Real Analysis and Applications

Fabio Silva Botelho

Real Analysis
and Applications

 Springer

Fabio Silva Botelho
Department of Mathematics
Federal University of Santa Catarina
Florianópolis, Santa Catarina, Brazil

ISBN 978-3-030-08750-0   ISBN 978-3-319-78631-5   (eBook)
https://doi.org/10.1007/978-3-319-78631-5

Printed on acid-free paper

This Springer imprint is published by the registered company Springer International Publishing AG part
of Springer Nature.
The registered company address is: Gewerbestrasse 11, 6330 Cham, Switzerland

Preface

This text develops basic and advanced concepts on real analysis in $\mathbb{R}$ and $\mathbb{R}^n$. In the first chapter, we present a careful construction of the real number system through the definition of cut. In the subsequent chapters, we present some definitions and results concerning metric spaces. We finish this study with a formal proof of the Arzela–Ascoli theorem.

Here we emphasize a standard study on real sequences and series is also presented. In these initial chapters, the results are in general not new and may be found in many other standard books, such as Walter Rudin [12] and Elon Lages Lima [9, 10], with the exception of the new proof for the intermediate value theorem and the rigorous study on the exponential functions and logarithms.

On the other hand, in the multi-variable part, we highlight the detailed proofs of the implicit function theorem for the vectorial case, which is performed through an application of the Banach fixed point theorem. This latter theorem is also the basis for an also detailed proof of the inverse function theorem for $\mathbb{R}^n$.

These proofs are rigorous, but we believe they are relatively easy to follow.

As an application of the implicit function theorem for the vectorial case, we develop our Lagrange multiplier result.

Once more the proofs are rigorous but very well developed and easy to follow.

Finally, in the last chapter, we present a study on surfaces in $\mathbb{R}^n$ which includes differential forms defined on such surfaces.

We prefer to use the word surface instead of manifold, since we are referring to a special class of manifolds properly specified.

In this context, we rigorously establish the definition of volume form and formally prove the algorithm to obtain the standard formulas.

The Stokes theorem in a more abstract fashion is also developed. In the final section we present an introduction to Riemannian geometry.

At this point we start a more detailed description of each chapter content.

Summary of Each Chapter Content

Chapter 1: Real Numbers

This chapter develops formal results relating the concept of real numbers. In the first sections we start with a study about the natural numbers. In the subsequent sections the definitions of integer and rational numbers are introduced. In the final sections the existence of the real field is proven through the definition of cut. We finish such a chapter with a study about countable and uncountable sets.

Chapter 2: Metric Spaces

In this chapter we present the main definitions and results related to metric spaces. Standard topics such as open, closed, and compact sets are developed in detail. In the final section we prove the Heine–Borel theorem.

Chapter 3: Real Sequences and Series

This chapter develops a standard study on real sequences and series. We develop in detail topics such as superior and inferior limits for bounded sequences and their applicability in the proof of root and ratio tests for convergence of series. The comparison criterion for series is also extensively addressed and applied to a great variety of situations.

Chapter 4: Real Function Limits

This chapter addresses the main results relating the concept of limits for one variable real functions. Topics such as subsequential limits and related cluster points are presented in detail. Other topics include the standard sandwich theorem and relating comparison results. We finish the chapter with a study on infinite limits and limits at infinity.

Chapter 5: Continuous Functions

In this chapter we present a study about continuity for one variable real functions. Standard topics such as the relations between continuity and compactness are developed extensively. We finish the chapter addressing the main definitions and results on uniform continuity.

Chapter 6: Derivatives

This chapter presents the main definitions and results related to derivatives for one variable real functions. Standard topics such as the derivative proprieties, the mean value theorem, and Taylor expansion are developed in detail. The inverse function theorem and related derivative for such a one real variable case is also addressed.

Chapter 7: The Riemann Integral

This chapter develops results concerning the Riemann integration of one variable real functions. The standard necessary and sufficient condition of zero Lebesgue measure for the set of discontinuities for the Riemann integrability of a one variable real function is addressed in detail. In the final sections we present a basic study on sequences and series of real functions.

Chapter 8: Differential Analysis in $\mathbb{R}^n$

This chapter starts with the basic definitions and results related to scalar functions in $\mathbb{R}^n$. In the first sections we address concepts such as limits, continuity, and differentiability. A study on optimality conditions for critical points is also developed. In the subsequent sections we address these same concepts of limits, continuity, and differentiability for vectorial functions in $\mathbb{R}^n$. In the final sections we develop detailed proofs of the implicit (scalar and vectorial cases) and inverse function theorems. Moreover, results concerning Lagrange Multipliers are also presented through an application of the implicit function theorem for the vectorial case. We finish the chapter with an introduction to differential geometry.

Chapter 9: Integration in $\mathbb{R}^n$

This chapter develops results concerning the Riemann integration of functions defined in $\mathbb{R}^n$. We address in detail the standard necessary and sufficient condition of zero Lebesgue measure for the set of discontinuities for the Riemann integrability of a scalar function defined in $\mathbb{R}^n$. Other topics such as change of variables are also presented in detail.

Chapter 10: Topics on Vector Calculus and Vector Analysis in $\mathbb{R}^n$

In this chapter we address the main definitions and results for vector analysis in $\mathbb{R}^n$. This part of the text comprises topics such as differential forms in surfaces in $\mathbb{R}^n$, including the volume form for a surface in $\mathbb{R}^n$, and the Green, Gauss, and Stokes theorems in both standard calculus and abstract versions. Indeed we develop an abstract version of the Stokes theorem and recover the classical Divergence and Stokes theorems from such a general approach. We finish the chapter with an introduction to Riemannian geometry.

Acknowledgments

I would like to express my gratitude to the Mathematics Department of Virginia Tech, USA, for the financial support received during my Ph.D. program in applied mathematics, developed from August 2006 to July 2009. Among the professors, I am especially grateful to Robert C. Rogers (Advisor), William Floyd (Elementary Real Analysis), Martin Day (Calculus of Variations), James Thomson (Real Analysis), and George Hagedorn (Functional Analysis) for the excellent lectured courses.

Florianópolis, Santa Catarina, Brazil   Fabio Silva Botelho
May 2017

Contents

Part I
One Variable Real Analysis

Chapter 1
Real Numbers

1.1 Introduction

In this chapter we present the construction of the real numbers. We start with the concept of sets and relations. In the final sections we introduce the concept of countable set and some concerned applications. The main references for this chapter are [9, 12].

1.2 Sets and Relations

By a set we shall understand a collection of objects (also called the set elements), without specifying a more formal definition. We shall describe and/or represent a set either by declaring a propriety satisfied by its elements or in a straightforward fashion, by specifying its elements.

For example, consider the set A, described by the propriety,

$$A = \{x \mid x \text{ is a month of the year with 31 days}\}.$$

Or specifying its elements,

$$A = \{\text{January, March, May, July, August, October, December}\}.$$

Given a set A, if x is in A (that is, if x is an element of A), we write $x \in A$. If x is not in A, we denote $x \notin A$.

Other examples of sets:

$$A = \{x \mid x \text{ is a country which shares a border with the USA}\},$$

© Springer International Publishing AG, part of Springer Nature 2018
F. S. Botelho, *Real Analysis and Applications*,
https://doi.org/10.1007/978-3-319-78631-5_1

hence

$$A = \{ \text{Canada, Mexico} \}.$$

$$B = \{x \mid 3/x \text{ is integer}\},$$

that is

$$B = \{3, -3, 1, -1\}.$$

1.2.1   Empty and Unitary Sets

If a set A has no elements, is said to be empty, and in such a case we denote $A = \emptyset$.
For example,

$$A = \{x \mid x \text{ is an American state which shares a border with England}\},$$

Hence $A = \emptyset$.
On the other hand, a set which has only one element is said to be an unitary one.
For example

$$A = \{x \mid 2x + 3 = 7\},$$

that is $A = \{2\}$ (unitary set).

Definition 1.2.1 Let A, B be sets. We say that A is contained in B if the following propriety holds:

$$\text{If } x \in A, \text{ then } x \in B.$$

In such a case we denote $A \subset B$.
Moreover, if $A \subset B$ and there exists $x \in B$ such that $x \notin A$ the inclusion is said to be proper.
Finally, if there exists $x \in A$ such that $x \notin B$ we say that A is not contained in B and denote $A \nsubseteq B$.

Remark 1.2.2 Let A, B be two sets. If $A \subset B$ and $B \subset A$ we say that A equals B and write $A = B$.

1.2.2   Properties of Inclusion

The inclusion has the following elementary properties:

Proposition 1.2.3 *Let A, B, C be sets. Thus,*

1. $\emptyset \subset A$,
2. $A \subset A$
3. *If $A \subset B$ and $B \subset C$, then $A \subset C$.*

1.2.3 Parts of a Set

Definition 1.2.4 Let A be a set. We define the set of parts of A, denoted by $\mathscr{P}(A)$ by

$$\mathscr{P}(A) = \{B \ : \ B \subset A\}.$$

Hence the set of parts of A is the set whose elements are the subsets of A.

1.2.4 Union of Sets

At this point we assume the set $A \subset U$, where U is a general class of sets. We call U the universal set of the class in question (Fig. 1.1).

Definition 1.2.5 (Union) Let A, B be sets. we define the union of A and B (Fig. 1.2), denoted by $A \cup B$, by

$$A \cup B = \{x \ : \ x \in A \text{ or } x \in B\}.$$

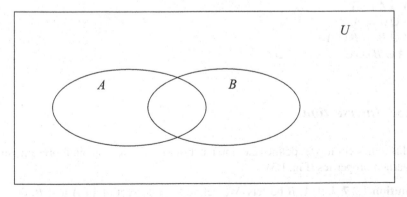

Fig. 1.1 Two sets A and B in a universal set U

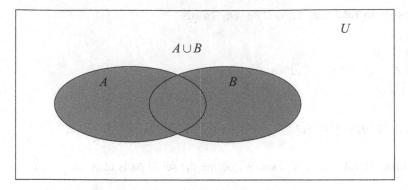

Fig. 1.2 Union of two sets A and B in a universal set U

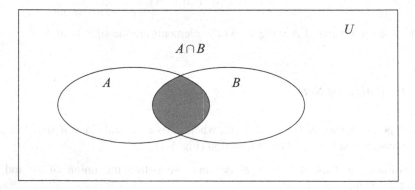

Fig. 1.3 Intersection of two sets A and B in a universal set U

Proposition 1.2.6 (Properties of Union) *Let A, B, C be sets. thus,*

1. $A \cup A = A$,
2. $A \cup \emptyset = A$,
3. $A \cup B = B \cup A$,
4. $(A \cup B) \cup C = A \cup (B \cup C)$.

1.2.5   Intersection

In this subsection we define the intersection between sets and present some concerned properties (Fig. 1.3).

Definition 1.2.7 Let A, B be sets. We define the intersection of A and B, denoted by $A \cap B$, by

$$A \cap B = \{x \ : \ x \in A \text{ and } x \in B\}.$$

Proposition 1.2.8 (Intersection Properties) *Let* A, B, C *be sets. Concerning intersection definition, the following properties hold:*

1. $A \cap A = A$,
2. $A \cap \emptyset = \emptyset$,
3. $A \cap B = B \cap A$,
4. $(A \cap B) \cap C = A \cap (B \cap C)$.

Definition 1.2.9 (Disjoint Sets) Let A, B be sets. If $A \cap B = \emptyset$, then A and B are said to be disjoint sets.

Proposition 1.2.10 (Properties Involving Unions and Intersections) *Let* A, B, C *be sets. The following proprieties are valid concerning unions and intersections:*

1. $A \cup (A \cap B) = A$,
2. $A \cap (A \cup B) = A$,
3. $A \cup (B \cap C) = (A \cup B) \cap (A \cup C)$,
4. $A \cap (B \cup C) = (A \cap B) \cup (A \cap C)$.

Proof We prove just property 3. Observe that $x \in A \cup (B \cap C)$ implies that $x \in A$ or $(x \in B$ and $x \in C)$

If $x \in A$, then $x \in (A \cup B)$ and $x \in (A \cup C)$ so that $x \in (A \cup B) \cap (A \cup C)$.
If $x \in B$ and $x \in C$, then $x \in (A \cup B)$ and $x \in (A \cup C)$, so that $x \in (A \cup B) \cap (A \cup C)$.

Hence, in any case

$$A \cup (B \cap C) \subset (A \cup B) \cap (A \cup C).$$

Conversely, assume $x \in (A \cup B) \cap (A \cup C)$. There are two cases to consider. $x \in A$ or $x \notin A$ If $x \in A$, then $x \in A \cup (B \cap C)$.

If $x \notin A$, as $x \in (A \cup B)$ we must have $x \in B$. Similarly as $x \in A \cup C$, we must have $x \in C$ so that $x \in B \cap C \subset A \cup (B \cap C)$.

Thus $(A \cup B) \cap (A \cup C) \subset A \cup (B \cap C)$.

We may conclude that $(A \cup B) \cap (A \cup C) = A \cup (B \cap C)$ which completes the proof.

We finish this section presenting the definition of difference between sets (Figs. 1.4 and 1.5).

Definition 1.2.11 (Difference) Let A, B be sets. We define A minus B, denoted by $A \setminus B$, by

$$A \setminus B = \{x \, : \, x \in A \text{ and } x \notin B\}.$$

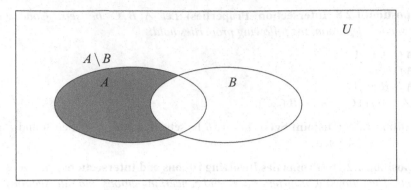

Fig. 1.4 Difference A minus B between the sets A and B

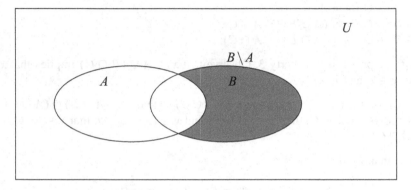

Fig. 1.5 Difference B minus A between the sets A and B

Hence we denote $A^c = U \setminus A$, that is,

$$A^c = \{x \in U \ : \ x \notin A\}.$$

We refer to A^c as the complement of A relating U.

1.3   Cartesian Product and Relations

Definition 1.3.1 (Cartesian Product, Relations) Let A, B be sets. We define the Cartesian product of A by B, denoted by $A \times B$ by:

$$A \times B = \{(x, y) \mid x \in A, \ y \in B\}.$$

Any subset $R \subset A \times B$ is said to be a relation from A to B.

Definition 1.3.2 (Function) Let A and B be two sets. A relation f from A to B is said to be a function if for each $x \in A$ there is one and only one $y \in B$ such that $(x, y) \in f$. In such a case we denote $f : A \to B$ and $y = f(x)$.

Definition 1.3.3 Let $f : A \to B$ be a function. We say that f is injective if the following propriety is valid:

$$\text{If } x_1, x_2 \in A \text{ and } x_1 \neq x_2 \text{ then } f(x_1) \neq f(x_2),$$

or equivalently

$$\text{If } x_1, x_2 \in A \text{ and } f(x_1) = f(x_2) \text{ then } x_1 = x_2.$$

Definition 1.3.4 A function $f : A \to B$ is said to be surjective if $R(f) = B$, where $R(f)$ denotes the range or image of f, defined by

$$R(f) = \{f(x) \ : \ x \in A\}.$$

Definition 1.3.5 Let $f : A \to B$ be a function. Let $E \subset B$. We define the inverse image of E, denoted by $f^{-1}(E)$, as

$$f^{-1}(E) = \{x \in A \mid f(x) \in E\}.$$

Definition 1.3.6 A function $f : A \to B$ is said to bijective if it is injective and surjective.

Definition 1.3.7 Let $f : A \to B$ be a bijective function. We define the inverse function relating f, denoted by

$$f^{-1} : B \to A,$$

through the relation,

$$x = f^{-1}(y) \text{ if, and only if, } y = f(x).$$

1.4 Natural Numbers

In the next sections we develop the formal concept of real numbers. We start studying the set of natural numbers. Afterwards we define the concept of rational numbers and then the concept of cut which is related to the formal definition of real number.

At this point, we consider a set denoted by $\mathbb{N}$, whose elements are not yet precisely specified and a function $s : \mathbb{N} \to \mathbb{N}$ which satisfies the following properties, known as the Peano axioms:

1. $s : \mathbb{N} \to \mathbb{N}$ is injective.
2. $\mathbb{N} - s(\mathbb{N}) = \{1\}$, where $1 \in \mathbb{N}$ is said to be the natural unit.
3. Induction principle:

>   If $X \subset \mathbb{N}$ and X satisfies the properties:

   (a) $1 \in X$,
   (b) if $n \in X$, then $s(n) \in X$,

then

$$X = \mathbb{N},$$

where $s(n)$ is also denoted by $s(n) = n + 1$ (we read: the successor of n is $n + 1$).
Through such a function we may identify the natural numbers as the set

$$\mathbb{N} = \{1, 1 + 1, (1 + 1) + 1, \ldots\}$$

which we may denote by

$$\mathbb{N} = \{1, 2, 3, \ldots\}.$$

1.4.1   A Sum in $\mathbb{N}$

Definition 1.4.1 We define a sum in $\mathbb{N}$ by:

$$m + n = s^n(m), \forall m, n \in \mathbb{N}$$

where, as above:

$$s(n) = n + 1.$$

Hence, for example:

$$
\begin{aligned}
m + 2 &= s^2(m) \\
&= s(s(m)) \\
&= s(m + 1) \\
&= (m + 1) + 1
\end{aligned}
\tag{1.1}
$$

Observe that

$$m + s(n) = m + (n + 1)$$

and

$$s(m + n) = (m + n) + 1.$$

Remark 1.4.2 For $s : \mathbb{N} \to \mathbb{N}$ we shall assume the following fundamental property:

$$s(m + n) = m + s(n), \ \forall m, n \in \mathbb{N},$$

that is,

$$(m + n) + 1 = m + (n + 1), \forall m, n \in \mathbb{N}. \tag{1.2}$$

The property indicated in (1.2) is called the fundamental sum associativity.

1.4.2 Properties of Sum in $\mathbb{N}$

Proposition 1.4.3 *The sum in $\mathbb{N}$ has the following properties:*

1. *Associativity:*
   $(m + n) + p = m + (n + p), \forall m, n, p \in \mathbb{N}$
2. *Commutativity:*
   $m + n = n + m, \forall m, n \in \mathbb{N}$
3. *Law of cancelation:*
   Let $m, n \in \mathbb{N}$. If there exists $p \in \mathbb{N}$ such that

$$m + p = n + p,$$

   then

$$m = n.$$

Proof (Associativity) Choose $m, n \in \mathbb{N}$ and define

$$B = \{p \in \mathbb{N} \mid (m + n) + p = m + (n + p)\}.$$

From (1.2) (the fundamental associativity) we have

$$(m + n) + 1 = m + (n + 1),$$

that is

$$1 \in B.$$

Suppose $p \in B$. Thus,

$$(m + n) + p = m + (n + p). \tag{1.3}$$

From (1.2) we have

$$(m + n) + (p + 1) = ((m + n) + p) + 1.$$

From this and (1.3) we obtain

$$(m + n) + (p + 1) = (m + (n + p)) + 1,$$

so that from (1.2) again we get

$$(m + n) + (p + 1) = m + ((n + p) + 1)$$
$$= m + (n + (p + 1)), \tag{1.4}$$

that is,

$$p + 1 \in B.$$

From the induction principle, we have

$$B = \mathbb{N}.$$

Being $m, n \in \mathbb{N}$ arbitrary, we may infer that

$$(m + n) + p = m + (n + p), \ \forall m, n, p \in \mathbb{N}.$$

Proof of Commutativity First we prove that

$$n + 1 = 1 + n, \ \forall n \in \mathbb{N}.$$

Define

$$B = \{n \in \mathbb{N} \mid n + 1 = 1 + n\}.$$

From

$$1 + 1 = 1 + 1$$

we obtain $1 \in B$. Suppose that $n \in B$, thus $n + 1 = 1 + n$. Therefore,

$$(n + 1) + 1 = (1 + n) + 1,$$

and from this and (1.2)

$$(n + 1) + 1 = 1 + (n + 1),$$

that is,

$$n + 1 \in B.$$

From the principle of induction, we get

$$B = \mathbb{N}.$$

Thus,

$$n + 1 = 1 + n, \quad \forall n \in \mathbb{N}. \tag{1.5}$$

Now choose $m \in \mathbb{N}$ and define

$$C = \{p \in \mathbb{N} \mid m + p = p + m\}.$$

From (1.5) we have that

$$1 \in C.$$

Suppose $p \in C$. Thus,

$$m + p = p + m.$$

Observe that from this, from the associativity, from (1.2) and (1.5) we may infer that

$$\begin{aligned}
m + (p + 1) &= (m + p) + 1 \\
&= (p + m) + 1 \\
&= 1 + (p + m) \\
&= (1 + p) + m \\
&= (p + 1) + m, \tag{1.6}
\end{aligned}$$

that is, $p + 1 \in C$. From the principle of induction

$$C = \mathbb{N}.$$

Hence, being $m \in \mathbb{N}$ arbitrary, we obtain

$$m + p = p + m, \ \forall m, p \in \mathbb{N}.$$

Proof of Law of Cancelation Let $m, n \in \mathbb{N}$ be such that there exists $p \in \mathbb{N}$ such that

$$m + p = n + p.$$

From the sum definition, we get:

$$s^p(m) = s^p(n). \tag{1.7}$$

Since $s : \mathbb{N} \to \mathbb{N}$ is injective, we have that $s^p : \mathbb{N} \to \mathbb{N}$ is injective, and thus from this and (1.7), we obtain:

$$m = n.$$

1.4.3 An Order Relation for $\mathbb{N}$

Let $m, n \in \mathbb{N}$ we denote $m < n$ if there exists $p \in \mathbb{N}$ such that

$$m + p = n,$$

that is,

$$s^p(m) = n.$$

If $m < n$ we also denote $n > m$ and $m \leq n$ will denote $m < n$ or $m = n$.

Proposition 1.4.4 *The following properties are valid for the order relation in $\mathbb{N}$:*

1. *Transitivity: If $m < p$ and $p < n$, then $m < n$.*
2. *Trichotomy: Given $m, n \in \mathbb{N}$, then exactly one of the three possibilities below indicated is valid:*

   (a) $m < n$
   (b) $n < m$
   (c) $m = n$.

Proof We prove just the transitivity, leaving the proof of trichotomy as exercise. Suppose $m, n, p \in \mathbb{N}$ and let be such that $m < p$ and $p < n$. Thus, there exist $t_1, t_2 \in \mathbb{N}$ such that

$$m + t_1 = p,$$

and

$$p + t_2 = n.$$

Hence,

$$(m + t_1) + t_2 = n,$$

that is,

$$m + (t_1 + t_2) = n,$$

so that

$$m < n.$$

The proof is complete.

1.4.4 The Multiplication in $\mathbb{N}$

Fix $m \in \mathbb{N}$. Define the function $f_m : \mathbb{N} \to \mathbb{N}$ by:

$$f_m(p) = p + m, \forall p \in \mathbb{N}.$$

Definition 1.4.5 Let $m, n \in \mathbb{N}$. We define the multiplication of m by n, denoted by $m \cdot n$, by

$$m \cdot n = f_m^{n-1}(m),$$

and in particular

$$m \cdot 1 = f_m^0(m) = m.$$

For example,

$$\begin{aligned}
m \cdot 3 &= f_m^{3-1}(m) \\
&= f_m^2(m) \\
&= f_m(f_m(m)) \\
&= f_m(m + m) \\
&= (m + m) + m. \qquad (1.8)
\end{aligned}$$

Observe that given $m, n \in \mathbb{N}$ we have:

$$m \cdot (n + 1) = f_m^n(m)$$
$$= f_m(f_m^{n-1}(m))$$
$$= f_m(m \cdot n)$$
$$= m \cdot n + m. \tag{1.9}$$

Hence,

$$m \cdot (n + 1) = m \cdot n + m, \quad \forall m, n \in \mathbb{N}. \tag{1.10}$$

1.4.5 Properties of Multiplication

Proposition 1.4.6 *The following properties hold for the multiplication in* $\mathbb{N}$*:*

1. *Distributivity:* $m \cdot (n + p) = m \cdot n + m \cdot p, \ \forall m, n, p \in \mathbb{N}$.
2. *Associativity:* $m \cdot (n \cdot p) = (m \cdot n) \cdot p, \ \forall m, n, p \in \mathbb{N}$.
3. *Commutativity:* $m \cdot n = n \cdot m, \ \forall m, n \in \mathbb{N}$.
4. *Cancelation law: Let* $m, n \in \mathbb{N}$. *if there exists* $p \in \mathbb{N}$ *such that*

$$m \cdot p = n \cdot p$$

then $m = n$.

Proof We shall prove just the distributivity. The remaining proofs are left as exercises. Choose $m, n \in \mathbb{N}$. Define B by

$$B = \{p \in \mathbb{N} \mid m \cdot (n + p) = m \cdot n + m \cdot p\}.$$

From (1.10) we have

$$m \cdot (n + 1) = m \cdot n + m = m \cdot n + m \cdot 1,$$

and hence

$$1 \in B.$$

Suppose $p \in B$. Thus $m \cdot (n + p) = m \cdot n + m \cdot p$. Observe that

$$m \cdot (n + (p + 1)) = m \cdot ((n + p) + 1)$$
$$= m \cdot (n + p) + m$$

$$= (m \cdot n + m \cdot p) + m$$
$$= m \cdot n + (m \cdot p + m)$$
$$= m \cdot n + m \cdot (p + 1), \qquad (1.11)$$

Hence $p + 1 \in B$. From the induction principle we obtain:

$$B = \mathbb{N}.$$

since $m, n \in \mathbb{N}$ arbitrary we may infer that

$$m \cdot (n + p) = m \cdot n + m \cdot p, \ \forall m, n, p \in \mathbb{N}.$$

The proof is complete.

Exercise 1.4.7 Prove the associativity, commutativity, and the cancelation law for the multiplication in $\mathbb{N}$.

1.4.6 The Well-Ordering Principle

In this section we formally prove that all subset of $\mathbb{N}$ has a minimum. We start with the definitions:

Definition 1.4.8 Let $X \subset \mathbb{N}$. We say that $p \in X$ is the minimum point of X if

$$p \leq n, \ \forall n \in X.$$

By analogy, we say that $q \in X$ is the maximum point of X if

$$n \leq q, \ \forall n \in X.$$

Let $n \in \mathbb{N}$. Denote $I_n = \{1, \ldots, n\}$.

Definition 1.4.9 A set A is said to be finite if there exists a bijection between A and I_n, for some $n \in \mathbb{N}$. In such a case we denote $I_n \sim A$ and say A has cardinality n. If $A \neq \emptyset$ is not finite, it is said to be infinite.

Theorem 1.4.10 (Well-Ordering Principle) *Let $X \subset \mathbb{N}$ such that $X \neq \emptyset$. Under such hypotheses there exists $p \in X$ such that $p \leq n, \forall n \in X$.*

Proof Suppose to obtain contradiction that for each $m \in X$ there exists $m_1 \in X$ such that $m_1 < m$. Choose $m_0 \in X$ (this is possible since $X \neq \emptyset$). Hence there exists $m_1 \in X \subset \mathbb{N}$ such that $m_1 < m_0$. And also, there exists $m_2 \in X \subset \mathbb{N}$ such that $m_2 < m_1 < m_0$. Proceeding in this fashion we may obtain

$$\cdots < m_n < m_{n-1} < \cdots < m_1 < m_0,$$

that is an infinite set of natural numbers smaller than m_0, which contradicts I_{m_0} to be finite.

The proof is complete.

1.5   The Set of Integers

In this section we develop some details concerning a set which is a natural extension of $\mathbb{N}$, called the set of integers and denoted by $\mathbb{Z}$.

Definition 1.5.1 (Integers) We define a set denoted by $\mathbb{Z}$ whose elements will be more precisely specified in the next lines, which together with the functions $s : \mathbb{Z} \rightarrow \mathbb{Z}$ and $a : \mathbb{Z} \rightarrow \mathbb{Z}$, satisfy the following properties:

1. s and a are injective and onto (bijections).
2. $\mathbb{N} \subset \mathbb{Z}$ and $s(\mathbb{N}) = \mathbb{N} \setminus \{1\}$, where $1 \in \mathbb{N}$ is the natural unit.
3. There exist elements denoted by $0 \in \mathbb{Z}$ and $-1 \in \mathbb{Z}$ such that $s(0) = 1$, and $s(-1) = 0$ where we denote $s(n) = n + 1, \forall n \in \mathbb{Z}$.
4. a is such that $a(0) = -1$, $a(1) = 0$, so that we denote $a(n) = n + (-1), \forall n \in \mathbb{Z}$.
5. We also denote $a^{n-1}(-1) = -n, \forall n \in \mathbb{N}$ so that we may express

$$\mathbb{Z} = \{0, n, -n \; : \; n \in \mathbb{N}\}.$$

6. (Induction principle for $\mathbb{Z}$)
   If $X \neq \emptyset$, $X \subset \mathbb{Z}$ and the following property is valid:

   - If $p \in X$, then $p + 1 \in X$ and $p + (-1) \in X$;

   we have $X = Z$.

Definition 1.5.2 (Sum for $\mathbb{Z}$) Given $m \in \mathbb{Z}$ and $n \in \mathbb{N}$ we define the sum of m and n, denoted by $m + n$, by

$$m + n = s^n(m).$$

We also define, the sum of m and $-n$, denoted by $m + (-n)$, by

$$m + (-n) = a^n(m).$$

Finally, we define $m + 0 = 0 + m = m, \forall m \in \mathbb{Z}$.

Remark 1.5.3 We have denoted $s(n) = n + 1$. Here we read, "the successor of n is $n + 1$."

Also $a(n) = n + (-1)$ and here we read "the antecessor of n is $n + (-1), \forall n \in \mathbb{Z}$."

Definition 1.5.4 (Multiplication for $\mathbb{Z}$) Given $m \in \mathbb{Z}$ and $n \in \mathbb{N}$ we define the multiplication of m by n denoted by $m \cdot n$, by

$$m \cdot n = f_m^{n-1}(m),$$

where

$$f_m(p) = p + m, \forall m \in \mathbb{Z}, p \in \mathbb{Z}.$$

Also, the multiplication of m by $-n$, denoted by $m \cdot (-n)$, is defined by

$$m \cdot (-n) = -f_m^{n-1}(m).$$

Remark 1.5.5 We assume the following fundamental associativity properties in $\mathbb{Z}$:

$$(m + n) + 1 = m + (n + 1), \forall m, n \in \mathbb{Z}, \tag{1.12}$$

$$(m + n) + (-1) = m + (n + (-1)), \forall m, n \in \mathbb{Z}. \tag{1.13}$$

Exercise 1.5.6 Prove the associativity, commutativity, and cancelation law for the sum and multiplication in $\mathbb{Z}$.

Definition 1.5.7 (Order for $\mathbb{Z}$) Given $m, n \in \mathbb{Z}$ we denote $m < n$, if there exists p in $\mathbb{N}$ such that $m + p = n$.

We may write $m \leq n$ to denote $m < n$ or $m = n$.

Exercise 1.5.8 Prove the transitivity and trichotomy for the order relation for $\mathbb{Z}$.

1.6 The Rational Set

We will define $\mathbb{Z}^* = \mathbb{Z} \setminus \{0\}$ and $\mathbb{Q} = \mathbb{Z} \times \mathbb{Z}^*$, where $\mathbb{Q}$ are said to be the set of rational numbers. Thus, $(m, n) \in \mathbb{Q}$ if and only if $m \in \mathbb{Z}$ and $n \in \mathbb{Z}^*$.

We denote $(m, n) = \frac{m}{n}$ so that $\mathbb{Q}$ contains a not relabeled natural set $\mathbb{N}^*$ where

$$\mathbb{N}^* = \left\{ (n, 1) = \frac{n}{1}, : n \in \mathbb{N} \right\}.$$

With such identification, we shall denote simply $\mathbb{N}^* = \mathbb{N}$.

Also we define $(m, n) = (m_1, n_1)$ if in a more usual sense,

$$\frac{m}{n} = \frac{m_1}{n_1},$$

that is,

$$m \cdot n_1 = n \cdot m_1.$$

Definition 1.6.1 (Sum and Multiplication for $\mathbb{Q}$) Given $x = (m, n) \in \mathbb{Q}$ and $y = (m_1, n_1) \in \mathbb{Q}$ we define the sum of x and y denoted by $x + y$, by

$$x + y = (m \cdot n_1 + n \cdot m_1, n \cdot n_1)$$

or in a more informal notation

$$x + y = \frac{m}{n} + \frac{m_1}{n_1} = \frac{m \cdot n_1 + m_1 \cdot n}{n \cdot n_1}.$$

And the multiplication $x \cdot y$ is defined by

$$x \cdot y = (m \cdot m_1, n \cdot n_1),$$

or in a more informal notation:

$$x \cdot y = \frac{m}{n} \cdot \frac{m_1}{n_1} = \frac{m \cdot m_1}{n \cdot n_1},$$

where in both definitions the multiplication between elements of $\mathbb{Z}$ is the usual previously defined.

1.7 Ordered Sets

In this section we introduce the definition of ordered set.

Definition 1.7.1 (Order Relation) Let A be a set. A relation in A which we denote by $<$ is said to be an order relation, if the following properties are satisfied:

1. If $x, y \in A$, then one and only one of the following possibilities holds:

   (a) $x < y$,
   (b) $x = y$,
   (c) $y < x$.

2. Given $x, y, z \in A$, if $x < y$ and $y < z$, then $x < z$.

Eventually, $x \le y$ denotes $x < y$ or $x = y$.

Definition 1.7.2 (Ordered Set) Any set for which is defined an ordered relation is said to be an ordered one.

Definition 1.7.3 (Upper Bound) Let S be an ordered set and suppose $E \subset S$. If there exists $\beta \in S$ such that $x \le \beta$, $\forall x \in E$, the set E is said to be upper bounded and β is said to be an upper bound for E. By analogy if there exists $\gamma \in S$ such that $\gamma \le x$, $\forall x \in E$, then E is said to be lower bounded and γ is said to be a lower bound for E.

Definition 1.7.4 (Supremum) Let S be an ordered set, suppose $E \subset S$ is upper bounded. If there exists $\alpha \in S$ such that

1. $x \leq \alpha$, $\forall x \in E$,
2. If $\gamma \in S$ and $\gamma < \alpha$, then γ is not an upper bound for E,

we say that α is the supremum of E and denote

$$\alpha = \sup E.$$

Definition 1.7.5 (Infimum) Let S an ordered set, suppose $E \subset S$ is lower bounded. If there exists $\beta \in S$ such that

1. $\beta \leq x$, $\forall x \in E$,
2. If $\gamma \in S$ and $\beta < \gamma$, then γ is not a lower bound for E,

we say that β is the infimum for E and we denote

$$\beta = \inf E.$$

Definition 1.7.6 (Property of the Least Upper Bound) Let S be an ordered set. We say that S has the property of the least upper bound if for each $E \subset S$ upper bounded, there exists $\alpha \in S$ such that

$$\sup E = \alpha.$$

Let us see the counter example $S = \mathbb{Q}$ and

$$E = \left\{ \frac{m}{n} \mid m, n \in \mathbb{N} \text{ and } \frac{m^2}{n^2} < 2 \right\}.$$

Observe that E has no supremum in $\mathbb{Q}$. Thus $\mathbb{Q}$ has not the property of the least upper bound. In fact if we redefine $S = \mathbb{R}$, we get $\sup E = \sqrt{2} \notin \mathbb{Q}$.

Theorem 1.7.7 *Let S an ordered set with the property of the least upper bound. Suppose $B \subset S$ ($B \neq \emptyset$) is lower bounded. Let $L \subset S$ the set of all lower bounds for B in S. Thus there exists $\alpha \in S$ such that*

$$\alpha = \sup L$$

and so that

$$\alpha = \inf B.$$

Proof From the hypotheses B is lower bounded, thus there exists $\beta \in S$ such that $\beta \leq x$, $\forall x \in B$. Hence $L \neq \emptyset$. Fix $x_0 \in B$. Thus, if $\beta_1 \in L$, then $\beta_1 \leq x$, $\forall x \in B$, and in particular

$$\beta_1 \leq x_0, \ \forall \beta_1 \in L.$$

Therefore $L \subset S$ is upper bounded. As S has the least upper bound property, there exists $\alpha \in S$ such that

$$\alpha = \sup L.$$

Let $\gamma \in S$. If $\gamma < \alpha$, then γ is not an upper bound for L, thus there exists $\beta_2 \in L$ such that

$$\gamma < \beta_2 \leq x, \ \forall x \in B.$$

Hence,

$$\text{if } \gamma < \alpha \text{ then } \gamma \notin B,$$

the contrapositive is

$$\text{if } x \in B \text{ then } x \geq \alpha.$$

We conclude that α is a lower bound for B. However if $\beta > \alpha = \sup L$, then β is not a lower bound for B. Hence α is the greatest lower bound for B, that is, $\alpha = \inf B$.
    The proof is complete.

1.8 Fields

We begin this section with the definition of field.

Definition 1.8.1 (Field) A field is a set which we shall denote by $\mathbb{F}$, for which are defined two operations, namely, a sum and a multiplication, defined in $\mathbb{F} \times \mathbb{F}$ and denoted by $(+)$ and $(\cdot)$ respectively, for which the following properties are valid:

1. **Sum properties**:

    (a) if $x, y \in \mathbb{F}$ then $x + y \in \mathbb{F}$,
    (b) $x + y = y + x, \ \forall x, y \in \mathbb{F}$,
    (c) $(x + y) + z = x + (y + z), \ \forall x, y, z \in \mathbb{F}$,
    (d) there exists an element denoted by $0 \in \mathbb{F}$ such that $x + 0 = x, \ \forall x \in \mathbb{F}$,
    (e) for each $x \in \mathbb{F}$ there exists an unique element $y \in \mathbb{F}$ such that $x + y = 0$, so that we denote $y = -x$.

2. **Multiplication properties**:

    (a) if $x, y \in \mathbb{F}$, then $x \cdot y \in \mathbb{F}$,
    (b) $x \cdot y = y \cdot x, \ \forall x, y \in \mathbb{F}$,
    (c) $(x \cdot y) \cdot z = x \cdot (y \cdot z), \ \forall x, y, z \in \mathbb{F}$

(d) there exists an element denoted by $1 \in \mathbb{F}$ such that $1 \neq 0$ and $1 \cdot x = x$, $\forall x \in \mathbb{F}$

(e) if $x \in \mathbb{F}$ and $x \neq 0$, then there exists a unique $y \in \mathbb{F}$ such that $x \cdot y = 1$, so that we denote $y = 1/x$.

3. **Distributive law**: $x \cdot (y + z) = x \cdot y + x \cdot z$, $\forall x, y, z \in \mathbb{F}$.

Proposition 1.8.2 *Let* $x, y, z \in \mathbb{F}$. *The sum properties imply the following results:*

1. *if* $x + y = x + z$, *then* $y = z$,
2. *if* $x + y = x$, *then* $y = 0$,
3. *if* $x + y = 0$, *then* $y = -x$.
4. $-(-x) = x$.

Proof We shall prove just the first property, leaving the remaining proofs as exercises.

Suppose that $x + y = x + z$. Thus,

$$(-x) + (x + y) = (-x) + (x + z),$$

that is,

$$(-x + x) + y = (-x + x) + z,$$

so that,

$$0 + y = 0 + z,$$

that is,

$$y = z.$$

The proof is complete.

Proposition 1.8.3 *Let* $\mathbb{F}$ *be a field. Thus,*

1. $0 \cdot x = 0$, $\forall x \in \mathbb{F}$,
2. *if* $x \neq 0$ *and* $y \neq 0$ *then* $x \cdot y \neq 0$,
3. $(-x) \cdot y = -(x \cdot y) = x \cdot (-y)$, $\forall x, y \in \mathbb{F}$,
4. $(-x) \cdot (-y) = x \cdot y$, $\forall x, y \in \mathbb{F}$.

Proof We shall prove just the first item, leaving the remaining proofs as exercises. Let $x \in \mathbb{F}$. Observe that $(0 + 0) = 0$, and thus

$$(0 + 0) \cdot x = 0 \cdot x$$

so that

$$0 \cdot x + 0 \cdot x = 0 \cdot x.$$

From this and the second item of Proposition 1.8.2, we obtain

$$0 \cdot x = 0.$$

The proof is complete.

Exercise 1.8.4 Prove the items not proven yet in Propositions 1.8.2 and 1.8.3.

Definition 1.8.5 (Ordered Field) An ordered field, which we shall also denote by $\mathbb{F}$, is a field for which is defined an order relation, denoted by $<$, so that the following properties are satisfied.

1. If $x, y \in \mathbb{F}$ and $x < y$, then $x + z < y + z$, $\forall z \in \mathbb{F}$.
2. If $x, y \in \mathbb{F}$, $x > 0$, $y > 0$, then $x \cdot y > 0$.

Moreover if $x > 0$ we say that x is positive and if $x < 0$ we say that x is negative.

Proposition 1.8.6 *Let $\mathbb{F}$ be an ordered field and let $x, y, z \in \mathbb{F}$. Thus, the following properties are valid:*

1. *If $x > 0$, then $-x < 0$ and the reciprocal is valid.*
2. *If $x > 0$ and $y < z$, then $xy < xz$.*
3. *If $x < 0$ and $y < z$, then $xy > xz$.*
4. *If $x \neq 0$, then $x^2 = x \cdot x > 0$ and in particular $1 = 1^2 > 0$.*
5. *If $0 < x < y$, then $0 < 1/y < 1/x$.*

Proof We shall prove just the first item leaving the remaining proofs as exercises. Suppose $x > 0$. From the definition of ordered field, we obtain:

$$x + (-x) > 0 + (-x) = (-x),$$

that is

$$0 > (-x).$$

Hence

$$-x < 0.$$

The proof is complete.

Exercise 1.8.7

Prove the items not yet proven in Proposition 1.8.6.
Prove that $\mathbb{Q}$ is a field.

1.9   The Real Field

Let $\mathbb{F}$ be an ordered field. We say that $A \subset \mathbb{F}$ is inductive if:

1. $1 \in A$,
2. if $x \in A$, then $x + 1 \in A$.

Definition 1.9.1 (Naturals in $\mathbb{F}$) We define the set of naturals in $\mathbb{F}$, denoted by $\mathbb{N}_{\mathbb{F}}$, as the intersection of all inductive sets contained in $\mathbb{F}$.

It may be easily verified that

$$\mathbb{N}_{\mathbb{F}} = \{1, 1 + 1, (1 + 1) + 1, \ldots\},$$

where we may use the usual notation, that is,

$$1 + 1 = 2,$$

$$(1 + 1) + 1 = 3, \text{ etc.}$$

From now on we shall assume that $\mathbb{N}_{\mathbb{F}}$ satisfy the Peano axioms and shall denote it simply by $\mathbb{N}$.

Definition 1.9.2 (Integers in $\mathbb{F}$) We define the set of integers in $\mathbb{F}$ denoted simply by $\mathbb{Z}$, by

$$\mathbb{Z} = \{0, x, -x \mid x \in \mathbb{N}\}.$$

We shall denote also,

$$\mathbb{Z}^+ = \mathbb{N},$$

and

$$\mathbb{Z}^- = \{-x \mid x \in \mathbb{N}\},$$

where $\mathbb{Z}^+$ is called the set of positive integers and $\mathbb{Z}^-$ is the set of negative integers.

Definition 1.9.3 (Rationales in $\mathbb{F}$) We define the rational set in $\mathbb{F}$, denoted by $\mathbb{Q}$, by

$$\mathbb{Q} = \left\{ m \cdot \frac{1}{n} \mid m \in \mathbb{Z} \text{ and } n \in \mathbb{N} \right\}.$$

1.9.1   Existence of the Real Field

Theorem 1.9.4 *There exists a field which in some sense contains the set* $\mathbb{Q}$ *above specified and which has the least upper bound property. Such a field will be denoted by* $\mathbb{R}_\mathbb{F}$ *or simply* $\mathbb{R}$*, and is said to be the real field associated with* $\mathbb{F}$.

Proof We shall divide the proof in several parts in order to make easier its understanding.

1. Cut definition:

   The elements of $\mathbb{R}$ will be some subsets of $\mathbb{Q}$, which we shall call cuts.

   Definition 1.9.5 (Cut) We say that α is a cut if $\alpha \subset \mathbb{Q}$ and the following properties are valid:

   $$\alpha \neq \emptyset \text{ and } \alpha \neq \mathbb{Q}. \tag{1.14}$$

   $$\text{If } p \in \alpha, \ q \in \mathbb{Q} \text{ and } q < p, \text{ then } q \in \alpha. \tag{1.15}$$

   $$\text{If } p \in \alpha, \text{ then } p < r \text{ for some } r \in \alpha. \tag{1.16}$$

   For example, we could denote the cut associated with the rational 1 by 1^*, where

   $$1^* = \{x \in \mathbb{Q} \mid x < 1\}.$$

   We may associate with any $r \in \mathbb{Q}$ the cut r^* where

   $$r^* = \{x \in \mathbb{Q} \mid x < r\}.$$

   In the next lines Latin letters $p, q, r, \cdot$ will denote rational numbers, whereas Greek letters $\alpha, \beta, \gamma, \cdot$ will denote cuts. It is worth emphasizing that:

   $$\text{If } p \in \alpha \text{ and } q \notin \alpha, \text{ then } p < q. \tag{1.17}$$

   $$\text{If } r \notin \alpha \text{ and } r < s, \text{ then } s \notin \alpha. \tag{1.18}$$

2. Order definition for $\mathbb{R}$

   Define

   $$\alpha < \beta \Leftrightarrow \alpha \subsetneq \beta,$$

   where we read: α is smaller than β if, and only if, α is properly contained in β.

   We are going to show such order is transitive: Suppose $\alpha < \beta$ and $\beta < \gamma$. Thus $\alpha \subsetneq \beta$ and $\beta \subsetneq \gamma$ so that $\alpha \subsetneq \gamma$, that is, $\alpha < \gamma$. Hence the order relation is transitive.

Observe also that given α, β at most one of the three possibilities holds:

$$\alpha < \beta \text{ or } \alpha = \beta \text{ or } \beta < \alpha.$$

Therefore, to show the above relation is in fact an order one, it suffices to show that at least one possibility holds.

Choose α, $\beta \in \mathbb{R}$. Suppose we do not have $\alpha < \beta$ and we do not have $\alpha = \beta$. Hence, α is not a subset of β, so that there exists $p \in \alpha$ such that $p \notin \beta$.

Let $q \in \beta$. From (1.15) we obtain $q < p$ and thus from $p \in \alpha$ and (1.15) we may conclude that $q \in \alpha$. Thus, if $q \in \beta$, then $q \in \alpha$ and since $\alpha \neq \beta$ we obtain $\beta \subsetneq \alpha$, that is, $\beta < \alpha$. Therefore, the concerned order relation is well defined for $\mathbb{R}$.

3. We are going to show that $\mathbb{R}$ has the least upper bound property.

Let $A \subset \mathbb{R}$, such that $A \neq \emptyset$ and A is upper bounded. Thus, there exists $\beta \in \mathbb{R}$ such that $\alpha \leq \beta$, $\forall \alpha \in A$.

Define $\gamma = \cup_{\alpha \in A} \alpha$.

Hence $p \in \gamma$ if, and only if, $p \in \alpha$ for some $\alpha \in A$. We shall prove that γ is a cut.

First we shall verify (1.14). Observe that since $A \neq \emptyset$ we have that there exists $\alpha_0 \in A$ such that $\alpha_0 \neq \emptyset$, that is, $\alpha_0 \subset \gamma$ and thus

$$\gamma \neq \emptyset.$$

On the other hand, since $\alpha < \beta$, $\forall \alpha \in A$, we have that $\alpha \subset \beta$, $\forall \alpha \in A$. Therefore $\gamma \subset \beta$, that is, $\gamma \neq \mathbb{Q}$. Hence (1.14) has been verified.

Let us verify now the properties (1.15) and (1.16).

Let $p \in \gamma$. Observe that $p \in \gamma$ implies that $p \in \alpha_1$ for some $\alpha_1 \in A$. Hence if $q < p$, then $p \in \alpha_1 \subset \gamma$, that is, $q \in \gamma$. Thus, γ satisfies (1.15). On the other hand, from (1.16) for α_1 there exists $r \in \alpha_1 \subset \gamma$ such that $r > p$. Hence, $r > p$ and $r \in \gamma$. Thus (1.16) is valid for γ and we may infer that γ is a cut.

Now, we are going to prove that $\gamma = \sup A$.

First, observe that

$$\alpha \subset \gamma, \ \forall \alpha \in A,$$

that is,

$$\alpha \leq \gamma, \ \forall \alpha \in A.$$

Hence γ is an upper bound for A.

Let $\delta \in \mathbb{R}$ such that $\delta < \gamma$. Thus

$$\delta \subsetneq \gamma$$

(here we mean $\delta \subset \gamma$ and $\delta \neq \gamma$), so that there exists $s \in \gamma$ such that $s \notin \delta$. As $s \in \gamma$, there exists $\alpha \in A$ such that $s \in \alpha$. Thus $s \in \alpha$ and $s \notin \delta$, that is $\delta < \alpha$ and $\alpha \in A$, which means that δ is not an upper bound for A. Therefore γ is the least upper bound for A, that is

$$\gamma = \sup A.$$

4. Definition of a sum for $\mathbb{R}$.

Definition 1.9.6 (Sum for $\mathbb{R}$) Let $\alpha, \beta \in \mathbb{R}$, we define the sum of α and β, denoted by $\alpha + \beta$, by:

$$\alpha + \beta = \{r + s \mid r \in \alpha, \ s \in \beta\}.$$

The zero element for $\mathbb{R}$, will be denoted by 0^*, where

$$0^* = \{x \in \mathbb{Q} \mid x < 0\}.$$

Let $\alpha > 0^*$ be a cut.
Define

$$\beta = \{p \in \mathbb{Q} \text{ such that}$$

$$\text{there exists } s < 0 \text{ such that}$$

$$p + q < s, \ \forall q \in \alpha\}. \tag{1.19}$$

We are going to show that β is a cut and

$$\alpha + \beta = 0^*.$$

Observe that, since $\alpha \neq \mathbb{Q}$, there exists $s_1 \in \mathbb{Q}$ such that $s_1 \notin \alpha$. In particular,

$$q < s_1 + 1/2, \ \forall q \in \alpha,$$

so that

$$q - s_1 - 1 < -1/2 < 0, \ \forall q \in \alpha.$$

From the definition of β we may conclude that $-s_1 - 1 \in \beta$, so that $\beta \neq \emptyset$.

Let $s_2 \in \mathbb{Q}$ be such that $s_2 > 0$. Choose $p \in \alpha$ such that $p > 0$.

Hence $s_2 + p > 0$ so that from the definition of β we may infer that $s_2 \notin \beta$, so that $\beta \neq \mathbb{Q}$.

So, we have verified the first item concerning the cut definition.

Let $p \in \beta$ and let $q \in \mathbb{Q}$ be such that $q < p$.

From the definition of β there exists $s < 0$ such that

$$p + r < s, \; \forall r \in \alpha.$$

In particular,

$$q + r < p + r < s, \; \forall r \in \alpha,$$

so that from the definition of β, $q \in \beta$.

Thus we have verified the second item concerning the cut definition.
Finally, also from $p \in \beta$, we recall that there exists $s < 0$ such that

$$p + q < s, \forall q \in \alpha.$$

Select $m \in \mathbb{N}$ such that $s + 1/m < 0$.
Hence,

$$p + 1/m + q < s + 1/m < 0, \; \forall q \in \alpha,$$

so that, from the definition of β,

$$r = p + 1/m \in \beta.$$

We have verified the third item concerning the cut definition so that we may conclude β is a cut.

Now we are going to prove that

$$\alpha + \beta = 0^*.$$

Let $p \in \alpha + \beta$ be given.
Hence, there exists $r \in \alpha$ and $t \in \beta$ such that

$$p = r + t.$$

From the definition of β there exists $s < 0$ such that $r + t < s$, so that

$$r + t = p \in 0^*.$$

Since $p \in \alpha + \beta$ was arbitrary, we may conclude that

$$\alpha + \beta \subset 0^*. \tag{1.20}$$

Reciprocally, let $u \in 0^*$ be given.

Hence $u \in \mathbb{Q}$ and $u < 0$. Define

$$A = \{p \in \mathbb{Q} \text{ such that } q + p < u, \forall q \in \alpha, \ q > 0\} \subset \beta.$$

Now we are going to show that A is nonempty.
Since α is a cut, there exists $r \in \mathbb{Q}$ such that $r > 0$ and $q < r, \forall q \in \alpha$.
In the next lines, for all $w \in \mathbb{Q}$, we generically denote

$$|w| = \begin{cases} w, & \text{if } w \geq 0, \\ -w, & \text{if } w < 0. \end{cases}$$

Thus, in particular $q < r + |u| + u, \forall q \in \alpha$, so that

$$q - r - |u| < u, \forall q \in \alpha,$$

and hence $-r - |u| \in A$, that is, $A \neq \emptyset$.
Define $r_3 = -r - |u|$, and select $m \in \mathbb{N}$ such that

$$u + |r_3|/m < 0.$$

Define also

$$k_0 = \max A_0,$$

where

$$A_0 = \{k \in \mathbb{N} \cup \{0\} \text{ such that } q + r_3 + k|r_3|/m < u, \ \forall q \in \alpha\}.$$

Choose $q_0 \in \alpha$. Observe that the collection $\{r_3 + k|r_3|/m\}$ defined through A_0 is such that $r_3 + k|r_3|/m \in \beta, \forall k \in A_0$, where β is a cut.
Hence, there exists $s_3 \in \mathbb{Q}$ such that

$$w < s_3, \forall w \in \beta,$$

so that if $k \in A_0$, then $q_0 + r_3 + k|r_3|/m < q_0 + s_3$.
Thus, A_0 is a bounded set of naturals. From this $k_0 = \max A_0$ is well defined.
Also, for some $q_1 \in \alpha$ we have,

$$u \leq q_1 + r_3 + (k_0 + 1)|r_3|/m < u + |r_3|/m < 0.$$

Observe that

$$q + r_3 + k_0|r_3|/m < u, \forall q \in \alpha,$$

so that

$$q + r_3 + (k_0 + 1)|r_3|/m < u + |r_3|/m < 0, \forall q \in \alpha.$$

From this we may conclude that

$$r_3 + (k_0 + 1)|r_3|/m \in \beta,$$

so that

$$q_1 + r_3 + (k_0 + 1)|r_3|/m \in \alpha + \beta.$$

From this and,

$$u \leq q_1 + r_3 + (k_0 + 1)|r_3|/m \in \alpha + \beta,$$

we obtain,

$$u \in \alpha + \beta, \forall u \in 0^*,$$

so that

$$0^* \subset \alpha + \beta.$$

Therefore, from this and (1.20) we have

$$\alpha + \beta = 0^*.$$

With such a result in mind, we denote,

$$-\alpha = \beta.$$

Remark 1.1 Similarly we define $-\alpha$ for the case $\alpha < 0^*$. We leave the details as an exercise.

Exercise 1.9.7 Complete the step 4 proving the sum axioms relating the field definition.

Exercise 1.9.8 Prove that,

$$\text{if } \alpha, \beta, \gamma \in \mathbb{R} \text{ and } \beta < \gamma \text{ then } \alpha + \beta < \alpha + \gamma.$$

5. Definition of a multiplication for $\mathbb{R}$.

Definition 1.9.9 (Multiplication for $\mathbb{R}$) Let $\alpha, \beta \in \mathbb{R}$ be such that $\alpha > 0^*$ and $\beta > 0^*$. We define the multiplication of α by β, denoted by $\alpha \cdot \beta$, by

$$\alpha \cdot \beta = \{p \in \mathbb{Q} \mid p < rs, \text{ for some } r \in \alpha, s \in \beta \text{ such that } r > 0, s > 0\}.$$

The unit in $\mathbb{R}$ will be denoted 1^* and is defined by:

$$1^* = \{x \in \mathbb{Q} \mid x < 1\}.$$

Finally, if $\alpha < 0^*$ and $\beta > 0^*$, we define

$$\alpha \cdot \beta = -[(-\alpha) \cdot \beta],$$

if $\alpha > 0^*$ and $\beta < 0^*$, we define

$$\alpha \cdot \beta = -[\alpha \cdot (-\beta)],$$

and if $\alpha < 0^*$ and $\beta < 0^*$, we define

$$\alpha \cdot \beta = [(-\alpha) \cdot (-\beta)],$$

with the last right side multiplications are performed as above specified for positive cuts.

Let $\alpha > 0^*$ be a cut.
    Define,

$$\beta = \{p \in \mathbb{Q} \text{ such that there exists}$$
$$r \in \mathbb{Q} \text{ such that } 0 < r < 1, \text{ for which } pq < r,$$
$$\forall q \in \alpha, \ q > 0\}. \tag{1.21}$$

We are going to show that β is a cut and

$$\alpha \cdot \beta = 1^*.$$

Pick $s_1 \in \mathbb{Q}$ such that

$$q < s_1, \forall q \in \alpha.$$

Select $r \in \mathbb{Q}$ such that $0 < r < 1$. Select $m \in \mathbb{N}$ such that

$$s_1/m < r.$$

Hence

$$(1/m)q < (1/m)s_1 < r < 1, \ \forall q \in \alpha,$$

so that $1/m \in \beta$. From this we may conclude that $\beta \neq \emptyset$.
    Select $p \in \alpha$ such that $p > 0$.

Define $q = p + 1/p \in \mathbb{Q}$ and observe that

$$pq = p^2 + 1 > 1,$$

so that from the definition of β, $q \notin \beta$ and thus $\beta \neq \mathbb{Q}$.

We have verified the first item concerning the cut definition.

Let $p \in \beta$ be such that $p > 0$ and let $q \in \mathbb{Q}$ be such that $0 < q < p$.

We are going to show that $q \in \beta$.

Firstly, observe that, from the definition of β, there exists $0 < s < 1$ such that

$$pr < s < 1, \forall r \in \alpha, \ r > 0.$$

Therefore,

$$qr < pr < s < 1, \forall r \in \alpha, \ r > 0,$$

so that from the β definition,

$$q \in \beta.$$

We have verified the second item concerning the cut definition.

For the same $p \in \beta$, $p > 0$ we recall again there exists $0 < s < 1$ such that

$$0 < pq < s < 1, \ \forall q \in \alpha, \ q > 0.$$

On the other hand, since α is a cut, we may select $r \in \mathbb{Q}$ such that,

$$q < r, \ \forall q \in \alpha.$$

Pick $m \in \mathbb{N}$ such that

$$0 < s + r/m < 1.$$

Hence,

$$(p + 1/m)q = pq + q/m < s + r/m < 1,$$

$\forall q \in \alpha, q > 0$, so that

$$p + 1/m \in \beta.$$

Therefore, we have verified the third item concerning the cut definition.

We may conclude that β is a cut such that clearly $\beta > 0^*$.

Now we are going to show that

$$\alpha \cdot \beta = 1^*.$$

Let $p \in \alpha \cdot \beta$ be such that $p > 0$.
Hence $p < rt$, for some $r \in \alpha$, $t \in \beta$, $r > 0, t > 0$.
From the β definition there exists $0 < s < 1$, $s \in \mathbb{Q}$ such that

$$rt < s < 1,$$

so that

$$rt \in 1^*,$$

and since 1^* is a cut and $p < rt$ we may infer that

$$p \in 1^*, \ \forall p \in \alpha \cdot \beta.$$

Thus,

$$\alpha \cdot \beta \subset 1^*.$$

Reciprocally, let $p \in 1^*$ be such that $p > 0$, so that $0 < p < 1$.
Define,

$$B_0 = \{q \in \mathbb{Q} \ : \ q > 0 \text{ and } qs < p < 1, \ \forall s \in \alpha, \ s > 0\}.$$

Clearly,

$$B_0 \subset \beta.$$

Since α is a cut, we may select $s_2 \in \mathbb{Q}$, $s_2 > 0$ such that

$$s < s_2, \forall s \in \alpha.$$

Choose $m \in \mathbb{N}$ such that

$$0 < s_2/m < \min\{p, (1 - p)/2\}.$$

Thus,

$$s_2/m < p, \ \ 2s_2/m + p < 1,$$

and

$$0 < s/m < s_2/m < p < 1, \ \forall s \in \alpha, \ s > 0,$$

so that $1/m \in B_0 \subset \beta$, that is, B_0 is nonempty.

Define

$$k_0 = \max B_1,$$

where

$$B_1 = \{k \in \mathbb{N} \,:\, 0 < (k/m)s < p, \forall s \in \alpha, \ s > 0\}.$$

Observe that the collection $\{k/m\}$ defined through B_1 is such that $k/m \in \beta$, $\forall k \in B_1$, where β is a cut.

Also, there exists $s_3 \in \mathbb{Q}$ such that

$$w < s_3, \forall w \in \beta,$$

so that if $k \in B_1$, then $k/m < s_3$.

From this, B_1 is a bounded set of natural numbers, so that we may conclude the maximum k_0 of B_1 is well defined.

From the definition of B_1, we may find $s_1 \in \alpha$ such that,

$$
\begin{aligned}
p &\le [(k_0+1)/m]s_1 < [(k_0+2)/m]s_1 \\
&= (k_0 s_1/m) + (2s_1/m) \\
&< p + \frac{2s_2}{m} \\
&< 1,
\end{aligned}
\tag{1.22}
$$

Observe that,

$$0 < [(k_0+2)/m]s < (k_0 s/m) + (2s_2/m) < p + (2s_2/m) < 1,$$

$\forall s \in \alpha, \ s > 0$.

From this we may infer that

$$(k_0+2)/m \in \beta,$$

and in particular,

$$p < [(k_0+2)/m]s_1,$$

so that from $s_1 \in \alpha$ we may conclude that

$$p \in \alpha \cdot \beta.$$

Since such a $p \in 1^*$ was arbitrary, we may infer that $1^* \subset \alpha \cdot \beta$, so that

$$1^* = \alpha \cdot \beta.$$

With such a result in mind, we denote,

$$\alpha^{-1} = \beta.$$

Exercise 1.9.10 Show that such a multiplication satisfies the field axioms for multiplication.

These last steps would complete the proof of existence of the field $\mathbb{R}$.

1.9.2 The Irrational Set

We have seen that we may associate with each rational $r \in \mathbb{Q}$ a cut

$$r^* = \{x \in \mathbb{Q} \mid x < r\}.$$

However, there are cuts that are not associated with any rational in such a fashion. For example, consider the cut

$$\alpha = \{m \cdot 1/n, \ m \in \mathbb{Z}, \ n \in \mathbb{N} \mid m^2/n^2 < 2\} \cup \{x \in \mathbb{Q} \mid x < 0\}.$$

Clearly we could denote

$$\alpha = [\sqrt{2}]^*.$$

This motivates the following definition.

Definition 1.9.11 We shall define the rational subset of $\mathbb{R}$, denoted by $\mathbb{Q}^*$, by,

$$\mathbb{Q}^* = \{r^* \mid r \in \mathbb{Q}\},$$

where

$$r^* = \{x \in \mathbb{Q} \mid x < r\}.$$

We shall define the irrational subset of $\mathbb{R}$, denoted by $\mathbb{I}^*$, by,

$$\mathbb{I}^* = \mathbb{R} \setminus \mathbb{Q}^*.$$

Exercise 1.9.12

Let $r \in \mathbb{Q}$, prove that $r^* = \{p \in \mathbb{Q} \mid p < r\}$ is a cut.

Let $r \in \mathbb{Q}$ be such that $r > 0$. Prove that $-(r^*) = (-r)^*$ where

$$(-r)^* = \{p \in \mathbb{Q} \ : \ p < -r\}$$

and where

$$-(r^*) = \{p \in \mathbb{Q} \ : \ \text{there exists } s \in \mathbb{Q}, \ s < 0 \text{ such that } p + q < s, \ \forall q \in r^*\}.$$

Let $\alpha \in \mathbb{Q}$, where $\alpha \neq 0$, and $\beta \in \mathbb{I}$. Assuming $\mathbb{Q}^*$ is a field, show that $\alpha + \beta$ and $\alpha \cdot \beta$ are irrationals.

Remark 1.9.13 Since there exists a well-established relation between $\mathbb{Q}$ and $\mathbb{Q}^*$, we shall denote $\mathbb{Q}^*$ simply by $\mathbb{Q}$. Thus, from now on we consider $\mathbb{R}$ and respective operations in a more usual sense.

1.9.3 Archimedean Property

Theorem 1.9.14 (Archimedean Property) *Let x, $y \in \mathbb{R}$ so that $x > 0$. Thus, there exists $n \in \mathbb{N}$ such that $nx > y$.*

Proof Suppose to obtain contradiction, that

$$nx \leq y, \forall n \in \mathbb{N}.$$

Thus

$$A = \{nx \mid n \in \mathbb{N}\},$$

is upper bounded by y, and since $A \subset \mathbb{R}$, there exists $\alpha \in \mathbb{R}$ such that

$$\alpha = \sup A. \tag{1.23}$$

From the definition of supremum, since $x > 0$ we have that $\alpha - x$ is not an upper bound for A. Therefore, there exists $m \in \mathbb{N}$ such that $\alpha - x < mx$.

Hence

$$\alpha < mx + x = (m + 1)x,$$

which contradicts (1.23).

The proof is complete.

Theorem 1.9.15 *Let x, $y \in \mathbb{R}$ such that $x < y$. Thus, there exists $p \in \mathbb{Q}$ such that*

$$x < p < y.$$

Proof Since $x < y$ we have $y - x > 0$. From Theorem 1.9.14, there exists $n \in \mathbb{N}$ such that

$$n(y - x) > 1.$$

Also from Theorem 1.9.14, there exist $m_1, m_2 \in \mathbb{N}$ such that

$$m_1 \cdot 1 > nx$$

and

$$m_2 \cdot 1 > -nx.$$

Thus,

$$-m_2 < nx < m_1.$$

Define

$$\tilde{m}_1 = \max\{z \in \mathbb{Z} \mid z \leq nx\},$$

and

$$\tilde{m}_2 = \min\{z \in \mathbb{Z} \mid z > nx\}.$$

Therefore

$$nx - \tilde{m}_1 < 1$$

and

$$\tilde{m}_2 - nx \leq 1,$$

so that

$$\tilde{m}_2 - \tilde{m}_1 < 2,$$

and since $\tilde{m}_1, \tilde{m}_2 \in \mathbb{Z}$ and $\tilde{m}_1 < \tilde{m}_2$, we obtain

$$\tilde{m}_2 = \tilde{m}_1 + 1.$$

Hence:

$$\tilde{m}_1 \leq nx < \tilde{m}_1 + 1 = \tilde{m}_2 \leq nx + 1 < ny,$$

that is,

$$nx < \tilde{m}_2 < ny,$$

so that

$$x < \frac{\tilde{m}_2}{n} < y.$$

Since

$$p = \frac{\tilde{m}_2}{n} \in \mathbb{Q},$$

the proof is complete.

Theorem 1.9.16 $\sqrt{2} \notin \mathbb{Q}$.

Proof Suppose, to obtain contradiction that $\sqrt{2} \in \mathbb{Q}$. Thus there exist $m \in \mathbb{Z}$ and $n \in \mathbb{N}$ such that $g.c.d.\{m, n\} = 1$ (here g.c.d. stands for the greatest common divisor) and

$$\sqrt{2} = \frac{m}{n}.$$

Hence,

$$2 = \frac{m^2}{n^2},$$

so that

$$2n^2 = m^2.$$

Therefore $2|m^2$ and since 2 is prime, $2|m$, so that $2^2|m^2$. Thus,

$$2\frac{n^2}{2^2} = \frac{m^2}{2^2} \in \mathbb{Z},$$

that is,

$$\frac{n^2}{2} \in \mathbb{Z}.$$

Thus, $2|n^2$ and since 2 is prime we have that $2|n$.

We have got $2|m$ and $2|n$, which contradicts $g.c.d.\{m, n\} = 1$. Thus,

$$\sqrt{2} \notin \mathbb{Q}.$$

Exercise 1.9.17 Prove that $\sqrt{11} \notin \mathbb{Q}$.

Exercise 1.9.18 Prove that $\sqrt{18} \notin \mathbb{Q}$.

Exercise 1.9.19 Prove that $\sqrt{88} \notin \mathbb{Q}$.

Exercise 1.9.20 Prove that $\sqrt{72} \notin \mathbb{Q}$.

Exercise 1.9.21 Let S be an ordered set and let $E \subset S$, where $E \neq \emptyset$. Suppose $\alpha \in S$ is a lower bound for E and $\beta \in S$ is an upper bound for E. Show formally that

$$\alpha \leq \beta.$$

At this point we define the modulus or absolute value of a real number

Definition 1.9.22 (Modulus of a Real Number) Let $x \in \mathbb{R}$. We define the modulus or absolute value of x, denoted by $|x|$, as

$$|x| = \begin{cases} x, & \text{if } x \geq 0, \\ -x, & \text{if } x < 0. \end{cases}$$

Exercise 1.9.23 About the modulus proprieties, prove that

1. $|x| \geq 0, \ \forall x \in \mathbb{R}$,
2. $|x| = 0$ if, and only if, $x = 0$,
3. $|x \cdot y| = |x||y|, \ \forall x, y \in \mathbb{R}$,
4. $|x|^2 = x^2, \ \forall x \in \mathbb{R}$,
5. $x \leq |x|, \forall x \in \mathbb{R}$,
6. Triangular inequality:

$$|x + y| \leq |x| + |y|, \ \forall x, y \in \mathbb{R},$$

7.

$$||x| - |y|| \leq |x - y|, \ \forall x, y \in \mathbb{R}.$$

8. For $a \geq 0$ we have

$$|x| \leq a \text{ if, and only if, } -a \leq x \leq a$$

and

$$|x| \geq a \text{ if, and only if, } x \geq a \text{ or } x \leq -a.$$

1.10   Finite, Infinite, Countable, and Uncountable Sets

Definition 1.10.1 Let A and B be two sets. We say that A and B have the same cardinality if there exists a bijection from A to B. In such a case, we say that A and B are equivalents and denote $A \sim B$.

We may define a relation, defining that A is related to B if $A \sim B$. We may show that such a relation is reflexive, symmetric, and transitive.

Observe that

- $A \sim A$ since $f : A \to A$ defined by $f(x) = x$ is a bijection.
- $A \sim B \Rightarrow B \sim A$, since if $f : A \to B$ is a bijection, then $f^{-1} : B \to A$ is a bijection as well.
- $A \sim B$ and $B \sim C \Rightarrow A \sim C$, since if $f : A \to B$ and $g : B \to C$ are bijections we have that $h = g \circ f : A \to C$ is a bijection from A to C.

We recall that for each $n \in \mathbb{N}$ we earlier had denoted:

$$I_n = \{1, 2, \ldots, n\}.$$

Definition 1.10.2 Let A be a nonempty set.

1. We say that A is finite if $A \sim I_n$, for some $n \in \mathbb{N}$.
2. If A is not finite it is said to be infinite.
3. We say that A is countable if it is finite or if there exists a bijection from A to $\mathbb{N}$, that is, $A \sim \mathbb{N}$.
4. If A is not countable it is said to be uncountable.

As an example, consider $A = \mathbb{Z}$, the set of integers. We may show that $\mathbb{Z}$ is countable. Consider a bijection $f : \mathbb{N} \to \mathbb{Z}$ given by:

$$f(n) = \begin{cases} n/2, & \text{if } n \text{ is even} \\ -(n-1)/2, & \text{if } n \text{ is odd,} \end{cases} \tag{1.24}$$

Clearly such a function is injective and surjective, so that

$$\mathbb{Z} = \{f(n) \mid n \in \mathbb{N}\} = \{1, 2, 3, 4, \ldots\} \cup \{0, -1, -2, -3, \ldots\}.$$

Hence $\mathbb{Z}$ is countable.

Definition 1.10.3 (Sequence) All function whose domain is $\mathbb{N}$ is said to be a sequence. Thus $f : \mathbb{N} \to A$ is a sequence in A. We also denote $f(n) = x_n$ or the sequence simply by $\{x_n\}$.

Theorem 1.10.4 *Let A be countable. Assume that $E \subset A$ and that E is infinite. Under such hypotheses E is countable.*

Proof By hypothesis, A is countable and infinite. Hence A may be expressed by a sequence of distinct elements, since $A \sim \mathbb{N}$, that is, $A = \{x_n\}_{n \in \mathbb{N}}$. Let n_1 be the smallest natural such that $x_{n_1} \in E$. Reasoning inductively, having $n_1 < n_2 < \cdots < n_{k-1}$ define n_k as the smallest integer greater than n_{k-1} such that $x_{n_k} \in E$. Define $f : \mathbb{N} \to E$ by

$$f(k) = x_{n_k}.$$

Being the elements of $\{x_n\}$ distinct, we have that f is injective. Let us show that f is also surjective. Let $x_j \in E$. Define k_0 as the greatest natural number such that $n_{k_0} < j$. Since $x_j \in E$ we obtain $n_{k_0+1} = j$, that is, $x_j \in \{x_{n_k}\}$, so that $E \subset \{x_{n_k}\}$. Since by definition $\{x_{n_k}\} \subset E$, we obtain $E = \{x_{n_k}\}$, so that f is surjective. The proof is complete.

Definition 1.10.5 Let A be a set of indices such that for each $\alpha \in A$ we associate an unique set denoted by E_α. The union of the sets E_α we shall denote by S so that

$$S = \cup_{\alpha \in A} E_\alpha.$$

Thus $x \in S \Leftrightarrow x \in E_\alpha$ for some $\alpha \in A$. If $A = \{1, \ldots, n\}$, we write

$$S = \cup_{i=1}^n E_i,$$

and if $A = \mathbb{N}$ we write

$$S = \cup_{n=1}^\infty E_n.$$

By analogy, the intersection between the sets E_α will be denoted by P, that is

$$P = \cap_{\alpha \in A} E_\alpha.$$

Thus

$$x \in P \Leftrightarrow x \in E_\alpha, \ \forall \alpha \in A.$$

If $A = \{1, \ldots, n\}$ we write

$$P = \cap_{i=1}^n E_i = E_1 \cap E_2 \cap \cdots \cap E_n.$$

If $A = \mathbb{N}$ we write,

$$P = \cap_{n=1}^\infty E_n.$$

Finally, if $A \cap B = \emptyset$ we say that A and B are disjoint.

Theorem 1.10.6 *Let* $\{E_n\}$ *be a sequence of countable sets. Under such hypotheses,* $S = \cup_{n=1}^{\infty} E_n$ *is countable.*

Proof We give just a sketch of the proof. Observe that for each $n \in \mathbb{N}$ E_n is countable, so that we may denote

$$E_n = \{x_{n_k}\}_{k \in \mathbb{N}}.$$

Hence,

$$E_1 = \{x_{11}, , x_{12}, x_{13}, \cdots\}$$

$$E_2 = \{x_{21}, , x_{22}, x_{23}, \cdots\}$$

$$E_3 = \{x_{31}, , x_{32}, x_{33}, \cdots\}$$

$$E_4 = \cdots \cdots \cdots \cdots \cdots$$

$$\cdots = \cdots \cdots \cdots \cdots \cdots$$

We may pass an arrow through x_{11} and define values for a function $f : \mathbb{N} \to S$ by setting $f(1) = x_{11}$. After that, we may pass a diagonal arrow from x_{21} to x_{12} and define $f(2) = x_{21}$, $f(3) = x_{12}$. We may pass a third arrow through x_{31}, x_{22}, and x_{13} and define $f(4) = x_{31}$, $f(5) = x_{22}$, $f(6) = x_{13}$. Proceeding in this fashion, we continue to pass diagonal arrows, associating a natural number through f, as an element of the table is touched by a concerned arrow. Observe that to each element $S = \cup_{n=1}^{\infty} E_n$ will be associated a natural number (Fig. 1.6).

If there exist repeated elements in the table above defined by S, we may infer that f will be a bijection between S and a subset T of $\mathbb{N}$. Hence, from the last theorem:

$$\mathbb{N} \sim T \sim S,$$

Thus

$$\mathbb{N} \sim S,$$

that is, $S = \cup_{n=1}^{\infty} E_n$ is countable.

Theorem 1.10.7 *Let* A *be a countable set. Then,* $A \times A$ *is a countable set.*

Proof The case in which A if finite is immediate. Thus assume A is infinite. Since A is countable we may denote $A = \{x_n\}$. Let $n \in \mathbb{N}$. Define $E_n = \{(x_n, x_m) \mid m \in \mathbb{N}\}$. Consider a bijection $f : E_n \to A$ defined by $f(x_n, x_m) = x_m$. Thus $E_n \sim A \sim \mathbb{N}$. Therefore, $E_n \sim \mathbb{N}$, that is E_n is countable, $\forall n \in \mathbb{N}$. Since

$$A \times A = \cup_{n=1}^{\infty} E_n$$

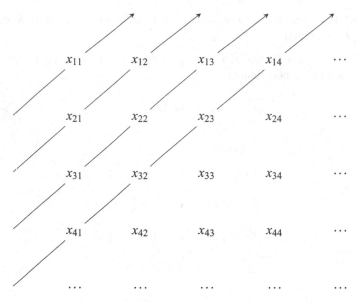

Fig. 1.6 Countability of a countable union of countable sets

from the last theorem we may conclude that $A \times A$ is countable.

The proof is complete.

Exercise 1.10.8 Show that if A and B are countable sets, then so is $A \times B$.

Exercise 1.10.9 Show that if A is countable and $n \in \mathbb{N}$, then $A^n = A \times A \times \cdots \times A$ (n times) is also countable, where we denote $A^2 = A \times A$ and $A^n = A^{n-1} \times A$.

Hint: Use induction.

Theorem 1.10.10 *The rational set $\mathbb{Q}$ is countable.*

Proof We have already proven that $\mathbb{Z}$ is countable and therefore from Theorem 1.10.7, $\mathbb{Z} \times \mathbb{Z}$ is countable. Consider the subset $A \subset \mathbb{Z} \times \mathbb{Z}$ defined by:

$$A = \{(m, n) \in \mathbb{Z} \times \mathbb{N} \mid g.c.d.(m, n) = 1\}.$$

Thus, since A is a subset of a countable set, it is also countable. Consider the function $f : A \to \mathbb{Q}$ defined by

$$f(m, n) = m/n.$$

Thus, f is a bijection from A to $\mathbb{Q}$. Therefore, $\mathbb{N} \sim A \sim \mathbb{Q}$, that is,

$$\mathbb{N} \sim \mathbb{Q}$$

so that $\mathbb{Q}$ is countable. The proof is complete.

Theorem 1.10.11 *The set A of all sequences whose elements are just 0 or 1 is uncountable.*

Proof Suppose, to obtain contradiction that A is countable. Hence, we may write $A = \{S_n\}_{n \in \mathbb{N}}$ where $S_n = \{s_{nk}\}_{k \in \mathbb{N}}$ and where $s_{nk} = 0$ or $s_{nk} = 1$, $\forall\, k, n \in \mathbb{N}$. Define the sequence $\tilde{S} = \{\tilde{s}_k\}$ by

$$\tilde{s}_k = \begin{cases} 1, & \text{if } s_{kk} = 0, \\[2mm] 0, & \text{if } s_{kk} = 1. \end{cases} \tag{1.25}$$

Thus $\tilde{S} \neq S_n$, $\forall n \in \mathbb{N}$. Hence, $\tilde{S} \notin A$. However, by its definition $\tilde{S} \in A$. We have got a contradiction. The proof is complete.

Corollary 1.10.12 $\mathbb{R}$ *is uncountable.*

Proof Consider the interval $[0, 0.2]$ and the decimal expansions of the form $0.\, x_1\, x_2 \cdots x_k \cdots$ where $x_k = 0$ or $x_k = 1$, $\forall k \in \mathbb{N}$. From the last theorem the collection of such numbers is uncountable. Therefore the interval $[0, 0.2]$ is uncountable so that $\mathbb{R} \supset [0, 0.2]$ is uncountable.

Exercise 1.10.13 Show that $\mathbb{R} \setminus \mathbb{Q}$, the set of irrationals, is uncountable.

Exercise 1.10.14 A complex number z is said to be algebraic if there exist $a_0, a_1, \ldots, a_n \in \mathbb{Z}$ not all zero, such that

$$a_0 z^n + a_1 z^{n-1} + \cdots + a_{n-1} z + a_n = 0.$$

Prove that the set of algebraic numbers is countable.

Exercise 1.10.15 Prove by induction that,

1.

$$1^2 + 3^2 + 5^2 + \cdots + (2n-1)^2 = \frac{n(2n+1)(2n-1)}{3}, \ \forall n \in \mathbb{N},$$

2.

$$1 \cdot 2^1 + 2 \cdot 2^2 + 3 \cdot 2^3 + \cdots + n \cdot 2^n = (n-1)2^{n+1} + 2 \ \forall n \in \mathbb{N},$$

3.

$$\frac{1}{1 \cdot 4} + \frac{1}{4 \cdot 7} + \frac{1}{7 \cdot 10} + \cdots + \frac{1}{(3n-2) \cdot (3n+1)} = \frac{n}{3n+1}, \ \forall n \in \mathbb{N},$$

4.

$$n^2 \leq n!, \ \forall n \geq 4,$$

5.

$$2^n > n^2, \ \forall n \geq 5,$$

6.

$$1 + x + x^2 + \cdots + x^n = \frac{1 - x^{n+1}}{1 - x}, \ \forall n \in \mathbb{N}, \ x \in \mathbb{R}, \ \text{such that } x \neq 1,$$

Exercise 1.10.16 Prove by induction that if A and B are square matrices such that

$$AB = BA,$$

then

$$AB^n = B^n A, \ \forall n \in \mathbb{N}.$$

Exercise 1.10.17 Let $f : \mathbb{N} \to \mathbb{N}$ be a function such that

$$f(n + m) = f(n) + f(m), \ \forall m, n \in \mathbb{N}.$$

1. Prove that

$$f(na) = nf(a), \ \forall n \in \mathbb{N}.$$

Hint: Fix $a \in \mathbb{N}$ and define

$$A = \{n \in \mathbb{N} \ : \ f(na) = nf(a)\},$$

and prove by induction that

$$A = \mathbb{N}.$$

2. Prove that there exists $b \in \mathbb{N}$ such that

$$f(n) = bn, \ \forall n \in \mathbb{N}.$$

Exercise 1.10.18 Let $x, y \in \mathbb{R}$ be such that

$$x < y + \varepsilon, \ \forall \varepsilon > 0.$$

Prove formally that

$$x \leq y.$$

Hint: Suppose, to obtain contradiction, that $x > y$.

Exercise 1.10.19 Let $a, b \in \mathbb{R}$ be such that $0 < a < b$. Prove by induction that

$$0 < a^n < b^n, \ \forall n \in \mathbb{N}.$$

Exercise 1.10.20 Let $a \in \mathbb{R}$ be such that $0 < a < 1$.
Let $\varepsilon > 0$. Prove that there exists $n_0 \in \mathbb{N}$ such that

$$0 < a^{n_0} < \varepsilon.$$

Show also that if $n > n_0$, then

$$0 < a^n < a^{n_0} < \varepsilon.$$

Exercise 1.10.21 Let $K \in \mathbb{R}$ be such that $K > 0$. Let $a \in \mathbb{R}$ be such that $a > 1$.
Prove formally that there exists $n_0 \in \mathbb{N}$ such that if $n > n_0$, then

$$a^n > K.$$

Exercise 1.10.22 Let $A, B \subset \mathbb{R}$ be nonempty upper bounded sets. Define

$$A + B = \{x + y \ : \ x \in A \text{ and } y \in B\}.$$

Show that $A + B$ is upper bounded and

$$\sup(A + B) = \sup A + \sup B.$$

Exercise 1.10.23 Let $X \subset \mathbb{R}$. A function $f : X \to \mathbb{R}$ is said to be upper bounded
if its range

$$f(X) = \{f(x) \ : \ x \in X\},$$

is upper bounded. In such a case we define the supremum of f on X by

$$\sup f = \sup\{f(x) \ : \ x \in X\}.$$

Given two functions $f, g : X \to \mathbb{R}$, the sum $(f + g) : X \to \mathbb{R}$ is defined by

$$(f + g)(x) = f(x) + g(x), \ \forall x \in X.$$

Prove that if $f, g : X \to \mathbb{R}$ are upper bounded, then so is $f + g$ and also prove that

$$\sup(f + g) \leq \sup f + \sup g.$$

Finally, give an example for which the strict inequality is valid.

Exercise 1.10.24 Let $A \subset \mathbb{R}$ be a nonempty bounded set. Define $-A$ by

$$-A = \{-x \; : \; x \in A\}.$$

Prove that

$$\inf A = -\sup(-A).$$

Exercise 1.10.25 Let $A \subset \mathbb{R}$ be a nonempty bounded set and let $c > 0$. Define cA by

$$cA = \{cx \; : \; x \in A\}.$$

show that

$$\sup(cA) = c \sup A,$$

and

$$\inf(cA) = c \inf A.$$

Exercise 1.10.26 Let $A, B \subset \mathbb{R}^+ = [0, +\infty)$ be nonempty bounded sets. Define

$$A \cdot B = \{xy \; : \; x \in A \text{ and } y \in B\}.$$

Prove that

$$\sup(A \cdot B) = \sup A \; \sup B$$

and

$$\inf(A \cdot B) = \inf A \; \inf B.$$

Exercise 1.10.27 Verify if the sets below indicated are countable or uncountable. Please justify your answers.

1. The set of all sequences having only 0 and 1 entries, with exactly 3 entries equal to 1.

2. For $k \in \mathbb{N}$, the set of all sequences having only 0 and 1 entries, with at most k entries equal to 1.
3.

$$\mathscr{A} = \{\{a_n\} \ : \ a_n \in \mathbb{N} \cup \{0\}, \ \text{such that } a_n = 0, \ \forall n \in \mathbb{N},$$

$$\text{except for a finite number of } n's\}.$$

4.

$$\mathscr{B} = \{\{a_n\} \ : \ a_n \in \mathbb{N} \text{ and } a_n \geq a_{n+1}, \forall n \in \mathbb{N}\}.$$

5.

$$\mathscr{C} = \{\{a_n\} \ : \ a_n \text{ is prime } \forall n \in \mathbb{N}\}.$$

6.

$$\mathscr{D} = \{\{a_n\} \ : \ a_n \in \mathbb{N} \text{ and } a_{n+1} \text{ is a multiple of } a_n, \ \forall n \in \mathbb{N}\}.$$

7.

$$\mathscr{E} = \{\{a_n\} \ : \ a_n \in \mathbb{N} \text{ and } a_{n+1} \text{ is a divisor of } a_n, \ \forall n \in \mathbb{N}\}.$$

8. The set of all polynomials in x with rational coefficients.
9. The set of all power series $\sum_{n=0}^{\infty} a_n x^n$, such that $a_n \in \mathbb{Z}, \ \forall n \in \mathbb{N} \cup \{0\}$.

Exercise 1.10.28 Prove that if a set B is countable and there exists an injective function

$$f : A \to B,$$

then A is countable.

Exercise 1.10.29 Let A be a countable set and B be a finite set. Constructing a bijection between $\mathbb{N}$ and $A \cup B$, show that $A \cup B$ is countable.

Chapter 2
Metric Spaces

2.1 Introduction

In this chapter we develop the main definitions and properties related to metric spaces. The main reference for this chapter is [12] where more details may be found.

Definition 2.1.1 (Metric Spaces) Let U be a set. We say that U is a metric space if it is possible to define a function $d : U \times U \to \mathbb{R}^+ = [0, +\infty)$ such that

1. $d(u, v) > 0$, if $u \neq v$ and $d(u, u) = 0, \forall u, v \in U$.
2. $d(u, v) = d(v, u), \forall u, v \in U$.
3. $d(u, v) \leq d(u, w) + d(w, v), \forall u, v, w \in U$.

The function d is called a metric for U so that we may denote the metric space in question by (U, d).

Examples 2.1.2

1. $U = \mathbb{R}$, that is the real set is a metric space with the metric $d : \mathbb{R} \times \mathbb{R} \to \mathbb{R}^+$ given by

$$d(u, v) = |u - v|, \forall u, v \in \mathbb{R}.$$

2. $U = \mathbb{R}^2$ is a metric space with the metric $d : \mathbb{R}^2 \times \mathbb{R}^2 \to \mathbb{R}^+$ given by

$$d(\mathbf{u}, \mathbf{v}) = \sqrt{(u_1 - v_1)^2 + (u_2 - v_2)^2}, \forall \mathbf{u}, \mathbf{v} \in \mathbb{R}^2,$$

where $\mathbf{u} = (u_1, u_2)$ and $\mathbf{v} = (v_1, v_2)$.

© Springer International Publishing AG, part of Springer Nature 2018
F. S. Botelho, *Real Analysis and Applications*,
https://doi.org/10.1007/978-3-319-78631-5_2

3. $U = \mathbb{R}^n$ is a metric space with the metric $d : \mathbb{R}^n \times \mathbb{R}^n \to \mathbb{R}^+$ given by

$$d(\mathbf{u}, \mathbf{v}) = \|\mathbf{u} - \mathbf{v}\| = \sqrt{(u_1 - v_1)^2 + \cdots + (u_n - v_n)^2}, \forall \mathbf{u}, \mathbf{v} \in \mathbb{R}^n,$$

where $\mathbf{u} = (u_1, \ldots, u_n)$ and $\mathbf{v} = (v_1, \ldots, v_n) \in \mathbb{R}^n$.

Let $\mathbf{u} \in \mathbb{R}^n$, we define an open ball with center $\mathbf{u}$ and radius r, denoted by $B_r(\mathbf{u})$ by

$$B_r(\mathbf{u}) = \{v \in \mathbb{R}^n \mid \|\mathbf{v} - \mathbf{u}\| < r\}.$$

By analogy, we define a closed ball with center $u \in \mathbb{R}^n$ and radius r, denoted by $\overline{B}_r(\mathbf{u})$ by

$$\overline{B}_r(\mathbf{u}) = \{v \in \mathbb{R}^n \mid \|\mathbf{v} - \mathbf{u}\| \leq r\}.$$

Definition 2.1.3 A set $E \subset \mathbb{R}^n$ is said to be convex if

$$\lambda \mathbf{u} + (1 - \lambda)\mathbf{v} \in E, \forall \mathbf{u}, \mathbf{v} \in E, \lambda \in [0, 1].$$

Exercise 2.1.4 Let $\mathbf{x} \in \mathbb{R}^n$ and $r > 0$. Prove that $B_r(\mathbf{x})$ is convex.

2.1.1   Some Fundamental Definitions

Definition 2.1.5 (Neighborhood) Let (U, d) be a metric space. Let $u \in U$ and $r > 0$. We define the neighborhood of center u and radius r, denoted by $V_r(u)$, by

$$V_r(u) = \{v \in U \mid d(u, v) < r\}.$$

Definition 2.1.6 (Limit Point) Let (U, d) be a metric space and $E \subset U$. A point $u \in U$ is said to be a limit point of E if for each $r > 0$ there exists $v \in V_r(u) \cap E$ such that $v \neq u$.

We shall denote by E' the set of all limit points of E.

Example 2.1.7 $U = \mathbb{R}^2$, $E = B_r(0)$. Thus $E' = \overline{B}_r(0)$.

Remark 2.1.8 In the next definitions U shall denote a metric space with a metric d.

Definition 2.1.9 (Isolated Point) Let $u \in E \subset U$. We say that u is an isolated point of E if it is not a limit point of E.

Example 2.1.10 $U = \mathbb{R}^2$, $E = B_1((0, 0)) \cup \{(3, 3)\}$. Thus $(3, 3)$ is an isolated point of E.

Definition 2.1.11 (Closed Set) Let $E \subset U$ and let E' be the set of limit points of E. We say that E is closed if $E \supset E'$.

Example 2.1.12 Let $U = \mathbb{R}^2$ and $r > 0$, thus $E = \overline{B}_r((0,0))$ is closed.

Definition 2.1.13 A point $u \in E \subset U$ is said to be an interior point of E if there exists $r > 0$ such that $V_r(u) \subset E$, where

$$V_r(U) = \{v \in U \mid d(u,v) < r\}.$$

Example 2.1.14 For $U = \mathbb{R}^2$, let $E = B_1((0,0)) \cup \{(3,3)\}$, for example $u = (0.25, 0.25)$ is an interior point of E, in fact, for $r = 0.5$, if $v \in V_r(u)$ then $d(u,v) < 0.5$ so that $d(v,(0,0)) \leq d((0,0),u) + d(u,v) \leq \sqrt{1/8} + 0.5 < 1$ that is, $v \in B_1((0,0))$ and thus $V_r(u) \subset B_1((0,0))$. We may conclude that u is an interior point of $B_1((0,0))$ and therefore an interior point of E is interior.

Definition 2.1.15 (Open Set) $E \subset U$ is said to be open if all its points are interior.

Example 2.1.16 For $u = \mathbb{R}^2$, the ball $B_1(0,0)$ is open.

Definition 2.1.17 Let $E \subset U$, we define its complement, denoted by E^c, by:

$$E^c = \{v \in U \mid v \notin E\}.$$

Definition 2.1.18 A set $E \subset U$ is said to be bounded if there exists $M > 0$ such that

$$\sup\{d(u,v) \mid u, v \in E\} \leq M.$$

Definition 2.1.19 A set $E \subset U$ is said to be dense in U if each point of U is either a point of E or it is a limit point of E, that is, $U = E \cup E'$.

Example 2.1.20 The set $\mathbb{Q}$ is dense in $\mathbb{R}$. Let $u \in \mathbb{R}$ and let $r > 0$. Thus, from Theorem 1.9.15 there exists $v \in \mathbb{Q}$ such that $u < v < u + r$, that is, $v \in \mathbb{Q} \cap V_r(u)$ and $v \neq u$, where $V_r(u) = (u - r, u + r)$. Therefore u is a limit point of $\mathbb{Q}$. Since $u \in \mathbb{R}$ is arbitrary, we may conclude that $\mathbb{R} \subset \mathbb{Q}'$, that is, $\mathbb{Q}$ is dense in $\mathbb{R}$.

Theorem 2.1.21 *Let (U,d) be a metric space. Let $u \in U$ and $r > 0$. Thus, $V_r(u)$ is open.*

Proof First we recall that

$$V_r(u) = \{v \in U \mid d(u,v) < r\}.$$

Let $v \in V_r(u)$. We have to show that v is an interior point of $V_r(u)$. Define $r_1 = r - d(u,v) > 0$. We shall show that $V_{r_1}(v) \subset V_r(u)$.

Let $w \in V_{r_1}(v)$, thus $d(v,w) < r_1$. Hence

$$d(u,w) \leq d(u,v) + d(v,w) < d(u,v) + r_1 = r.$$

Therefore $w \in V_r(u), \forall w \in V_{r_1}(v)$, that is $V_{r_1}(v) \subset V_r(u)$, so that we may conclude that v is an interior point of $V_r(u), \forall v \in V_r(u)$, that is, $V_r(u)$ is open.

The proof is complete.

Theorem 2.1.22 *Let u be a limit point of $E \subset U$, where (U, d) is a metric space. Thus, each neighborhood of u has an infinite number of points of E, distinct from u.*

Proof Suppose to obtain contradiction, that there exists $r > 0$ such that $V_r(u)$ has a finite number of points of E distinct from u. Let $\{v_1, \ldots, v_n\}$ be such points of $V_r(u) \cap E$ distinct from u. Choose $0 < r_1 < \min\{d(u, v_1), d(u, v_2), \ldots, d(u, v_n)\}$. Hence $V_{r_1}(u) \subset V_r(u)$ and $v_i \notin V_{r_1}(u), \forall i \in \{1, 2, \ldots, n\}$. Therefore either $V_{r_1}(u) \cap E = \{u\}$ or $V_{r_1} \cap E = \emptyset$, which contradicts the fact that u is a limit point of E.

The proof is complete.

Corollary 2.1.23 *Let $E \subset U$ be a finite set. Then E has no limit points.*

2.1.2 Properties of Open and Closed Sets in a Metric Space

In this section we present some basic properties of open and closed sets.

Proposition 2.1.24 *Let $\{E_\alpha, \ \alpha \in L\}$ be a collection of sets. Thus,*

$$(\cup_{\alpha \in L} E_\alpha)^c = \cap_{\alpha \in L} E_\alpha^c.$$

Proof Observe that

$$
\begin{aligned}
u \in (\cup_{\alpha \in L} E_\alpha)^c &\Leftrightarrow u \notin \cup_{\alpha \in L} E_\alpha \\
&\Leftrightarrow u \notin E_\alpha, \forall \alpha \in L \\
&\Leftrightarrow u \in E_\alpha^c, \forall \alpha \in L \\
&\Leftrightarrow u \in \cap_{\alpha \in L} E_\alpha^c.
\end{aligned}
\tag{2.1}
$$

Exercise 2.1.25 Prove that

$$(\cap_{\alpha \in L} E_\alpha)^c = \cup_{\alpha \in L} E_\alpha^c.$$

Theorem 2.1.26 *Let (U, d) be a metric space and $E \subset U$. Thus, E is open if and only if E^c is closed.*

Proof Suppose E^c is closed. Choose $u \in E$, thus $u \notin E^c$ and therefore u is not a limit point E^c. Hence there exists $r > 0$ such that $V_r(u) \cap E^c = \emptyset$. Hence, $V_r(u) \subset E$, that is, u is an interior point of $E, \forall u \in E$, so that E is open.

Reciprocally, suppose E is open. Let $u \in (E^c)'$. Thus for each $r > 0$ there exists $v \in V_r(u) \cap E^c$ such that $v \neq u$, so that

$$V_r(u) \nsubseteq E, \ \forall r > 0.$$

Therefore u is not an interior point of E. Since E is open we have that $u \notin E$, that is, $u \in E^c$. Hence $(E^c)' \subset E^c$, that is, E^c is closed.

The proof is complete.

Corollary 2.1.27 *Let (U, d) be a metric space, $F \subset U$ is closed if and only if F^c is open.*

Theorem 2.1.28 *Let (U, d) be a metric space.*

1. *If $G_\alpha \subset U$ and G_α is open $\forall \alpha \in L$ then*

$$\cup_{\alpha \in L} G_\alpha$$

is open.

2. *If $F_\alpha \subset U$ and F_α is closed $\forall \alpha \in L$ then*

$$\cap_{\alpha \in L} F_\alpha$$

is closed.

3. *If $G_1, \ldots, G_n \subset U$ and G_i is open $\forall i \in \{1, \ldots, n\}$ then*

$$\cap_{i=1}^{n} G_i$$

is open.

4. *If $F_1, \ldots, F_n \subset U$ and F_i is closed $\forall i \in \{1, \ldots, n\}$ then*

$$\cup_{i=1}^{n} F_i$$

is closed.

Proof

1. Let $G_\alpha \subset U$, where G_α is open $\forall \alpha \in L$. Let $u \in \cup_{\alpha \in L} G_\alpha$. Thus $u \in G_{\alpha_0}$ for some $\alpha_0 \in L$. Since G_{α_0} is open, there exists $r > 0$ such that $V_r(u) \subset G_{\alpha_0} \subset \cup_{\alpha \in L} G_\alpha$. Hence, u is an interior point, $\forall u \in \cup_{\alpha \in L} G_\alpha$. Thus $\cup_{\alpha \in L} G_\alpha$ is open.

2. Let $F_\alpha \subset U$, where F_α is closed $\forall \alpha \in L$. Thus, F_α^c is open $\forall \alpha \in L$. From the last item, we have $\cup_{\alpha \in L} F_\alpha^c$ is open so that

$$\cap_{\alpha \in L} F_\alpha = \left(\cup_{\alpha \in L} F_\alpha^c \right)^c$$

is closed.

3. Let $G_1, \ldots, G_n \subset U$ be open sets. Let

$$u \in \cap_{i=1}^{n} G_i.$$

Thus,

$$u \in G_i, \forall i \in \{1, \ldots, n\}.$$

Since G_i is open, there exists $r_i > 0$ such that $V_{r_i}(u) \subset G_i$.

Define $r = \min\{r_1, \ldots, r_n\}$. Hence, $V_r(u) \subset V_{r_i}(u) \subset G_i, \forall i \in \{1, \ldots, n\}$ and therefore

$$V_r(u) \subset \cap_{i=1}^n G_i.$$

This means that u is an interior point of $\cap_{i=1}^n G_i$, and being $u \in \cap_{i=1}^n G_i$ arbitrary we obtain that $\cap_{i=1}^n G_i$ is open.

4. Let $F_1, \ldots, F_n \subset U$ be closed sets. Thus, $F_1^c, \ldots, F_n^c$ are open. Thus, from the last item, we obtain:

$$\cap_{i=1}^n F_i^c$$

is open, so that

$$\cup_{i=1}^n F_i = \left(\cap_{i=1}^n F_i^c \right)^c$$

is closed.

The proof is complete.

Exercise 2.1.29 Let (U, d) be a metric space and let $u_0 \in U$. Show that $A = \{u_0\}$ is closed. Let $B = \{u_1, \ldots, u_n\} \subset U$. Show that B is closed.

Definition 2.1.30 (Closure) Let (U, d) be a metric space and let $E \subset U$. Denote the set of limit points of E by E'. We define the closure of E, denoted by $\overline{E}$, by:

$$\overline{E} = E \cup E'.$$

Examples 2.1.31

1. Let $U = \mathbb{R}^2$, $E = B_1(0, 0)$, we have that $E' = \overline{B}_1(0, 0)$, so that in this example $\overline{E} = E \cup E' = E'$.
2. Let $U = \mathbb{R}$, $A = \{1/n : n \in \mathbb{N}\}$, we have that $A' = \{0\}$, and thus $\overline{A} = A \cup A' = A \cup \{0\}$.

Theorem 2.1.32 *Let (U, d) be a metric space and $E \subset U$. Thus,*

1. *$\overline{E}$ is closed.*
2. *$E = \overline{E} \Leftrightarrow E$ is closed.*
3. *If $F \supset E$ and F is closed, then $F \supset \overline{E}$.*

Proof

1. Observe that $\overline{E} = E \cup E'$. Let $u \in \overline{E}^c$. Thus $u \notin E$ and $u \notin E'$ (u is not a limit point of E). Therefore, there exists $r > 0$ such that $V_r(u) \cap E = \emptyset$, that is, $V_r(u) \subset E^c$, thus, u is an interior point of E^c.

   We shall prove that $V_r(u) \cap \overline{E} = \emptyset$. Let $v \in V_r(u)$ and define $r_1 = r - d(u, v) > 0$. We shall show that

   $$V_{r_1}(v) \subset V_r(u).$$

   Let $w \in V_{r_1}(v)$, thus $d(v, w) < r_1$ and therefore

   $$d(u, w) \le d(u, v) + d(v, w) < d(u, v) + r_1 = r,$$

   that is, $w \in V_r(u)$. Hence,

   $$V_{r_1}(v) \subset V_r(u),$$

   and thus v is not a limit point of E, that is, $v \in \overline{E}^c$, $\forall v \in V_r(u)$. Thus, $V_r(u) \subset \overline{E}^c$ which means that u is an interior point of $\overline{E}^c$, so that $\overline{E}^c$ is open, and hence $\overline{E}$ is closed.

2. Observe that $E \subset \overline{E} = E \cup E'$. Suppose that E is closed. Thus $E \supset E'$, that is $E \supset E \cup E' = \overline{E}$. Hence $E = \overline{E}$. Suppose $E = \overline{E}$. From the last item $\overline{E}$ is closed, and thus E is closed.

3. Let F be a closed set such that $F \supset E$. Thus, $F' \supset E'$.

   Hence

   $$F = \overline{F} = F \cup F' \supset E \cup E' = \overline{E}.$$

The proof is complete.

Exercises 2.1.33

1. In the proof of the last theorem we have used a result which now is requested to be proven in an exercise form.

   Let U be a metric space. Assume $A \subset B \subset U$. Show that $A' \subset B'$.

2. Let U be a metric space and let $A, B \subset U$. Show that

   $$A' \cup B' = (A \cup B)'.$$

3. Let U be a metric space and let $E \subset U$. Show that E' is closed.

4. Let $B_1, B_2, \ldots$ be subsets of a metric space U.

   (a) Show that if

   $$B_n = \cup_{i=1}^{n} B_i \text{ then } \overline{B}_n = \cup_{i=1}^{n} \overline{B}_i.$$

(b) Show that if

$$B = \cup_{i=1}^{\infty} B_i \text{ then } \overline{B} \supset \cup_{i=1}^{\infty} \overline{B}_i.$$

5. Let U be a metric space and let $E \subset U$. Recall that the interior of E, denoted by E°, is defined as the set of all interior points of E.

   (a) Show that E° is open.
   (b) Show that E is open, if and only if, $E = E^\circ$.
   (c) Show that if $G \subset E$ and G is open, then $G \subset E^\circ$.
   (d) Prove that $(E^\circ)^c - \overline{E^c}$.
   (e) Do E and $\overline{E}$ have always the same interior? If not, present a counter example.
   (f) Do E and E^0 have always the same closure? If not, present a counter example.

6. Prove that $\mathbb{Q}$, the rational set, has empty interior.
7. Prove that $\mathbb{I}$, the set of irrationals, has empty interior.
8. Prove that given $x, y \in \mathbb{R}$ such that $x < y$, there exists $\alpha \in \mathbb{I}$, such that

$$x < \alpha < y.$$

9. Prove that $\mathbb{I}$ is dense in $\mathbb{R}$.
      Hint: Prove that

$$x \in \mathbb{I}', \ \forall x \in \mathbb{R},$$

   where $\mathbb{I}'$ denotes the set of limit of points of $\mathbb{I}$.
10. Let $B \subset \mathbb{R}$ be an open set. Show that for all $x \in \mathbb{R}$ the set

$$x + B = \{x + y \mid y \in B\}$$

   is open.
11. Let $A, B \subset \mathbb{R}$ be open sets. Show that the set

$$A + B = \{x + y \ : \ x \in A \text{ and } y \in B\},$$

   is open.
12. Let $B \subset \mathbb{R}$ be an open set. Show that for all $x \in \mathbb{R}$ such that $x \neq 0$ the set

$$x \cdot B = \{x \cdot y \mid y \in B\}$$

   is open.

13. Let $A, B \subset \mathbb{R}$, show that

    (a)

$$(A \cap B)^\circ = A^\circ \cap B^\circ,$$

    (b)

$$(A \cup B)^\circ \supset A^\circ \cup B^\circ,$$

    and give an example in which the inclusion is proper.

14. Let $A \subset \mathbb{R}$ be an open set and $a \in A$. Prove that $A \setminus \{a\}$ is open.
15. Let $A, B \subset \mathbb{R}$. Prove that:

    (a) $\overline{A \cup B} = \overline{A} \cup \overline{B}$,
    (b) $\overline{A \cap B} \subset \overline{A} \cap \overline{B}$, and give an example for which the last inclusion is proper.

16. Show that a set A is dense in $\mathbb{R}$ if, and only if, A^c has empty interior.
17. Let $F \subset \mathbb{R}$ be a closed set and let $x \in F$. Show that x is an isolated point of F if, and only if, $F \setminus \{x\}$ is closed.
18. Show that if $A \subset \mathbb{R}$ is uncountable, then so is A'.
19. Show that if $A \subset \mathbb{R}$, then $\overline{A} \setminus A'$ is countable.
20. Let U be a metric space and let $A \subset U$ be an open set. Assume $a_1, \ldots, a_n \in A$. Prove that $A \setminus \{a_1, \ldots, a_n\}$ is open.
21. Let U be a metric space, let $A \subset U$ be an open set, and let $F \subset U$ be a closed one.

    Show that $A \setminus F$ is open and $F \setminus A$ is closed.
22. Let $A \subset \mathbb{R}$ be an uncountable set. Prove that $A \cap A' \neq \emptyset$.

2.1.3 Compact Sets

Definition 2.1.34 (Open Covering) Let (U, d) be a metric space. We say that a collection of sets $\{G_\alpha, \alpha \in L\} \subset U$ is an open covering of $A \subset U$ if

$$A \subset \cup_{\alpha \in L} G_\alpha$$

and G_α is open, $\forall \alpha \in L$.

Definition 2.1.35 (Compact Set) Let (U, d) be a metric space and $K \subset U$. We say that K is compact if each open covering $\{G_\alpha, \ \alpha \in L\}$ of K admits a finite sub-covering. That is, if $K \subset \cup_{\alpha \in L} G_\alpha$, and G_α is open $\forall \alpha \in L$, then there exist $\alpha_1, \alpha_2, \ldots, \alpha_n \in L$ such that $K \subset \cup_{i=1}^n G_{\alpha_i}$.

Theorem 2.1.36 *Let (U, d) be a metric space. Let $K \subset U$ where K is compact. Under such hypotheses, K is closed.*

Proof Let us show that K^c is open. Let $u \in K^c$, let us generically denote in this proof $V_r(u) = V(u, r)$.

For each $v \in K$ we have $d(u, v) > 0$. Define $r_v = d(u, v)/2$. Thus,

$$V(u, r_v) \cap V(v, r_v) = \emptyset, \forall v \in K. \tag{2.2}$$

Observe that

$$\cup_{v \in K} V(v, r_v) \supset K.$$

since K is compact, there exist $v_1, \ldots, v_n \in K$ such that

$$K \subset \cup_{i=1}^n V(v_i, r_{v_i}). \tag{2.3}$$

Define $r_0 = \min\{r_{v_1}, \ldots, r_{v_n}\}$, thus

$$V(u, r_0) \subset V(u, r_{v_i}), \forall i \in \{1, \ldots, n\},$$

so that from this and (2.2) we get

$$V(u, r_0) \cap V(v_i, r_{v_i}) = \emptyset, \forall i \in \{1, 2, \ldots, n\}.$$

Hence,

$$V(u, r_0) \cap \left(\cup_{i=1}^n V(v_i, r_{v_i})\right) = \emptyset.$$

From this and (2.3) we obtain, $V(u, r_0) \cap K = \emptyset$, that is $V(u, r_0) \subset K^c$. Therefore u is an interior point of K^c and being $u \in K^c$ arbitrary, K^c is open so that K is closed.

The proof is complete.

Theorem 2.1.37 *Let (U, d) be a metric space. Assume $F \subset K \subset U$, where K is compact and F is closed.*

Under such hypotheses, F is compact.

Proof Let $\{G_\alpha, \alpha \in L\}$ be an open covering of F, that is

$$F \subset \cup_{\alpha \in L} G_\alpha.$$

Observe that $U = F \cup F^c \supset K$, and thus,

$$F^c \cup (\cup_{\alpha \in L} G_\alpha) \supset K.$$

Therefore, since F^c is open $\{F^c, G_\alpha, \alpha \in L\}$ is an open covering of K, and since K is compact, there exist $\alpha_1, \ldots, \alpha_n \in L$ such that

$$F^c \cup G_{\alpha_1} \cup \cdots \cup G_{\alpha_n} \supset K \supset F.$$

Therefore

$$G_{\alpha_1} \cup \cdots \cup G_{\alpha_n} \supset F,$$

so that F is compact.

Exercise 2.1.38 Show that if F is closed and K is compact, then $F \cap K$ is compact.

Theorem 2.1.39 *If $\{K_\alpha, \ \alpha \in L\}$ is a collection of compact sets in a metric space (U, d) such that the intersection of each finite subcollection is nonempty, then*

$$\cap_{\alpha \in L} K_\alpha \neq \emptyset.$$

Proof Suppose, to obtain contradiction, that

$$\cap_{\alpha \in L} K_\alpha = \emptyset. \tag{2.4}$$

Fix $\alpha_0 \in L$ and denote $L_1 = L \setminus \{\alpha_0\}$. From (2.4) we obtain

$$K_{\alpha_0} \cap \left(\cap_{\alpha \in L_1} K_\alpha\right) = \emptyset.$$

Hence

$$K_{\alpha_0} \subset \left(\cap_{\alpha \in L_1} K_\alpha\right)^c,$$

that is,

$$K_{\alpha_0} \subset \cup_{\alpha \in L_1} K_\alpha^c.$$

Since, K_{α_0} is compact and K_α^c is open, $\forall \alpha \in L$, there exist $\alpha_1, \alpha_2, \ldots, \alpha_n \in L_1$ such that

$$K_{\alpha_0} \subset \cup_{j=1}^n K_{\alpha_j}^c = \left(\cap_{j=1}^n K_{\alpha_j}\right)^c,$$

therefore,

$$K_{\alpha_0} \cap \left(\cap_{j=1}^n K_{\alpha_j}\right) = K_{\alpha_0} \cap K_{\alpha_1} \cap \cdots \cap K_{\alpha_n} = \emptyset,$$

which contradicts the hypotheses. The proof is complete.

Corollary 2.1.40 *Let (U, d) be a metric space. If $\{K_n, \ n \in \mathbb{N}\} \subset U$ is a sequence of compact nonempty sets such that $K_n \supset K_{n+1}, \forall n \in \mathbb{N}$, then $\cap_{n=1}^\infty K_n \neq \emptyset$.*

Theorem 2.1.41 *Let (U, d) be a metric space. If $E \subset K \subset U$, K is compact and E is infinite, then E has at least one limit point in K.*

Proof Suppose, to obtain contradiction, that no point of K is a limit point of E. Then, for each $u \in K$ there exists $r_u > 0$ such that $V(u, r_u)$ has at most one point of E, namely, u if $u \in E$. Observe that $\{V(u, r_u), \ u \in K\}$ is an open covering of K and therefore of E. Since each $V(u, r_u)$ has at most one point of E which is infinite, no finite sub-covering (relating the open cover in question) covers E, and hence no finite sub-covering covers $K \supset E$, which contradicts the fact that K is compact. This completes the proof.

Theorem 2.1.42 *Let $\{I_n\}$ be a sequence of bounded closed nonempty real intervals, such that $I_n \supset I_{n+1}, \forall n \in \mathbb{N}$. Thus, $\cap_{n=1}^{\infty} I_n \neq \emptyset$.*

Proof Let $I_n = [a_n, b_n]$ and let $E = \{a_n, \ n \in \mathbb{N}\}$. Thus, $E \neq \emptyset$ and E is upper bounded by b_1. Let $x = \sup E$.

Observe that, given $m, n \in \mathbb{N}$ we have that

$$a_n \leq a_{n+m} \leq b_{n+m} \leq b_m,$$

so that

$$\sup_{n \in \mathbb{N}} a_n \leq b_m, \forall m \in \mathbb{N},$$

that is, $x \leq b_m, \forall m \in \mathbb{N}$. Hence,

$$a_m \leq x \leq b_m, \forall m \in \mathbb{N},$$

that is,

$$x \in [a_m, b_m], \ \forall m \in \mathbb{N},$$

so that $x \in \cap_{m=1}^{\infty} I_m$.

The proof is complete.

Theorem 2.1.43 *Let $I = [a, b] \subset \mathbb{R}$ be a bounded closed nonempty real interval. thus, I is compact.*

Proof Observe that if $x, y \in [a, b]$, then $|x - y| \leq (b - a)$. Suppose there exists an open covering of I, denoted by $\{G_\alpha, \ \alpha \in L\}$ for which there is no finite sub-covering.

Let $c = (a + b)/2$. Thus, either $[a, c]$ or $[c, b]$ has no finite sub-covering related to $\{G_\alpha, \ \alpha \in L\}$. Denote such an interval by I_1. Dividing I_1 into two connected closed subintervals of same size, we get an interval I_2 for which there is no finite sub-covering related to $\{G_\alpha, \ \alpha \in L\}$.

Proceeding in this fashion, we may obtain a sequence of closed intervals $\{I_n\}$ such that

1. $I_n \supset I_{n+1}, \forall n \in \mathbb{N}$.
2. No finite subcollection of $\{G_\alpha, \alpha \in L\}$ covers $I_n, \ \forall n \in \mathbb{N}$.
3. If $x, y \in I_n$, then $|x - y| \leq 2^{-n}(b - a)$.

From the last theorem, there exists $x^* \in \mathbb{R}$ such that $x^* \in \cap_{n=1}^{\infty} I_n \subset I \subset \cup_{\alpha \in L} G_\alpha$. Hence, there exists $\alpha_0 \in L$ such that $x^* \in G_{\alpha_0}$. Since G_{α_0} is open, there exists $r > 0$ such that

$$V_r(x^*) = (x^* - r, x^* + r) \subset G_{\alpha_0}.$$

Choose $n_0 \in \mathbb{N}$ such that

$$2^{-n_0}(b - a) < r/2.$$

Hence, since $x^* \in I_{n_0}$, if $y \in I_{n_0}$ then from item 3 above, $|y - x^*| \le 2^{-n_0}(b - a) < r/2$, that is $y \in V_r(x^*) \subset G_{\alpha_0}$.

Therefore

$$y \in I_{n_0} \Rightarrow y \in G_{\alpha_0},$$

so that $I_{n_0} \subset G_{\alpha_0}$, which contradicts the item 2 indicated above.

The proof is complete.

Theorem 2.1.44 (Heine–Borel) *Let $E \subset \mathbb{R}$, thus the following three properties are equivalent.*

1. *E is closed and bounded.*
2. *E is compact.*
3. *Each infinite subset of E has a limit point in E.*

Proof

- 1 implies 2: Let $E \subset \mathbb{R}$ be closed and bounded. Thus, since E is bounded there exists $[a, b]$ a bounded closed interval such that $E \subset [a, b]$. From the last theorem $[a, b]$ is compact and since E is closed, from Theorem 2.1.37 we may infer that E is compact.
- 2 implies 3: This follows from Theorem 2.1.41.
- 3 implies 1: We prove the contrapositive, that is, the negation of 1 implies the negation of 3.

  The negation of 1 is: E is not bounded or E is not closed. If $E \subset \mathbb{R}$ is not bounded, for each $n \in \mathbb{N}$ there exists $x_n \in E$ such that $|x_n| > n + |x_{n-1}| \ge n$. Hence $\{x_n\}$ has no limit points so that we have got the negation of 3.

  On the other hand, suppose E is not closed. Thus there exists $x_0 \in \mathbb{R}$ such that $x_0 \in E'$ and $x_0 \notin E$.

  Since $x_0 \in E'$, for each $n \in \mathbb{N}$ there exists $x_n \in E$ such that $|x_n - x_0| < 1/n$ ($x_n \in V_{1/n}(x_0)$).

  Let $y \in E$, we are going to show that y is not limit point $\{x_n\} \subset E$. Observe that,

  $$|x_n - y| \ge |x_0 - y| - |x_n - x_0|$$

$$> |x_0 - y| - 1/n$$

$$> |x_0 - y|/2 > 0 \qquad\qquad (2.5)$$

for all n sufficiently big.

Hence y is not a limit point of $\{x_n\}$, $\forall y \in E$. Therefore $\{x_n\} \subset E$ is an infinite set with no limit point in E.

In any case, we have got the negation of 3. This completes the proof.

Exercises 2.1.45

1. Let U be a metric space and let $\{K_\lambda, \ \lambda \in L\}$ be a collection of compact sets, such that $K_\lambda \subset U$, $\forall \lambda \in L$. Prove that $\cap_{\lambda \in L} K_\lambda$ is compact.
2. Let U be a metric space and let $K_1, K_2, \ldots, K_n \subset U$ be compact sets. Prove that

$$\cup_{j=1}^{n} K_j$$

is compact.

Theorem 2.1.46 (Weierstrass) *Any real set which is bounded and infinite has a limit point in* $\mathbb{R}$.

Proof Let $E \subset \mathbb{R}$ be a bounded infinite set. Thus, there exists $r > 0$ such that $E \subset [-r, r] = I_r$. Since E is infinite and I_r is compact, from Theorem 2.1.41, E has a limit point in $I_r \subset \mathbb{R}$. The proof is complete.

Chapter 3
Real Sequences and Series

3.1 Introduction

This chapter develops a standard study on real sequences and series. We develop in detail topics such as superior and inferior limits for bounded sequences and their applicability in the proof of root and ratio tests for convergence of series. The comparison criterion for series is also extensively addressed and applied to a great variety of situations.

The main references for this chapter are [9, 12].

3.2 Real Sequences

We start with the formal definition of sequence.

Definition 3.2.1 (Sequence) A real sequence is a real function whose domain is $\mathbb{N}$, the natural set. Denoting such a function by $f : \mathbb{N} \to \mathbb{R}$ we also denote $f(n) = x_n$ and the sequence simply by $\{x_n\}$.

Definition 3.2.2 (Subsequence) Let $\{x_n\}$ be a sequence. Given the set $\{n_1, n_2, \ldots\}$ $= \{n_k, \ k \in \mathbb{N}\} \subset \mathbb{N}$ such that $n_1 < n_2 < n_3 < \cdots < n_k < \cdots$, the subsequence $\{x_{n_k}\}$ of $\{x_n\}$ is defined by

$$\tilde{f}(k) = x_{n_k}, \ \forall k \in \mathbb{N}.$$

© Springer International Publishing AG, part of Springer Nature 2018
F. S. Botelho, *Real Analysis and Applications*,
https://doi.org/10.1007/978-3-319-78631-5_3

Definition 3.2.3 (Limit of a Real Sequence) Let $\{x_n\} \subset \mathbb{R}$. We say that $a \in \mathbb{R}$ is the limit of $\{x_n\}$ as n goes to infinity if, for each $\varepsilon > 0$, there exists $n_0 \in \mathbb{N}$ such that if $n > n_0$, then

$$|x_n - a| < \varepsilon.$$

In such a case we write,

$$\lim_{n \to \infty} x_n = a$$

and we say that $\{x_n\}$ converges to a as $n \to \infty$. If a sequence does not converge to a real number, it is said to be divergent.

Theorem 3.2.4 (Limit Uniqueness) *Let $\{x_n\} \subset \mathbb{R}$. Assume $\lim_{n \to \infty} x_n = a \in \mathbb{R}$ and $\lim_{n \to \infty} x_n = b \in \mathbb{R}$.*

Under such hypotheses, $a = b$.

Proof Let $\varepsilon > 0$.
From

$$\lim_{n \to \infty} x_n = a,$$

there exists $n_1 \in \mathbb{N}$ such that if $n > n_1$, then

$$|x_n - a| < \varepsilon/2.$$

From

$$\lim_{n \to \infty} x_n = b,$$

there exists $n_2 \in \mathbb{N}$ such that if $n > n_2$, then

$$|x_n - b| < \varepsilon/2.$$

Choose $n_0 > \max\{n_1, n_2\}$, thus

$$\begin{aligned}
|b - a| &= |b - x_{n_0} + x_{n_0} - a| \\
&\leq |b - x_{n_0}| + |x_{n_0} - a| \\
&< \varepsilon/2 + \varepsilon/2 = \varepsilon.
\end{aligned} \tag{3.1}$$

Hence $|b - a| < \varepsilon$, $\forall \varepsilon > 0$, so that $|b - a| = 0$ that is, $b = a$. The proof is complete.

Theorem 3.2.5 *Let $\{x_n\} \subset \mathbb{R}$. Suppose $\lim_{n \to \infty} x_n = a \in \mathbb{R}$. Under such hypotheses, for all subsequence $\{x_{n_k}\}$ we have*

$$\lim_{k \to \infty} x_{n_k} = a.$$

Proof Let $\{x_{n_k}\}$ a subsequence of $\{x_n\} \subset \mathbb{R}$, where

$$\lim_{n \to \infty} x_n = a \in \mathbb{R}.$$

Let $\varepsilon > 0$. Thus there exists $n_0 \in \mathbb{N}$ such that if $n > n_0$, then $|x_n - a| < \varepsilon$.

Choose $k_0 \in \mathbb{N}$ such that $n_{k_0} > n_0$. Thus if $k > k_0$, we have $n_k > n_{k_0} > n_0$ and therefore

$$|x_{n_k} - a| < \varepsilon, \text{ if } k > k_0.$$

Hence

$$\lim_{k \to \infty} x_{n_k} = 0.$$

Theorem 3.2.6 *All real convergent sequence is bounded.*

Proof Let $\{x_n\} \subset \mathbb{R}$ be such that $\lim_{n \to \infty} x_n = a \in \mathbb{R}$.

For $\varepsilon = 1$ there exists $n_0 \in \mathbb{N}$ such that if $n > n_0$, then

$$|x_n - a| < \varepsilon = 1.$$

Hence $x_n \in (a - 1, a + 1)$, if $n > n_0$. Define $F = \{x_1, x_2, \ldots, x_{n_0}, a - 1, a + 1\}$, $c = \min F$ and $d = \max F$. Clearly $x_n \in [c, d], \forall n \in \mathbb{N}$. Therefore $\{x_n\}$ is bounded.

The proof is complete.

Remark 3.2.7 The reciprocal is false. For example $\{x_n\} = \{0, 1, 0, 1, \ldots\}$ is bounded but not convergent.

Definition 3.2.8 (Monotone Sequences) Let $\{x_n\} \subset \mathbb{R}$.

1. If $x_n \le x_{n+1}, \forall n \in \mathbb{N}$, we say that such a sequence is nondecreasing.
2. If $x_n < x_{n+1}, \forall n \in \mathbb{N}$, we say that such a sequence is increasing.
3. If $x_n \ge x_{n+1}, \forall n \in \mathbb{N}$, we say that such a sequence is nonincreasing.
4. If $x_n > x_{n+1}, \forall n \in \mathbb{N}$, we say that such a sequence is decreasing.

If $\{x_n\}$ is either nondecreasing, increasing, nonincreasing, or decreasing is said to be monotonic.

Theorem 3.2.9 *All real, monotonic, bounded sequence is convergent.*

Proof Suppose $\{x_n\}$ is nondecreasing, that is, $x_n \le x_{n+1}, \forall n \in \mathbb{N}$, and bounded (other cases may be dealt similarly).

Thus $\{x_n\}$ is upper bounded. Hence there exists $a \in \mathbb{R}$ such that $a = \sup\{x_n\}$.

Let $\varepsilon > 0$, since $a - \varepsilon < a = \sup\{x_n\}$, $a - \varepsilon$ is not an upper bound for $\{x_n\}$. Therefore there exists $n_0 \in \mathbb{N}$ such that $a - \varepsilon < x_{n_0} \le a$, and since $\{x_n\}$ is nondecreasing, if $n > n_0$ we have

$$a - \varepsilon < x_{n_0} \leq x_n \leq a < a + \varepsilon,$$

that is, $|x_n - a| < \varepsilon,$ if $n > n_0$. Thus, $\lim_{n \to \infty} x_n = a$. The proof is complete.

Corollary 3.2.10 *Let $\{x_n\} \subset \mathbb{R}$ be a monotonic sequence which has a convergent subsequence. Thus, $\{x_n\}$ is convergent.*

Proof Suppose $\{x_n\}$ is nondecreasing, other cases are dealt similarly. Suppose that

$$\lim_{k \to \infty} x_{n_k} = a \in \mathbb{R}.$$

From the last theorem

$$a = \sup_{k \in \mathbb{N}} \{x_{n_k}\}.$$

Fix $n \in \mathbb{N}$. Thus there exists $k_0 \in \mathbb{N}$ such that $n_{k_0} > n$, so that

$$x_n \leq x_{n_{k_0}} \leq a,$$

and hence

$$\sup_{n \in \mathbb{N}} \{x_n\} \leq a.$$

On the other hand, since $\{x_{n_k}\} \subset \{x_n\}$, we have

$$\sup_{n \in \mathbb{N}} \{x_n\} \geq \sup_{k \in \mathbb{N}} \{x_{n_k}\} = a.$$

Therefore,

$$\lim_{n \to \infty} x_n = \sup_{n \in \mathbb{N}} \{x_n\} = \sup_{k \in \mathbb{N}} \{x_{n_k}\} = a.$$

The proof is complete.

Exercises 3.2.11

1. Let $a \in \mathbb{R}$ be such that $|a| < 1$. Define $\{x_n\} \subset \mathbb{R}$ by

$$x_n = 1 + a + a^2 + \cdots + a^n, \ \forall n \in \mathbb{N}.$$

  Under such assumptions, prove that

$$\lim_{n \to \infty} x_n = \frac{1}{1 - a}.$$

3.3 Proprieties of Limits

We start with the following theorem.

Theorem 3.3.1 *Let $\{x_n\}$ and $\{y_n\} \subset \mathbb{R}$. Assume $\lim_{n\to\infty} x_n = 0$, and $\{y_n\}$ is bounded.*

Under such hypotheses,

$$\lim_{n\to\infty} x_n y_n = 0.$$

Proof From the hypotheses, there exists $c > 0$ such that

$$|y_n| \le c, \forall n \in \mathbb{N}.$$

Let $\varepsilon > 0$. Since $\lim_{n\to\infty} x_n = 0$ there exists $n_0 \in \mathbb{N}$ such that if $n > n_0$, then

$$|x_n| < \varepsilon/c.$$

Hence:

$$|x_n y_n| = |x_n||y_n| < \frac{\varepsilon}{c} c = \varepsilon, \text{ if } n > n_0.$$

Thus,

$$\lim_{n\to\infty} x_n y_n = 0.$$

The proof is complete.

Exercise 3.3.2 Let $x \in \mathbb{R}$. Show that

$$\lim_{n\to\infty} \frac{\sin(nx)}{n} = 0.$$

Theorem 3.3.3 *Let $\{x_n\}$ and $\{y_n\} \subset \mathbb{R}$ be sequences such that*

$$\lim_{n\to\infty} x_n = a \in \mathbb{R}$$

and

$$\lim_{n\to\infty} y_n = b \in \mathbb{R}.$$

Under such hypotheses we have,

1. $\lim_{n\to\infty} \alpha x_n = \alpha a, \forall \alpha \in \mathbb{R}.$
2. $\lim_{n\to\infty}(x_n + y_n) = a + b,$

3. $\lim_{n \to \infty} x_n y_n = ab$,
4. $\lim_{n \to \infty} x_n / y_n = a/b, \quad if \, b \neq 0$.

Proof We prove just the items 2, 3, and 4 leaving the proof of item 1 as an exercise.
Let $\varepsilon > 0$.

1. From $\lim_{n \to \infty} x_n = a$, there exists $n_1 \in \mathbb{N}$ such that if $n > n_1$, then

$$|x_n - a| < \varepsilon/2.$$

From $\lim_{n \to \infty} y_n = b$, there exists $n_2 \subset \mathbb{N}$ such that if $n > n_2$, then $|y_n - b| < \varepsilon/2$. Define $n_3 = \max\{n_1, n_2\}$. Thus, if $n > n_3$, then

$$|(x_n + y_n) - (a + b)| \leq |x_n - a| + |y_n - b| < \varepsilon/2 + \varepsilon/2 = \varepsilon.$$

therefore

$$\lim_{n \to \infty} (x_n + y_n) = a + b.$$

2. Observe that $|x_n y_n - ab| = |x_n y_n - x_n b + x_n b - ab| \leq |x_n||y_n - b| + |x_n - a||b|$.
Observe also that since $\{x_n\}$ is convergent, there exists $K > 0$ such that $|x_n| < K, \forall n \in \mathbb{N}$. From $\lim_{n \to \infty} y_n = b$, there exists $n_1 \in \mathbb{N}$ such that if $n > n_1$, then

$$|y_n - b| < \frac{\varepsilon}{2K}.$$

From $\lim_{n \to \infty} x_n = a$ there exists $n_2 \in \mathbb{N}$ such that if $n > n_2$, then

$$|x_n - a| < \frac{\varepsilon}{2(|b| + 1)}.$$

Thus, defining $n_3 = \max\{n_1, n_2\}$, if $n > n_3$ we have

$$\begin{aligned}
|x_n y_n - ab| &= |x_n y_n - x_n b + x_n b - ab| \\
&\leq |x_n||y_n - b| + |x_n - a||b| \\
&< K\frac{\varepsilon}{2K} + |b|\frac{\varepsilon}{2(|b| + 1)} \\
&< \varepsilon.
\end{aligned} \tag{3.2}$$

Thus,

$$\lim_{n \to \infty} x_n y_n = ab.$$

3. Assume $b \neq 0$ where

$$\lim_{n \to \infty} y_n = b.$$

Firstly we are going to prove that

$$\lim_{n \to \infty} \frac{1}{y_n} = \frac{1}{b}.$$

Observe that

$$\left| \frac{1}{y_n} - \frac{1}{b} \right| = \frac{|y_n - b|}{|y_n||b|}. \tag{3.3}$$

Let $\varepsilon_1 = \frac{|b|}{2} > 0$.
From $\lim_{n \to \infty} y_n = b$, there exists $n_1 \in \mathbb{N}$ such that if $n > n_1$, then

$$|y_n - b| < \varepsilon_1 = \frac{|b|}{2}.$$

Thus,

$$||y_n| - |b|| \leq |y_n - b| < \frac{|b|}{2},$$

so that

$$-\frac{|b|}{2} < |y_n| - |b|,$$

that is,

$$|y_n| > \frac{|b|}{2},$$

and hence

$$0 \leq \frac{1}{|y_n|} < \frac{2}{|b|}. \tag{3.4}$$

Let a new $\varepsilon > 0$ be given.
From $\lim_{n \to \infty} y_n = b$, there exists $n_2 \in \mathbb{N}$ such that if $n > n_2$, then

$$|y_n - b| < \varepsilon \frac{|b|^2}{2}. \tag{3.5}$$

Define $n_0 = \max\{n_1, n_2\}$.

Therefore, if $n > n_0$, from (3.3), (3.4), and (3.5) we obtain

$$\left| \frac{1}{y_n} - \frac{1}{b} \right| = \frac{|y_n - b|}{|y_n||b|}$$

$$\leq \frac{\varepsilon|b|^2}{2} \frac{2}{|b|\,|b|}$$

$$= \varepsilon. \tag{3.6}$$

From this, we may infer that

$$\lim_{n \to \infty} \frac{1}{y_n} = \frac{1}{b}.$$

From this and the item 3, we may finally obtain

$$\lim_{n \to \infty} x_n \frac{1}{y_n} = \lim_{n \to \infty} x_n \lim_{n \to \infty} \frac{1}{y_n} = \frac{a}{b}.$$

The proof is complete.

Theorem 3.3.4 (Sign Keeping) *Let $\{x_n\} \subset \mathbb{R}$ be a sequence such that*

$$\lim_{n \to \infty} x_n = a \in \mathbb{R}$$

where $a > 0$. Thus, there exists $n_0 \in \mathbb{N}$ such that if $n > n_0$, then $x_n > 0$.

Proof Let $\varepsilon = a/2$. From

$$\lim_{n \to \infty} x_n = a,$$

there exists $n_0 \in \mathbb{N}$ such that if $n > n_0$, then

$$|x_n - a| < \varepsilon = a/2.$$

Thus,

$$-a/2 < x_n - a < a/2,$$

that is,

$$x_n > a - a/2 = a/2 > 0,$$

if $n > n_0$.

The proof is complete.

Corollary 3.3.5 *Let $\{x_n\}$ and $\{y_n\} \subset \mathbb{R}$ be convergent real sequences. Thus, if*

$$x_n \leq y_n, \forall n \in \mathbb{N},$$

then

$$\lim_{n \to \infty} x_n \leq \lim_{n \to \infty} y_n.$$

Proof Exercise.

Corollary 3.3.6 *Let $\{x_n\} \subset \mathbb{R}$ be a real convergent sequence. If*

$$x_n \geq a, \forall n \in \mathbb{N},$$

then

$$\lim_{n \to \infty} x_n \geq a.$$

Theorem 3.3.7 *Let $\{x_n\}$, $\{y_n\}$ and $\{z_n\} \subset \mathbb{R}$ be real sequences such that*

$$x_n \leq y_n \leq z_n, \ \forall n \in \mathbb{N}.$$

Assume

$$\lim_{n \to \infty} x_n = a \ and \ \lim_{n \to \infty} z_n = a.$$

Under such hypotheses,

$$\lim_{n \to \infty} y_n = a.$$

Proof Let $\varepsilon > 0$. From

$$\lim_{n \to \infty} x_n = a$$

there exists $n_1 \in \mathbb{N}$ such that if $n > n_1$, then

$$|x_n - a| < \varepsilon.$$

From

$$\lim_{n \to \infty} z_n = a,$$

there exists $n_2 \in \mathbb{N}$ such that if $n > n_2$, then

$$|z_n - a| < \varepsilon.$$

Defining $n_3 = \max\{n_1, n_2\}$, we have that if $n > n_3$, then

$$a - \varepsilon < x_n \leq y_n \leq z_n < a + \varepsilon,$$

that is,

$$|y_n - a| < \varepsilon,$$

if $n > n_3$.

Therefore,

$$\lim_{n \to \infty} y_n = a.$$

The proof is complete.

Theorem 3.3.8 *$a \in \mathbb{R}$ is the limit of a real subsequence of $\{x_n\} \subset \mathbb{R}$ if, and only if, for each $\varepsilon > 0$ and $k \in \mathbb{N}$, there exists $n > k$, $n \in \mathbb{N}$ such that $x_n \in (a - \varepsilon, a + \varepsilon)$.*

Proof Necessity:

Let $\{x_n\} \subset \mathbb{R}$ and let $\{x_{n_k}\}$ be a subsequence such that

$$\lim_{k \to \infty} x_{n_k} = a.$$

Let $\varepsilon > 0$. Thus, there exists $k_0 \in \mathbb{N}$ such that if $k > k_0$, then

$$|x_{n_k} - a| < \varepsilon.$$

Thus, $x_{n_k} \in (a - \varepsilon, a + \varepsilon)$, if $k > k_0$. Fix $k_1 \in \mathbb{N}$. Let $k_2 \in \mathbb{N}$ be such that $k_2 > \max\{k_0, k_1\}$.

Hence $n_{k_2} \geq k_2 > k_1$ and

$$x_{n_{k_2}} \in (a - \varepsilon, a + \varepsilon)..$$

Therefore the condition in question is necessary.

Sufficiency:

Suppose for each $\varepsilon > 0$ and each $k > 0$, there exists $n > k$ such that $|x_n - a| < \varepsilon$.

Let $\varepsilon = 1$, thus there exists $n_1 > 1$ such that $x_{n_1} \in (a - 1, a + 1)$.

For $\varepsilon = 1/2$ from the hypotheses, there exists $n_2 > n_1$ and $x_{n_2} \in (a - 1/2, a + 1/2)$.

Hence, inductively for $\varepsilon = 1/k$ there exists $n_k > n_{k-1} > n_{k-2} > \cdots > n_1$ such that $x_{n_k} \in (a - 1/k, a + 1/k)$, $\forall k \in \mathbb{N}$.

Let us now show that

$$\lim_{k \to \infty} x_{n_k} = a.$$

Let $\varepsilon > 0$.
Thus there exists $k_0 \in \mathbb{N}$ such that

$$1/k_0 < \varepsilon.$$

Therefore, if $k > k_0$ we obtain,

$$x_{n_k} \in (a - 1/k_0, a + 1/k_0) \subset (a - \varepsilon, a + \varepsilon).$$

Thus,

$$|x_{n_k} - a| < \varepsilon, \ \forall k > k_0,$$

so that,

$$\lim_{k \to \infty} x_{n_k} = a.$$

The proof is complete.

Definition 3.3.9 (Subsequential Limit) We say that $a \in \mathbb{R}$ is a subsequential limit for a sequence $\{x_n\} \subset \mathbb{R}$, if a is the limit of a subsequence of $\{x_n\}$.

3.4 Superior and Inferior Limits for a Bounded Sequence

Definition 3.4.1 Let $\{x_n\} \subset \mathbb{R}$ be a real bounded sequence, that is, assume that there exist $\alpha, \beta \in \mathbb{R}$ such that,

$$\alpha \leq x_n \leq \beta, \forall n \in \mathbb{N}.$$

Define $X_n = \{x_n, x_{n+1}, x_{n+2}, \ldots\}, \ \forall n \in \mathbb{N}$.
Thus

$$[\alpha, \beta] \supset X_1 \supset X_2 \supset \cdots \supset X_n \supset \cdots$$

Define $a_n = \inf X_n$ and $b_n = \sup X_n, \ \forall n \in \mathbb{N}$
Observe that

$$\alpha \leq a_1 \leq a_2 \leq \cdots \leq a_n \leq \cdots \leq b_n \leq b_{n-1} \leq \cdots \leq b_2 \leq b_1 \leq \beta.$$

Hence $\{a_n\}$ is monotonic and bounded, therefore convergent. We define the inferior limit of $\{x_n\}$, denoted by $a \in \mathbb{R}$, by

$$a = \lim_{n \to \infty} a_n = \sup_{n \in \mathbb{N}}\{a_n\}.$$

We also denote

$$a = \liminf_{n \to \infty} x_n.$$

Observe that

$$\liminf_{n \to \infty} x_n = \lim_{n \to \infty} \left(\inf\{x_n, x_{n+1}, \ldots\}\right).$$

Thus,

$$\liminf_{n \to \infty} x_n = \lim_{n \to \infty} \left(\inf\{x_k \; : \; k \geq n\}\right) = \sup_{n \in \mathbb{N}} \left(\inf\{x_k \; : \; k \geq n\}\right).$$

On the other hand $\{b_n\}$ is also monotone and bounded, therefore convergent.
Since $\{b_n\}$ is nondecreasing and bounded, we may define the superior limit of $\{x_n\}$, denoted by b, by

$$b = \lim_{n \to \infty} b_n = \inf_{n \in \mathbb{N}}\{b_n\}.$$

We also denote

$$b = \limsup_{n \to \infty} x_n.$$

Observe that,

$$\limsup_{n \to \infty} x_n = \lim_{n \to \infty} \left(\sup\{x_n, x_{n+1}, \ldots\}\right).$$

thus,

$$\limsup_{n \to \infty} x_n = \lim_{n \to \infty} \left(\sup\{x_k \; : \; k \geq n\}\right) = \inf_{n \in \mathbb{N}} \left(\sup\{x_k \; : \; k \geq n\}\right).$$

Theorem 3.4.2 *Let $\{x_n\} \subset \mathbb{R}$ be a bounded sequence. Thus $\liminf_{n \to \infty} x_n$ is the smallest subsequential limit of $\{x_n\}$ and $\limsup_{n \to \infty} x_n$ is the largest one.*

Proof First we show that $\liminf_{n \to \infty} x_n$ is a subsequential limit.
Define

$$a = \liminf_{n \to \infty} x_n.$$

Observe that

$$a = \lim_{n \to \infty} a_n,$$

where

$$a_n = \inf\{x_n, x_{n+1}, \ldots\}.$$

Let $\varepsilon > 0$ and $n_0 \in \mathbb{N}$. We may obtain $n_1 > n_0$ such that

$$a - \varepsilon < a_{n_1} < a + \varepsilon.$$

Thus, since $a + \varepsilon$ is not a lower bound for $\{x_{n_1}, x_{n_1+1}, \ldots\}$ there exists $n \geq n_1 > n_0$, such that

$$a - \varepsilon < a_{n_1} \leq x_n < a + \varepsilon.$$

Thus, for all $n_0 \in \mathbb{N}$ we may obtain $n > n_0$ such that

$$x_n \in (a - \varepsilon, a + \varepsilon).$$

From Theorem 3.3.8 there exists a subsequence which converges to a. Thus, a is a subsequential limit.

Let $c < a$, we shall show that c is not a subsequential limit.

Since $a = \lim_{n \to \infty} a_n$ and $c < a$, for $\varepsilon = a - c$, there exists $n_0 \in \mathbb{N}$ such that if $n > n_0$, then

$$a - a_n < \varepsilon = a - c,$$

so that

$$c < a_n, \ \forall n > n_0.$$

Choose $n_1 > n_0$, thus $c < a_{n_1} \leq a_n \leq a, \forall n > n_1$, and in particular, defining $\delta = a_{n_1} - c$, we obtain,

$$c + \delta = a_{n_1} \leq x_n, \forall n > n_1.$$

Thus $x_n \notin (c - \delta, c + \delta), \forall n > n_1$ so that by Theorem 3.3.8, c is not a subsequential limit.

The proof concerning the superior limit is similar and it is left as an exercise.

Corollary 3.4.3 *All bounded sequence $\{x_n\} \subset \mathbb{R}$ has a convergent subsequence.*

Proof Just observe that from the last theorem $a = \liminf_{n\to\infty} x_n$ is a subsequential limit of $\{x_n\}$.

Corollary 3.4.4 *A bounded sequence $\{x_n\} \subset \mathbb{R}$ is convergent if and only if*

$$\liminf_{n\to\infty} x_n = \limsup_{n\to\infty} x_n.$$

Proof Suppose

$$\lim_{n\to\infty} x_n = a.$$

Since a is the unique subsequential limit, we have

$$\liminf_{n\to\infty} x_n = \limsup_{n\to\infty} x_n = \lim_{n\to\infty} x_n = a.$$

Reciprocally, suppose

$$\liminf_{n\to\infty} x_n = \limsup_{n\to\infty} x_n = 0.$$

Denoting

$$a_n = \inf\{x_n, x_{n+1}, \ldots\},$$

and

$$b_n = \sup\{x_n, x_{n+1}, \ldots\},$$

we have

$$\lim_{n\to\infty} a_n = a = \lim_{n\to\infty} b_n.$$

Let $\varepsilon > 0$ be given. Thus, there exists $n_1 \in \mathbb{N}$ such that if $n > n_1$ we have that

$$a - \varepsilon < a_n < a + \varepsilon.$$

And also, there exists $n_2 \in \mathbb{N}$ such that if $n > n_2$, then

$$a - \varepsilon < b_n < a + \varepsilon.$$

Thus, if $n > \max\{n_1, n_2\} \equiv n_0$ we have,

$$a - \varepsilon < a_n \le x_n \le b_n < a + \varepsilon.$$

Thus,

$$\lim_{n \to \infty} x_n = a.$$

The proof is complete.

Exercise 3.4.5 Let $\{x_n\} \subset \mathbb{R}$ be a bounded sequence. Let $c \in \mathbb{R}$ such that $c < \liminf_{n \to \infty} x_n = 0$.
Show that there exists $n_0 \in \mathbb{N}$ such that if $n > n_0$, then $x_n > c$.

3.5 Real Cauchy Sequences

Definition 3.5.1 A sequence $\{x_n\} \subset \mathbb{R}$ is said to be a Cauchy one if for each $\varepsilon > 0$ there exists $n_0 > 0$ such that if $m > n_0$ and $n > n_0$, then

$$|x_m - x_n| < \varepsilon.$$

Exercise 3.5.2 Show that every real convergent sequence is a Cauchy one.

Exercise 3.5.3 Show that every real Cauchy sequence in bounded.

Theorem 3.5.4 *Every Cauchy sequence $\{x_n\} \subset \mathbb{R}$ is convergent.*

Proof Let $\{x_n\} \subset \mathbb{R}$ be a Cauchy sequence.
Let $\varepsilon > 0$. Thus there exists $n_0 \in \mathbb{N}$ such that if $m, n > n_0$ then

$$-\varepsilon < x_n - x_m < \varepsilon.$$

Thus,

$$-\varepsilon \le \limsup_{n \to \infty}\{x_n\} - x_m \le \varepsilon.$$

From this we get,

$$-\varepsilon \le x_m - \limsup_{n \to \infty}\{x_n\} \le \varepsilon,$$

so that,

$$-\varepsilon \le \liminf_{m \to \infty}\{x_m\} - \limsup_{n \to \infty}\{x_n\} \le \varepsilon.$$

Since $\varepsilon > 0$ is arbitrary and $\{x_n\}$ is bounded, we obtain,

$$\liminf_{m \to \infty}\{x_m\} = \limsup_{n \to \infty}\{x_n\} = \lim_{n \to \infty} x_n.$$

3.6   A Special Class of Cauchy Sequences

Let $0 \le \lambda < 1$. Suppose that $\{x_n\} \subset \mathbb{R}$ is such that,

$$|x_{n+2} - x_{n+1}| \le \lambda|x_{n+1} - x_n|, \forall n \in \mathbb{N}.$$

We shall prove that $\{x_n\}$ is Cauchy sequence so that it is convergent. Observe that

$$|x_3 - x_2| \le \lambda|x_2 - x_1|$$
$$|x_4 - x_3| \le \lambda|x_3 - x_2| \le \lambda^2|x_2 - x_1|$$
$$\cdots \quad \cdots \quad \cdots$$
$$|x_{n+1} - x_n| \le \lambda^{n-1}|x_2 - x_1|. \tag{3.7}$$

Thus, for $n, p \in \mathbb{N}$ we have that

$$|x_{n+p} - x_n| = |x_{n+p} - x_{n+p-1} + x_{n+p-1} - x_{n+p-2} + x_{n+p-1} + \cdots - x_{n+1}$$
$$+ x_{n+1} - x_n|$$
$$\le |x_{n+p} - x_{n+p-1}| + |x_{n+p-1} - x_{n+p-2}| + \cdots + |x_{n+1} - x_n|$$
$$\le (\lambda^{n+p-2} + \lambda^{n+p-3} + \cdots + \lambda^{n-1})|x_2 - x_1|. \tag{3.8}$$

Therefore,

$$|x_{n+p} - x_n| \le \lambda^{n-1}(\lambda^{p-1} + \lambda^{p-2} + \cdots + 1)|x_2 - x_1|$$
$$\le \frac{\lambda^{n-1}(1 - \lambda^p)}{1 - \lambda}|x_2 - x_1|$$
$$\le \frac{\lambda^{n-1}}{1 - \lambda}|x_2 - x_1|. \tag{3.9}$$

Observe that

$$\lim_{n \to \infty} \frac{\lambda^{n-1}}{1 - \lambda}|x_2 - x_1| = 0.$$

Let $\varepsilon > 0$. Thus there exists $n_0 \in \mathbb{N}$ such that if $n > n_0$, then

$$\frac{\lambda^{n-1}}{1 - \lambda}|x_2 - x_1| < \varepsilon.$$

From this and (3.9), we get

$$|x_{n+p} - x_n| < \varepsilon, \ \text{if } n > n_0.$$

Hence, for $m = p + n$ we obtain

$$|x_m - x_n| < \varepsilon, \ \text{if } m > n > n_0.$$

Thus, $\{x_n\}$ is a Cauchy sequence, therefore it is convergent.

3.7 Infinite Limits

In this short section we establish the definitions of infinite limits.

Definition 3.7.1 (Infinite Limit) Let $\{x_n\} \subset \mathbb{R}$ be a real sequence.

We say that the limit of x_n as n goes to infinity is plus infinite $(+\infty)$, as for each $A > 0$ there exists $n_0 \in \mathbb{N}$ such that if $n > n_0$, then $x_n > A$.

In such a case we denote:

$$\lim_{n \to \infty} x_n = +\infty.$$

We say that the limit of x_n as n goes to infinity is minus infinite $(-\infty)$, as for each $B < 0$ there exists $n_0 \in \mathbb{N}$ such that if $n > n_0$, then $x_n < B$.

In such a case we denote:

$$\lim_{n \to \infty} x_n = -\infty.$$

3.8 Numerical Series

A numerical series is an infinite sum of the form

$$\sum_{n=1}^{\infty} a_n$$

where $\{a_n\}$ is a real sequence.

The sequence $\{s_n\}$ where

$$s_n = \sum_{k=1}^{n} a_k = a_1 + a_2 + \cdots + a_n,$$

is called the sequence of partial sums relating the series in question.

If there exists $s \in \mathbb{R}$ such that

$$s = \lim_{n \to \infty} s_n = \lim_{n \to \infty} \sum_{k=1}^{n} a_k,$$

we say that the series converges to s and write

$$\sum_{n=1}^{\infty} a_n = s.$$

If the sequence of partial sums is not convergent, we say that $\sum_{n=1}^{\infty} a_n$ is divergent.

Theorem 3.8.1 *If $\sum_{n=1}^{\infty} a_n$ is convergent then*

$$\lim_{n \to \infty} a_n = 0.$$

Proof Suppose

$$\lim_{n \to \infty} \sum_{k=1}^{n} a_k = \lim_{n \to \infty} s_n = s \in \mathbb{R}.$$

Observe that

$$s_n - s_{n-1} = a_n,$$

Hence,

$$\lim_{n \to \infty} a_n = \lim_{n \to \infty} (s_n - s_{n-1}) = \lim_{n \to \infty} s_n - \lim_{n \to \infty} s_{n-1} = s - s = 0.$$

The proof is complete.

Remark 3.8.2 The reciprocal of this last theorem is false. We shall prove that

$$\sum_{n=1}^{\infty} \frac{1}{n} = +\infty$$

despite the fact that

$$\lim_{n \to \infty} \frac{1}{n} = 0.$$

Consider the function $f(x) = 1/x$ defined on $[1, +\infty)$.

From its graph, we get,

$$\int_1^\infty 1/x \, dx < \sum_{n=1}^\infty 1/n.$$

Thus, from a elementary calculus result we have,

$$\int_1^\infty 1/x \, dx = \lim_{t \to \infty} \int_1^t 1/x \, dx = \lim_{t \to \infty} (\ln x)_1^t = \lim_{t \to \infty} \ln(t) = +\infty.$$

Therefore,

$$\sum_{n=1}^\infty 1/n = +\infty.$$

Another proof not involving integrals may be obtained observing that, for

$$s_n = 1 + \frac{1}{2} + \frac{1}{3} + \cdots \frac{1}{n},$$

we have

$$\begin{aligned}
s_{2^n} &= 1 + \frac{1}{2} + \left(\frac{1}{3} + \frac{1}{4}\right) \\
&\quad + \left(\frac{1}{5} + \frac{1}{6} + \frac{1}{7} + \frac{1}{8}\right) + \cdots + \left(\frac{1}{2^{n-1}+1} + \cdots \frac{1}{2^n}\right) \\
&> 1 + \frac{1}{2} + \frac{2}{4} + \frac{4}{8} + \cdots + \frac{2^{n-1}}{2^n} \\
&= 1 + n\frac{1}{2} \to +\infty, \text{ as } n \to +\infty,
\end{aligned} \tag{3.10}$$

so that

$$\lim_{n \to +\infty} s_n = \lim_{n \to \infty} s_{2^n} = \sum_{n=1}^\infty \frac{1}{n} = +\infty.$$

Remark 3.8.3 The series

$$\sum_{n=1}^\infty \frac{1}{n(n+1)}$$

is convergent.

Observe that

$$\frac{1}{n(n+1)} = \frac{1}{n} - \frac{1}{n+1}.$$

Thus,

$$s_n = \sum_{k=1}^{n} \frac{1}{k(k+1)} = 1 - \frac{1}{2} + \left(\frac{1}{2} - \frac{1}{3}\right)$$

$$+ \left(\frac{1}{3} - \frac{1}{4}\right) + \cdots + \left(\frac{1}{n} - \frac{1}{n+1}\right)$$

$$= 1 - \frac{1}{n+1}. \tag{3.11}$$

Therefore,

$$\lim_{n \to \infty} s_n = \lim_{n \to \infty} \left(1 - \frac{1}{n+1}\right) = 1,$$

so that

$$\sum_{n=1}^{\infty} \frac{1}{n(n+1)} = 1.$$

Through such a result we may show that the series

$$\sum_{n=1}^{\infty} \frac{1}{n^2}$$

is also convergent.

Observe that

$$\frac{1}{(n+1)^2} < \frac{1}{n(n+1)}, \forall n \in \mathbb{N}.$$

Therefore,

$$s_n = \sum_{k=1}^{n} \frac{1}{(k+1)^2} < \sum_{k=1}^{n} \frac{1}{k(k+1)} \le \sum_{n=1}^{\infty} \frac{1}{n(n+1)} = 1.$$

From this we may conclude that the sequence of partial sums is bounded and monotonic, therefore it is convergent.

Hence, the series

$$\sum_{n=1}^{\infty} 1/n^2$$

is convergent.

3.8.1 Comparison Criterion

Theorem 3.8.4 (Comparison Criterion) *Let*

$$\sum_{n=1}^{\infty} a_n \text{ and } \sum_{n=1}^{\infty} b_n,$$

be series such that $a_n, b_n \geq 0, \forall n \in \mathbb{N}$.
Suppose there exists $c > 0$ and $n_0 \in \mathbb{N}$ such that

$$a_n \leq cb_n, \forall n > n_0.$$

Assume also that $\sum_{n=1}^{\infty} b_n$ is convergent.
Under such hypotheses, $\sum_{n=1}^{\infty} a_n$ is also convergent.

Proof Observe that defining $s_n = \sum_{k=n_0+1}^{n} a_k, n > n_0$, we have

$$s_n \leq \sum_{k=n_0+1}^{n} cb_k = c \sum_{k=n_0+1}^{n} b_k \leq c \sum_{n=1}^{\infty} b_n \in \mathbb{R}.$$

Thus $\{s_n\}_{n>n_0}$ is a bounded and monotonic sequence, therefore there exists $s \in \mathbb{R}$ such that

$$s = \lim_{n \to \infty} s_n = \sum_{n=n_0+1}^{\infty} a_n.$$

Hence,

$$\sum_{n=1}^{\infty} a_n = a_1 + a_2 + \cdots + a_{n_0} + s \in \mathbb{R}.$$

The proof is complete.

Theorem 3.8.5 (Cauchy Criterion for Series) *In order that a real series $\sum_{n=1}^{\infty} a_n$ be convergent is necessary and sufficient that, for each $\varepsilon > 0$, we may find $n_0 \in \mathbb{N}$ such that*

$$|a_{n+1} + a_{n+2} + \cdots + a_{n+p}| < \varepsilon, \forall n > n_0, \ p \in \mathbb{N}.$$

Proof Observe that $|a_{n+1} + a_{n+2} + \cdots + a_{n+p}| = |s_{n+p} - s_n|$, where $s_n = \sum_{k=1}^{n} a_n$.

Hence, the condition above stated is equivalent to $\{s_n\}$ be a Cauchy sequence, which is equivalent to $\{s_n\}$ be convergent.

The proof is complete.

Definition 3.8.6 A series $\sum_{n=1}^{\infty} a_n$ is said to be absolutely convergent if

$$\sum_{n=1}^{\infty} |a_n|$$

is convergent.

Exercises 3.8.7

1. Prove that if $\sum_{n=1}^{\infty} a_n$ is absolutely convergent, then it is convergent.
2. Let $\sum_{n=1}^{\infty} b_n$ be a convergent series such that $b_n \geq 0, \forall n \in \mathbb{N}$. Prove that if there exists $c > 0$ and $n_0 \in \mathbb{N}$ such that $|a_n| \leq cb_n, \forall n > n_0$, then $\sum_{n=1}^{\infty} a_n$ is absolutely convergent.
3. Prove that if there exist $0 \leq c < 1$, $K > 0$, and $n_0 \in \mathbb{N}$ such that

$$|a_n| < Kc^n, \forall n > n_0,$$

then

$$\sum_{n=1}^{\infty} a_n$$

is absolutely convergent.

Theorem 3.8.8 (The Root Test) *Let $\{a_n\} \subset \mathbb{R}$ be a real sequence. Assume*

$$\limsup_{n \to \infty} \sqrt[n]{|a_n|} = c,$$

where $0 \leq c < 1$. Under such hypotheses

$$\sum_{n=1}^{\infty} |a_n|$$

is convergent.

Proof Define

$$b_n = \sup\{\sqrt[n]{|a_n|},\ \sqrt[n+1]{|a_{n+1}|},\ \cdots\}$$

Thus

$$c = \lim_{n \to \infty} b_n.$$

Define $\varepsilon = (1 - c)/2$.
Hence, there exists $n_0 \in \mathbb{N}$ such that if $n > n_0$, then

$$|b_n - c| < \varepsilon.$$

In particular,

$$\sqrt[n]{|a_n|} \leq b_n < c + \varepsilon = c + (1 - c)/2 = 1/2 + c/2 = c_0 < 1.$$

Therefore

$$|a_n| \leq c_0^n,\quad \forall n > n_0.$$

Since $0 \leq c_0 < 1$ we have that

$$\sum_{n=1}^{\infty} c_0^n$$

is convergent, so that from the comparison criterion we may conclude that

$$\sum_{n=1}^{\infty} |a_n|$$

is also convergent.

The proof is complete.

Exercise 3.8.9 Prove that if $\liminf_{n \to \infty} \sqrt[n]{|a_n|} > 1$, then

$$\sum_{n=1}^{\infty} a_n$$

is divergent.

Theorem 3.8.10 (The Ratio Test) *Let $\sum_{n=1}^{\infty} a_n$ be a series such that $a_n \neq 0,\ \forall n \in \mathbb{N}$.*

Let $\sum_{n=1}^{\infty} b_n$ be a convergent series such that $b_n > 0$, $\forall n \in \mathbb{N}$. Assume there exists $n_0 \in \mathbb{N}$ such that

$$\frac{|a_{n+1}|}{|a_n|} \leq \frac{b_{n+1}}{b_n}, \quad \forall n > n_0.$$

Under such hypotheses,

$$\sum_{n-1}^{\infty} |a_n|$$

is convergent.

Proof Choose $n > n_0$.
   Thus

$$\frac{|a_{n_0+2}|}{|a_{n_0+1}|} \leq \frac{b_{n_0+2}}{b_{n_0+1}}, \frac{|a_{n_0+3}|}{|a_{n_0+2}|} \leq \frac{b_{n_0+3}}{b_{n_0+2}}, \ldots, \frac{|a_n|}{|a_{n-1}|} \leq \frac{b_n}{b_{n-1}},$$

so that

$$\frac{|a_{n_0+2}|}{|a_{n_0+1}|} \cdot \frac{|a_{n_0+3}|}{|a_{n_0+2}|} \cdot \ldots \cdot \frac{|a_n|}{|a_{n-1}|}$$

$$\leq \frac{|b_{n_0+2}|}{|b_{n_0+1}|} \cdot \frac{|b_{n_0+3}|}{|b_{n_0+2}|} \cdot \ldots \cdot \frac{|b_n|}{|a_{n-1}|}. \tag{3.12}$$

Therefore,

$$\frac{|a_n|}{|a_{n_0+1}|} \leq \frac{|b_n|}{|b_{n_0+1}|},$$

that is,

$$|a_n| \leq c|b_n|, \quad \text{if } n > n_0,$$

where

$$c = \frac{|a_{n_0+1}|}{|b_{n_0+1}|}.$$

Since $\sum_{n=1}^{\infty} b_n$ is convergent, from the comparison criterion so is $\sum_{n=1}^{\infty} |a_n|$. The proof is complete.

Exercises 3.8.11

1. Prove that if there exists $0 < c < 1$ and $n_0 \in \mathbb{N}$ such that

$$\frac{|a_{n+1}|}{|a_n|} < c, \forall n \geq n_0,$$

then $\sum_{n=1}^{\infty} |a_n|$ is convergent.
2. Prove that if

$$\limsup_{n \to \infty} \frac{|a_{n+1}|}{|a_n|} = c < 1,$$

then $\sum_{n=1}^{\infty} |a_n|$ is convergent.
3. Prove that if

$$\liminf_{n \to \infty} \frac{|a_{n+1}|}{|a_n|} > 1,$$

then

$$\sum_{n=1}^{\infty} a_n$$

is divergent.

Theorem 3.8.12 *Let $\{a_n\}$ be a bounded sequence, where $a_n > 0$, $\forall n \in \mathbb{N}$.*
Under such hypotheses

$$\liminf_{n \to \infty} \frac{a_{n+1}}{a_n} \leq \liminf_{n \to \infty} \sqrt[n]{a_n} \leq \limsup_{n \to \infty} \sqrt[n]{a_n} \leq \limsup_{n \to \infty} \frac{a_{n+1}}{a_n}.$$

Proof We shall prove just that

$$\limsup_{n \to \infty} \sqrt[n]{a_n} \leq \limsup_{n \to \infty} \frac{a_{n+1}}{a_n},$$

since the second inequality is immediate and the first one may be proven in a similar
fashion.

Suppose, to obtain contradiction, that

$$\limsup_{n \to \infty} \frac{a_{n+1}}{a_n} < \limsup_{n \to \infty} \sqrt[n]{a_n}.$$

Hence, there exists $c > 0$ such that

$$\limsup_{n \to \infty} \frac{a_{n+1}}{a_n} < c < \limsup_{n \to \infty} \sqrt[n]{a_n}. \tag{3.13}$$

Let

$$d = \limsup_{n \to \infty} \frac{a_{n+1}}{a_n} < c.$$

Define

$$b_n = \sup \left\{ \frac{a_{n+1}}{a_n}, \frac{a_{n+2}}{a_{n+1}}, \dots \right\}.$$

Thus

$$d = \lim_{n \to \infty} b_n.$$

Let $\varepsilon = (c - d)/2$. Hence there exists $n_0 \in \mathbb{N}$ such that if $n > n_0$, then

$$|b_n - d| < \varepsilon = (c - d)/2,$$

so that

$$b_n < (c + d)/2 \equiv c_0 < c,$$

that is,

$$0 < \frac{a_{n+1}}{a_n} \le b_n < c, \text{ if } n > n_0.$$

Hence, for $n > p > n_0 + 1$ we have

$$\frac{a_{p+1}}{a_p} < c, \frac{a_{p+2}}{a_{p+1}} < c, \dots, \frac{a_n}{a_{n-1}} < c,$$

and thus,

$$\frac{a_{p+1}}{a_p} \frac{a_{p+2}}{a_{p+1}} \cdots \frac{a_n}{a_{n-1}} < c^{n-p},$$

that is,

$$\frac{a_n}{a_p} < c^{n-p}, \forall n > p > n_0 + 1.$$

Hence,

$$\sqrt[n]{a_n} < \sqrt[n]{\frac{a_p}{c^p}} c.$$

Thus,

$$\limsup_{n\to\infty} \sqrt[n]{a_n} \le \lim_{n\to\infty} \sqrt[n]{\frac{a_p}{c^p}}c = c,$$

that is,

$$\limsup_{n\to\infty} \sqrt[n]{a_n} \le c,$$

which contradicts (3.13).

The proof is complete.

Exercises 3.8.13

1. Calculate the limits below indicated and prove formally that your result is right.

   (a)
   $$\lim_{n\to\infty} \frac{3n-9}{2-12n},$$

   (b)
   $$\lim_{n\to\infty} \frac{-4n}{2-7n},$$

   (c)
   $$\lim_{n\to\infty} \frac{2n}{n+4\sqrt{n}},$$

   (d)
   $$\lim_{n\to\infty} \frac{1}{\sqrt{n+1}},$$

   (e)
   $$\lim_{n\to\infty} \sqrt{n+10}-\sqrt{n},$$

   (f)
   $$\lim_{n\to\infty} \frac{3n-9}{2-12n},$$

2. Prove that

$$\lim_{n \to \infty} a_n = 0,$$

where

$$a_n = \frac{7^n}{n!}.$$

Hint: Prove that

$$0 < a_n < \frac{7^7}{7!} \frac{7}{n}, \text{ if } n > 7.$$

3. Prove that

$$\lim_{n \to \infty} a_n = +\infty,$$

where

$$a_n = n(\sqrt{n+1} - \sqrt{n}), \forall n \in \mathbb{N}.$$

4. Let $a, b > 0$ and let

$$c_n = \sqrt[n]{a^n + b^n}, \ \forall n \in \mathbb{N}.$$

Show that

$$\lim_{n \to \infty} c_n = \max\{a, b\}.$$

5. Let $a_1, \dots, a_k > 0$ and let

$$c_n = \sqrt[n]{a_1^n + \dots + a_k^n}, \ \forall n \in \mathbb{N}.$$

Show that

$$\lim_{n \to \infty} c_n = \max\{a_1, \dots, a_k\}.$$

6. Given $A, B \subset \mathbb{R}$, we define the distance between A and B, denoted by $d(A, B)$, as

$$d(A, B) = \inf\{|u - v| \ : \ u \in A \text{ and } v \in B\}.$$

Let $K, F \subset \mathbb{R}$ be sets such that K is compact, F is closed, and

$$K \cap F = \emptyset.$$

Prove that $d(K, F) > 0$ and there exist $u_0 \in K$ and $v_0 \in F$ such that

$$d(K, F) = |u_0 - v_0|.$$

7. Let $\{x_n\}$ be such that $x_1 = 1$ and

$$x_{n+1} = 3 + \frac{3}{x_n + 10}, \quad \forall n \in \mathbb{N}.$$

(a) Show that there exists $0 < c < 1$, such that

$$|x_{n+2} - x_{n+1}| \le c|x_{n+1} - x_n|, \quad \forall n \in \mathbb{N},$$

and conclude that $\{x_n\}$ is convergent.

(b) Calculate

$$\lim_{n \to \infty} x_n.$$

8. Let $\{x_n\} \subset \mathbb{R}$ be a sequence such that $x_1 = 1$ and

$$x_{n+1} = 3 + \frac{x_n}{15} + \frac{1}{x_n + 10}, \quad \forall n \in \mathbb{N}.$$

(a) Show there exists $0 < c < 1$, such that

$$|x_{n+2} - x_{n+1}| \le c|x_{n+1} - x_n|, \quad \forall n \in \mathbb{N},$$

and conclude that $\{x_n\}$ is convergent.

(b) Calculate

$$\lim_{n \to \infty} x_n.$$

9. Let $\{x_n\} \subset \mathbb{R}$ be defined by $x_1 = \sqrt{2}$ and

$$x_{n+1} = \sqrt{2 + x_n}, \forall n \in \mathbb{N}.$$

Show that $\{x_n\}$ is convergent and calculate its limit.
Hint: Firstly prove by induction that $\sqrt{2} \le x_n \le 2, \forall n \in \mathbb{N}$.

10. Let $\{x_n\} \subset \mathbb{R}$ be defined by $x_1 = \sqrt{a}$ where $a > 0$ and

$$x_{n+1} = \sqrt{a + x_n}, \forall n \in \mathbb{N}.$$

Show that $\{x_n\}$ is convergent and calculate its limit.

11. Let $\{a_n\}$ be such that $a_n > 0$ and

$$\lim_{n \to \infty} \frac{a_{n+1}}{a_n} = \alpha < 1.$$

Show that there exists $0 < c < 1$ and $n_0 \in \mathbb{N}$ such that

$$a_{n+1} < c a_n, \quad \forall n > n_0.$$

Prove also that, in such a case,

$$\lim_{n \to \infty} a_n = 0.$$

12. Let $a > 0$. Using the results of last item, prove that

$$\lim_{n \to \infty} \frac{a^n}{n!} = 0.$$

Using the same results, prove also that,

$$\lim_{n \to \infty} \frac{n!}{n^n} = 0.$$

13. Let $\{a_n\} \subset \mathbb{R}$ be a decreasing sequence. Suppose $\sum_{n=1}^{\infty} a_n$ is convergent. Under such hypotheses, prove that

$$\lim_{n \to \infty} n \, a_n = 0.$$

14. Let $a \in \mathbb{R}$ be such that $0 \leq |a| < 1$. Using either the ratio or root test, show that

$$\sum_{n=1}^{\infty} n |a|^n$$

is convergent.

15. Let $\sum_{n=1}^{\infty} a_n$ and $\sum_{n=1}^{\infty} b_n$ be series such that $a_n > 0$, $\forall n \in \mathbb{N}$ and

$$\lim_{n \to \infty} \frac{a_n}{b_n} = 5.$$

(a) Show there exists $n_0 \in \mathbb{N}$ such that if $n > n_0$, then $b_n > 0$.
(b) Show that $\sum_{n=1}^{\infty} a_n$ is convergent if and only if $\sum_{n=1}^{\infty} b_n$ is convergent.

16. Let $\sum_{n=1}^{\infty} a_n$ and $\sum_{n=1}^{\infty} b_n$ be series such that $b_n > 0$, $\forall n \in \mathbb{N}$ and

$$\lim_{n \to \infty} \frac{a_n}{b_n} = c > 0, \text{ where } c \in \mathbb{R}.$$

  (a) Show there exists $n_0 \in \mathbb{N}$ such that if $n > n_0$, then $a_n > 0$.
  (b) Show that $\sum_{n=1}^{\infty} a_n$ is convergent if and only if $\sum_{n=1}^{\infty} b_n$ is convergent.

17. Let $\{a_n\} \subset \mathbb{R}$ and $\{b_n\} \subset \mathbb{R}$ be sequences such that

$$\lim_{n \to \infty} \frac{a_n}{b_n} = -4.$$

  Prove that $\sum_{n=1}^{\infty} |a_n|$ is convergent if, and only if, $\sum_{n=1}^{\infty} |b_n|$ is convergent.

18. Let $\{a_n\}$ be a sequence such that

$$\lim_{n \to \infty} a_n = a \in \mathbb{R}.$$

  Show that

$$\lim_{N \to \infty} \frac{\sum_{n=1}^{N} a_n}{N} = a.$$

19. Let $\{y_n\} \subset \mathbb{R}$ be such that $y_n > 0$, $\forall n \in \mathbb{N}$. Suppose that

$$\sum_{n=1}^{\infty} y_n = +\infty.$$

  Suppose also that $\{x_n\} \subset \mathbb{R}$ is such that

$$\lim_{n \to \infty} \frac{x_n}{y_n} = a.$$

  Prove that

$$\lim_{n \to \infty} c_n = a,$$

  where

$$c_n = \frac{x_1 + x_2 + \cdots + x_n}{y_1 + y_2 + \cdots + y_n}, \quad \forall n \in \mathbb{N}.$$

20. Let $\{x_n\} \subset \mathbb{R}$ and $\{y_n\} \subset \mathbb{R}$ be sequences such that

$$\lim_{n \to +\infty} x_n = a \in \mathbb{R}$$

and

$$\lim_{n \to \infty} \frac{x_n}{y_n} = b \in \mathbb{R},$$

where $b \neq 0$.
    Prove that

$$\lim_{n \to \infty} c_n = b,$$

where

$$c_n = \frac{x_1 + x_2 + \cdots + x_n}{y_1 + y_2 + \cdots + y_n}, \quad \forall n \in \mathbb{N}.$$

Chapter 4
Real Function Limits

4.1 Introduction

This chapter addresses the main results relating the concept of limits for one variable real functions. Topics such as subsequential limits and related cluster points are presented in detail. Other topics include the standard sandwich theorem and relating comparison results. We finish the chapter with a study on infinite limits and limits at infinity.

The main reference for this chapter is [9]. We start by introducing the formal definition of limit.

4.2 Some Preliminary Definitions and Results

At this point, we remark that for this chapter and the subsequent ones, in general $f : X \to \mathbb{R}$ will denote a real function whose domain is $X \subset \mathbb{R}$, unless otherwise specified.

Definition 4.2.1 (Limit) Let $X \subset \mathbb{R}$ and $a \in X'$. Given $f : X \to \mathbb{R}$, we say that $L \in \mathbb{R}$ is the limit of $f(x)$ as x goes to a, if for each $\varepsilon > 0$ there exists $\delta > 0$ such that if $x \in X$ and $0 < |x - a| < \delta$, then $|f(x) - L| < \varepsilon$.

In such a case we write

$$\lim_{x \to a} f(x) = L.$$

Here we present a result concerning the limit uniqueness.

Theorem 4.2.2 (Limit Uniqueness) *Let $X \subset \mathbb{R}$ and $a \in X'$. For a function $f : X \to \mathbb{R}$, assume*

© Springer International Publishing AG, part of Springer Nature 2018
F. S. Botelho, *Real Analysis and Applications*,
https://doi.org/10.1007/978-3-319-78631-5_4

$$\lim_{x \to a} f(x) = L_1 \in \mathbb{R}$$

and

$$\lim_{x \to a} f(x) = L_2 \in \mathbb{R}.$$

Under such hypotheses,

$$L_1 = L_2.$$

Proof Let $\varepsilon > 0$. From $\lim_{x \to a} f(x) = L_1$, there exists $\delta_1 > 0$ such that if $x \in X$ and $0 < |x - a| < \delta_1$, then $|f(x) - L_1| < \varepsilon/2$. From $\lim_{x \to a} f(x) = L_2$, there exists $\delta_2 > 0$ such that if $x \in X$ and $0 < |x - a| < \delta_2$, then $|f(x) - L_2| < \varepsilon/2$.

Choose $\tilde{x}$ such that $\tilde{x} \in X$ and $0 < |\tilde{x} - a| < \min\{\delta_1, \delta_2\}$.

Thus,

$$\begin{aligned}
|L_1 - L_2| &= |L_1 - f(\tilde{x}) + f(\tilde{x}) - L_2| \\
&\leq |L_1 - f(\tilde{x})| + |f(\tilde{x}) - L_2| \\
&\leq \varepsilon/2 + \varepsilon/2 = \varepsilon.
\end{aligned} \tag{4.1}$$

Therefore, $|L_1 - L_2| < \varepsilon$, and since $\varepsilon > 0$ is arbitrary, we obtain,

$$L_1 = L_2.$$

4.2.1   Some Examples of Limits

The next example concerns the quadratic function.

Proposition 4.2.3 *Let* $f : \mathbb{R} \to \mathbb{R}$ *be defined by*

$$f(x) = ax^2 + bx + c, \ \forall x \in \mathbb{R},$$

where $a, b, c \in \mathbb{R}, a \neq 0$.

Under such assumptions, we have,

$$\lim_{x \to x_0} f(x) = f(x_0), \ \forall x_0 \in \mathbb{R}.$$

Proof Choose $x_0 \in \mathbb{R}$. We are going to show that

$$\lim_{x \to x_0} ax^2 + bx + c = ax_0^2 + bx_0 + c.$$

Observe that,

$$|ax^2 + bx + c - (ax_0^2 + bx_0 + c)|$$
$$= |a(x^2 - x_0^2) + b(x - x_0)|$$
$$\le |a||x - x_0||x + x_0| + |b||x - x_0|. \tag{4.2}$$

Let $0 < \delta \le 1$.

Thus, if $|x - x_0| < \delta \le 1$, then $-1 < x - x_0 < 1$, that is $-1 + x_0 < x < 1 + x_0$, so that,

$$-1 - 2|x_0| \le -1 + 2x_0 < x + x_0 < 1 + 2x_0 \le 1 + 2|x_0|,$$

that is,

$$|x + x_0| < 1 + 2|x_0|. \tag{4.3}$$

Summarizing if $|x - x_0| < \delta \le 1$, then from (4.2) and (4.3), we have,

$$|ax^2 + bx + c - (ax_0^2 + bx_0 + c)|$$
$$\le |a||x - x_0||x + x_0| + |b||x - x_0|$$
$$< |a|\delta(1 + 2|x_0|) + |b|\delta$$
$$= [|a|(1 + 2|x_0|) + |b|]\delta(\equiv \varepsilon). \tag{4.4}$$

Hence, let $\varepsilon > 0$ be given.

From (4.4), for

$$\delta = \min\left\{1, \frac{\varepsilon}{|a|(1 + 2|x_0|) + |b|}\right\},$$

we have, if $|x - x_0| < \delta$, then

$$|ax^2 + bx + c - (ax_0^2 + bx_0 + c)| < [|a|(1 + 2|x_0|) + |b|]\delta \le \varepsilon.$$

More formally, we could write:

For each $\varepsilon > 0$, there exists

$$\delta = \min\left\{1, \frac{\varepsilon}{|a|(1 + 2|x_0|) + |b|}\right\} > 0,$$

such that if $|x - x_0| < \delta$, then

$$|ax^2 + bx + c - (ax_0^2 + bx_0 + c)| < [|a|(1 + 2|x_0|) + |b|]\delta \le \varepsilon,$$

so that

$$\lim_{x \to x_0} ax^2 + bx + c = ax_0^2 + bx_0 + c.$$

The proof is complete.

Proposition 4.2.4 *Let* $f : \mathbb{R} \setminus \{-d/c\} \to \mathbb{R}$ *be such that*

$$f(x) = \frac{ax + b}{cx + d}, \forall x \neq -d/c,$$

where $a, b, c, d \in \mathbb{R}$, $ad + bc \neq 0$ *and* $c \neq 0$.
  Let $x_0 \in \mathbb{R}$ *be such that* $x_0 \neq -d/c$.
  Under such statements and assumptions, we have

$$\lim_{x \to x_0} f(x) = \frac{ax_0 + b}{cx_0 + d}.$$

Proof Observe that,

$$\left| \frac{ax + b}{cx + d} - \frac{ax_0 + b}{cx_0 + d} \right|$$

$$= \left| \frac{(ax + b)(cx_0 + d) - (ax_0 + b)(cx + d)}{(cx + d)(cx_0 + d)} \right|$$

$$= \left| \frac{(ad + bc)(x - x_0)}{(cx + d)(cx_0 + d)} \right|$$

$$= \frac{|ad + bc| \, |x - x_0|}{|cx + d| \, |cx_0 + d|} \tag{4.5}$$

From the last proposition with $a = 0$ we may easily obtain

$$\lim_{x \to x_0} cx + d = cx_0 + d.$$

From this, specifically for $\varepsilon_0 = |cx_0 + d|/2 > 0$, there exists $\delta_1 > 0$ such that if $|x - x_0| < \delta_1$, then

$$|cx + d - cx_0 + d| < \varepsilon_0 = |cx_0 + d|/2.$$

In particular

$$-|cx_0 + d|/2 < |cx + d| - |cx_0 + d|,$$

so that

$$|cx + d| > |cx_0 + d|/2,$$

and therefore,

$$\frac{1}{|cx + d|} < \frac{2}{|cx_0 + d|},$$

if $|x - x_0| < \delta_1$.

From this and (4.5), we obtain, if $|x - x_0| < \delta_1$, then

$$\left| \frac{ax + b}{cx + d} - \frac{ax_0 + b}{cx_0 + d} \right| = \frac{|ad + bc||x - x_0|}{|cx + d||cx_0 + d|}$$

$$\leq \frac{2|ad + bc||x - x_0|}{|cx_0 + d|^2}. \tag{4.6}$$

Let $\varepsilon > 0$ be given and let $\delta_2 > 0$ be such that

$$\frac{2|ad + bc|\delta_2}{|cx_0 + d|^2} = \varepsilon,$$

that is,

$$\delta_2 = \frac{|cx_0 + d|^2 \varepsilon}{2|ad + bc|},$$

and define $\delta = \min\{\delta_1, \delta_2\}$.

Thus, if $|x - x_0| < \delta$, from (4.6), we obtain,

$$\left| \frac{ax + b}{cx + d} - \frac{ax_0 + b}{cx_0 + d} \right| \leq \frac{2|ad + bc||x - x_0|}{|cx_0 + d|^2}$$

$$< \frac{2|ad + bc|\delta}{|cx_0 + d|^2}$$

$$< \frac{2|ad + bc|\delta_2}{|cx_0 + d|^2}$$

$$= \varepsilon. \tag{4.7}$$

From this we may infer that

$$\lim_{x \to x_0} \frac{ax + b}{cx + d} = \frac{ax_0 + b}{cx_0 + d}.$$

The proof is complete.

The next result asserts that if the limit of a function exists at some point, such a function keeps bounded in a neighborhood of the concerned point.

Theorem 4.2.5 *Let $X \subset \mathbb{R}$, $f : X \to \mathbb{R}$, and $a \in X'$. If*

$$\lim_{x \to a} f(x)$$

exists, then there exist $A > 0$ and $\delta > 0$ such that if $x \in X$ and $0 < |x - a| < \delta$, then $|f(x)| < A$.

Proof Suppose

$$\lim_{x \to a} f(x) = L \in \mathbb{R}.$$

For $\varepsilon = 1$, there exists $\delta > 0$ such that if $x \in X$ and $0 < |x - a| < \delta$, then $|f(x) - L| < \varepsilon = 1$, that is,

$$|f(x)| - |L| < 1,$$

and hence,

$$|f(x)| < 1 + |L| \equiv A, \ \text{if } x \in X \text{ and } 0 < |x - a| < \delta.$$

At this point we present some types of limit comparison theorems.

Theorem 4.2.6 (Sandwich Theorem) *Let $X \subset \mathbb{R}$, $f, g, h : X \to \mathbb{R}$, and $a \in X'$. Assume there exists $\delta_0 > 0$ such that if $x \in X$ and $0 < |x - a| < \delta_0$, then*

$$f(x) \le g(x) \le h(x).$$

Suppose also that

$$\lim_{x \to a} f(x) = \lim_{x \to a} h(x) = L \in \mathbb{R}.$$

Under such hypotheses, we have,

$$\lim_{x \to a} g(x) = L.$$

Proof Let $\varepsilon > 0$. From

$$\lim_{x \to a} f(x) = L,$$

there exists $\delta_1 > 0$ such that if $x \in X$ and $0 < |x - a| < \delta_1$, then

$$|f(x) - L| < \varepsilon.$$

From

$$\lim_{x \to a} h(x) = L,$$

there exists $\delta_2 > 0$ such that if $x \in X$ and $0 < |x - a| < \delta_2$, then

$$|h(x) - L| < \varepsilon.$$

Define $\delta = \min\{\delta_0, \delta_1, \delta_2\}$, thus if $x \in X$ and $0 < |x - a| < \delta$, then

$$L - \varepsilon < f(x) \leq g(x) \leq h(x) < L + \varepsilon,$$

that is,

$$|g(x) - L| < \varepsilon, \ \text{ if } x \in X \text{ and } 0 < |x - a| < \delta.$$

Thus,

$$\lim_{x \to a} g(x) = L.$$

The proof is complete.

Theorem 4.2.7 *Let* $X \subset \mathbb{R}$, $a \in X'$, *and let* $f, g : X \to \mathbb{R}$ *be functions. Assume*

$$\lim_{x \to a} f(x) = L \in \mathbb{R}$$

and

$$\lim_{x \to a} g(x) = M \in \mathbb{R}$$

where $L < M$.

Under such hypotheses, there exists $\delta > 0$ *such that if* $x \in X$ *and* $0 < |x - a| < \delta$, *then* $f(x) < g(x)$.

Proof For $\varepsilon = (M - L)/2$, from

$$\lim_{x \to a} f(x) = L,$$

there exists $\delta_1 > 0$ such that if $x \in X$ and $0 < |x - a| < \delta_1$, then

$$|f(x) - L| < \varepsilon = (M - L)/2,$$

that is,

$$f(x) < (M - L)/2 + L = (M + L)/2.$$

By analogy, from

$$\lim_{x \to a} g(x) = M,$$

there exists $\delta_2 > 0$ such that if $x \in X$ and $0 < |x - a| < \delta_2$, then

$$|g(x) - M| < \varepsilon = (M - L)/2,$$

that is,

$$-(M - L)/2 < g(x) - M,$$

so that

$$g(x) > (L + M)/2.$$

Define $\delta = \min\{\delta_1, \delta_2\}$. Thus, from above if $x \in X$ and $0 < |x - a| < \delta$, then

$$f(x) < (M + L)/2 < g(x).$$

This completes the proof.

Exercises 4.2.8

1. Let $f : X \to \mathbb{R}$ be a function. Prove that if $\lim_{x \to a} f(x) = L > 0$, then there exists $\delta > 0$ such that if $x \in X$ and $0 < |x - a| < \delta$, then

$$f(x) > 0.$$

2. Let $f, g : X \to \mathbb{R}$ be real functions. Prove that if $f(x) \le g(x), \forall x \in X$ such that $x \ne a$,

$$\lim_{x \to a} f(x) = L$$

and

$$\lim_{x \to a} g(x) = M,$$

then

$$L \le M.$$

Theorem 4.2.9 *Let* $X \subset \mathbb{R}$, $f : X \to \mathbb{R}$ *and* $a \in X'$. *In order that we have*

$$\lim_{x \to a} f(x) = L \in \mathbb{R},$$

it is necessary and sufficient that we have

$$\lim_{n \to \infty} f(x_n) = L$$

for each sequence $\{x_n\} \subset X \setminus \{a\}$ *such that*

$$\lim_{n \to \infty} x_n = a.$$

Proof Suppose $\lim_{x \to a} f(x) = L$. Let $\{x_n\} \subset X \setminus \{a\}$ be such that

$$\lim_{n \to \infty} x_n = a.$$

Let $\varepsilon > 0$. Thus, there exists $\delta > 0$ such that if $x \in X$ and $0 < |x - a| < \delta$, then $|f(x) - L| < \varepsilon$.

From

$$\lim_{n \to \infty} x_n = a,$$

there exists $n_0 \in \mathbb{N}$ such that if $n > n_0$, then

$$0 < |x_n - a| < \delta,$$

so that

$$|f(x_n) - L| < \varepsilon, \text{ if } n > n_0.$$

Thus,

$$\lim_{n \to \infty} f(x_n) = L.$$

For the reciprocal, we prove the contrapositive. Suppose we do not have

$$\lim_{x \to a} f(x) = L.$$

Thus, there exists $\varepsilon > 0$ such that for each $n \in \mathbb{N}$, we may obtain $x_n \in X$ such that

$$0 < |x_n - a| < 1/n,$$

and

$$|f(x_n) - L| \geq \varepsilon.$$

Hence,

$$\lim_{n \to \infty} x_n = a$$

and we do not have

$$\lim_{n \to \infty} f(x_n) = L.$$

The proof is complete.

Corollary 4.2.10 *For*

$$\lim_{x \to a} f(x)$$

to exist, it is necessary and sufficient that

$$\lim_{n \to \infty} f(x_n)$$

exists and such a limit be independent of the sequence $\{x_n\} \subset X \setminus \{a\}$ such that

$$\lim_{n \to \infty} x_n = a.$$

Corollary 4.2.11 *For*

$$\lim_{x \to a} f(x)$$

to exist, it suffices that

$$\lim_{n \to \infty} f(x_n)$$

exists for each sequence $\{x_n\} \subset X \setminus \{a\}$ such that

$$\lim_{n \to \infty} x_n = a.$$

Proof Suppose that

$$\lim_{n \to \infty} f(x_n)$$

exists for each sequence $\{x_n\} \in X \setminus \{a\}$ such that

$$\lim_{n \to \infty} x_n = a.$$

We claim that such a limit does not depend on the sequence $\{x_n\}$. To obtain contradiction, suppose we have $\{x_n\} \subset X \setminus \{a\}$ and $\{y_n\} \in X \setminus \{a\}$ such that

$$\lim_{n \to \infty} x_n = a$$

and

$$\lim_{n \to \infty} y_n = a$$

and,

$$\lim_{n \to \infty} f(x_n) = L \neq M = \lim_{n \to \infty} f(y_n).$$

Consider the sequence $\{z_n\} \subset X \setminus \{a\}$, such that $z_{2n} = x_{2n}$ and $z_{2n-1} = y_{2n-1}$. Thus,

$$\lim_{n \to \infty} z_n = a,$$

and

$$\lim_{n \to \infty} f(z_{2n}) = L \neq M = \lim_{n \to \infty} f(z_{2n-1}).$$

Hence

$$\lim_{n \to \infty} f(z_n)$$

does not exists, which contradicts the hypotheses.

Therefore

$$\lim_{n \to \infty} f(x_n)$$

does not depend on the sequence $\{x_n\}$ in question, and hence, from Theorem 4.2.9,

$$\lim_{x \to a} f(x)$$

does exist.

Theorem 4.2.12 (Limit Properties) *Let $X \subset \mathbb{R}$, $a \in X'$, $f, g : X \to \mathbb{R}$. If*

$$\lim_{x \to a} f(x) = L \in \mathbb{R}$$

and

$$\lim_{x \to a} g(x) = M \in \mathbb{R},$$

then

1. $\lim_{x \to a} \alpha f(x) = \alpha L,\ \forall \alpha \in \mathbb{R}$,
2. $\lim_{x \to a} (f(x) + g(x)) = L + M$,
3. $\lim_{x \to a} (f(x) \cdot g(x)) = L \cdot M$,
4. If $M \neq 0$, then $\lim_{x \to a} (f(x)/g(x)) = L/M$.

Proof We shall prove just 3 and 4, leaving the proof of remaining items as exercises.
Observe that

$$|f(x)g(x) - LM| = |f(x)g(x) - f(x)M + f(x)M - LM|$$
$$\leq |f(x)||g(x) - M| + |M||f(x) - L|. \tag{4.8}$$

Let $\varepsilon > 0$.
From

$$\lim_{x \to a} f(x) = L$$

there exists $\delta_1 > 0$ and $A > 0$ such that if $x \in X$ and $0 < |x - a| < \delta_1$, then $|f(x)| < A$.

And also, there exists $\delta_2 > 0$ such that $x \in X$ and $0 < |x - a| < \delta_2$, then

$$|f(x) - L| < \frac{\varepsilon}{2|M| + 1}.$$

From

$$\lim_{x \to a} g(x) = M$$

there exists $\delta_3 > 0$ such that if $x \in X$ and $0 < |x - a| < \delta_3$, then

$$|g(x) - M| < \frac{\varepsilon}{2A}.$$

Thus, defining $\delta = \min\{\delta_1, \delta_2, \delta_3\}$ we have that if $x \in X$ and $0 < |x - a| < \delta$, then

$$|f(x)g(x) - LM| = |f(x)g(x) - f(x)M + f(x)M - LM|$$
$$\leq |f(x)||g(x) - M| + |M||f(x) - L|$$
$$< A\frac{\varepsilon}{2A} + \frac{\varepsilon|M|}{2|M| + 1}$$
$$< \varepsilon. \tag{4.9}$$

Thus,

$$\lim_{x \to a} f(x)g(x) = LM.$$

For 4, assume $M \neq 0$.
We prove first that

$$\lim_{x \to a} \frac{1}{g(x)} = \frac{1}{M}.$$

Observe that

$$\left| \frac{1}{g(x)} - \frac{1}{M} \right| = \frac{|g(x) - M|}{|g(x)||M|}.$$

Let $\varepsilon_1 = \frac{|M|}{2}$.
From

$$\lim_{x \to a} g(x) = M,$$

there exists $\delta_1 > 0$ such that if $x \in X$ and $0 < |x - a| < \delta_1$, then

$$|g(x) - M| < \varepsilon_1 = \frac{|M|}{2},$$

so that,

$$-\frac{|M|}{2} < |g(x)| - |M| < \frac{|M|}{2},$$

and hence,

$$|g(x)| > \frac{|M|}{2},$$

and thus,

$$\frac{1}{|g(x)|} < \frac{2}{|M|}.$$

Let a new $\varepsilon > 0$ be given. Also, from

$$\lim_{x \to a} g(x) = M,$$

there exists $\delta_2 > 0$ such that if $x \in X$ and $0 < |x - a| < \delta_2$, then

$$|g(x) - M| < \varepsilon \frac{|M|^2}{2}.$$

Thus, defining $\delta = \min\{\delta_1, \delta_2\}$, we have, if $x \in X$ and $0 < |x - a| < \delta$, then

$$\left| \frac{1}{g(x)} - \frac{1}{M} \right| = \frac{|g(x) - M|}{|g(x)||M|} < \frac{\varepsilon |M|^2}{2} \frac{2}{|M||M|} = \varepsilon.$$

Therefore,

$$\lim_{x \to a} \frac{1}{g(x)} = \frac{1}{M}.$$

From this and item 3 we finally obtain,

$$\lim_{x \to a} \frac{f(x)}{g(x)} = \lim_{x \to a} f(x) \lim_{x \to a} \frac{1}{g(x)} = \frac{L}{M}.$$

The proof is complete.

As a first application of such results, we have,

Proposition 4.2.13 *Let $x_0 \in \mathbb{R}$. Thus,*

$$\lim_{x \to x_0} x^n = x_0^n, \ \forall n \in \mathbb{N}.$$

Proof We prove the result by induction.
Define

$$B = \{ n \in \mathbb{N} \ : \ \lim_{x \to x_0} x^n = x_0^n \}.$$

From

$$\lim_{x \to x_0} x = x_0,$$

we obtain,

$$1 \in B.$$

Suppose $n \in B$. Thus,

$$\lim_{x \to x_0} x^n = x_0^n,$$

so that from this and the limit properties, we get

$$\lim_{x \to x_0} x^{n+1} = \lim_{x \to x_0} x^n \cdot x$$

$$= \lim_{x \to x_0} x^n \cdot \lim_{x \to x_0} x$$

$$= x_0^n \cdot x_0$$

$$= x_0^{n+1}. \tag{4.10}$$

From this $n + 1 \in B$.

We have got, $B \subset \mathbb{N}$, $1 \in B$ and if $n \in B$, then $n + 1 \in B$. From the induction principle,

$$B = \mathbb{N},$$

so that

$$\lim_{x \to x_0} x^n = x_0^n, \quad \forall n \in \mathbb{N}.$$

The proof is complete.

We finish this section with a important definition to be addressed in the subsequent sections.

Definition 4.2.14 Let $X \subset \mathbb{R}$ and let $f : X \to \mathbb{R}$ be a function.

1. We say that f is increasing, if the following property is satisfied.
   If $x, y \in X$ and $x < y$, then $f(x) < f(y)$.
2. We say that f is nondecreasing, if the following property is satisfied.
   If $x, y \in X$ and $x < y$, then $f(x) \leq f(y)$.
3. We say that f is decreasing, if the following property is satisfied.
   If $x, y \in X$ and $x < y$, then $f(x) > f(y)$.
4. We say that f is nonincreasing, if the following property is satisfied.
   If $x, y \in X$ and $x < y$, then $f(x) \geq f(y)$.

Finally, if f is either increasing, nondecreasing, decreasing, or nonincreasing, it is said to be monotonic.

4.3 Composite Functions, Limits

We start this section with a theorem dealing with limits for composite functions.

Theorem 4.3.1 *Let $X, Y \subset \mathbb{R}$, $f : X \to \mathbb{R}$, $g : Y \to \mathbb{R}$ be such that $f(X) \subset Y$. Let $a \in X'$ and $b \in Y' \cap Y$. Assume*

$$\lim_{x \to a} f(x) = b$$

$$\lim_{y \to b} g(y) = c$$

and $g(b) = c$.

Under such hypotheses,

$$\lim_{x \to a} g(f(x)) = c = g(b).$$

Proof Let $\varepsilon > 0$. From $\lim_{y \to b} g(y) = g(b) = c$ we have that there exists $\eta > 0$ such that if $|y - b| < \eta$, then $|g(y) - c| < \varepsilon$.

From

$$\lim_{x \to a} f(x) = b,$$

for such a $\eta > 0$, there exists $\delta > 0$ such that if $x \in X$ and $0 < |x - a| < \delta$, then $|f(x) - b| < \eta$. Hence, $|g(f(x)) - c| < \varepsilon$ if $x \in X$ and $0 < |x - a| < \delta$.

Therefore,

$$\lim_{x \to a} g(f(x)) = c = g(b).$$

4.4   One-Sided Limits

We shall denote by X'_+ the set of cluster points on the right, relating X, that is, $a \in X'_+$ if, and only if, for each $\delta > 0$, there exists $x \in X \cap (a, a + \delta)$.

By analogy, $a \in X'_-$ if, and only if, for each $\delta > 0$, there exists $x \in X \cap (a - \delta, a)$.

Definition 4.4.1 We say that $L_1 \in \mathbb{R}$ is the one-sided right-hand limit for $f(x)$ as x goes to $a \in X'+$, if for each $\varepsilon > 0$, there exists $\delta > 0$ such that if $x \in X$ and $0 < x - a < \delta$, then $|f(x) - L_1| < \varepsilon$.

In such a case, we write

$$\lim_{x \to a^+} f(x) = L_1.$$

By analogy, for $a \in X'_-$, we define the one-sided left-hand limit. Hence, we write,

$$\lim_{x \to a^-} f(x) = L_2 \in \mathbb{R},$$

as, for each $\varepsilon > 0$, there exists $\delta > 0$ such that if $x \in X$ and $a - \delta < x < a$, then $|f(x) - L_2| < \varepsilon$.

Theorem 4.4.2 *Suppose $X \subset \mathbb{R}$, $f : X \to \mathbb{R}$ and $a \in X'_+ \cap X'_-$. We have*

$$\lim_{x \to a} f(x) = L \in \mathbb{R},$$

if, and only if, both one-sided limits exist and be equal to L, that is,

$$\lim_{x \to a^-} f(x) = L = \lim_{x \to a^+} f(x).$$

Proof Let

$$\lim_{x \to a} f(x) = L$$

and $a \in X'_+ \cap X'_-$. Obviously,

$$\lim_{x \to a^-} f(x) = L = \lim_{x \to a^+} f(x).$$

Reciprocally, suppose

$$\lim_{x \to a^-} f(x) = L = \lim_{x \to a^+} f(x).$$

Let $\varepsilon > 0$. From

$$\lim_{x \to a^+} f(x) = L,$$

there exists $\delta_1 > 0$ such that if $x \in X \cap (a, a + \delta_1)$, then

$$|f(x) - L| < \varepsilon.$$

From

$$\lim_{x \to a^-} f(x) = L,$$

there exists $\delta_2 > 0$ such that if $x \in X \cap (a - \delta_2, a)$, then

$$|f(x) - L| < \varepsilon.$$

Thus, defining $\delta = \min\{\delta_1, \delta_2\}$ we have that if $x \in (X \setminus \{a\}) \cap (a - \delta, a + \delta)$, then $|f(x) - L| < \varepsilon$.

Therefore,

$$\lim_{x \to a} f(x) = L.$$

Theorem 4.4.3 *Let* $X \subset \mathbb{R}$ *and let* $f : X \to \mathbb{R}$ *be a bounded and monotonic function.*

Suppose also $a \in X'_+$ *and* $b \in X'_-$.

Under such hypotheses, both one side limits in a^+ *and* b^- *exist, that is, there exist* $L, M \in \mathbb{R}$ *such that*

$$L = \lim_{x \to a^+} f(x),$$

and

$$M = \lim_{x \to b^-} f(x).$$

Proof Suppose f is nondecreasing (other cases may be dealt similarly). Let

$$L = \inf\{f(x) \ : \ x \in X, \ x > a\}.$$

We shall show that

$$L = \lim_{x \to a^+} f(x).$$

Let $\varepsilon > 0$. $L + \varepsilon$ is not a lower bound for the set

$$\{f(x) \ : \ x \in X, x > a\}.$$

Hence, there exists $y \in X$ such that $y > a$ and

$$L \leq f(y) < L + \varepsilon.$$

Define $\delta = y - a$. Since f is nondecreasing, if $x \in X$ and $a < x < y = a + \delta$, then

$$L \leq f(x) \leq f(y) < L + \varepsilon.$$

Thus, if $x \in X$ and $a < x < a + \delta$ we have $|f(x) - L| < \varepsilon$.

Therefore,

$$\lim_{x \to a^+} f(x) = L.$$

The case relating the limit on the left may be proven similarly.

4.5 Cluster Values for a Function: Superior and Inferior Limits

Definition 4.5.1 Let $X \subset \mathbb{R}$, $f : X \to \mathbb{R}$ and $a \in X'$. Such a function f is said to be bounded on a neighborhood V_δ of a, if there exists $K > 0$ and $\delta > 0$ such that

$$|f(x)| < K, \forall x \in (X \setminus \{a\}) \cap (a - \delta, a + \delta) \equiv V_\delta.$$

Definition 4.5.2 Let $X \subset \mathbb{R}$ and let $f : X \to \mathbb{R}$ be a real function. A real number c is said to be a cluster value for f at the point $a \in X'$, if there exists a sequence $\{x_n\} \subset X \setminus \{a\}$ such that

$$\lim_{n \to \infty} x_n = a$$

and

$$\lim_{n \to \infty} f(x_n) = c.$$

Theorem 4.5.3 *Let $X \subset \mathbb{R}$ and let $f : X \to \mathbb{R}$ be a real function. Thus, $c \in \mathbb{R}$ is a cluster value for f at the point $a \in X'$ if, and only if, for each $\delta > 0$, $c \in \overline{f(V_\delta)}$.*

Proof Let $c \in \mathbb{R}$ be a cluster value for f at $a \in X'$. Thus, there exists $\{x_n\} \subset X \setminus \{a\}$ such that

$$\lim_{n \to \infty} x_n = a$$

and

$$\lim_{n \to \infty} f(x_n) = c.$$

Let $\delta > 0$, thus there exists $n_0 \in \mathbb{N}$ such that if $n > n_0$, then

$$x_n \in (X \setminus \{a\}) \cap (a - \delta, a + \delta) \equiv V_\delta.$$

Therefore, since

$$c = \lim_{n \to \infty} f(x_n)$$

and

$$\{f(x_n)\}_{n>n_0} \subset f(V_\delta),$$

we have,

$$c \in \overline{f(V_\delta)}.$$

Reciprocally, assume $c \in \overline{f(V_\delta)}$, $\forall \delta > 0$.
Thus,

$$c \in \overline{f(V_{1/n})}, \ \forall n \in \mathbb{N}.$$

Hence, for each $n \in \mathbb{N}$ there exists $x_n \in V_{1/n}$ such that

$$|f(x_n) - c| < 1/n.$$

Thus, $\{x_n\} \subset X \setminus \{a\}$,

$$\lim_{n \to \infty} x_n = a$$

and

$$\lim_{n \to \infty} f(x_n) = c,$$

that is, c is a cluster value for f at $a \in X'$.
   The proof is complete.

   We denote by $CV(f; a)$ the set of cluster values for f at a.

Corollary 4.5.4 $CV(f; a) = \cap_{\delta>0}\overline{f(V_\delta)}$.

Corollary 4.5.5 $CV(f; a) = \cap_{n=1}^{\infty}\overline{f(V_{1/n})}$.

Proof Obviously

$$CV(f; a) = \cap_{\delta>0}\overline{f(V_\delta)} \subset \cap_{n=1}^{\infty}\overline{f(V_{1/n})}.$$

Reciprocally, let

$$c \in \cap_{n=1}^{\infty}\overline{f(V_{1/n})}.$$

In particular,

$$c \in \overline{f(V_{1/n})}, \ \forall n \in \mathbb{N}.$$

Let $\delta > 0$. Let $\tilde{n} \in \mathbb{N}$ be such that

$$1/\tilde{n} < \delta.$$

Thus,

$$c \in \overline{f(V_{1/\tilde{n}})} \subset \overline{f(V_\delta)}.$$

Since $\delta > 0$ is arbitrary, we obtain,

$$c \in \cap_{\delta>0}\overline{f(V_\delta)} = CV(f; a).$$

Therefore,

$$\cap_{n=1}^{\infty}\overline{f(V_{1/n})} \subset \cap_{\delta>0}\overline{f(V_\delta)},$$

so that

$$\cap_{n=1}^{\infty}\overline{f(V_{1/n})} = \cap_{\delta>0}\overline{f(V_\delta)} = CV(f; a).$$

Corollary 4.5.6 *The set of cluster values for f at $a \in X'$ is closed. If f is bounded in a neighborhood V_δ of $a \in X'$, then such a set is compact and nonempty.*

Proof Observe that

$$CV(f; a) = \cap_{\delta>0}\overline{f(V_\delta)},$$

is closed, as an intersection of closed sets. We denote $K_n = \overline{f(V_{1/n})}$. Thus, assuming f is bounded in a neighborhood V_δ of $a \in X'$, there exists $n_0 \in \mathbb{N}$ such that $\overline{f(V_{1/(n_0)})}$ is bounded (just take $n_0 \in \mathbb{N}$ such that $n_0 > 1/\delta$) and hence K_{n_0} is compact. Observe that

$$CV(f; a) = \cap_{n>n_0} K_n = \cap_{n>n_0}\overline{f(V_{1/n})},$$

is compact as an intersection of compact sets. Since $K_{n+1} \subset K_n, \forall n > n_0$, from Corollary 2.1.40, we obtain that $CV(f; a)$ is nonempty.

The proof is complete.

Example 4.5.7 Let $f : \mathbb{R} \setminus \{0\} \to \mathbb{R}$ where

$$f(x) = \sin\left(\frac{1}{x}\right), \quad \forall x \neq 0.$$

We shall show that $CV(f; 0) = [-1, 1]$.
Indeed, since

$$|\sin(1/x)| \leq 1, \forall x \neq 0,$$

we obtain

$$CV(f; 0) \subset [-1, 1]. \tag{4.11}$$

Let $y \in [-1, 1]$. Thus we may find $x_0 \in [0, 2\pi]$ such that

$$\sin(x_0) = y.$$

Define

$$x_n = \frac{1}{x_0 + 2n\pi}, \quad \forall n \in \mathbb{N}.$$

Therefore,

$$\lim_{n \to \infty} x_n = 0,$$

and

$$\begin{aligned}
\lim_{n \to \infty} \sin(1/x_n) &= \lim_{n \to \infty} \sin(x_0 + 2n\pi) \\
&= \lim_{n \to \infty} y \\
&= y, \tag{4.12}
\end{aligned}$$

so that

$$y \in CV(f; 0), \quad \forall y \in [-1, 1].$$

Hence

$$[-1, 1] \subset CV(f; 0),$$

and thus from this and (4.11), we obtain

$$CV(f; 0) = [-1, 1].$$

This completes the example.

4.5.1   Superior and Inferior Limits for a Function at a Point

For f bounded in a neighborhood V_δ of $a \in X'$, we define the superior limit of f at a, as the greatest cluster value for f at a.

Thus, we write

$$L = \limsup_{x \to a} f(x),$$

to express that $L \in \mathbb{R}$ is the superior limit in question.

By analogy, we define the inferior limit of f at a as the smallest cluster value for f at a.

Denoting such an inferior limit by $l \in \mathbb{R}$, we write:

$$l = \liminf_{x \to a} f(x).$$

Finally, we write

$$\limsup_{x \to a} f(x) = +\infty,$$

to denote that f is upper unbounded in each neighborhood V_δ of a.

On the other hand, we write $c \in CV(f; +\infty)$ as there exists $\{x_n\} \subset X$ such that

$$\lim_{n \to \infty} x_n = +\infty,$$

and

$$c = \lim_{n \to +\infty} f(x_n).$$

We also denote

$$\limsup_{x \to +\infty} f(x) = +\infty,$$

if there exists a sequence $\{x_n\} \subset X$ such that

$$\lim_{n \to \infty} x_n = +\infty$$

and

$$\lim_{n \to \infty} f(x_n) = +\infty.$$

Similar notations may be defined for

$$\liminf_{x \to a} f(x) = -\infty,$$

$$CV(f; -\infty),$$

$$\liminf_{x \to -\infty} f(x) = -\infty,$$

$$\liminf_{x \to +\infty} f(x) = -\infty,$$

and

$$\limsup_{x \to -\infty} f(x) = +\infty.$$

Now consider again a function $f : X \subset \mathbb{R} \to \mathbb{R}$ bounded in a neighborhood $V_{\delta_1} \equiv (X \setminus \{a\}) \cap (a - \delta_1, a + \delta_1)$ of $a \in X'$. Thus, there exists $\delta_0 > 0$ such that $f(V_{\delta_0})$ is a bounded set.

Therefore, $f(V_\delta)$ is bounded, $\forall \delta \in (0, \delta_0]$. On the interval $(0, \delta_0]$ we shall define the functions $\delta \mapsto L_\delta$ and $\delta \mapsto l_\delta$ where

$$L_\delta = \sup\{f(V_\delta)\} = \sup_{x \in V_\delta}\{f(x)\}$$

and

$$l_\delta = \inf\{f(V_\delta)\} = \inf_{x \in V_\delta}\{f(x)\}.$$

Observe that

$$l_{\delta_0} \leq l_\delta \leq L_\delta \leq L_{\delta_0}, \ \forall \delta \in (0, \delta_0].$$

Observe also that if

$$0 < \delta' \leq \delta'' \leq \delta_0,$$

then

$$V_{\delta'} \subset V_{\delta''}$$

and therefore

$$l_{\delta''} \leq l_{\delta'}$$

and

$$L_{\delta'} \leq L_{\delta''}.$$

Hence, l_δ is a nonincreasing (monotonic) function of δ whereas L_δ is a nondecreasing one.

From Theorem 4.4.2 the following limits exist

$$\lim_{\delta \to 0^+} l_\delta$$

and

$$\lim_{\delta \to 0^+} L_\delta$$

as lateral limits on the right of 0.

Theorem 4.5.8 *Let $f : X \to \mathbb{R}$ be a bounded function in a neighborhood $V_{\delta_0} = (X \setminus \{a\}) \cap (a - \delta_0, a + \delta_0)$ of $a \in X'$, where $\delta_0 > 0$.*
  Under such hypotheses,

$$\limsup_{x \to a} f(x) = \lim_{\delta \to 0^+} L_\delta,$$

and

$$\liminf_{x \to a} f(x) = \lim_{\delta \to 0^+} l_\delta,$$

where,

$$L_\delta = \sup_{x \in V_\delta} \{f(x)\}$$

and

$$l_\delta = \inf_{x \in V_\delta} \{f(x)\}.$$

Proof We denote,

$$L = \limsup_{x \to a} f(x)$$

and

$$L_0 = \lim_{\delta \to 0^+} L_\delta.$$

Since L is the largest cluster value for f at a,

$$L \in \overline{f(V_\delta)}, \ \forall \delta > 0.$$

Therefore,

$$L \le \sup f(V_\delta) = L_\delta,$$

and thus,

$$L \le \lim_{\delta \to 0^+} L_\delta = L_0.$$

To show that $L_0 \le L$, it suffices to prove that L_0 is a cluster value for f at $a \in X'$.

Observe that

$$L_0 = \lim_{n \to \infty} L_{1/n}.$$

Observe also that

$$L_{1/n} = \sup_{x \in V_{1/n}} f(x).$$

Hence, for each $n \in \mathbb{N}$ there exists $x_n \in V_{1/n}$ such that

$$L_{1/n} - 1/n < f(x_n) \le L_{1/n}.$$

Therefore, $|x_n - a| < 1/n$, that is,

$$\lim_{n \to \infty} x_n = a,$$

and

$$\lim_{n \to \infty} f(x_n) = \lim_{n \to \infty} L_{1/n} = L_0.$$

We may conclude that L_0 is a cluster value for f at a, that is $L_0 \le L$ and thus $L = L_0$.

For the case $\liminf f(x)$ the proof is similar and left as an exercise.

Example 4.5.9 Consider $f : \mathbb{R} \to \mathbb{R}$ where

$$f(x) = \begin{cases} -x^2 + 1, & \text{if } x < 1, \\ 10, & \text{if } x = 1, \\ 2x + 3, & \text{if } x > 1. \end{cases} \tag{4.13}$$

We shall calculate $\limsup_{x \to 1} f(x)$ and $\liminf_{x \to 1} f(x)$.

Observe that for $\delta > 0$ sufficiently small and $V_\delta = (1 - \delta, 1 + \delta) \setminus \{1\}$, we have

$$L_\delta = \sup f(V_\delta) = f(1 + \delta) = 2(1 + \delta) + 3,$$

and

$$l_\delta = \inf f(V_\delta) = 0,$$

so that

$$\limsup_{x \to 1} f(x) = \lim_{\delta \to 0^+} L_\delta$$
$$= \lim_{\delta \to 0^+} 2(1 + \delta) + 3$$
$$= 5, \qquad (4.14)$$

and

$$\liminf_{x \to 1} f(x) = \lim_{\delta \to 0^+} l_\delta$$
$$= \lim_{\delta \to 0^+} 0$$
$$= 0. \qquad (4.15)$$

The example is complete.

4.6   Infinite Limits

In this section we study infinite limits. We start with the following definition.

Definition 4.6.1 Let $X \subset \mathbb{R}$ be a nonempty set. Let $f : X \to \mathbb{R}$ be a function and let $a \in X'$.

We say that the limit of f as x approaches $a \in X'$ is plus infinity $(+\infty)$, if for each $A > 0$, there exists $\delta > 0$ such that if $x \in X$ and $0 < |x - a| < \delta$, then $f(x) > A$.

In such a case, we write,

$$\lim_{x \to a} f(x) = +\infty.$$

Let us see the example, $f : \mathbb{R} \setminus \{a\} \to \mathbb{R}$, where

$$f(x) = \frac{1}{(x - a)^2}, \forall x \in \mathbb{R} \setminus \{a\}.$$

Let $A > 0$ be given.

Thus for $x \neq a$, we have,

$$\frac{1}{(x-a)^2} > A \Leftrightarrow |x-a|^2 < 1/A$$

$$\Leftrightarrow |x-a| < \frac{1}{\sqrt{A}} (\equiv \delta). \qquad (4.16)$$

Hence, if $x \neq a$ and

$$|x-a| < \delta = \frac{1}{\sqrt{A}},$$

then

$$\frac{1}{(x-a)^2} > A.$$

Therefore, we have satisfied the limit definition, that is, given $A > 0$, we have obtained the corresponding δ concerning the definition in question.

Summarizing, for each $A > 0$, there exists $\delta = 1/\sqrt{A} > 0$ such that if $0 < |x-a| < \delta$, then

$$f(x) = \frac{1}{|x-a|^2} > A,$$

so that,

$$\lim_{x \to a} f(x) = +\infty.$$

Similarly, we write

$$\lim_{x \to a} f(x) = -\infty,$$

if for each $B < 0$, there exists $\delta > 0$ such that if $x \in X$ and $0 < |x-a| < \delta$, then

$$f(x) < B.$$

In such a context, we may define the one-sided limits as well.

Definition 4.6.2 (One-Sided Infinite Limits) Let $X \subset \mathbb{R}$ be a nonempty set. Let $f : X \to \mathbb{R}$ be a function and let $a \in X'_+$.

We say that the limit of f as x approaches $a \in X'_+$ on the right is plus infinite $(+\infty)$, if for each $A > 0$, there exists $\delta > 0$ such that if $x \in X$ and $0 < x - a < \delta$, then $f(x) > A$.

In such a case we write,

$$\lim_{x \to a^+} f(x) = +\infty.$$

Similarly, we say that the limit of f as x approaches $a \in X'_+$ on the right is minus infinite $(-\infty)$, if for each $B < 0$, there exists $\delta > 0$ such that if $x \in X$ and $0 < x - a < \delta$, then $f(x) < B$.

In such a case we write,

$$\lim_{x \to a^+} f(x) = -\infty.$$

Definition 4.6.3 Let $X \subset \mathbb{R}$ be a nonempty set. Let $f : X \to \mathbb{R}$ be a function and let $a \in X'_-$.

We say that the limit of f as x approaches $a \in X'_-$ on the left is plus infinite $(+\infty)$, if for each $A > 0$, there exists $\delta > 0$ such that if $x \in X$ and $0 < a - x < \delta$, then $f(x) > A$.

In such a case we write,

$$\lim_{x \to a^-} f(x) = +\infty.$$

Similarly, we say that the limit of f as x approaches $a \in X'_-$ on the left is minus infinite $(-\infty)$, if for each $B < 0$, there exists $\delta > 0$ such that if $x \in X$ and $0 < a - x < \delta$, then $f(x) < B$.

In such a case we write,

$$\lim_{x \to a^-} f(x) = -\infty.$$

Let us see the example where

$$f : \mathbb{R} \setminus \{1\} \to \mathbb{R},$$

is given by

$$f(x) = \frac{1}{x - 1}, \forall x \in \mathbb{R} \setminus \{1\}.$$

We have that

$$\lim_{x \to 1^+} f(x) = +\infty.$$

To show this, let $A > 0$.

Observe that,

$$x > 1 \text{ and } \frac{1}{x-1} > A \Leftrightarrow 0 < x - 1 < \frac{1}{A} (\equiv \delta). \qquad (4.17)$$

Hence, for each $A > 0$, there exists $\delta = 1/A > 0$ such that if $0 < x - 1 < \delta = 1/A$, then

$$\frac{1}{x-1} > A,$$

so that we may write,

$$\lim_{x \to 1^+} f(x) = +\infty.$$

Similarly, we may prove that

$$\lim_{x \to 1^-} f(x) = -\infty.$$

Indeed, let $B < 0$ be given. Thus,

$$x < 1 \text{ and } \frac{1}{x-1} < B \Leftrightarrow 0 < 1 - x < \frac{-1}{B} (\equiv \delta). \qquad (4.18)$$

Hence, for each $B < 0$, there exists $\delta = -1/B > 0$ such that if $0 < 1 - x < \delta = -1/B$, then

$$\frac{1}{x-1} < B$$

so that we may write,

$$\lim_{x \to 1^-} f(x) = -\infty.$$

Theorem 4.6.4 *Let $X \subset \mathbb{R}$ be a nonempty set and let $a \in X'$. Let $f, g : X \to \mathbb{R}$ be functions such that there exist $\delta_1 > 0$ and $c > 0$ such that*

$$f(x) > c, \forall x \in X \text{ such that } 0 < |x - a| < \delta_1,$$

there exists $\delta_2 > 0$ such that

$$g(x) > 0, \forall x \in X \text{ such that } 0 < |x - a| < \delta_2,$$

and

$$\lim_{x \to a} g(x) = 0.$$

Under such hypotheses, we have

$$\lim_{x \to a} \frac{f(x)}{g(x)} = +\infty.$$

Proof Let $A > 0$ be given.
For $\varepsilon = c/A$, from

$$\lim_{x \to a} g(x) = 0,$$

there exists $\delta_3 > 0$ such that if $x \in X$ and $0 < |x - a| < \delta_3$, then $|g(x)| < \varepsilon = c/A$.
Define $\delta = \min\{\delta_1, \delta_2, \delta_3\}$.
Thus, if $x \in X$ and $0 < |x - a| < \delta$, then

$$g(x) = |g(x)| < c/A,$$

and

$$f(x) > c > 0,$$

so that

$$\frac{f(x)}{g(x)} > \frac{A}{c}c = A.$$

Thus, we may conclude that

$$\lim_{x \to a} \frac{f(x)}{g(x)} = +\infty.$$

Theorem 4.6.5 *Let $X \subset \mathbb{R}$ be a nonempty set and let $a \in X'$. Let $f, g : X \to \mathbb{R}$ be functions such that there exists $\delta_1 > 0$ and $c > 0$ such that*

$$|f(x)| < c, \forall x \in X \text{ such that } 0 < |x - a| < \delta_1,$$

and

$$\lim_{x \to a} g(x) = +\infty.$$

Under such hypotheses, we have

$$\lim_{x \to a} \frac{f(x)}{g(x)} = 0.$$

Proof Let $\varepsilon > 0$ be given.

Let $A = c/\varepsilon > 0$. From

$$\lim_{x \to a} g(x) = +\infty$$

there exists $\delta_2 > 0$ such that if $x \in X$ and $0 < |x - a| < \delta_2$, then

$$g(x) > A = c/\varepsilon.$$

Define $\delta = \min\{\delta_1, \delta_2\}$.

Thus, if $x \in X$ and $0 < |x - a| < \delta$, then

$$0 < \frac{1}{g(x)} < \varepsilon/c,$$

and

$$|f(x)| < c,$$

so that

$$\left| \frac{f(x)}{g(x)} - 0 \right| = \frac{|f(x)|}{|g(x)|} < \frac{\varepsilon}{c} c = \varepsilon.$$

Thus, we may conclude that

$$\lim_{x \to a} \frac{f(x)}{g(x)} = 0.$$

The proof is complete.

4.7   Real Limits at Infinity

We start with a formal definition of real limit at infinity.

Definition 4.7.1 Let $X \subset \mathbb{R}$ be a nonempty set which is upper unbounded.

Let $f : X \to \mathbb{R}$ be a function.

We say that the limit as of f as x goes to plus infinity $(x \to +\infty)$ is $L \in \mathbb{R}$, if for each $\varepsilon > 0$ there exists $A > 0$ such that if $x \in X$ and $x > A$, then

$$|f(x) - L| < \varepsilon.$$

In such a case we write,

$$\lim_{x \to +\infty} f(x) = L \in \mathbb{R}.$$

Definition 4.7.2 Let $X \subset \mathbb{R}$ be a nonempty set which is lower unbounded.
Let $f : X \to \mathbb{R}$ be a function.
We say that the limit as of f as x goes to minus infinite $(x \to -\infty)$ is $L \in \mathbb{R}$, if
for each $\varepsilon > 0$ there exists $B < 0$ such that if $x \in X$ and $x < B$, then

$$|f(x) - L| < \varepsilon.$$

In such a case we write,

$$\lim_{x \to -\infty} f(x) = L \in \mathbb{R}.$$

Let us see the example where $f : \mathbb{R} \setminus \{-3/2\} \to \mathbb{R}$ is given by

$$f(x) = \frac{-5x + 2}{2x + 3}.$$

We are going to formally show that

$$\lim_{x \to +\infty} f(x) = -\frac{5}{2}.$$

Let $\varepsilon > 0$ be given.
Observe that,

$$x > 0 \text{ and } \left| \frac{-5x + 2}{2x + 3} - \left(-\frac{5}{2} \right) \right| < \varepsilon$$

$$\Leftrightarrow x > 0 \text{ and } \left| \frac{-5x + 2}{2x + 3} + \frac{5}{2} \right| < \varepsilon$$

$$\Leftrightarrow x > 0 \text{ and } \left| \frac{(-5x + 2)2 + 5(2x + 3)}{(2x + 3)2} \right| < \varepsilon$$

$$\Leftrightarrow x > 0 \text{ and } \left| \frac{19}{(2x + 3)2} \right| < \varepsilon$$

$$\Leftrightarrow x > 0 \text{ and } \left| \frac{(2x + 3)2}{19} \right| > \frac{1}{\varepsilon}$$

$$\Leftrightarrow x > 0 \text{ and } |(2x + 3)| > \frac{19}{2\varepsilon}$$

$$\Leftrightarrow x > 0 \text{ and } 2x + 3 > \frac{19}{2\varepsilon}$$

$$\Leftrightarrow x > \max\left\{0, \frac{19}{4\varepsilon} - \frac{3}{2}\right\} (\equiv A). \qquad (4.19)$$

Thus, if $x > A = \max\left\{0, \frac{19}{4\varepsilon} - \frac{3}{2}\right\}$, then

$$\left|\frac{-5x + 2}{2x + 3} - \left(-\frac{5}{2}\right)\right| < \varepsilon.$$

Hence, we have satisfied the concerning limit definition. Formally, for each $\varepsilon > 0$ there exists $A = \max\left\{1, \frac{19}{4\varepsilon} - \frac{3}{2}\right\}$ such that if $x > A$, then

$$|f(x) - (-5/2)| < \varepsilon,$$

so that

$$\lim_{x \to +\infty} f(x) = -\frac{5}{2}.$$

4.8   Infinite Limits at Infinity

We finish this chapter by presenting a study about infinite limits at infinity.

We start with the following definition.

Definition 4.8.1 Let $X \subset \mathbb{R}$ be an upper unbounded set.

Let $f : X \to \mathbb{R}$ be a function.

We say that the limit of f as x goes to plus infinity ($x \to +\infty$) is plus infinite ($+\infty$), if for each $A > 0$ there exists $B > 0$ such that if $x \in X$ and $x > B$, then $f(x) > A$.

In such a case we write

$$\lim_{x \to +\infty} f(x) = +\infty.$$

Similarly, we say that the limit of f as x goes to plus infinity ($x \to +\infty$) is minus infinite ($-\infty$), if for each $A < 0$ there exists $B > 0$ such that if $x \in X$ and $x > B$, then $f(x) < A$.

In such a case we write

$$\lim_{x \to +\infty} f(x) = -\infty.$$

Definition 4.8.2 Let $X \subset \mathbb{R}$ be a lower unbounded set.

Let $f : X \to \mathbb{R}$ be a function.

We say that the limit of f as x goes to minus infinity $(x \to -\infty)$ is plus infinite $(+\infty)$, if for each $A > 0$ there exists $B < 0$ such that if $x \in X$ and $x < B$, then $f(x) > A$.

In such a case we write

$$\lim_{x \to -\infty} f(x) = +\infty.$$

Similarly, we say that the limit of f as x goes to minus infinity $(x \to -\infty)$ is minus infinite $(-\infty)$, if for each $A < 0$ there exists $B < 0$ such that if $x \in X$ and $x < B$, then $f(x) < A$.

In such a case we write

$$\lim_{x \to -\infty} f(x) = -\infty.$$

Consider the example in which $f : \mathbb{R} \to \mathbb{R}$ is given by

$$f(x) = x^7 + 10x^3 + 10x \sin(x).$$

We are going to formally prove that

$$\lim_{x \to -\infty} f(x) = -\infty.$$

Observe that for $x < 0$, we have

$$x^7 + 10x^3 + 10x \sin(x) \leq x^7 + 10x^3 - 10x. \tag{4.20}$$

On the other hand,

$$x < 0 \text{ and } 10x^3 - 10x < 5x^3 \Leftrightarrow x < 0 \text{ and } 5x^3 - 10x < 0$$
$$\Leftrightarrow x < 0 \text{ and } 5x(x^2 - 2) < 0$$
$$\Leftrightarrow x < 0 \text{ and } x^2 - 2 > 0$$
$$\Leftrightarrow x < 0 \text{ and } x^2 > 2$$
$$\Leftrightarrow x < -\sqrt{2}. \tag{4.21}$$

So, from this and (4.20) we obtain,

$$x < -\sqrt{2} \text{ implies that } x^7 + 10x^3 + 10x \sin(x) \leq x^7 + 10x^3 - 10x < x^7 + 5x^3. \tag{4.22}$$

Observe that,

$$x < 0 \text{ and } x^7 + 5x^3 < \frac{x^7}{2} \Leftrightarrow x < 0 \text{ and } \frac{x^7}{2} + 5x^3 < 0$$

$$\Leftrightarrow x < 0 \text{ and } x^3(\frac{x^4}{2} + 5) < 0$$

$$\Leftrightarrow x < 0 \text{ and } \frac{x^4}{2} + 5 > 0$$

$$\Leftrightarrow x < 0. \tag{4.23}$$

From this and (4.22), we obtain,

$$x < -\sqrt{2} \text{ implies that } x^7 + 10x^3 + 10x \sin(x) \le x^7 + 10x^3 - 10x < x^7 + 5x^3 < \frac{x^7}{2}. \tag{4.24}$$

Let $A < 0$.
Thus,

$$x < 0 \text{ and } \frac{x^7}{2} < A \Leftrightarrow x < \sqrt[7]{2A}. \tag{4.25}$$

From this and (4.24), we obtain

$$\text{if } x < \min\{-\sqrt{2}, \sqrt[7]{2A}\} \equiv B, \text{ then } x^7 + 10x^3 + 10x \sin(x) < \frac{x^7}{2} < A.$$

Thus, we have satisfied the concerning limit definition. More formally, we could write:
For each $A < 0$ there exists $B = \min\{-\sqrt{2}, \sqrt[7]{2A}\}$ such that if $x < B$, then

$$x^7 + 10x^3 + 10x \sin(x) < A,$$

so that

$$\lim_{x \to -\infty} x^7 + 10x^3 + 10x \sin(x) = -\infty.$$

Exercises 4.8.3

1. Let $X \subset \mathbb{R}$ be a nonempty set and let $a \in X'$. Let $f, g : X \to \mathbb{R}$ be functions such that

$$\lim_{x \to a} f(x) = c < 0,$$

there exists $\delta_2 > 0$ such that

$$g(x) > 0, \forall x \in X \text{ such that } 0 < |x - a| < \delta_2,$$

and

$$\lim_{x \to a} g(x) = 0.$$

Upon such hypotheses, show that

$$\lim_{x \to a} \frac{f(x)}{g(x)} = -\infty.$$

2. Let $X \subset \mathbb{R}$ be a nonempty set.
   Let $g : X \to \mathbb{R}$ be a function and let $a \in X'$.
   We shall denote

$$\lim_{x \to a} g(x) = 0^+,$$

if

$$\lim_{x \to a} g(x) = 0,$$

and there exists $\delta_0 > 0$ such that if $x \in X$ and $0 < |x - a| < \delta_0$, then $g(x) > 0$.
Similarly, we shall denote

$$\lim_{x \to a} g(x) = 0^-,$$

if

$$\lim_{x \to a} g(x) = 0,$$

and there exists $\delta_0 > 0$ such that if $x \in X$ and $0 < |x - a| < \delta_0$, then $g(x) < 0$.

(a) Let $f : X \to \mathbb{R}$ be such that

$$\lim_{x \to a} f(x) = c > 0,$$

and let $g : X \to \mathbb{R}$ be such that

$$\lim_{x \to a} g(x) = 0^+.$$

Prove formally that

$$\lim_{x \to a} \frac{f(x)}{g(x)} = +\infty.$$

(b) Let $f : X \to \mathbb{R}$ be such that

$$\lim_{x \to a} f(x) = c < 0,$$

and let $g : X \to \mathbb{R}$ be such that

$$\lim_{x \to a} g(x) = 0^+.$$

Prove formally that

$$\lim_{x \to a} \frac{f(x)}{g(x)} = -\infty.$$

(c) Let $f : X \to \mathbb{R}$ be such that

$$\lim_{x \to a} f(x) = c > 0,$$

and let $g : X \to \mathbb{R}$ be such that

$$\lim_{x \to a} g(x) = 0^-.$$

Prove formally that

$$\lim_{x \to a} \frac{f(x)}{g(x)} = -\infty.$$

(d) Let $f : X \to \mathbb{R}$ be such that

$$\lim_{x \to a} f(x) = c < 0,$$

and let $g : X \to \mathbb{R}$ be such that

$$\lim_{x \to a} g(x) = 0^-.$$

Prove formally that

$$\lim_{x \to a} \frac{f(x)}{g(x)} = +\infty.$$

3. Let $X \subset \mathbb{R}$ be a nonempty set. Let $f, g : X \to \mathbb{R}$ be functions such that

$$\lim_{x \to a} f(x) = c > 0,$$

and

$$\lim_{x \to a} g(x) = -\infty.$$

(a) Show formally that

$$\lim_{x \to a} f(x) + g(x) = -\infty.$$

(b) Show formally that

$$\lim_{x \to a} f(x)g(x) = -\infty.$$

(c) Show formally that

$$\lim_{x \to a} f(x)/g(x) = 0.$$

4. Let $f : \mathbb{R} \setminus \{0\} \to \mathbb{R}$ be defined by

$$f(x) = e^{1/x}.$$

Prove that

$$\lim_{x \to 0^+} f(x) = +\infty$$

and

$$\lim_{x \to 0^-} f(x) = 0.$$

5. Let $f : \mathbb{R} \setminus \{0\} \to \mathbb{R}$ be defined by

$$f(x) = \frac{1}{1 + e^{1/x}}.$$

Prove that

$$\lim_{x \to 0^+} f(x) = 0$$

and

$$\lim_{x \to 0^-} f(x) = 1.$$

6. Let $f : \mathbb{R} \to \mathbb{R}$ be defined by

$$f(x) = x + \frac{x}{2}\sin(x), \ \forall x \in \mathbb{R}.$$

Prove formally that

$$\lim_{x \to +\infty} f(x) = +\infty$$

and

$$\lim_{x \to -\infty} f(x) = -\infty.$$

7. Let $f : \mathbb{R} \to \mathbb{R}$ be defined by

$$f(x) = x^2 + 10x\sin(x), \ \forall x \in \mathbb{R}.$$

Prove formally that

$$\lim_{x \to +\infty} f(x) = +\infty$$

and

$$\lim_{x \to -\infty} f(x) = +\infty.$$

8. Let $f : \mathbb{R} \to \mathbb{R}$ be defined by

$$f(x) = x^3 + x^2 + 10x\sin(x), \ \forall x \in \mathbb{R}.$$

Prove formally that

$$\lim_{x \to +\infty} f(x) = +\infty$$

and

$$\lim_{x \to -\infty} f(x) = -\infty.$$

9. Let $X \subset \mathbb{R}$ and let $f : X \to \mathbb{R}$ be a monotonic function such that $f(X) \subset [a, b]$. Prove that if $f(X)$ is dense in $[a, b]$, then for each $c \in X'_+ \cap X'_-$ we have that

$$\lim_{x \to c^-} f(x) = \lim_{x \to c^+} f(x).$$

Also prove that if $c \in X$, then

$$\lim_{x \to c} f(x) = f(c).$$

10. Let $X \subset \mathbb{R}$, let $f : X \to \mathbb{R}$ be a monotonic function, and let $a \in X'$. Suppose there exists a sequence $\{x_n\} \subset X$ such that $x_n > a$, $\forall n \in \mathbb{N}$,

$$\lim_{n \to \infty} x_n = a.$$

and

$$\lim_{n \to \infty} f(x_n) = L \in \mathbb{R}.$$

Upon such hypotheses, show that

$$\lim_{x \to a^+} f(x) = L.$$

11. Let $f : [a, b] \to \mathbb{R}$ be a monotonic function. Define

$$A = \{c \in (a, b) : \lim_{x \to c^-} f(x) \neq \lim_{x \to c^+} f(x)\}.$$

Prove that A is countable.

12. Prove that

$$\lim_{x \to +\infty} \ln(x) = +\infty$$

and that

$$\lim_{x \to 0+} \ln(x) = -\infty.$$

Let $a > 1$. Prove that

$$\lim_{x \to +\infty} a^x = +\infty,$$

and

$$\lim_{x \to -\infty} a^x = 0,$$

13. Let $p : \mathbb{R} \to \mathbb{R}$ be a real polynomial function, that is,

$$p(x) = \sum_{j=0}^{n} a_j x^j = a_0 + a_1 x + a_2 x^2 + \cdots + a_n x^n.$$

Assume $n \geq 1$ and $a_n > 0$. Prove that

$$\lim_{x \to \infty} p(x) = +\infty,$$

and if n is even, then

$$\lim_{x \to -\infty} p(x) = +\infty$$

if n is odd, then

$$\lim_{x \to -\infty} p(x) = -\infty.$$

14. Obtain the cluster values for the function

$$f : \mathbb{R} \setminus \{0\} \to \mathbb{R}$$

at the point $x = 0$, where

$$f(x) = \frac{\sin(1/x)}{1 + e^{1/x}}.$$

15. For each $x \in \mathbb{R}$, denote by $[x]$ the greatest integer smaller or equal to x. Let $a, b \in \mathbb{R}$ be such that $a > 0$ and $b > 0$. Prove that

$$\lim_{x \to 0^+} \frac{x}{a} \left[\frac{b}{x} \right] = \frac{b}{a}$$

and

$$\lim_{x \to 0^+} \frac{b}{x} \left[\frac{x}{a} \right] = 0.$$

Show also that

$$\lim_{x \to 0^-} \frac{x}{a} \left[\frac{b}{x} \right] = \frac{b}{a}.$$

and

$$\lim_{x \to 0^-} \frac{b}{x} \left[\frac{x}{a} \right] = +\infty.$$

- In the next exercises, assume always $X \subset \mathbb{R}$.

16. Let f be a bounded function on a neighborhood $V_{\delta_0} = (X \setminus \{a\}) \cap (a - \delta_0, a + \delta_0)$ of $a \in X'$, where $\delta_0 > 0$. Prove that for each $\varepsilon > 0$ there exists $\delta > 0$ such that if $x \in X$ and $0 < |x - a| < \delta$, then

$$l - \varepsilon < f(x) < L + \varepsilon,$$

where

$$l = \liminf_{x \to a} f(x)$$

and

$$L = \limsup_{x \to a} f(x).$$

17. Let f be a bounded function on a neighborhood $V_{\delta_0} = (X \setminus \{a\}) \cap (a - \delta_0, a + \delta_0)$ of $a \in X'$, where $\delta_0 > 0$. Show that

$$\lim_{x \to a} f(x)$$

exists if, and only if, f has only one cluster value at a.

18. Let $f, g : X \to \mathbb{R}$ be bounded functions on a neighborhood of $a \in X'$. Show that

(a)

$$\limsup_{x \to a}(f(x) + g(x)) \leq \limsup_{x \to a} f(x) + \limsup_{x \to a} g(x),$$

(b)

$$\liminf_{x \to a}(f(x) + g(x)) \geq \liminf_{x \to a} f(x) + \liminf_{x \to a} g(x),$$

(c)

$$\limsup_{x \to a}[-f(x)] = -\liminf_{x \to a} f(x),$$

(d)

$$\liminf_{x \to a}[-f(x)] = -\limsup_{x \to a} f(x).$$

19. Let $f : [0, +\infty) \to \mathbb{R}$ be a bounded function on each bounded interval. Suppose

$$\lim_{x \to +\infty} (f(x+1) - f(x)) = \alpha \in \mathbb{R}.$$

Under such hypotheses, show that

$$\lim_{x \to +\infty} \frac{f(x)}{x} = \alpha.$$

Chapter 5
Continuous Functions

5.1 Introduction

This chapter presents a study about continuity for one variable real functions. Standard topics such as the relations between continuity and compactness are developed extensively. We finish the chapter addressing the main definitions and results on uniform continuity.

For such a chapter the main references are [9, 12].

We start with the formal definition of continuity.

Definition 5.1.1 (Continuous Function) Let $f : X \subset \mathbb{R} \to \mathbb{R}$ be a function and let $a \in X$. We say that f is continuous at a if for each $\varepsilon > 0$ there exists $\delta > 0$ such that if $x \in X$ and $|x - a| < \delta$, then $|f(x) - f(a)| < \varepsilon$.

Intuitively, this means that as x approaches a through the domain, $f(x)$ approaches $f(a)$.

Remark 5.1.2

1. If $a \in X$ is an isolated point of X, then all real function $f : X \to \mathbb{R}$ is continuous at a. Indeed, let $\varepsilon > 0$. Thus we may obtain $\delta > 0$ such that $X \cap (a - \delta, a + \delta) = \{a\}$. Hence, if $x \in X$ and $|x - a| < \delta$ we necessarily have $x = a$ and therefore

$$|f(x) - f(a)| = |f(a) - f(a)| = 0 < \varepsilon.$$

2. As f is continuous on all its domain, we simply say that f is continuous.
3. Let $a \in X' \cap X$. Thus f is continuous at a if, and only if,

$$\lim_{x \to a} f(x) = f(a).$$

© Springer International Publishing AG, part of Springer Nature 2018
F. S. Botelho, *Real Analysis and Applications*,
https://doi.org/10.1007/978-3-319-78631-5_5

Observe that any $f : \mathbb{Z} \to \mathbb{R}$ is continuous, since each point of $\mathbb{Z}$ is isolated. Similarly, for $X = \{1, 1/2, 1/3, \ldots, 1/n, \ldots\}$, any function whose domain is X is continuous. On the other hand, if

$$X = \{1/n \,:\, n \in \mathbb{N}\} \cup \{0\},$$

then a function $f : X \to \mathbb{R}$ is continuous if, and only if,

$$\lim_{n \to \infty} f(1/n) = f(0).$$

5.2 Some Preliminary Results

Theorem 5.2.1 *Let $f : X \to \mathbb{R}$ be a continuous function and $a \in X' \cap X$. Under such hypotheses f is bounded in a neighborhood of $a \in X' \cap X$. That is, there exist $\delta > 0$ and $A > 0$ such that $|f(x)| < A$, $\forall x \in V_\delta = X \cap (a - \delta, a + \delta)$.*

Exercise 5.2.2 Prove Theorem 5.2.1.

Theorem 5.2.3 *Let $f : X \to \mathbb{R}$ and $a \in X' \cap X$ where $f(a) > 0$. Under such hypotheses, there exists $\delta > 0$ such that if $x \in X$ and $|x - a| < \delta$, then $f(x) > 0$.*

Proof Let $\varepsilon = f(a)/2 > 0$.
   From

$$\lim_{x \to a} f(x) = f(a),$$

there exists $\delta > 0$ such that if $|x - a| < \delta$, then $|f(x) - f(a)| < \varepsilon = f(a)/2$.
   In particular

$$-\varepsilon = -f(a)/2 < f(x) - f(a),$$

that is

$$f(x) > f(a)/2 > 0, \ \text{if } x \in X \text{ and } |x - a| < \delta.$$

Theorem 5.2.4 *Let $X \subset \mathbb{R}$ be an open set. Let $f : X \to \mathbb{R}$ be a function.*
   Under such hypotheses, f is continuous if, and only if, $f^{-1}(B)$ is open whenever $B \subset \mathbb{R}$ is open.

Proof Assume f is continuous. Let $B \subset \mathbb{R}$ be an open set. We are going to show that $f^{-1}(B)$ is open, where

$$f^{-1}(B) = \{x \in X \,:\, f(x) \in B\}.$$

If $f^{-1}(B) = \emptyset$ the results follow immediately, since $\emptyset$ is open.
Thus, suppose $f^{-1}(B) \neq \emptyset$.
Let $x_0 \in f^{-1}(B)$. Hence, $f(x_0) \in B$.
Since B is open, there exists $\varepsilon > 0$ such that

$$V_\varepsilon(f(x_0)) = (f(x_0) - \varepsilon, f(x_0) + \varepsilon) \subset B.$$

Since f is continuous in x_0, there exists $\delta_1 > 0$ such that if $x \in X$ and $|x - x_0| < \delta_1$, then

$$|f(x) - f(x_0)| < \varepsilon. \tag{5.1}$$

On the other hand, since X is open, there exists $\delta_2 > 0$ such that

$$V_{\delta_2}(x_0) = (x_0 - \delta_2, x_0 + \delta_2) \subset X.$$

From this and (5.1), defining $\delta = \min\{\delta_1, \delta_2\}$, we have that if $|x - x_0| < \delta$, then

$$|f(x) - f(x_0)| < \varepsilon.$$

Therefore

$$f(x) \in (f(x_0) - \varepsilon, f(x_0) + \varepsilon) = V_\varepsilon(f(x_0)) \subset B,$$

so that

$$x \in f^{-1}(B), \ \forall x \in V_\delta(x_0),$$

that is,

$$V_\delta(x_0) \subset f^{-1}(B).$$

From this we may infer that x_0 is an interior point, $\forall x_0 \in f^{-1}(B)$, so that $f^{-1}(B)$ is open.
Reciprocally, assume $f^{-1}(B)$ is open whenever $B \subset \mathbb{R}$ is open.
Let $x_0 \in X$ and let $\varepsilon > 0$.
Observe that

$$V_\varepsilon(f(x_0)) = (f(x_0) - \varepsilon, f(x_0) + \varepsilon)$$

is open, so that, from the hypotheses,

$$f^{-1}(V_\varepsilon(f(x_0)))$$

is open.

Observe also that since $f(x_0) \in V_\varepsilon(f(x_0))$ we obtain

$$x_0 \in f^{-1}(V_\varepsilon(f(x_0))$$

so that there exists $\delta > 0$ such that

$$V_\delta(x_0) \subset f^{-1}(V_\varepsilon(f(x_0))).$$

Hence, if $|x - x_0| < \delta$, then $x \in f^{-1}(V_\varepsilon(f(x_0)))$ so that

$$f(x) \in V_\varepsilon(f(x_0)),$$

that is,

$$|f(x) - f(x_0)| < \varepsilon.$$

From this we may infer that f is continuous at x_0, $\forall x_0 \in X$, so that f is continuous.

The proof is complete.

At this point we propose some closely related exercises.

Exercises 5.2.5

1. Let $f : \mathbb{R} \to \mathbb{R}$ be a continuous function. Let $c \in \mathbb{R}$ and define

$$A = \{x \in \mathbb{R} : f(x) = c\}.$$

Show that A is closed.

2. Let $f, g : \mathbb{R} \to \mathbb{R}$ be continuous functions. Define

$$A = \{x \in \mathbb{R} : f(x) = g(x)\}.$$

Show that A is closed.

3. Let $f, g : X \to \mathbb{R}$ be continuous functions at $a \in X'$. Show that if $f(a) < g(a)$, then there exists $\delta > 0$ such that if $x \in X$ and $|x - a| < \delta$, then $f(x) < g(x)$.

4. Let $f : X \to \mathbb{R}$ be a continuous function and $a \in X'$. Show that if $f(a) < K$, then there exists $\delta > 0$ such that if $x \in X$ and $|x - a| < \delta$, then $f(x) < K$.

5. Let $f, g : \mathbb{R} \to \mathbb{R}$ be continuous functions. Define

$$A = \{x \in \mathbb{R} : f(x) > g(x)\}.$$

Show that A is open.

6. Let $f : X \to \mathbb{R}$ be a continuous function where $X \subset \mathbb{R}$ is an open set. Let $K \in \mathbb{R}$.

   Show that $A_K = \{x \in X : f(x) < K\}$ is open.

7. Let $f : \mathbb{R} \to \mathbb{R}$ be a continuous function. Let $K \in \mathbb{R}$.
   Show that $B_K = \{x \in \mathbb{R} : f(x) \leq K\}$ is closed.

5.2.1 Some Properties of Continuous Functions

The proof of the next theorems are very similar to those of limit theory.

Theorem 5.2.6 *Let $f, g : X \to \mathbb{R}$ be a continuous functions at $a \in X$. Under such hypotheses, αf, $\forall \alpha \in \mathbb{R}$, $f + g$, fg, and if $g(a) \neq 0$, f/g, are continuous at a.*

Proof The proof results from the limit properties.

Theorem 5.2.7 *A composition of two continuous functions is continuous. More specifically, let $X, Y \subset \mathbb{R}$ and let $f : X \to \mathbb{R}$, $g : Y \to \mathbb{R}$ be functions such that $f(X) \subset Y$. Let $a \in X' \cap X$ and $b \in Y' \cap Y$. Assume*

$$\lim_{x \to a} f(x) = f(a) \equiv b$$

$$\lim_{y \to b} g(y) = g(b) \equiv c.$$

Under such hypotheses,

$$\lim_{x \to a} g(f(x)) = c = g(b).$$

Proof The proof is almost the same as that of Theorem 4.3.1 and will not be repeated here.

Theorem 5.2.8 *Let $X \subset \bigcup_{\lambda \in L} A_\lambda$, where $A_\lambda \subset \mathbb{R}$ is open, $\forall \lambda \in L$. Let $f : X \to \mathbb{R}$ be such that, for each $\lambda \in L$ the restriction $f|_{(A_\lambda \cap X)}$ is continuous.*
   Under such hypotheses, f is continuous.

Proof Let $a \in X' \cap X$. We shall show that f is continuous at a.

Let $\varepsilon > 0$. From the hypotheses, there exists $\lambda_0 \in L$ such that $a \in A_{\lambda_0}$. Since A_{λ_0} is open, there exists $\delta_1 > 0$ such that $(a - \delta_1, a + \delta_1) \subset A_{\lambda_0}$. Since, $f|_{(A_{\lambda_0} \cap X)}$ is continuous, there exists $\delta_2 > 0$ such that if $x \in A_{\lambda_0} \cap X$ and $|x - a| < \delta_2$, then

$$|f(x) - f(a)| < \varepsilon.$$

Define $\delta = \min\{\delta_1, \delta_2\}$. Thus, if $x \in X$ and $|x - a| < \delta = \min\{\delta_1, \delta_2\}$, then $x \in A_{\lambda_0} \cap X$ and also $|f(x) - f(a)| < \varepsilon$.
   Thus, f is continuous at a.

5.3 Discontinuities

Definition 5.3.1 Let $f : X \to \mathbb{R}$. We say that $a \in X$ is a point of discontinuity of f, if f is not continuous at a. Thus, if $a \in X$ is a point of discontinuity, there exists $\varepsilon_0 > 0$ such that for each $\delta > 0$, there exists $x_\delta \in X$ such that $0 < |x_\delta - a| < \delta$ and $|f(x_\delta) - f(a)| \geq \varepsilon_0$.

Exercise 5.3.2 Show that $f : \mathbb{R} \to \mathbb{R}$ is discontinuous on all the real set, where

$$f(x) = \begin{cases} 0, & \text{if } x \in \mathbb{Q} \\ 1, & \text{if } x \in \mathbb{R} \setminus \mathbb{Q}. \end{cases} \tag{5.2}$$

5.4 Types of Discontinuities

Definition 5.4.1 Let $f : X \to \mathbb{R}$. We say that f has a first kind discontinuity at $a \in X$ if f is discontinuous and also at such a point, if $a \in X'_+$, then the one-sided limit

$$\lim_{x \to a^+} f(x)$$

exists and if $a \in X'_-$, then

$$\lim_{x \to a^-} f(x)$$

exists.

Remark 5.4.2 In order that f be discontinuous at $a \in X$ it is necessary that $a \in X'$. We also emphasize that if $a \in X'_+$ and $a \notin X'_-$, then in order that the discontinuity be of first kind it is just necessary that

$$\lim_{x \to a^+} f(x)$$

exists. A similar remark is valid for the case in which $a \in X'_-$ and $a \notin X'_+$.

Definition 5.4.3 A discontinuity that is not of first kind is said to be of second one. Thus either $a \in X'_+$ and $\lim_{x \to a^+} f(x)$ does not exist, or $a \in X'_-$ and $\lim_{x \to a^-} f(x)$ does not exist.

Resulting from theorem of existence of one-sided limits for monotonic functions, we may obtain the following result:

Theorem 5.4.4 *A monotonic function $f : X \to \mathbb{R}$ has no discontinuities of second kind.*

Theorem 5.4.5 *Let* $X \subset \mathbb{R}$ *and let* $f : X \to \mathbb{R}$ *be a monotonic function. Then if* $f(X)$ *is dense in some interval* I, *then* f *is continuous.*

Proof For each $a \in X'_+$ we shall denote

$$f(a+) = \lim_{x \to a^+} f(x),$$

and for each $a \in X'_-$ we denote

$$f(a-) = \lim_{x \to a^-} f(x).$$

Suppose f is nondecreasing (another cases may be dealt similarly). Let $a \in X'$, thus $a \in X'_+$ or $a \in X'_-$. Suppose that $a \in X'_+$. We shall show that $f(a) = f(a+)$.

Observe that

$$f(a+) = \inf\{f(x) \mid x \in X, x > a\}$$

and also if $a < x$ and $x \in X$ then $f(a) \leq f(x)$.

Thus,

$$f(a) \leq \inf\{f(x) \mid x \in X, x > a\} = f(a+).$$

Suppose, to obtain contradiction, that $f(a) < f(a+)$.

Since $a \in X'_+$, there exists $x > a$ such that $f(x) \geq f(a+)$. Assume I is an interval which contains $f(X)$ and such that $f(X)$ is dense on I.

Since $f(X)$ is dense on I, we have that $I \supset (f(a), f(a+))$. However, there is no points of X such that $f(a) < f(x) < f(a+)$, since if $x \leq a$, then $f(x) \leq f(a)$, and if $x > a$, then $f(x) \geq f(a+)$. This contradicts $f(X)$ to be dense on I. Thus, $f(a) = f(a+)$.

Similarly, we may show that if $a \in X'_-$, then $f(a-) = f(a)$.

Hence, f is continuous at $a \in X'$.

The proof of the next corollary results directly of this last theorem.

Corollary 5.4.6 *Let* $f : X \to \mathbb{R}$ *be a monotonic function such that* $f(X)$ *is an interval.*

Under such hypotheses, f *is continuous.*

5.5 Continuous Functions on Compact Sets

Theorem 5.5.1 *Let* $f : X \to \mathbb{R}$ *be a continuous function. If* $X \subset \mathbb{R}$ *is compact, then* $f(X)$ *is compact.*

Proof Let $\{y_n\} \subset f(X)$. Thus there exists a sequence $\{x_n\} \subset X$ such that $f(x_n) = y_n$, $\forall n \in \mathbb{N}$. Since X is compact, there exists a subsequence $\{x_{n_k}\}$ of $\{x_n\}$ and $x_0 \in X$ such that $\lim_{k \to \infty} x_{n_k} = x_0$, and thus, since f is continuous,

$$\lim_{k\infty} y_{n_k} = \lim_{k \to \infty} f(x_{n_k}) = f(x_0) \equiv y_0 \in f(X).$$

Thus, for any sequence $\{y_n\} \subset f(X)$ we may obtain a convergent subsequence with limit in $f(X)$.

From the Heine–Borel theorem, $f(X)$ is compact.

Corollary 5.5.2 (Weierstrass) *Let* $f : X \to \mathbb{R}$ *be a continuous function where* $X \subset \mathbb{R}$ *is compact. Under such hypotheses, f attains its extremal points on X, that is, there exist $x_1, x_2 \in X$ such that $f(x_1) \le f(x) \le f(x_2), \forall x \in X$.*

Proof From the last theorem $f(X)$ is compact. Therefore, it is bounded and closed.

Let $y_2 = \sup f(X)$. Thus for each $n \in \mathbb{N}$ there exists $y_2 - 1/n < y_n \le y_2$. Hence y_2 is a limit point of $f(X)$. Since $f(X)$ is closed, we obtain $y_2 \in f(X)$ and thus, there exists $x_2 \in X$ such that $f(x_2) = y_2 = \sup f(X)$.

Similarly, we may prove that there exists $x_1 \in X$ such that $f(x_1) = \inf f(X)$.

5.6   The Intermediate Value Theorem

Theorem 5.6.1 *Let* $f : [a, b] \to \mathbb{R}$ *be a continuous function. Define* $g : [a, b] \to \mathbb{R}$ *by*

$$g(x) = \max\{f(t) \quad t \in [a, x]\}.$$

Then g is nondecreasing and continuous on $[a, b]$.

Proof Clearly g is nondecreasing. Let $x \in (a, b)$.

We shall show that

$$\lim_{y \to x} g(y) = g(x).$$

Let $\varepsilon > 0$ be given. Since

$$\lim_{y \to x} f(y) = f(x),$$

there exists $\delta > 0$ such that if $x \le y < x + \delta$, then

$$|f(y) - f(x)| < \varepsilon.$$

Choose y such that $x < y < x + \delta$. Hence,

$$g(y) = \max\{f(t) : t \in [a, y]\}.$$

There are two possibilities, either such a max is attained in $[a, x]$ and in such a case $g(y) = g(x)$, or the max is attained at some $t \in (x, y]$. Since, $x < t \leq y < x + \delta$ we obtain

$$|g(y) - g(x)| = g(y) - g(x) \leq g(y) - f(x) = f(t) - f(x) < \varepsilon,$$

that is, in any case

$$|g(y) - g(x)| < \varepsilon, \ \forall y \in (x, x + \delta).$$

Since $\varepsilon > 0$ is arbitrary, we may conclude that

$$\lim_{y \to x^+} g(y) = g(x).$$

Similarly, we may prove that

$$\lim_{y \to x^-} g(y) = g(x),$$

and conclude that

$$\lim_{y \to x} g(y) = g(x).$$

Finally, the case in which either $x = a$ or $x = b$ may be dealt similarly with one-sided limits.

This completes the proof.

Theorem 5.6.2 *Let $f : [a, b] \to \mathbb{R}$ be a continuous function on $[a, b]$. Suppose that $f(a) < f(b)$. Let $d \in \mathbb{R}$ be such that*

$$f(a) < d < f(b).$$

Under such hypotheses, there exists $c \in (a, b)$ such that

$$f(c) = d.$$

Proof Define $g : [a, b] \to \mathbb{R}$ by

$$g(x) = \max\{f(t) \ : \ t \in [a, x]\}, \forall x \in [a, b].$$

Hence, from the last theorem g is continuous and nondecreasing on $[a, b]$. Define

$$A = \{x \in [a, b] \ : \ g(x) < d\},$$

and

$$B = \{x \in [a, b] \; : \; g(x) > d\}.$$

Observe that $a \in A$ and $b \in B$ so that $A \neq \emptyset$ and $B \neq \emptyset$.

Define $\alpha = \sup A$ and $\beta = \inf B$. We claim that $\alpha \leq \beta$.

Suppose, to obtain contradiction, that $\beta < \alpha$. Let $\varepsilon = \frac{\alpha - \beta}{4}$.

Thus, there exists $x_1 \in A$ and $x_2 \in B$ such that $x_1 > \alpha - \varepsilon$ and $x_2 < \beta + \varepsilon$, so that $-x_2 > -\beta - \varepsilon$ and hence

$$x_1 - x_2 > \alpha - \beta - 2\varepsilon = \alpha - \beta - \frac{\alpha - \beta}{2} = \frac{\alpha - \beta}{2} > 0.$$

Therefore $x_2 < x_1$ and since g is nondecreasing, we obtain

$$g(x_2) \leq g(x_1) < d$$

which contradicts

$$x_2 \in B.$$

Thus $\alpha \leq \beta$.

Suppose $\alpha < \beta$. Choose $c_1 \in \mathbb{R}$ such that

$$\alpha < c_1 < \beta.$$

Since $c_1 > \alpha$, $c_1 \notin A$ so that $g(c_1) \geq d$. Similarly, since $c_1 < \beta$, $c_1 \notin B$ so that $g(c_1) \leq d$.

Hence

$$d \leq g(c_1) \leq d$$

so that

$$g(c_1) = d.$$

From the Weierstrass Theorem, there exists $c \in [0, c_1]$ such that

$$f(c) = \max\{f(x), \; x \in [a, c_1]\} = g(c_1) = d,$$

that is, $f(c) = d$.

The other possibility is $\alpha = \beta$, that is

$$\sup A = \inf B = \alpha$$

Thus there exists $\{x_n\} \subset A$ be such that $x_n \to \alpha$ as $n \to \infty$. By continuity,

$$g(\alpha) = \lim_{n \to \infty} g(x_n) \le d.$$

Similarly there exists $\{y_n\} \subset B$ be such that $y_n \to \alpha$ as $n \to \infty$. By continuity,

$$g(\alpha) = \lim_{n \to \infty} g(y_n) \ge d.$$

Hence

$$d \le g(\alpha) \le d,$$

so that

$$g(\alpha) = d.$$

From the Weierstrass Theorem, there exists $c \in [a, \alpha]$ such that

$$f(c) = \max\{f(x), \ x \in [a, \alpha]\} = g(\alpha) = d,$$

that is $f(c) = d$.

This completes the proof.

We recall that a set $I \subset \mathbb{R}$ is said to be an interval if

1. $I \ne \emptyset$,
2. if $a, b \in I$ and $a < c < b$, then $c \in I$.

Corollary 5.6.3 *Let $f : I \to \mathbb{R}$ be a continuous function on the interval I. Under such hypotheses $f(I)$ is an interval.*

Proof The proof results from the intermediate value theorem and from the definition of interval. We do not give the details here.

Theorem 5.6.4 *Let $f : I \to \mathbb{R}$ be a continuous injective function, defined on the interval I. Under such hypotheses, f is monotonic, its range $J = f(I)$ is an interval and its inverse is continuous $f^{-1} : J \to I$ is continuous.*

Proof To conclude that f is monotonic, it suffices to show that f is monotonic on each closed interval $[a, b] \subset I$.

So, we may assume $I = [a, b]$. Since f is injective, we have that $f(a) \ne f(b)$. Assume $f(a) < f(b)$ (the case $f(a) > f(b)$ may be dealt similarly)

Since we have assumed $f(a) < f(b)$ we have to show that f is increasing.

Suppose, to obtain contradiction, that there exist $x, y \in [a, b]$ such that $x < y$ and $f(x) > f(y)$.

There are two possibilities to be considered, namely,

1. $f(a) < f(y)$
2. $f(a) > f(y)$

In the first case $f(a) < f(y) < f(x)$. From the intermediate value theorem, there exists $c \in (a, x)$ such that $f(c) = f(y)$, which contradicts f to be injective.

In the second case we have $f(y) < f(a) < f(b)$. Also from the intermediate value theorem there exists $c \in (y, b)$ such that $f(c) = f(a)$, which also contradicts f to be injective.

Therefore, f increasing. From the corollary of the intermediate value theorem we obtain that $J = f(I)$ is an interval. Since f is monotonic and injective, we have that $f^{-1} : J \to I$ is also monotonic and since $f^{-1}(J) = I$ is an interval, from Theorem 5.4.5, f^{-1} is also continuous.

5.6.1 Applications of the Intermediate Value Theorem

We start with the following proposition.

Proposition 5.6.5 *Let $a, b \in \mathbb{R}$ be such that $0 \le a < b$. Under such hypotheses $0 \le a^n < b^n, \forall n \in \mathbb{N}$.*

Proof The proof is left as an exercise. As a hint, define

$$A = \{n \in \mathbb{N} \text{ such that } 0 \le a^n < b^n\}$$

and prove that $A = \mathbb{N}$.

Proposition 5.6.6 *Let $n \in \mathbb{N}$ and let $y > 0$. Under such hypotheses there exists $x > 0$ such that*

$$x^n = y,$$

so that we denote

$$x = y^{1/n}.$$

Proof Observe that for $x_0 = 0$ and $x_1 = (y + 1)$, we have

$$x_0 = 0 = 0^n < y < y + 1 < (y + 1)^n = x_1^n. \tag{5.3}$$

On the other hand, we have already proven that

$$\lim_{x \to x_0} x^n = x_0^n, \ \forall x_0 \in \mathbb{R},$$

so that $f : \mathbb{R} \to \mathbb{R}$ where

$$f(x) = x^n, \ \forall x \in \mathbb{R}$$

is a continuous function.

From this, (5.3) and the intermediate value theorem, there exists $x \in (x_0, x_1)$ such that

$$x^n = y,$$

so that we denote,

$$x = y^{1/n}.$$

This completes the proof.

Proposition 5.6.7 *Let $n \in \mathbb{N}$ be an odd natural number and let $y \in \mathbb{R}$. Under such hypotheses there exists $x \in \mathbb{R}$ such that*

$$x^n = y,$$

so that we denote

$$x = y^{1/n}.$$

Proof The case in which $y \geq 0$ has already been dealt in the last proposition.
  Thus, let us assume $y < 0$.
  Hence for $x_0 = y - 1$ and $x_1 = 0$ we have,

$$x_0^n = (y - 1)^n \leq (y - 1) < y < 0 = 0^n = x_1^n. \tag{5.4}$$

On the other hand, as we have already seen, the function $f : \mathbb{R} \to \mathbb{R}$ where

$$f(x) = x^n, \ \forall x \in \mathbb{R}$$

is continuous.

From this, (5.4) and the intermediate value theorem, there exists $x \in (x_0, x_1)$ such that

$$x^n = y,$$

so that we denote,

$$x = y^{1/n}.$$

This completes the proof.

Remark 5.1 Indeed, if n is odd, such a function $f(x) = x^n$ is strictly increasing so that $x \in \mathbb{R}$ such that $x^n = y$ is unique, $\forall y \in \mathbb{R}$. If n is even $f(x) = x^n$ is an even function, but it is strictly increasing on $[0, +\infty)$ so that for each $y > 0$ there exists a unique $x > 0$ such that $x^n = y$. However in such a case we have also $(-x)^n = y$.

At this point we present some concerning examples of continuous functions.

Proposition 5.6.8 *Let* $f : [0, +\infty) \to \mathbb{R}$ *be defined by*

$$f(x) = \sqrt{x}.$$

Under such assumptions, we have,

$$\lim_{x \to x_0} f(x) = f(x_0), \ \forall x_0 > 0,$$

and

$$\lim_{x \to 0^+} f(x) = 0,$$

so that f *is a continuous function.*

Proof Choose $x_0 > 0$. We are going to show that

$$\lim_{x \to x_0} \sqrt{x} = \sqrt{x_0}.$$

Let $\varepsilon > 0$ be given. Define

$$\varepsilon_1 = \min\{\sqrt{x_0}/2, \varepsilon\}.$$

Thus,

$$
\begin{aligned}
&|\sqrt{x} - \sqrt{x_0}| < \varepsilon_1 \\
&\Leftrightarrow -\varepsilon_1 < \sqrt{x} - \sqrt{x_0} < \varepsilon_1 \\
&\Leftrightarrow \sqrt{x_0} - \varepsilon_1 < \sqrt{x} < \sqrt{x_0} + \varepsilon_1 \\
&\Leftrightarrow (\sqrt{x_0} - \varepsilon_1)^2 < x < (\sqrt{x_0} + \varepsilon_1)^2 \\
&\Leftrightarrow -\delta_1 \equiv (\sqrt{x_0} - \varepsilon_1)^2 - x_0 < x - x_0 < (\sqrt{x_0} + \varepsilon_1)^2 - x_0 \equiv \delta_2. \quad (5.5)
\end{aligned}
$$

So, defining $\delta = \min\{\delta_1, \delta_2\}$, if $0 < |x - x_0| < \delta$, we have,

$$-\delta_1 \le -\delta < x - x_0 < \delta \le \delta_2,$$

so that from this and (5.6), we obtain

$$|\sqrt{x} - \sqrt{x_0}| < \varepsilon_1 \le \varepsilon.$$

Formally, we could write:
For each $\varepsilon > 0$, defining $\varepsilon_1 = \min\{\sqrt{x_0}/2, \varepsilon\}$, there exists $\delta = \min\{\delta_1, \delta_2\}$, where $\delta_1 = x_0 - (\sqrt{x_0} - \varepsilon_1)^2$, and $\delta_2 = (\sqrt{x_0} + \varepsilon_1)^2 - x_0$, such that if $0 <$

$|x - x_0| < \delta$, then

$$|\sqrt{x} - \sqrt{x_0}| < \varepsilon_1 \leq \varepsilon,$$

so that

$$\lim_{x \to x_0} \sqrt{x} = \sqrt{x_0}.$$

Finally, let us consider the case in which $x_0 = 0$.
Observe that

$$0 < \sqrt{x} < \varepsilon \Leftrightarrow 0 < x < \varepsilon^2 (\equiv \delta).$$

Thus, more formally, we could write.
For each $\varepsilon > 0$ there exists $\delta = \varepsilon^2 > 0$ such that if $0 < x < \delta = \varepsilon^2$, then

$$\sqrt{x} < \varepsilon,$$

so that

$$\lim_{x \to 0^+} \sqrt{x} = 0.$$

The proof is complete.

Proposition 5.6.9 *Let $n \in \mathbb{N}$ be an even natural number. Let $f : [0, +\infty) \to \mathbb{R}$ be defined by*

$$f(x) = \sqrt[n]{x}.$$

Under such assumptions, we have,

$$\lim_{x \to x_0} f(x) = f(x_0), \ \forall x_0 > 0,$$

and

$$\lim_{x \to 0^+} f(x) = 0.$$

Thus, f is a continuous function.

Proof Choose $x_0 > 0$. We are going to show that

$$\lim_{x \to \infty} \sqrt[n]{x} = \sqrt[n]{x_0}.$$

Let $\varepsilon > 0$ be given. Define

$$\varepsilon_1 = \min\{\sqrt[n]{x_0}/2, \varepsilon\}.$$

Thus,

$$|\sqrt[n]{x} - \sqrt[n]{x_0}| < \varepsilon_1$$

$$\Leftrightarrow -\varepsilon_1 < \sqrt[n]{x} - \sqrt[n]{x_0} < \varepsilon_1$$

$$\Leftrightarrow \sqrt[n]{x_0} - \varepsilon_1 < \sqrt[n]{x} < \sqrt[n]{x_0} + \varepsilon_1$$

$$\Leftrightarrow (\sqrt[n]{x_0} - \varepsilon_1)^n < x < (\sqrt[n]{x_0} + \varepsilon_1)^n$$

$$\Leftrightarrow -\delta_1 \equiv (\sqrt[n]{x_0} - \varepsilon_1)^n - x_0 < x - x_0 < (\sqrt[n]{x_0} + \varepsilon_1)^n - x_0 \equiv \delta_2. \quad (5.6)$$

So, defining $\delta = \min\{\delta_1, \delta_2\}$, if $0 < |x - x_0| < \delta$, we have,

$$-\delta_1 \le -\delta < x - x_0 < \delta \le \delta_2,$$

so that from this and (5.6), we obtain

$$|\sqrt[n]{x} - \sqrt[n]{x_0}| < \varepsilon_1 \le \varepsilon.$$

Formally, we could write:
For each $\varepsilon > 0$, defining $\varepsilon_1 = \min\{\sqrt[n]{x_0}/2, \varepsilon\}$, there exists $\delta = \min\{\delta_1, \delta_2\}$, where $\delta_1 = x_0 - (\sqrt[n]{x_0} - \varepsilon_1)^n$, and $\delta_2 = (\sqrt[n]{x_0} + \varepsilon_1)^n - x_0$, such that if $0 < |x - x_0| < \delta$, then

$$|\sqrt[n]{x} - \sqrt[n]{x_0}| < \varepsilon_1 \le \varepsilon,$$

so that

$$\lim_{x \to x_0} \sqrt[n]{x} = \sqrt[n]{x_0}.$$

Finally, let us consider the case in which $x_0 = 0$.
Observe that

$$0 < \sqrt[n]{x} < \varepsilon \Leftrightarrow 0 < x < \varepsilon^n (\equiv \delta).$$

Thus, more formally, we could write.
For each $\varepsilon > 0$ there exists $\delta = \varepsilon^n > 0$ such that if $0 < x < \delta = \varepsilon^n$, then

$$0 < \sqrt[n]{x} < \varepsilon,$$

so that

$$\lim_{x \to 0^+} \sqrt[n]{x} = 0.$$

The proof is complete.

Proposition 5.6.10 *Let* $n \in \mathbb{N}$ *be an odd natural number. Let* $f : \mathbb{R} \to \mathbb{R}$ *be defined by*

$$f(x) = \sqrt[n]{x}.$$

Under such assumptions, we have,

$$\lim_{x \to x_0} f(x) = f(x_0), \ \forall x_0 \in \mathbb{R}.$$

Proof The cases in which $x_0 > 0$ and $x_0 = 0^+$ the proofs are completely analogous to those of the last proposition.

Thus, let us choose $x_0 < 0$.

Hence, from the last proposition, we may write,

$$\begin{aligned}
\lim_{x \to x_0} \sqrt[n]{x} &= \lim_{-x \to x_0} \sqrt[n]{-x} \\
&= \lim_{x \to -x_0} \sqrt[n]{-x} \\
&= \lim_{x \to -x_0} -\sqrt[n]{x} \\
&= - \lim_{x \to -x_0} \sqrt[n]{x} \\
&= - \sqrt[n]{-x_0} \\
&= \sqrt[n]{x_0}.
\end{aligned} \tag{5.7}$$

5.7 Uniform Continuity

Definition 5.7.1 Let $f : X \to \mathbb{R}$ be a continuous function. We say that f is uniformly continuous if for each $\varepsilon > 0$ there exists $\delta > 0$ such that if $x, y \in X$ and $|x - y| < \delta$, then $|f(x) - f(y)| < \varepsilon$.

Theorem 5.7.2 *Let* $f : X \to \mathbb{R}$ *be a uniform continuous function. Let* $\{x_n\}$ *be a Cauchy sequence in X. Under such hypotheses,* $\{f(x_n)\}$ *is also a Cauchy sequence.*

Proof Let $\varepsilon > 0$. Thus, there exists $\delta > 0$ such that if $x, y \in X$ and $|x - y| < \delta$, then $|f(x) - f(y)| < \varepsilon$. Since $\{x_n\}$ is Cauchy sequence, there exists $n_0 \in \mathbb{N}$ such that if $m, n > n_0$, then $|x_n - x_m| < \delta$ so that $|f(x_n) - f(x_m)| < \varepsilon$. Thus, $\{f(x_n)\}$ is a Cauchy sequence. This completes the proof.

Corollary 5.7.3 *Let* $f : X \to \mathbb{R}$ *be a uniformly continuous function. Under such hypothesis, for each* $a \in X'$ *the*

$$\lim_{x \to a} f(x).$$

exists.

Exercise 5.7.4
Prove the Corollary 5.7.3.

Theorem 5.7.5 *Let $X \subset \mathbb{R}$ be a compact set. Let $f : X \to \mathbb{R}$ be a continuous function. Then, f is also uniformly continuous.*

Proof Let $\varepsilon > 0$. Let $x \in X$. Since f is continuous, there exists $\delta_x > 0$ such that if $y \in X$ and $|y - x| < 2\delta_x$, then $|f(x) - f(y)| < \varepsilon/2$. Denoting $I_x = (x - \delta_x, x + \delta_x)$, we have,

$$X \subset \cup_{x \in X} I_x.$$

Since X is compact, there exist $x_1, x_2, \ldots, x_n \in X$ such that

$$X \subset \cup_{i=1}^{n} I_{x_i}.$$

Define $\delta = \min\{\delta_1, \ldots, \delta_n\}$.
Let $x, y \in X$ be such that $|x - y| < \delta$.
Hence $x \in I_{x_j}$ for some $j \in \{1, \ldots, n\}$ so that

$$|x - x_j| < \delta_{x_j}.$$

Thus, $|f(x) - f(x_j) < \varepsilon/2$. And also,

$$
\begin{aligned}
|y - x_j| &= |y - x + x - x_j| \\
&\leq \delta + \delta_{x_j} < 2\delta_{x_j}
\end{aligned}
\tag{5.8}
$$

so that $|f(y) - f(x_j)| < \varepsilon/2$. Finally

$$
\begin{aligned}
|f(x) - f(y)| &= |f(y) - f(x_j) + f(x_j) - f(x)| \\
&\leq |f(y) - f(x_j)| + |f(x) - f(x_j)| \\
&\leq \varepsilon/2 + \varepsilon/2 = \varepsilon, \; \forall x, y \in X, \; \text{such that } |x - y| < \delta.
\end{aligned}
\tag{5.9}
$$

The proof is complete.

5.8   Uniformly Continuous Extensions

Theorem 5.8.1 *Let $X \subset \mathbb{R}$ and let $f : X \to \mathbb{R}$ be a uniformly continuous function. Under such hypotheses, f admits a uniformly continuous extension $\phi : \overline{X} \to \mathbb{R}$. Moreover, ϕ is the only continuous extension of f to $\overline{X}$.*

Proof Observe that $\overline{X} = X \cup X'$. For each $x' \in X'$, define $\phi(x') = \lim_{x \to x'} f(x)$.
Such a limit always exists, from the Corollary of Theorem 4.4.3. As f is continuous,
if $x' \in X \cap X'$, then $\phi(x') = f(x')$.
   Thus, we define $\phi(x) = f(x), \forall x \in X$.
   Let $\overline{x} \in \overline{X}$, then if $\{x_n\} \subset X$ is such that $\lim_{n \to \infty} x_n = \overline{x}$, then

$$\phi(\overline{x}) = \lim_{n \to \infty} \phi(x_n).$$

We shall show that ϕ is uniformly continuous.
   Let $\varepsilon > 0$. Since f is uniformly continuous, there exists $\delta > 0$ such that if
$x, y \in X$ and $|x - y| < \delta$, then $|f(x) - f(y)| < \varepsilon/2$.
   Thus, if $\overline{x}, \overline{y} \in \overline{X}$ and

$$|\overline{x} - \overline{y}| < \delta/2,$$

there exist $\{x_n\}, \{y_n\} \subset X$ such that

$$\lim_{n \to \infty} x_n = \overline{x}$$

and

$$\lim_{n \to \infty} y_n = \overline{y}.$$

Therefore, there exist $n_1 \in \mathbb{N}$ and $n_2 \in \mathbb{N}$ such that if $n > n_1$, then

$$|x_n - \overline{x}| < \delta/4,$$

and if $n > n_2$, then

$$|y_n - \overline{y}| < \delta/4.$$

Thus, if $n > \max\{n_1, n_2\} \equiv n_0$, then

$$\begin{aligned}
|x_n - y_n| &= |x_n - \overline{x} + \overline{x} - \overline{y} + \overline{y} - y_n| \\
&\leq |x_n - \overline{x}| + |\overline{x} - \overline{y}| + |y_n - \overline{y}| \\
&< \delta/4 + \delta/2 + \delta/4 = \delta.
\end{aligned} \tag{5.10}$$

Therefore $|f(x_n) - f(y_n)| < \varepsilon/2$, if $n > n_3$. Thus,

$$|\phi(\overline{x}) - \phi(\overline{y})| = \lim_{n \to \infty} |f(x_n) - f(y_n)| \leq \varepsilon/2 < \varepsilon.$$

Hence, $\phi : \overline{X} \to \mathbb{R}$ is uniformly continuous.

Let us now verify the uniqueness of ϕ.

Suppose that $\psi : \overline{X} \to \mathbb{R}$ is another extension of f.

For $\overline{x} \in \overline{X}$ there exists $\{x_n\} \subset X$ such that

$$\lim_{n \to \infty} x_n = \overline{x}.$$

Thus,

$$\begin{aligned}
\psi(\overline{x}) &= \psi\left(\lim_{n \to \infty} x_n\right) \\
&= \lim_{n \to \infty} \psi(x_n) \\
&= \lim_{n \to \infty} f(x_n) \\
&= \lim_{n \to \infty} \phi(x_n) \\
&= \phi\left(\lim_{n \to \infty} x_n\right) \\
&= \phi(\overline{x}).
\end{aligned} \tag{5.11}$$

Therefore,

$$\psi(\overline{x}) = \phi(\overline{x}), \ \forall \overline{x} \in \overline{X}.$$

The proof is complete.

Exercises 5.8.2

1. Let $f, g : \mathbb{R} \to \mathbb{R}$ be continuous functions.
   Define

   $$A = \{x \in \mathbb{R} \ : \ f(x) \le g(x)\}.$$

   Show that A is closed.

2. Let $X \subset \mathbb{R}$ and let $f : X \to \mathbb{R}$ be a continuous function. Assume $B \subset \mathbb{R}$ is an open set.
   Show that there exists an open set $A \subset \mathbb{R}$ such that $f^{-1}(B) = A \cap X$, where

   $$f^{-1}(B) = \{x \in X \ : \ f(x) \in B\}.$$

3. Let $X \subset \mathbb{R}$ and let $f : X \to \mathbb{R}$ be a function.
   Assume that for each open set $B \subset \mathbb{R}$ there exists an open set $A \subset \mathbb{R}$ such that $f^{-1}(B) = A \cap X$.
   Under such assumptions, prove that f is continuous on X.

4. Let $X \subset \mathbb{R}$ be a closed set and let $f : X \to \mathbb{R}$ be a continuous function. Let $F \subset \mathbb{R}$ be a closed set.
   Show $f^{-1}(F)$ is closed.

5. Let $X \subset \mathbb{R}$ be a nonempty set and let $f : X \to \mathbb{R}$ be a continuous function. Let $F \subset \mathbb{R}$ be a closed set.
   Show that there exists a closed $F_1 \subset \mathbb{R}$ such that $f^{-1}(F) = F_1 \cap X$.

6. Let $f : \mathbb{R} \to \mathbb{R}$ be a continuous function such that

$$\lim_{x \to +\infty} f(x) = \lim_{x \to -\infty} f(x) = +\infty.$$

Show that there exists $x_0 \in \mathbb{R}$ such that

$$f(x_0) \le f(x), \ \forall x \in \mathbb{R}.$$

7. Let $f : \mathbb{R} \to \mathbb{R}$ be a continuous function. Assume that for all open $A \subset \mathbb{R}$, we have that $f(A)$ is open.
   Show that f is injective.

8. Let $X \subset \mathbb{R}$ be a closed set and let $f : X \to \mathbb{R}$ be a function. Show that f is continuous if and only if $f^{-1}(F)$ is closed for each closed set $F \subset \mathbb{R}$.

9. Let $f : X \to \mathbb{R}$ be a real function such that there exists $K > 0$ such that

$$|f(x) - f(y)| \le K|y - x|, \ \forall x, y \in X.$$

Show that f is uniformly continuous on X.

10. Let $Z \subset \mathbb{R}$ be a nonempty set. Define $f : \mathbb{R} \to \mathbb{R}$ by

$$f(x) = \inf\{|x - z| \ : \ z \in Z\}.$$

Show

$$|f(x) - f(y)| \le |x - y|, \ \forall x, y \in \mathbb{R}.$$

Conclude that f is uniformly continuous on $\mathbb{R}$.

11. Let $a > 0$. Show that the function $f : [0, +\infty) \to [0, +\infty)$ defined by $f(x) = \sqrt[n]{x}$ is Lipschitzian on $[a, +\infty)$, with Lipschitz constant $c = 1/(n\sqrt[n]{a^{n-1}})$. Show f is uniformly continuous on $[0, +a]$. Conclude, justifying your answer, that f is uniformly continuous on $[0, +\infty)$.

12. Let $f, g : X \to \mathbb{R}$ be uniformly continuous functions on X. Show that

$$(\alpha f + \beta g) : X \to \mathbb{R}$$

is uniformly continuous on X, $\forall \alpha, \beta \in \mathbb{R}$, where

$$(\alpha f + \beta g)(x) = \alpha f(x) + \beta g(x), \ \forall x \in X.$$

13. Let $f, g : X \to \mathbb{R}$ be uniformly continuous functions on X. Show that $h : X \to \mathbb{R}$ is uniformly continuous, where

    (a)

    $$h(x) = |f(x) + g(x)|, \forall x \in X,$$

    (b)

    $$h(x) = \gamma f(x) + |\alpha f(x) - \beta g(x)|, \forall x \in X, \alpha, \beta, \gamma \in \mathbb{R},$$

    (c)

    $$h(x) = \max\{f(x), g(x)\}, \forall x \in X,$$

    Hint: $\max\{a, b\} = \frac{a+b}{2} + \frac{|a-b|}{2}, \forall a, b \in \mathbb{R}.$

    (d)

    $$h(x) = \min\{f(x), g(x)\}, \ \forall x \in X.$$

    Hint: $\min\{a, b\} = \frac{a+b}{2} - \frac{|a-b|}{2}, \forall a, b \in \mathbb{R}.$

5.9 Exponentials and Logarithms

In this section we formally develop theoretical results concerning the construction of exponential and logarithmic functions.

5.9.1 Introduction and Preliminary Results

We start with some preliminary results.

Proposition 5.9.1 *Let $a, b \in \mathbb{R}$ be such that $a > b > 0$. Under such hypotheses, we have*

$$a^n > b^n > 0, \forall n \in \mathbb{N}.$$

Proof Define $B = \{n \in \mathbb{N} : a^n > b^n\}$. From $a > b$ we have $1 \in B$. Suppose $n \in B$. Thus, $a^n > b^n$, so that $a^n - b^n > 0$, and thus $(a^n - b^n)a > 0$, so that

$$a^{n+1} > b^n a > b^n b = b^{n+1}.$$

Hence $n + 1 \in B$.
From the induction principle, $B = \mathbb{N}$.
The proof is complete.

Proposition 5.9.2 *Let $a \in \mathbb{R}$ be such that $a > 1$.*
Under such hypotheses, we have $a^{1/n} > 1$, $\forall n \in \mathbb{N}$.

Proof Suppose, to obtain contradiction, there exists $n_0 \in \mathbb{N}$ such that $a^{1/n_0} \leq 1$. From the last proposition, we obtain,

$$a = (a^{1/n_0})^{n_0} \leq 1^{n_0} = 1,$$

which contradicts the hypotheses.
The proof is complete.

Proposition 5.9.3 *Let $a, b \in \mathbb{R}$ be such that $a > b > 0$. Then*

$$a^{1/n} > b^{1/n}, \ \forall n \in \mathbb{N}.$$

Proof From the hypotheses $a/b > 1$, so that from the last proposition

$$(a/b)^{1/n} > 1, \ \forall n \in \mathbb{N}$$

so that

$$a^{1/n} > b^{1/n}, \ \forall n \in \mathbb{N}.$$

Proposition 5.9.4 *Let $a, b \in \mathbb{R}$ be such that $a > b > 0$. Then*

$$a^{m/n} > b^{m/n}, \ \forall m, n \in \mathbb{N}.$$

Proof Let $m, n \in \mathbb{N}$. From the last proposition

$$a^{1/n} \geq b^{1/n},$$

and from this and Proposition 5.9.1, we obtain

$$a^{m/n} > b^{m/n}, \ \forall m, n \in \mathbb{N}.$$

Proposition 5.9.5 *Let $a \in \mathbb{R}$ be such that $a > 0$ and $a \neq 1$. Under such hypotheses*

$$\lim_{n \to \infty} a^{1/n} = 1.$$

Proof Suppose first $a > 1$, the case $0 < a < 1$ may be dealt similarly.
From Proposition 5.9.3 we have $a^{1/n} > 1$, $\forall n \in \mathbb{N}$.

If $m, n \in \mathbb{N}$ are such that $m > n$, then $1/m < 1/n$ so that $1/n - 1/m > 0$, and thus, from the last proposition $a^{1/n-1/m} = a^{(m-n)/(mn)} > 1^{(m-n)/(mn)} = 1$, that is,

$$a^{1/n}/a^{1/m} > 1,$$

so that $a^{1/n} > a^{1/m}$.

Thus $\{a^{1/n}\}$ is a decreasing bounded below by 1 sequence, so that there exists $b \geq 1$, such that $b = \lim_{n \to \infty} a^{1/n}$.

Observe that in particular

$$\lim_{n \to \infty} a^{1/(n(n+1))} = b,$$

that is

$$
\begin{aligned}
b &= \lim_{n \to \infty} a^{1/(n(n+1))} \\
&= \lim_{n \to \infty} a^{[1/n - 1/(n+1)]} \\
&= \frac{\lim_{n \to \infty} a^{1/n}}{\lim_{n \to \infty} a^{1/(n+1)}} \\
&= b/b \\
&= 1,
\end{aligned}
$$

that is,

$$\lim_{n \to \infty} a^{1/n} = 1.$$

Proposition 5.9.6 *Let $\{r_n\} \subset \mathbb{Q}$ be such that $\lim_{n \to +\infty} r_n = 0$ and let $a \in \mathbb{R}$ be such that $a > 0$. Under such hypotheses,*

$$\lim_{n \to \infty} a^{r_n} = 1.$$

Proof Assume first $a > 1$ and $r_n \geq 0, \forall n \in \mathbb{N}$ (the case $0 < a < 1$ may be dealt similarly).

Let $\varepsilon > 0$. From $\lim_{n \to +\infty} a^{1/n} = 1$, there exists $n_0 \in \mathbb{N}$ such that if $n > n_0$, then $1 \leq a^{1/n} < 1 + \varepsilon$. Choose $n_1 > n_0$. Thus,

$$1 \leq a^{1/n_1} < 1 + \varepsilon.$$

Since $\lim_{n \to +\infty} r_n = 0$, there exists $n_2 \in \mathbb{N}$ such that if $n > n_2$ then

$$0 < r_n < 1/n_1.$$

Thus

$$1 \leq a^{r_n} < a^{1/n_1} < 1 + \varepsilon,$$

if $n > n_2$, so that

$$\lim_{n \to +\infty} a^{r_n} = 1.$$

For the general case, for notation convenience, denote $\lim_{k \to \infty} r_k = 0$. Fix $k_1 \in \mathbb{N}$. Hence, there exists $k_0 \in \mathbb{N}$ such that if $k > k_0$, we may find $n_k \in \mathbb{N}$ such that $|r_k| < 1/k_1$. Hence

$$0 < r_k + 1/k_1 < 2/k_1, \quad \text{if } k > k_0.$$

Thus,

$$1 \leq \limsup_{k \to +\infty} a^{(r_k + 1/k_1)} \leq a^{2/k_1},$$

so that

$$\limsup_{k \to +\infty} a^{r_k} = \limsup_{k \to +\infty} a^{(r_k + 1/k_1 - 1/k_1)}$$

$$= \limsup_{k \to +\infty} a^{(r_k + 1/k_1)} a^{-1/k_1}$$

$$\leq \frac{a^{2/k_1}}{a^{1/k_1}} = a^{1/k_1}, \; \forall k_1 \in \mathbb{N}. \tag{5.12}$$

from this and from

$$a^{-1/k_1} \leq a^{(r_k + 1/k_1)} a^{-1/k_1}, \; \forall k > k_0$$

we obtain

$$a^{-1/k_1} \leq \liminf_{k \to \infty} a^{r_k} \leq \limsup_{k \to \infty} a^{r_k} \leq a^{1/k_1}, \forall k_1 \in \mathbb{N}.$$

Letting $k_1 \to \infty$, we finally get

$$\lim_{k \to \infty} a^{r_k} = 1.$$

The proof is complete.

5.9.2   The Main Definitions and Results

Definition 5.9.7 Let $a \in \mathbb{R}$ be such that $a > 0$ and $a \neq 1$. We define $a^0 = 1$. Thus

$$a^0 = 1 = \lim_{n \to +\infty} a^{r_n},$$

for each sequence $\{r_n\} \subset \mathbb{Q}$ such that

$$\lim_{n \to +\infty} r_n = 0.$$

Let $x \in \mathbb{R}$. We define

$$a^x = \lim_{n \to \infty} a^{r_n}$$

where $\{r_n\} \subset \mathbb{Q}$ is any sequence such that

$$\lim_{n \to +\infty} r_n = x.$$

We shall show that a^x is well defined, $\forall x \in \mathbb{R}$.
Let $x \in \mathbb{R}$. Assume that

$$\lim_{n \to +\infty} r_n = \lim_{n \to \infty} s_n = x.$$

Since $\{s_n\}$ is convergent, it is also bounded, so that there exist $M > 0$, such that $|a^{s_n}| < M, \forall n \in \mathbb{N}$.
Thus,

$$\lim_{n \to +\infty} |a^{r_n} - a^{s_n}| \leq \lim_{n \to +\infty} |a^{s_n}||a^{r_n - s_n} - 1|$$

$$\leq M \lim_{n \to +\infty} |a^{r_n - s_n} - 1|$$

$$= 0. \tag{5.13}$$

On the other hand, since $\{r_n\}$ is convergent, it is bounded, so that $\{a^{r_n}\}$ is bounded.
Thus, there exists a subsequence $\{n_k\}$ and $b \in \mathbb{R}$ such that

$$\lim_{n \to \infty} a^{r_{n_k}} = b.$$

Suppose, to obtain contradiction, we do not have

$$\lim_{n \to \infty} a^{r_n} = b.$$

Hence, there exists $\varepsilon_0 > 0$ such that for each $k \in \mathbb{N}$ we may find $\tilde{n}_k > k$ such that

$$|a^{r_{\tilde{n}_k}} - b| \geq \varepsilon_0. \tag{5.14}$$

Since $r_{\tilde{n}_k} \to x$, as $k \to \infty$, we must have from (5.13)

$$\lim_{k \to \infty} (a^{r_{n_k}} - a^{r_{\tilde{n}_k}}) = 0,$$

so that $\lim_{n \to \infty} a^{r_{\tilde{n}_k}} = b$, which contradicts (5.14).

Hence

$$\lim_{n \to +\infty} a^{r_n} = b,$$

so that

$$\lim_{n \to \infty} a^x = b.$$

Thus, a^x is well defined.

Proposition 5.9.8 *Let $a \in \mathbb{R}$ be such that $a > 1$. Then $x < y$ if and only if $a^x < a^y$.*

Proof Let $x < y$. Let $\{r_n\} \subset \mathbb{Q}$ and $\{s_n\} \subset \mathbb{Q}$ be such that $r_n \to x$ and $s_n \to y$, as $n \to +\infty$.

Observe that $s_n - r_n \to y - x > 0$, hence there exists $n_0 \in \mathbb{N}$ such that if $n > n_0$ then $s_n - r_n > (y - x)/2 > b > 0$, for some $b \in \mathbb{Q}$.

Hence $a^{s_n - r_n} \geq a^b > 1$, $\forall n > n_0$, and thus

$$a^{y-x} = \lim_{n \to \infty} a^{s_n - r_n} \geq a^b > 1,$$

so that $a^{y-x} > 1$, that is $a^y / a^x > 1$, which means $a^y > a^x$.

Reciprocally, assume $a^x < a^y$. Suppose that $x \geq y$. In such a case from above $a^y \leq a^x$, a contradiction.

Thus, we have obtained,

$x < y$ if, and only if, $a^x < a^y$. $\qquad \blacksquare$

Remark 5.9.9 Similarly we may prove that if $0 < a < 1$ we have that $x < y$ if and only if $a^x > a^y$. We leave the details of such a proof as an exercise.

Definition 5.9.10 (Logarithms) Let $a \in \mathbb{R}$ be such that $a > 0$ and $a \neq 1$. Let $x > 0$. If $y \in \mathbb{R}$ is such that $x = a^y$ we denote,

$$y = \log_a(x),$$

so that

$$y = \log_a(x) \Leftrightarrow x = a^y, \forall x > 0, \; y \in \mathbb{R}$$

and in such case we say that y is the logarithm of x relating the basis a.

Proposition 5.9.11 *Let $a \in \mathbb{R}$ be such that $a > 0$ and $a \neq 1$. Then,*

$$\lim_{x \to 0} a^x = 1.$$

Proof To simplify the analysis we assume $a > 1$ (the case $0 < a < 1$ may be dealt similarly). Let $\varepsilon > 0$. Let $\{s_n\} \in \mathbb{Q}^+ \setminus \{0\}$ be such that $s_n \to 0$ as $n \to +\infty$. From above $a^{s_n} \to 1$, as $n \to +\infty$. Thus, there exists $n_0 \in \mathbb{N}$ such that if $n > n_0$, then $1 \leq a^{s_n} < 1 + \varepsilon$. Choose $n_1 > n_0$. Thus, if $0 < x < s_{n_1} \equiv \delta$, then

$$1 \leq a^x < a^{s_{n_1}} < 1 + \varepsilon.$$

Thus $\lim_{x \to 0^+} a^x = 1$. Similarly, we may show that $\lim_{x \to 0^-} a^x = 1$. This completes the proof.

Proposition 5.9.12 *Let $a \in \mathbb{R}$ be such that $a > 0$ and $a \neq 1$. Let $x_0 \in \mathbb{R}$. Then,*

$$\lim_{x \to x_0} a^x = a^{x_0}.$$

Proof Assume $x_0 \neq 0$ (the case $x_0 = 0$ has been dealt in the last proposition).

Observe that $x - x_0 \to 0$, as $x \to x_0$. From the last proposition, $a^{x - x_0} \to 1$, as $x \to x_0$.

Thus

$$\lim_{x \to x_0} a^x = \lim_{x \to x_0} a^{(x - x_0 + x_0)}$$

$$= \lim_{x \to x_0} [a^{(x - x_0)} a^{x_0}]$$

$$= [\lim_{x \to x_0} [a^{(x - x_0)}] a^{x_0}$$

$$= 1 a^{x_0}$$

$$= a^{x_0} \tag{5.15}$$

The proof is complete.

Proposition 5.9.13 (Existence of Logarithms) *Let $a \in \mathbb{R}$ be such that $a > 0$ and $a \neq 1$. Let $x > 0$. Then there exists $y \in \mathbb{R}$ such that $x = a^y$ so that $y = \log_a(x)$.*

Proof Suppose $a > 1$. The case $0 < a < 1$ may be dealt similarly.

Observe that $\lim_{n \to +\infty} a^{-n} = 0$ and $\lim_{n \to +\infty} a^n = +\infty$.

Then, there exists $n_0, n_1 \in \mathbb{N}$ such that $a^{-n_0} < x < a^{n_1}$. Since from the last proposition f is continuous, where $f(y) = a^y, \forall y \in \mathbb{R}$, from the intermediate value theorem, there exists $y \in (-n_0, n_1)$ such that $a^y = x$.

The proof is complete.

5.9.3 On the Fundamental Exponential Limit

In this section we shall prove in detail the fundamental exponential limit, namely

$$\lim_{x \to 0} \frac{e^x - 1}{x} = 1.$$

Such a limit has great applicability in analysis, including the calculus of derivatives. We will start with the following remark:

Remark 5.9.14 Consider the sequence $\{a_n\}$ where

$$a_n = 1 + \frac{1}{1!} + \frac{1}{2!} + \cdots + \frac{1}{n!}.$$

Such a sequence is increasing and

$$0 \le a_n < 1 + \frac{1}{1!} + \frac{1}{2} + \frac{1}{2 \cdot 2} + \frac{1}{2 \cdot 2^2} + \cdots + \frac{1}{2 \cdot 2^{n-2}}$$

$$< 3, \quad \forall n \in \mathbb{N}. \tag{5.16}$$

Hence $\{a_n\}$ is increasing and bounded, therefore, it is convergent.

On the other hand, let $\{b_n\} \subset \mathbb{R}$ be such that

$$b_n = \left(1 + \frac{1}{n}\right)^n$$

$$= 1 + n\frac{1}{n} + \frac{n(n-1)}{2}\frac{1}{n^2} + \cdots + \frac{n(n-1)\cdots 2}{n!}\frac{1}{n^n}$$

$$= 1 + 1 + \frac{1}{2!}\left(1 - \frac{1}{n}\right) + \frac{1}{3!}\left(1 - \frac{1}{n}\right)\left(1 - \frac{2}{n}\right)$$

$$+ \cdots + \frac{1}{n!}\left(1 - \frac{1}{n}\right)\left(1 - \frac{2}{n}\right)\cdots\left(1 - \frac{n-1}{n}\right)$$

$$< a_n. \tag{5.17}$$

Thus,

$$2 < b_n \le a_n < 3, \quad \forall n \in \mathbb{N}.$$

Thus, b_n is also increasing and bounded, therefore, it is convergent.
We define $e \in \mathbb{R}$ by

$$e = \lim_{n \to +\infty} b_n,$$

that is,

$$e = \lim_{n \to +\infty} \left(1 + \frac{1}{n}\right)^n.$$

We shall call $e \in \mathbb{R}$ the natural basis.
Thus $2 \le e \le 3$.

Proposition 5.9.15 *About the natural basis e, we have*

$$\lim_{x \to +\infty} \left(1 + \frac{1}{x}\right)^x = e.$$

Proof Let $x > 1$ and let $n \in \mathbb{N}$ be such that

$$n \le x \le n + 1.$$

Hence,

$$1 + \frac{1}{n+1} \le 1 + \frac{1}{x} \le 1 + \frac{1}{n},$$

so that

$$\alpha_n \equiv \left(1 + \frac{1}{n+1}\right)^n \le \left(1 + \frac{1}{x}\right)^x \le \left(1 + \frac{1}{n}\right)^{n+1} \equiv \beta_n.$$

Observe that

$$\lim_{n \to \infty} \alpha_n = \lim_{n \to +\infty} \left(1 + \frac{1}{n+1}\right)^{n+1-1}$$

$$= \lim_{n \to \infty} \left(1 + \frac{1}{n+1}\right)^{n+1} \lim_{n \to \infty} \left(1 + \frac{1}{n+1}\right)^{-1}$$

$$= e \cdot 1$$

$$= e. \tag{5.18}$$

Also,

$$\lim_{n \to \infty} \beta_n = \lim_{n \to +\infty} \left(1 + \frac{1}{n}\right)^{n+1}$$

$$= \lim_{n \to \infty} \left(1 + \frac{1}{n}\right)^n \lim_{n \to \infty} \left(1 + \frac{1}{n}\right)$$

$$= e \cdot 1$$

$$= e. \tag{5.19}$$

Summarizing,

$$\lim_{n \to +\infty} \alpha_n = \lim_{n \to +\infty} \beta_n = e.$$

Let $\varepsilon > 0$. From $\lim_{n \to +\infty} \alpha_n = e$ there exists $n_0 \in \mathbb{N}$ such that if $n > n_0$, then

$$|\alpha_n - e| < \varepsilon.$$

From $\lim_{n \to +\infty} \beta_n = e$ there exists $n_1 \in \mathbb{N}$ such that if $n > n_1$, then

$$|\beta_n - e| < \varepsilon.$$

Define $A = \max\{n_0 + 1, n_1 + 1\}$. For $x > A$, let $n_2 \in \mathbb{N}$ be such that

$$n_2 \le x \le n_2 + 1.$$

Thus $n_2 > \max\{n_0, n_1\}$, so that

$$e - \varepsilon < \alpha_{n_2} \le \left(1 + \frac{1}{x}\right)^x \le \beta_{n_2} < e + \varepsilon,$$

and hence,

$$\left|\left(1 + \frac{1}{x}\right)^x - e\right| < \varepsilon, \quad \text{if } x > A.$$

Therefore,

$$\lim_{x \to +\infty} \left(1 + \frac{1}{x}\right)^x = e.$$

The proof is complete.

Now we present the main result in this section, namely, the fundamental exponential limit. From the results of last chapter, observe that the logarithmic function is continuous on $(0, +\infty)$, since it is the inverse one of an exponential function, which is injective, monotonic, and continuous on $\mathbb{R}$.

Proposition 5.9.16 *About the exponential function with the natural basis e, we have:*

$$\lim_{x \to 0} \frac{e^x - 1}{x} = 1.$$

Proof Let $x \in \mathbb{R}$. Define $y = e^x$, so that $x = \ln(y)$.
   Hence

$$\frac{e^x - 1}{x} = \frac{y - 1}{\ln(y)}.$$

On the other hand

$$\frac{y - 1}{\ln(y)} = \frac{1}{\frac{1}{y-1} \ln y} = \frac{1}{\ln \left[y^{\frac{1}{y-1}} \right]},$$

Define $z = y - 1$, so that $y = 1 + z$,

$$\frac{e^x - 1}{x} = \frac{y - 1}{\ln(y)}$$

$$= \frac{1}{\frac{1}{y-1} \ln y} = \frac{1}{\ln \left[y^{\frac{1}{y-1}} \right]}$$

$$= \frac{1}{\ln \left((1 + z)^{1/z} \right)}. \tag{5.20}$$

Observe that

$$x \to 0^+ \Leftrightarrow y = e^x \to 1^+ \Leftrightarrow z = y - 1 \to 0^+.$$

Thus, denoting $w = 1/z$, from the last theorem,

$$\lim_{z \to 0^+} (1 + z)^{1/z} = \lim_{w \to +\infty} \left(1 + \frac{1}{w} \right)^w = e,$$

so that

$$\lim_{z \to 0^+} \frac{1}{\ln\left((1+z)^{1/z}\right)} = \frac{1}{\ln\left(\lim_{z \to 0^+}(1+z)^{1/z}\right)}$$

$$= \frac{1}{\ln(e)}$$

$$= 1/1$$

$$= 1. \tag{5.21}$$

Since $z \to 0^+ \Leftrightarrow x \to 0^+$, we obtain, from this, and (5.20),

$$\lim_{x \to 0^+} \frac{e^x - 1}{x} = \lim_{z \to 0^+} \frac{1}{\ln\left((1+z)^{1/z}\right)} = 1.$$

Moreover

$$\lim_{x \to 0^-} \frac{e^x - 1}{x} = \lim_{x \to 0^+} \frac{e^{-x} - 1}{-x}$$

$$= \lim_{x \to 0^+} \frac{1 - e^x}{-xe^x}$$

$$= \lim_{x \to 0^+} \frac{e^x - 1}{x} \lim_{x \to 0^+} \frac{1}{e^x}$$

$$= 1. \tag{5.22}$$

From these last two one-sided limits we may infer that

$$\lim_{x \to 0} \frac{e^x - 1}{x} = 1.$$

The proof is complete.

Exercise 5.9.17

1. Prove that

$$\lim_{h \to 0} \frac{e^{a(x+h)} - e^{ax}}{h} = ae^{ax}, \ \forall a, x \in \mathbb{R}.$$

5.10 The Trigonometric Functions

In this section we present a study on the trigonometric functions. We start by defining the complex set.

Definition 5.10.1 We define the complex set denoted by $\mathbb{C}$ by

$$\mathbb{C} = \{(a, b) \; : \; a \in \mathbb{R}, \; b \in \mathbb{R}\}.$$

For such a set we define a sum operation, denoted by $(+) : \mathbb{C} \times \mathbb{C} \to \mathbb{C}$, by,

$$(a, b) + (c, d) = (a + c, b + d), \forall (a, b), (c, d) \in \mathbb{C},$$

and a multiplication, denoted by $(\cdot) : \mathbb{C} \times \mathbb{C} \to \mathbb{C}$, by

$$(a, b) \cdot (c, d) = (ac - bd, bc + ad), \;\; \forall (a, b), (c, d) \in \mathbb{C}.$$

Finally, we may denote,

$$(\alpha, 0) \cdot (a, b) = (\alpha a, \alpha b),$$

simply by $\alpha(a, b), \forall (a, b) \in \mathbb{C}, \alpha \in \mathbb{R}$.

In particular, we have

$$(0, 1) \cdot (0, 1) = (-1, 0) = -(1, 0),$$

so that denoting $(0, 1) = i$ and $(1, 0) = 1$, we have

$$i^2 = -1,$$

and

$$z = (a, b) = a(1, 0) + b(0, 1) = a1 + bi = a + bi.$$

Thus, we could justify the multiplication definition as specified in the next lines. Observe that, for

$$z_1 = (a, b) = a + bi,$$

$$z_2 = (c, d) = c + di,$$

we have, in a more usual fashion,

$$\begin{aligned}
(z_1 \cdot z_2 &= (a + bi) \cdot (c + di) \\
&= ac + adi + bci + adi^2 \\
&= (ac - bd) + (bc + ad)i \\
&= (ac - bd)(1, 0) + (bc + ad)(0, 1) \\
&= (ac - bd, bc + ad).
\end{aligned} \tag{5.23}$$

At this point we highlight that $\mathbb{C}$ is a field with such sum and multiplication operations.

Moreover, given $z = a + bi \in \mathbb{C}$, we denote,

$$R_e[z] = a,$$

the real part of z, and,

$$Im[z] = b$$

the imaginary part of z.

We also define the conjugate of $z = a + bi \in \mathbb{C}$, denoted by $\overline{z} \in \mathbb{C}$ by

$$\overline{z} = a - bi.$$

Finally, the modulus of $z = a + bi$, denoted by $|z|$, is defined by,

$$|z| = \sqrt{z \cdot \overline{z}}.$$

Observe that

$$z \cdot \overline{z} = (a + bi) \cdot (a - bi) = a^2 - i^2 b^2 = a^2 + b^2,$$

so that

$$|z| = \sqrt{a^2 + b^2}.$$

At this point we define the exponential of a complex number z.

Definition 5.10.2 Given $z \in \mathbb{C}$, we define the exponential of z, denoted by e^z, by

$$e^z = \sum_{k=0}^{\infty} \frac{z^k}{k!},$$

where $k! = 1 \cdot 2 \cdots k, \forall k \in \mathbb{N}$ and $0! = 1! = 1$.

Observe that, denoting

$$a_n = \frac{z^n}{n!},$$

for $z \neq 0$, we have,

$$\frac{a_{n+1}}{a_n} = \frac{z^{n+1}}{(n+1)!} \frac{n!}{z^n} = \frac{z}{n+1}.$$

Hence,

$$\lim_{n \to \infty} \frac{|a_{n+1}|}{|a_n|} = \lim_{n \to \infty} \frac{|z|}{n+1} = 0 < 1.$$

Thus, from this and the ratio test, the series relating $Re[e^z]$ and $Im[e^z]$ are converging, so that e^z represents a well-defined convergent series, $\forall z \in \mathbb{C}$.

Proposition 5.10.3 *Let $z_1, z_2 \in \mathbb{C}$. Thus,*

$$e^{z_1 + z_2} = e^{z_1} \cdot e^{z_2},$$

and,

$$e^{z_1 - z_2} = \frac{e^{z_1}}{e^{z_2}}.$$

Proof Observe that

$$
\begin{aligned}
e^{z_1 + z_2} &= \sum_{n=0}^{\infty} \frac{(z_1 + z_2)^n}{n!} \\
&= \sum_{n=0}^{\infty} \left(\sum_{k=0}^{n} \binom{n}{k} \frac{z_1^{n-k} z_2^k}{n!} \right) \\
&= \sum_{n=0}^{\infty} \left(\sum_{k=0}^{n} \frac{n!}{(n-k)!k!n!} z_1^{n-k} z_2^k \right) \\
&= \sum_{n=0}^{\infty} \left(\sum_{k=0}^{n} \frac{z_1^{n-k} z_2^k}{(n-k)!k!} \right) \\
&= \left(\sum_{n=0}^{\infty} \frac{z_1^n}{n!} \right) \left(\sum_{n=0}^{\infty} \frac{z_2^n}{n!} \right) \\
&= e^{z_1} \cdot e^{z_2}.
\end{aligned}
\tag{5.24}
$$

From this,

$$1 = e^0 = e^{z_1 - z_1} = e^{z_1} \cdot e^{-z_1},$$

so that

$$e^{-z_1} = \frac{1}{e^{z_1}},$$

and thus,

$$e^{z_1 - z_2} = e^{z_1} \cdot e^{-z_2} = \frac{e^{z_1}}{e^{z_2}}.$$

This completes the proof.

Remark 5.2 Observe that for $t \in \mathbb{R}$ we have

$$e^{it} = \sum_{k=0}^{\infty} \frac{(it)^k}{k!},$$

so that so that

$$\overline{e^{it}} = e^{-it}.$$

Thus,

$$\begin{aligned}
|e^{it}|^2 &= e^{it} \cdot \overline{e^{it}} \\
&= e^{it} \cdot e^{-it} \\
&= e^{it-it} \\
&= e^0 \\
&= 1.
\end{aligned} \tag{5.25}$$

Hence, $|e^{it}| = 1, \ \forall t \in \mathbb{R}$.
Observe also that,

$$e^{it} = R_e[e^{it}] + Im[e^{it}]i,$$

so that, at this point we define

$$\cos(t) = R_e[e^{it}],$$

and

$$\sin(t) = I_m[e^{it}].$$

Thus,

$$e^{it} = \cos(t) + i\sin(t), \forall t \in \mathbb{R}.$$

Therefore, we may write,

$$1 = |e^{it}|^2 = R_e[e^{it}]^2 + I_m[e^{it}]^2,$$

so that,

$$[\cos(t)]^2 + [\sin(t)]^2 = 1, \forall t \in \mathbb{R}.$$

Also,

$$\sin(0) = I_m[e^{i0}] = I_m(1) = 0,$$

and

$$\cos(0) = R_e[e^{i0}] = R_e[1] = 1.$$

Proposition 5.10.4 *Let $x, y \in \mathbb{R}$.*
  Thus,

1. $\cos(x + y) = \cos x \cos y - \sin x \sin y,$
2. $\sin(x + y) = \sin x \cos y + \cos x \sin y,$
3. $\cos(x - y) = \cos x \cos y + \sin x \sin y,$
4. $\sin(x - y) = \sin x \cos y - \cos x \sin y.$

Proof Observe that

$$\begin{aligned}
\cos(x + y) &= R_e[e^{(x+y)i}] \\
&= R_e[e^{ix} \cdot e^{iy}] \\
&= R_e[(\cos x + i \sin x)(\cos y + i \sin y)] \\
&= R_e[\cos x \cos y + i \sin x \cos y + i \cos x \sin y - \sin x \sin y] \\
&= \cos x \cos y - \sin x \sin y. \quad\quad (5.26)
\end{aligned}$$

Similarly,

$$\begin{aligned}
\sin(x + y) &= I_m[e^{(x+y)i}] \\
&= I_m[e^{ix} \cdot e^{iy}] \\
&= I_m[(\cos x + i \sin x)(\cos y + i \sin y)] \\
&= I_m[\cos x \cos y + i \sin x \cos y + i \cos x \sin y - \sin x \sin y] \\
&= \sin x \cos y + \cos x \sin y. \quad\quad (5.27)
\end{aligned}$$

Now, observe that

$$
\begin{aligned}
\cos(-x) &= R_e[e^{i(-x)}] \\
&= R_e[e^{-ix}] \\
&= R_e[\overline{[e^{ix}]}] \\
&= R_e[e^{ix}] \\
&= \cos(x), \forall x \in \mathbb{R}.
\end{aligned}
\tag{5.28}
$$

Similarly,

$$
\begin{aligned}
\sin(-x) &= I_m[e^{i(-x)}] \\
&= I_m[e^{-ix}] \\
&= I_m[\overline{[e^{ix}]}] \\
&= -I_m[e^{ix}] \\
&= -\sin(x), \forall x \in \mathbb{R}.
\end{aligned}
\tag{5.29}
$$

From (5.26) and (5.28),

$$
\begin{aligned}
\cos(x - y) &= \cos x \cos(-y) - \sin x \sin(-y) \\
&= \cos x \cos y + \sin x \sin y.
\end{aligned}
\tag{5.30}
$$

Finally, from (5.27) and (5.29),

$$
\begin{aligned}
\sin(x - y) &= \sin x \cos(-y) + \cos x \sin(-y) \\
&= \sin x \cos y - \cos x \sin y.
\end{aligned}
\tag{5.31}
$$

The proof is complete.

Proposition 5.10.5 *The following equalities hold,*

$$
\sin a - \sin b = 2 \cos\left(\frac{a+b}{2}\right) \sin\left(\frac{a-b}{2}\right),
$$

$$
\cos a - \cos b = -2 \sin\left(\frac{a+b}{2}\right) \sin\left(\frac{a-b}{2}\right),
$$

$\forall a, b \in \mathbb{R}.$

Proof Choose $a, b \in \mathbb{R}$. Define $x, y \in \mathbb{R}$ be such that

$$
x + y = a \text{ and } x - y = b
$$

so that

$$x = \frac{a+b}{2} \text{ and } y = \frac{a-b}{2}.$$

From the last proposition, we have,

$$\sin(x+y) - \sin(x-y) = 2\cos x \sin y,$$

so that,

$$\sin a - \sin b = 2\cos\left(\frac{a+b}{2}\right)\sin\left(\frac{a-b}{2}\right).$$

Similarly, also from the last proposition,

$$\cos(x+y) - \cos(x-y) = -2\sin x \sin y,$$

so that,

$$\cos a - \cos b = -2\sin\left(\frac{a+b}{2}\right)\sin\left(\frac{a-b}{2}\right).$$

Since a, b were arbitrary, the proof is complete.

Theorem 5.10.6 (Fundamental Trigonometric Limit) *The following equalities hold,*

1. $\lim_{x\to 0} \sin x = 0$,
2. $\lim_{x\to 0} \cos x = 1$.
3.

$$\lim_{x\to 0} \frac{\sin x}{x} = 1.$$

This last limit is known as the fundamental trigonometric limit.

Proof Consider Fig. 5.1 indicated, where generically we denote $\overline{AB}$ as the measure of the segment from A to B.

Hence, for Fig. 5.1, as indicated,

$$\overline{0A} = 1,$$

$$\overline{BC} = \sin x,$$

$$\overline{0C} = \cos x.$$

Fig. 5.1 Trigonometric circle
where $\overline{0A} = 1$, $\overline{BC} = \sin x$,
and $\overline{0C} = \cos x$

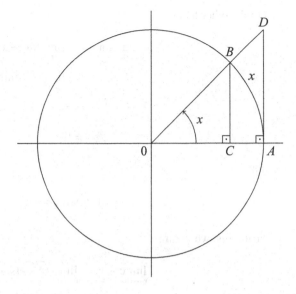

Observe that the area of triangle $0AB$ is smaller or equal to that of the area of circular sector $(0AB)$ that is,

$$0 \leq \frac{\overline{0A} \cdot \overline{BC}}{2} \leq \frac{\overline{0A} \cdot x}{2},$$

so that

$$0 \leq \frac{1 \cdot \sin x}{2} \leq \frac{1 \cdot x}{2},$$

and thus,

$$0 \leq \sin x \leq x, \forall x \geq 0.$$

Since

$$\lim_{x \to 0^+} x = 0,$$

from the Sandwich theorem we obtain,

$$\lim_{x \to 0^+} \sin x = 0.$$

On the other hand,

$$\lim_{x \to 0^-} \sin x = \lim_{x \to 0^+} \sin(-x)$$

$$= \lim_{x \to 0^+} [-\sin(x)]$$

$$= = -\lim_{x \to 0^+} \sin(x)$$

$$= 0. \tag{5.32}$$

so that,

$$\lim_{x \to 0} \sin x = 0.$$

From this, we obtain

$$\lim_{x \to 0} \cos x = \lim_{x \to 0} \sqrt{1 - [\sin x)]^2}$$

$$= \sqrt{1 - [\lim_{x \to 0} \sin x)]^2}$$

$$= \sqrt{1 - 0^2}$$

$$= 1. \tag{5.33}$$

On the other hand, also from the figure, we have the area of triangle $0AB$ is smaller than the area of the circular sector $0AB$, which is smaller than the area of the triangle $0AD$, so that

$$\frac{\sin x}{2} \le \frac{x}{2} \le \frac{\overline{AD}}{2}.$$

However, from the similarity between triangles $0BC$ and OAD we obtain,

$$\frac{\overline{AD}}{1} = \frac{\sin x}{\cos x},$$

that is

$$\overline{AD} = \frac{\sin x}{\cos x}$$

so that,

$$\sin x \le x \le \frac{\sin x}{\cos x}, \quad \forall 0 \le x < \frac{\pi}{2}.$$

Hence,

$$1 \le \frac{x}{\sin x} \le \frac{1}{\cos x},$$

that is,

$$\cos x \le \frac{\sin x}{x} \le 1,$$

$\forall x > 0$ sufficiently small.
Since

$$\lim_{x \to 0^+} \cos x = 1,$$

from these last lines and the Sandwich theorem, we obtain,

$$\lim_{x \to 0^+} \frac{\sin x}{x} = 1.$$

On the other hand,

$$\lim_{x \to 0^-} \frac{\sin x}{x} = = \lim_{x \to 0^+} \frac{\sin(-x)}{-x}$$
$$= \lim_{x \to 0^+} \frac{-\sin(x)}{-x}$$
$$= \lim_{x \to 0^+} \frac{\sin x}{x}$$
$$= 1. \tag{5.34}$$

so that we may infer that

$$\lim_{x \to 0} \frac{\sin x}{x} = 1.$$

This completes the proof.

At this point we highlight we could define two trigonometric functions $f : \mathbb{R} \to \mathbb{R}$ and $g : \mathbb{R} \to \mathbb{R}$ by,

$$f(x) = \sin(x), \forall x \in \mathbb{R},$$

ands

$$g(x) = \cos(x), \forall x \in \mathbb{R}.$$

In the next result we prove such functions are continuous.

Theorem 5.10.7 *The following equalities hold,*

$$\lim_{x \to x_0} = \sin(x) = \sin(x_0),$$

and

$$\lim_{x \to x_0} \cos(x) = \cos(x_0),$$

$\forall x_0 \in \mathbb{R}.$

Proof Choose $x_0 \in \mathbb{N}.$

Observe that from Proposition 5.10.5, we have,

$$\sin x - \sin x_0 = 2 \sin \left(\frac{x - x_0}{2} \right) \cos \left(\frac{x + x_0}{2} \right),$$

so that, for $x \neq x_0$, we obtain,

$$0 \leq |\sin x - \sin x_0|$$

$$= 2 \left| \sin \left(\frac{x - x_0}{2} \right) \right| \left| \cos \left(\frac{x + x_0}{2} \right) \right|$$

$$\leq 2 \left| \sin \left(\frac{x - x_0}{2} \right) \right|$$

$$= 2 \left| \sin \left(\frac{x - x_0}{2} \right) \right| |x - x_0|/|x - x_0|$$

$$= \left| \frac{\sin[(x - x_0)/2]}{(x - x_0)/2} \right| |x - x_0|. \tag{5.35}$$

Observe that,

$$\lim_{x \to x_0} \left| \frac{\sin[(x - x_0)/2]}{(x - x_0)/2} \right| |x - x_0|$$

$$\left| \lim_{x \to x_0} \frac{\sin[(x - x_0)/2]}{(x - x_0)/2} \right| \lim_{x \to x_0} |x - x_0|$$

$$1 \cdot 0$$

$$= 0, \tag{5.36}$$

From this, (5.35) and the Sandwich theorem, we obtain

$$\lim_{x \to x_0} |\sin x - \sin x_0| = 0,$$

so that

$$\lim_{x \to x_0} \sin x = \sin x_0.$$

The proof that

$$\lim_{x \to x_0} \cos x = \cos x_0$$

is left as an exercise. This would complete the proof.

Theorem 5.10.8 *The following equalities hold,*

$$\lim_{h \to 0} \frac{\sin(x + h) - \sin(x)}{h} = \cos x,$$

and

$$\lim_{h \to 0} \frac{\cos(x + h) - \cos(x)}{h} = -\sin x, \forall x \in \mathbb{R}.$$

Proof Choose $x \in \mathbb{R}$. Observe that, from Proposition 5.10.5, we have

$$\frac{\sin(x + h) - \sin(x)}{h} = \frac{2 \sin(h/2)}{h} \cos[(2x + h)/2]$$

$$= \frac{\sin(h/2)}{h/2} \cos[(2x + h)/2], \qquad (5.37)$$

so that

$$\lim_{h \to 0} \frac{\sin(x + h) - \sin(x)}{h} = \lim_{h \to 0} \left(\frac{2 \sin(h/2)}{h} \cos[(2x + h)/2] \right)$$

$$= \lim_{h \to 0} \frac{\sin(h/2)}{h/2} \lim_{h \to 0} \cos(x + h/2)$$

$$= 1 \cos x$$

$$= \cos x. \qquad (5.38)$$

The proof that

$$\lim_{h \to 0} \frac{\cos(x + h) - \cos(x)}{h} = -\sin x$$

is left as an exercise.

Since $x \in \mathbb{R}$ was arbitrary, this would complete the proof.

Chapter 6
Derivatives

6.1 Introduction

This chapter presents the main definitions and results related to derivatives for one variable real functions. Standard topics such as the derivative proprieties, the mean value theorem, and Taylor expansion are developed in detail. The inverse function theorem and related derivative formula for such a one real variable case is also addressed extensively.

The main references for this chapter are [9, 12].

6.2 First Definitions

We start by stating the first formal definitions concerning derivatives.

Definition 6.2.1 (Derivative) Let $f : [a, b] \to \mathbb{R}$ be a function. Let $x \in (a, b)$. We define the derivative of f at x denoted by $f'(x)$, by

$$f'(x) = \lim_{h \to 0} \frac{f(x + h) - f(x)}{h},$$

if such a limit exists, so that in such a case, we say that f is differentiable at x. Also, we define the one-sided derivatives $f'(a+)$ and $f'(b-)$, as

$$f'(a+) = \lim_{h \to 0^+} \frac{f(a + h) - f(a)}{h}$$

© Springer International Publishing AG, part of Springer Nature 2018
F. S. Botelho, *Real Analysis and Applications*,
https://doi.org/10.1007/978-3-319-78631-5_6

and

$$f'(b-) = \lim_{h \to 0^-} \frac{f(b+h) - f(b)}{h},$$

if such limits exist.

Theorem 6.2.2 *Let $f : [a, b] \to \mathbb{R}$ be a differentiable function in $x \in [a, b]$.*
  Under such hypotheses, f is continuous at x.

Proof Assume first $x \in (a, b)$. Observe that

$$f(x+h) - f(x) = \left(\frac{f(x+h) - f(x)}{h}\right) \cdot h \to f'(x) \cdot 0 = 0,$$

as $h \to 0$.
  The case in which $x = a$ or $x = b$ may be dealt similarly with one-sided limits.

Theorem 6.2.3 *Let $f, g : [a, b] \to \mathbb{R}$ be differentiable functions at $x \in [a, b]$.*
Under such hypotheses $f + g$, $f \cdot g$, are differentiable at x. Moreover, if $g(x) \neq 0$,
then f/g is also differentiable at x.
  Finally, we have:

1. $(f + g)'(x) = f'(x) + g'(x)$,
2. $(fg)'(x) = f'(x)g(x) + f(x)g'(x)$,
3.

$$(f/g)'(x) = \frac{f'(x)g(x) - f(x)g'(x)}{g(x)^2},$$

  if $g(x) \neq 0$.

Proof

1. This is left as an exercise.
2. Assume $x \in (a, b)$. Define $w = fg$.
      Thus,

$$\frac{w(x+h) - w(x)}{h} = \frac{f(x+h)g(x+h) - f(x)g(x)}{h}$$

$$= \frac{f(x+h)g(x+h) - f(x)g(x+h) + f(x)g(x+h) - f(x)g(x)}{h}$$

$$= \frac{(f(x+h) - f(x))}{h}g(x+h) + f(x)\frac{g(x+h) - g(x)}{h}$$

$$\to f'(x)g(x) + f(x)g'(x), \text{ as } h \to 0. \tag{6.1}$$

3. Define $w = f/g$. Thus,

$$\frac{w(x+h) - w(x)}{h} = \frac{\frac{f(x+h)}{g(x+h)} - \frac{f(x)}{g(x)}}{h}$$

$$= \frac{f(x+h)g(x) - f(x)g(x+h)}{g(x)g(x+h)h}$$

$$= \frac{f(x+h)g(x) - f(x)g(x) + f(x)g(x) - f(x)g(x+h)}{g(x)g(x+h)h}$$

$$= \frac{\frac{f(x+h)-f(x)}{h}g(x) - f(x)\frac{g(x+h)-g(x)}{h}}{g(x+h)g(x)}$$

$$\rightarrow \frac{f'(x)g(x) - f(x)g'(x)}{g(x)^2}, \text{ as } h \to 0. \tag{6.2}$$

The cases in which $x = a$ or $x = b$ may be dealt similarly with one-sided limits.

This completes the proof.

Theorem 6.2.4 (Chain Rule) *Suppose* $f : [a, b] \to \mathbb{R}$ *is continuous on* $[a, b]$. *Assume* $f'(x)$ *exists at* $x \in (a, b)$. *Assume also that* $g : I \to \mathbb{R}$ *is differentiable at* $y = f(x) \in I$, *where* I *is an open interval supposed to be such that* $I \supset f([a, b])$. *Under such hypotheses,* $w = (g \circ f) : [a, b] \to \mathbb{R}$, *where*

$$w(t) = (g \circ f)(t) = g(f(t)), \forall t \in [a, b],$$

is differentiable at x *and*

$$w'(x) = g'(f(x))f'(x).$$

Proof Observe that

$$f(x+h) - f(x) = f'(x)h + u(h)h,$$

where

$$u(h) = \frac{f(x+h) - f(x)}{h} - f'(x) \to 0, \text{ as } h \to 0.$$

On the other hand, denoting $y = f(x)$ we have,

$$g(s) - g(y) = g'(y)(s - y) + v(s)(s - y)$$

where

$$v(s) = \frac{g(s) - g(y)}{s - y} - g'(y) \to 0, \text{ as } s \to y.$$

In particular as $h \to 0$ we have that

$$f(x + h) \to f(x) = y,$$

so that

$$v(f(x + h)) \to 0, \text{ as } h \to 0.$$

Thus,

$$\frac{w(x + h) - w(x)}{h} = \frac{g(f(x + h)) - g(f(x))}{h}$$

$$= \frac{(g'(f(x)) + v(f(x + h)))(f(x + h) - f(x))}{h}$$

$$= \frac{(g'(f(x)) + v(f(x + h)))(f'(x)h + u(h)h)}{h}$$

$$= (g'(f(x)) + v(f(x + h)))(f'(x) + u(h))$$

$$\to g'(f(x))f'(x), \text{ as } h \to 0. \tag{6.3}$$

The proof is complete.

Definition 6.2.5 (Local Minimum) Let $X \subset \mathbb{R}$ be a nonempty set and let $f : X \to \mathbb{R}$ be a function. We say that $x_0 \in X$ is a point of local minimum for f in X if there exists $\delta > 0$ such that

$$f(x) \geq f(x_0), \ \forall x \in X \cap V_\delta(x_0),$$

where $V_\delta(x_0) = (x_0 - \delta, x_0 + \delta)$.
  Similarly we may define a point of local maximum.

Definition 6.2.6 (Global Minimum) Let $X \subset \mathbb{R}$ be a nonempty set and let $f : X \to \mathbb{R}$ be a function. We say that $x_0 \in X$ is a point of global minimum for f in X if

$$f(x) \geq f(x_0), \ \forall x \in X.$$

Similarly we may define a point of global maximum.

Theorem 6.2.7 *Let $x_0 \in (a, b)$ be a point of local minimum for $f : [a, b] \to \mathbb{R}$. Under such hypotheses, if $f'(x_0)$ exists, we have,*

$$f'(x_0) = 0.$$

Proof Choose $\delta > 0$ such that $(x_0 - \delta, x_0 + \delta) \subset [a, b]$. Thus, there exists $0 < \delta_1 < \delta$ such that if $|h| < \delta_1$, then

$$f(x_0 + h) \geq f(x_0),$$

so that if $0 < h < \delta_1$, then

$$\frac{f(x_0 + h) - f(x_0)}{h} \geq 0.$$

Letting $h \to 0^+$, we obtain

$$f'(x_0) \geq 0.$$

Similarly, if $-\delta_1 < h < 0$, then

$$f(x_0 + h) - f(x_0) \geq 0$$

and

$$\frac{f(x_0 + h) - f(x_0)}{h} \leq 0.$$

Letting $h \to 0^-$ we get

$$f'(x_0) \leq 0$$

so that

$$f'(x_0) = 0.$$

The proof is complete.

Remark 6.2.8 Similarly the result is also valid if f attains a local maximum at x_0, that is, $f'(x_0) = 0$.

6.3   The Mean Value Theorem

Theorem 6.3.1 *Let $f : [a, b] \to \mathbb{R}$ be a continuous function on $[a, b]$ and differentiable on (a, b). Thus, there exists $x \in (a, b)$ such that*

$$f(b) - f(a) = f'(x)(b - a).$$

Proof Define $g : [a, b] \to \mathbb{R}$ by

$$g(x) = (f(x) - f(a))(b - a) - (f(b) - f(a))(x - a).$$

Thus,

$$g(a) = g(b) = 0.$$

We may conclude that either $g(x) = 0$ on $[a, b]$ or if not, since g is continuous on $[a, b]$, it necessarily attains a local minimum or a local maximum at some $x \in (a, b)$.

From the last theorem, we have $g'(x) = 0$, so that

$$f'(x)(b - a) - (f(b) - f(a)) = 0,$$

that is,

$$f(b) - f(a) = f'(x)(b - a).$$

The proof is complete.

As an immediate application of the mean value theorem, we have the following result.

Proposition 6.3.2 *Let $f : [a, b] \to \mathbb{R}$ be a continuous function on $[a, b]$ and differentiable on (a, b).*

Under such hypotheses,

1. *If $f'(x) > 0$ on (a, b), then f is increasing.*
2. *If $f'(x) \geq 0$ on (a, b), then f is nondecreasing.*
3. *If $f'(x) < 0$ on (a, b), then f is decreasing.*
4. *If $f'(x) \leq 0$ on (a, b), then f is nonincreasing.*

Proof We shall prove only the item 1. The proofs of the remaining items are left as an exercise.

Thus, assume $f'(x) > 0$ on (a, b).

Let $x, y \in [a, b]$ be such that $x < y$.

From the mean value theorem, there exists $z \in (x, y)$ such that

$$f(y) - f(x) = f'(z)(y - x) > 0.$$

Therefore $f(x) < f(y)$, for all $x, y \in [a, b]$ such that $x < y$, so that f is increasing.

The proof is complete.

6.4 Higher Order Derivatives

Definition 6.4.1 Let $f : [a, b] \to \mathbb{R}$ be such that $f'(t)$ exists in a neighborhood of x. We define the second order derivative of f at x, denoted by $f''(x)$, by

$$f''(x) = \lim_{h \to 0} \frac{f'(x + h) - f'(x)}{h},$$

if such a limit exists. And inductively, if the derivative of order $n - 1$ of f at x, denoted by $f^{(n-1)}(t)$, exists in a neighborhood of x, then we define the derivative of order n of f at x, denoted by $f^{(n)}(x)$, by

$$f^{(n)}(x) = \lim_{h \to 0} \frac{f^{(n-1)}(x + h) - f^{(n-1)}(x)}{h},$$

if such a limit exists.

Definition 6.4.2 Let $X \subset \mathbb{R}$ be an open set. A function $f : X \to \mathbb{R}$ is said to be of $C^m(X)$ class (or simply of C^m class) if f and its derivatives of order up to m are continuous on X. If f is of C^m class, for all $m \in \mathbb{N}$, we say that f is of C^∞ class.

At this point we present the very important Taylor theorem.

Theorem 6.4.3 (Taylor Polynomial Approximation) *Suppose* $f : [a, b] \to \mathbb{R}$ *is such that* $f^{(n-1)}$ *is continuous on* $[a, b]$ *and that* $f^{(n)}$ *exists in* (a, b) *for some* $n \in \mathbb{N}$. *Let* $x_0 \in (a, b)$. *Define*

$$P(x) = \sum_{j=0}^{n-1} \frac{f^{(j)}(x_0)}{j!}(x - x_0)^j, \quad \forall x \in [a, b].$$

Let $x \in (a, b) \setminus \{x_0\}$. *Under such hypotheses, there exists* $\tilde{x}$ *between* x_0 *and* x *such that*

$$f(x) = P(x) + \frac{f^{(n)}(\tilde{x})(x - x_0)^n}{n!}.$$

Proof Define $M \in \mathbb{R}$ by

$$M = \frac{f(x) - P(x)}{(x - x_0)^n},$$

that is,

$$f(x) = P(x) + M(x - x_0)^n.$$

Define

$$g(t) = f(t) - P(t) - M(t - x_0)^n, \quad \forall t \in [a, b].$$

Observe that $g(x_0) = f(x_0) - P(x_0) = 0$, and

$$g(x) = f(x) - P(x) - M(x - x_0)^n = 0.$$

From the mean value theorem, there exists x_1 between x and x_0 such that $g'(x_1) = 0$. Observe also that $g'(x_0) = f'(x_0) - P'(x_0) = 0$.

Thus, by the mean value theorem, there exists x_2 between x_0 and x_1 such that

$$g''(x_2) = 0.$$

Proceeding inductively in this fashion, since $g(x_0) = g'(x_0) = \cdots = g^{(n-1)}(x_0) = 0$, after n steps we may find x_n between x_0 and x_{n-1} such that

$$g^{(n)}(x_n) = 0.$$

Denoting $x_n = \tilde{x}$ we have that $\tilde{x} \in (a, b)$ and

$$g^{(n)}(\tilde{x}) = f^{(n)}(\tilde{x}) - n!M = 0,$$

that is,

$$M = \frac{f^{(n)}(\tilde{x})}{n!},$$

so that

$$f(x) = P(x) + \frac{f^{(n)}(\tilde{x})(x - x_0)^n}{n!}.$$

This completes the proof.

6.5 Derivative of Inverse Function, One-Dimensional Case

In this section we obtain the derivative for the inverse of a differentiable function of one variable.

We start with the following preliminary result.

Theorem 6.5.1 *Let $f : [a, b] \to [a, b]$ be a function such that*

$$|f(x) - f(y)| \le \lambda|x - y|, \ \forall x, y \in [a, b],$$

for some $0 \le \lambda < 1$.

Remark 6.1 In such a case we say that f is a contractor function.

Under such hypotheses, there exists one and only one x_0 in $[a, b]$ such that

$$f(x_0) = x_0.$$

Such a $x_0 \in [a, b]$ is said to be a fixed point for f.

Proof Choose $x_1 \in [a, b]$ and define $\{x_n\} \subset [a, b]$ by

$$x_{n+1} = f(x_n), \ \forall n \in \mathbb{N}.$$

Hence,

$$|x_{n+1} - x_{n+1}| = |f(x_{n+1}) - f(x_n)|$$
$$\leq \lambda |x_{n+1} - x_n|, \ \forall n \in \mathbb{N}. \tag{6.4}$$

Since, $0 \leq \lambda < 1$, from the exposed at Sect. 3.6, $\{x_n\}$ is a Cauchy sequence, so that there exists $x_0 \in [a, b]$ such that

$$x_0 = \lim_{n \to \infty} x_n = x_0.$$

From the Lipschitz continuity of f, we get,

$$x_0 = \lim_{n \to} x_{n+1}$$
$$= \lim_{n \to \infty} f(x_n)$$
$$= f(x_0). \tag{6.5}$$

so that

$$f(x_0) = x_0,$$

that is, x_0 is a fixed point for f.

Now suppose $y_0 \in [a, b]$ is such that

$$f(y_0) = y_0.$$

Thus,

$$|x_0 - y_0| = |f(x_0) - f(y_0)| \leq \lambda |x_0 - y_0|,$$

so that

$$(1 - \lambda)|x_0 - y_0| \leq 0.$$

From this, since $1 - \lambda > 0$, we obtain,

$$|x_0 - y_0| = 0,$$

so that $y_0 = x_0$.

From this we may conclude that $x_0 \in [a, b]$ such that

$$f(x_0) = x_0$$

is unique.

The proof is complete.

Corollary 6.5.2 *Let $f : [a, b] \to [a, b]$ be a differentiable function such that*

$$\sup_{z \in [a,b]} |f'(z)| \leq \lambda < 1.$$

Under such assumptions, there exists one and only one $x_0 \in [a, b]$ such that

$$f(x_0) = x_0.$$

Proof Let $x, y \in [a, b]$ be such that $x \neq y$.

From the mean value theorem, there exists z between x and y such that

$$f(x) - f(y) = f'(z)(x - y),$$

so that

$$|f(x) - f(y)| \leq \sup_{z \in [a,b]} |f'(z)||x - y| \leq \lambda|x - y|.$$

Since $x, y \in [a, b]$ are arbitrary we may infer that f is a contractor function, so that from the last theorem there exists one and only one $x_0 \in [a, b]$ such that

$$f(x_0) = x_0.$$

This completes the proof.

Theorem 6.5.3 (Inverse Function Theorem for One Real Variable) *Let $f :$ $(a, b) \to \mathbb{R}$ be a C^1 class function and let $x_0 \in (a, b)$ be such that*

$$f'(x_0) \neq 0.$$

Under such hypotheses, there exist $\delta > 0$ and $\varepsilon > 0$ such that for each $y \in$ $(f(x_0) - \varepsilon, f(x_0) + \varepsilon) \equiv V$, there exists a unique $x \in (x_0 - \delta, x_0 + \delta) \equiv U$ such that $y = f(x)$, so that we denote,

$$x = f^{-1}(y) \equiv g(y).$$

Moreover,

$$g'(y) = \frac{1}{f'(g(y))}, \quad \forall y \in V.$$

We remark that g is a local inverse for f about $f(x_0)$.

Proof Denote $\lambda = f'(x_0)$ and with no loss in generality, assume $\lambda > 0$.

Since f' is continuous, there exists $\delta_0 > 0$ such that if $x \in (x_0 - \delta_0, x_0 + \delta_0)$, then $f'(x) > 0$.

Also, there exists $0 < \delta < \delta_0$ such that if $x \in [x_0 - \delta, x_0 + \delta] = \overline{U}$, then

$$|f'(x) - f'(x_0)| < \frac{\lambda}{2}.$$

Define $\varepsilon = \frac{\delta\lambda}{2}$ and let

$$y \in (f(x_0) - \varepsilon, f(x_0) + \varepsilon) \equiv V.$$

Define also $\phi : \mathbb{R} \to \mathbb{R}$ by

$$\phi(x) = x + \frac{1}{\lambda}(y - f(x)).$$

Hence for $x \in \overline{U}$ we have,

$$\phi'(x) = 1 - \frac{f'(x)}{\lambda}$$

$$= \frac{\lambda}{\lambda} - \frac{f'(x)}{\lambda}$$

$$= \frac{1}{\lambda}(\lambda - f'(x))$$

$$= \frac{1}{\lambda}(f'(x_0) - f'(x)), \tag{6.6}$$

so that

$$|\phi'(x)| \le \frac{1}{\lambda}|f'(x_0) - f'(x)| \le \frac{1}{\lambda}\frac{\lambda}{2} = \frac{1}{2}.$$

Hence,

$$|\phi'(x)| \le \frac{1}{2}, \ \forall x \in \overline{U}.$$

Let $x \in \overline{U}$ be given.
Hence,

$$|\phi(x) - x_0| = |\phi(x) - \phi(x_0) + \phi(x_0) - x_0|$$

$$\le \frac{1}{2}|x - x_0| + |\phi(x_0) - x_0|$$

$$\le \frac{\delta}{2} + \frac{1}{\lambda}|y - f(x_0)|$$

$$< \frac{\delta}{2} + \frac{\varepsilon}{\lambda} = \frac{\delta}{2} + \frac{\delta}{2}$$

$$= \delta. \tag{6.7}$$

Hence,

$$|\phi(x) - x_0| < \delta, \ \forall x \in \overline{U},$$

so that $\phi(\overline{U}) \subset U$.

From this and $|\phi'(x)| \leq \frac{1}{2}, \ \forall x \in \overline{U}$ we may infer that ϕ is a contractor function. Hence, there exists a unique not relabeled $x \in U$ such that

$$\phi(x) = x,$$

so that from $\phi(x) = x + \frac{1}{\lambda}(y - f(x))$, we have that $y - f(x) = 0$, so that $y = f(x)$.

Here we denote, $x = f^{-1}(y) \equiv g(y)$.

At this point we are going to show that g is continuous.

Let $y \in V$.

Let $\{y_n\} \subset \mathbb{R}$ be such that

$$\lim_{n \to \infty} y_n = y.$$

There is no loss of generality in assuming $\{y_n\} \subset V$.

We recall that there exists a unique $x \in U$ such that $y = f(x)$ so that

$$x = f^{-1}(y) = g(y).$$

Suppose, to obtain contradiction, we do not have

$$\lim_{n \to \infty} g(y_n) = x.$$

Thus, denoting $x_n = g(y_n)$, there exists $\varepsilon_0 > 0$ such that for each $k \in \mathbb{N}$ there exists $n_k > k, n_k \in \mathbb{N}$, such that

$$|x_{n_k} - x| \geq \varepsilon_0.$$

Observe that $\{x_{n_k}\} \subset U$ and such a set is bounded. So there exists a subsequence $\{x_{(n_k)_j}\}$ of $\{x_{n_k}\}$, which we shall denote simply by $\{x_{n_j}\}$, and $\tilde{x} \neq x$ such that

$$\lim_{j \to \infty} x_{n_j} = \tilde{x} \neq x.$$

Observe that

$$\lim_{j \to \infty} f(x_{n_j}) = f(\tilde{x})$$

$$= \lim_{j \to \infty} y_{n_j}$$

$$= y. \tag{6.8}$$

Hence, we have got,

$$y = f(x) = f(\tilde{x})$$

and

$$x \neq \tilde{x}.$$

This contradicts the uniqueness of x.
Hence,

$$\lim_{n \to \infty} g(y_n) = g(y).$$

Since $\{y_n\} \subset \mathbb{R}$ is arbitrary, we may infer that g is continuous at y, $\forall y \in V$.
Finally, let a new $y \in V$ be given and let $h \in \mathbb{R}$ be such that $h \neq 0$ and $y + h \in V$
Hence, there exist $x, x_1 \in U$ such that

$$y = f(x), \text{ so that } x = g(y),$$

and

$$y + h = f(x_1), \text{ so that } x_1 = g(y + h).$$

From the mean value theorem, there exists ξ between x and x_1 such that

$$f(x_1) - f(x) = f'(\xi)(x_1 - x),$$

so that

$$y + h - y = h = f'(\xi)(g(y + h) - g(y)),$$

and therefore,

$$\frac{g(y + h) - g(y)}{h} = \frac{1}{f'(\xi)}.$$

From the continuity of g,

$$g(y + h) \to g(y), \text{ as } h \to 0,$$

so that

$$x_1 \to x, \text{ as } h \to 0.$$

Hence,

$$\xi \to x, \text{ as } h \to 0,$$

so that

$$f'(\xi) \to f'(x), \text{ as } h \to 0.$$

Thus,

$$g'(y) = \lim_{h \to 0} \frac{g(y+h) - g(y)}{h}$$

$$= \lim_{h \to 0} \frac{1}{f'(\xi)}$$

$$= \frac{1}{f'(x)} = \frac{1}{f'(g(y))}. \tag{6.9}$$

The proof is complete.

Example 6.5.4 Consider the example in which $f : \mathbb{R} \to \mathbb{R}$ is given point-wisely by

$$f(x) = e^x.$$

Let $x \in \mathbb{R}$.
Hence,

$$f'(x) = \lim_{h \to 0} \frac{e^{x+h} - e^x}{h}$$

$$= \lim_{h \to 0} e^x \frac{e^h - 1}{h}$$

$$= e^x \lim_{h \to 0} \frac{e^h - 1}{h}$$

$$= e^x 1$$

$$= e^x. \tag{6.10}$$

Thus,

$$\frac{de^x}{dx} = e^x, \ \forall x \in \mathbb{R},$$

so that, by induction, we may easily obtain,

$$\frac{d^n e^x}{dx^n} = e^x, \ \forall x \in \mathbb{R}, \forall n \in m\mathbb{N},$$

and in particular

$$f^{(n)}(0) = 1, \ \forall n \in \mathbb{N}.$$

Choose $x \in \mathbb{R}$. From the Taylor's series about $x_0 = 0$ we obtain

$$f(x) = \sum_{k=1}^{n} f^{(k)}(0) \frac{x^k}{k!} + f^{(n+1)}(\tilde{x}) \frac{x^{n+1}}{n+1},$$

where $\tilde{x} \in \mathbb{R}$ is between 0 and x.

Thus,

$$e^x = \sum_{k=1}^{n} \frac{x^k}{k!} + e^{\tilde{x}} \frac{x^{n+1}}{(n+1)!}. \tag{6.11}$$

Observe that

$$|e^{\tilde{x}}| \le e^{|x|},$$

and

$$\left| \frac{x^{n+1}}{(n+1)!} \right| \le \frac{|x|^{n+1}}{(n+1)!}$$

$$= \frac{|x|}{1} \cdots \frac{|x|}{(n+1)}. \tag{6.12}$$

If $|x| < 1$ define $k_0 = 0$ and if $|x| \ge 1$ define

$$k_0 = \max \left\{ k \in \mathbb{N} \ : \ \frac{|x|}{k} \ge 1 \right\}.$$

Thus, if $k > k_0$ we have

$$\frac{|x|}{k} < 1.$$

Therefore, we have,

$$\frac{|x|^{n+1}}{(n+1)!} = \left(\frac{|x|}{1} \cdots \frac{|x|}{k_0} \right) \frac{|x|}{k_0+1} \cdots \frac{|x|}{(n+1)}$$

$$\le \left(\frac{|x|}{1} \cdots \frac{|x|}{k_0} \right) \frac{|x|}{(n+1)}$$

$$\to 0, \text{ as } n \to \infty. \tag{6.13}$$

Hence

$$e^{\tilde{x}} \frac{|x|^{n+1}}{(n+1)!} \le e^{|x|} \frac{|x|^{n+1}}{(n+1)!} \to 0, \text{ as } n \to \infty.$$

From this and (6.11), we obtain

$$\left| e^x - \sum_{k=0}^{n} \frac{x^k}{k!} \right| \leq e^{|x|} \frac{|x|^{n+1}}{(n+1)!} \to 0, \text{ as } n \to \infty.$$

From such results, we may write

$$e^x = \lim_{n \to \infty} \sum_{k-0}^{n} \frac{x^k}{k!}$$

$$= \sum_{n=0}^{\infty} \frac{x^n}{n!}. \tag{6.14}$$

Observe that

$$e^x = y \Leftrightarrow x = \ln(y), \forall x \in \mathbb{R}, \ y > 0,$$

so that $g : (0, +\infty) \to \mathbb{R}$ given point-wisely by,

$$g(y) = \ln(y),$$

is the inverse function of f, so that,

$$g(y) = f^{-1}(y), \ \forall y > 0.$$

From the inverse function theorem

$$\frac{dg(y)}{dy} = \frac{1}{f'(g(y))},$$

where $f'(x) = e^x$, so that

$$f'(g(y)) = f'(\ln(y)) = e^{\ln(y)} = y.$$

Summarizing,

$$\frac{dg(y)}{dy} = \frac{1}{y},$$

that is,

$$\frac{d \ln(y)}{dy} = \frac{1}{y}, \forall y > 0.$$

Example 6.5.5 Consider the example in which $f : (0, +\infty) \to \mathbb{R}$ is defined by

$$f(x) = x^r, \ \forall x > 0,$$

where

$$r \in \mathbb{R} \setminus \{0\}.$$

Fix $x > 0$. Observe that

$$x^r = e^{\ln(x^r)} = e^{r \ln(x)}.$$

Hence, from the last example and the chain rule,

$$
\begin{aligned}
\frac{dx^r}{dx} &= \frac{d[e^{r \ln(x)}]}{dx} \\
&= e^{r \ln(x)} \frac{d[r \ln(x)]}{dx} \\
&= x^r r \frac{d \ln(x)}{dx} \\
&= x^r r \frac{1}{x} \\
&= r x^{r-1}.
\end{aligned}
\tag{6.15}
$$

Summarizing,

$$\frac{dx^r}{dx} = r x^{r-1}, \forall x > 0.$$

Example 6.5.6 Similarly, consider the example in which $f : \mathbb{R} \to \mathbb{R}$ is defined point-wisely by,

$$f(x) = x^n,$$

where $n \in \mathbb{N}$.

Fix $x \in \mathbb{R}$. Thus,

$$
\begin{aligned}
\frac{dx^n}{dx} &= \lim_{h \to 0} \frac{(x+h)^n - x^n}{h} \\
&= \lim_{h \to 0} \frac{\sum_{k=0}^{n} \binom{n}{k} x^{(n-k)} h^k - x^n}{h}
\end{aligned}
$$

$$= \lim_{h \to 0} \frac{x^n + nx^{n-1}h + \mathcal{O}(h^2) - x^n}{h}$$

$$= \lim_{h \to 0} \left(nx^{n-1} + \frac{\mathcal{O}(h^2)}{h} \right)$$

$$= nx^{n-1}, \tag{6.16}$$

where

$$\mathcal{O}(h^2) = \sum_{k=2}^{n} \binom{n}{k} x^{n-k} h^k.$$

Summarizing,

$$\frac{dx^n}{dx} = nx^{n-1}, \forall x \in \mathbb{R}, \ n \in \mathbb{N}.$$

Let $f : \mathbb{R} \to \mathbb{R}$ be such that

$$f(x) = x^n, \forall x \in \mathbb{R},$$

where $n \in \mathbb{N}$ is odd.

Observe that,

$$y = x^n \Leftrightarrow x = y^{1/n}, \forall x, y \in \mathbb{R}.$$

Hence, $g : \mathbb{R} \to \mathbb{R}$ given point-wisely by

$$g(y) = y^{1/n},$$

is the inverse function of f, that is, $g = f^{-1}$.

Fix $y \in \mathbb{R}$. From the inverse function theorem for one variable,

$$\frac{dg(y)}{dy} = \frac{1}{f'(g(y))},$$

so that

$$\frac{dy^{1/n}}{dy} = \frac{1}{nx^{n-1}} \Big|_{x=y^{1/n}}$$

$$= \frac{1}{n[y^{1/n}]^n}$$

$$= \frac{1}{n} y^{1/n-1}. \tag{6.17}$$

Summarizing,

$$\frac{dx^{1/n}}{dx} = \frac{1}{n}x^{1/n-1}, \ \forall x \in \mathbb{R}, \ n \text{ odd.}$$

Similarly, we may obtain,

$$\frac{dx^{1/n}}{dx} = \frac{1}{n}x^{1/n-1}, \ \forall x > 0, \ n \in \mathbb{N}.$$

Example 6.5.7 In this example we consider the derivatives of trigonometric functions.

Theorem 6.5.8 *Let* $f, g : \mathbb{R} \to \mathbb{R}$ *defined by*

$$f(x) = \sin(x), \forall x \in \mathbb{R},$$

and

$$g(x) = \cos(x), \forall x \in \mathbb{R}.$$

Let $x \in \mathbb{R}$. *Under such assumptions we have,*

$$\lim_{h \to 0} \frac{\sin(x + h) - \sin(x)}{h} = \cos x,$$

so that

$$f'(x) = \cos(x)$$

and

$$\lim_{h \to 0} \frac{\cos(x + h) - \cos(x)}{h} = -\sin x, \forall x \in \mathbb{R},$$

so that

$$g'(x) = -\sin(x),$$

$\forall x \in \mathbb{R}.$

Proof Observe that, from Proposition 5.10.5, we have

$$\frac{\sin(x + h) - \sin(x)}{h} = \frac{2\sin(h/2)}{h}\cos[(2x + h)/2]$$

$$= \frac{\sin(h/2)}{h/2}\cos[(2x + h)/2], \qquad (6.18)$$

so that

$$f'(x) = \lim_{h \to 0} \frac{\sin(x+h) - \sin(x)}{h}$$

$$= \lim_{h \to 0} \left(\frac{2\sin(h/2)}{h} \cos[(2x+h)/2] \right)$$

$$= \lim_{h \to 0} \frac{\sin(h/2)}{h/2} \lim_{h \to 0} \cos(x+h/2)$$

$$= 1 \cos x$$

$$= \cos x. \tag{6.19}$$

Also from Proposition 5.10.5, we have

$$\frac{\cos(x+h) - \cos(x)}{h} = -2\frac{\sin(h/2)}{h} \sin[(2x+h)/2]$$

$$= -\frac{\sin(h/2)}{h/2} \sin[(2x+h)/2], \tag{6.20}$$

so that

$$g'(x) = \lim_{h \to 0} \frac{\cos(x+h) - \cos(x)}{h}$$

$$= -\lim_{h \to 0} \frac{\sin(h/2)}{h/2} \lim_{h \to 0} \cos(x+h/2)$$

$$= -1 \sin x$$

$$= -\sin x. \tag{6.21}$$

The proof is complete.

Example 6.5.9 As an application of derivatives, we shall prove that

$$\lim_{n \to \infty} \sqrt[n]{n} = 1.$$

First define $f : [1, +\infty) \to \mathbb{R}$ by

$$f(x) = x^{1/x}.$$

Hence,

$$\ln f(x) = \ln(x^{1/x}) = \frac{1}{x} \ln(x).$$

Thus,

$$[\ln(f(x))]' = \frac{f'(x)}{f(x)} = \left(\frac{1}{x}\ln(x)\right)' = \frac{-1}{x^2}\ln(x) + \frac{1}{x^2},$$

so that

$$f'(x) = f(x)\frac{1}{x^2}(1 - \ln(x)).$$

Therefore,

$$f'(x) < 0 \text{ if and only if } 1 - \ln(x) < 0,$$

which means $\ln(x) > 1$, so that $x = e^{\ln(x)} > e^1 = e$.

Hence, if $x > e$, then $f'(x) < 0$.

From this and Proposition 6.3.2, we obtain that $f(x) = x^{1/x} = \sqrt[x]{x}$ is decreasing on $[e, +\infty)$, so that

$$\sqrt[n]{n}$$

is decreasing from the third term and on.

Observe that $\sqrt[n]{n} \geq 1$, $\forall n \in \mathbb{N}$. Therefore $\{\sqrt[n]{n}\}$ is monotone from the third term and bounded, so that it is convergent.

Let $\alpha \geq 1$ be such that

$$\alpha = \lim_{n\to\infty} \sqrt[n]{n}.$$

Thus,

$$\alpha = \lim_{n\to\infty} \sqrt[n]{n} = \lim_{n\to\infty} \sqrt[2n]{2n}.$$

Hence,

$$\begin{aligned}
\alpha^2 &= \lim_{n\to\infty} [(2n)^{1/(2n)}]^2 \\
&= \lim_{n\to\infty} (2n)^{1/(n)} \\
&= \lim_{n\to\infty} 2^{1/n} \lim_{n\to\infty} n^{1/n} \\
&= \lim_{n\to\infty} n^{1/n} \\
&= \alpha.
\end{aligned} \tag{6.22}$$

Therefore,

$$\alpha^2 = \alpha.$$

Since $\alpha \geq 1$, we obtain $\alpha = 1$.

Exercises 6.5.10

1. Let $I \subset \mathbb{R}$ be an interval. Let $f : I \to \mathbb{R}$ be a differentiable function on I. Assume $f'(x) = 0, \; \forall x \in I$. Under such hypotheses, prove that

$$f(x) = c, \; \forall x \in I,$$

for some $c \in \mathbb{R}$.

2. Let $f : I \to \mathbb{R}$ be a differentiable function on the interval I. Assume that

$$|f'(x)| \le c,$$

$\forall x \in I$. Under such hypotheses, prove that

$$|f(x) - f(y)| \le c|y - x|, \; \forall x, y \in I.$$

3. Let $f, g : [a, b] \to \mathbb{R}$ be continuous functions on $[a, b]$ and differentiable on (a, b). Show that there exists $c \in (a, b)$ such that

$$(f(b) - f(a))g'(c) = (g(b) - g(a))f'(c).$$

4. Let $f : [0, +\infty) \to \mathbb{R}$ be a differentiable function and assume

$$\lim_{x \to +\infty} f'(x) = \alpha \in \mathbb{R}.$$

Under such hypotheses, prove that, for each $c > 0$, we have

$$\lim_{x \to +\infty} [f(x + c) - f(x)] = c\alpha,$$

and also,

$$\lim_{x \to +\infty} \frac{f(x)}{x} = \alpha.$$

5. Let $f : [0, +\infty) \to \mathbb{R}$ be a function twice differentiable. Assume f'' is bounded on $[0, +\infty)$ and that

$$\lim_{x \to +\infty} f(x) = c \in \mathbb{R}.$$

Under such hypotheses, prove that

$$\lim_{x \to +\infty} f'(x) = 0.$$

6. Let $f : [a, b] \to \mathbb{R}$ be a continuous function on $[a, b]$ and C^2 class function on (a, b). Let $c \in (a, b)$. Define $\eta : (a, b) \to \mathbb{R}$ by

$$\eta(x) = \frac{f(x) - f(c)}{x - c},$$

if $x \neq c$ and

$$\eta(c) = f'(c).$$

Under such hypotheses, prove that

$$\eta'(c) = \frac{f''(c)}{2}.$$

Chapter 7
The Riemann Integral

7.1 Introduction

This chapter develops results concerning the Riemann integration of one variable real functions. The standard necessary and sufficient condition of zero Lebesgue measure for the set of discontinuities for the Riemann integrability of a one variable real function is addressed in detail. In the final sections we present a basic study on sequences and series of real functions.

The main references for this chapter are again [9, 12, 13].

7.2 First Definitions and Results

We shall start with the definition of partition.

Definition 7.2.1 (Partition and Respective Norm) Let $[a, b] \subset \mathbb{R}$ be a closed interval. We define a partition of $[a, b]$, denoted by P, as

$$P = \{x_0 = a, x_1, x_2, \ldots, x_n = b\},$$

where $x_{i-1} < x_i, \ \forall i \in \{1, \ldots, n\}$.

Moreover, denoting $\Delta x_i = x_i - x_{i-1}, \ \forall i \in \{1, \ldots, n\}$, we define the norm of P, denoted by $|P|$, as

$$|P| = \max\{\Delta x_i, \ i \in \{1, \ldots, n\}\}.$$

Definition 7.2.2 (Lower and Upper Sums) Let $f : [a, b] \to \mathbb{R}$ be such that $|f(x)| < K, \ \forall x \in [a, b]$, for some $K > 0$. Let $P = \{x_0 = a, x_1, x_2, \ldots, x_n = b\}$ be a partition of $[a, b]$. We define the upper sum of f related to P, denoted by S_f^P,

© Springer International Publishing AG, part of Springer Nature 2018
F. S. Botelho, *Real Analysis and Applications*,
https://doi.org/10.1007/978-3-319-78631-5_7

by

$$S_f^P = \sum_{i=1}^{n} M_i \Delta x_i,$$

where

$$M_i = \sup\{f(x), \ x \in [x_{i-1}, x_i]\}.$$

Similarly, we define the lower sum of f related to P, denoted by s_f^P, by

$$s_f^P = \sum_{i=1}^{n} m_i \Delta x_i,$$

where

$$m_i = \inf\{f(x), \ x \in [x_{i-1}, x_i]\}.$$

Here, as above we denote $\Delta x_i = x_i - x_{i-1}, \forall i \in \{1, \dots, n\}$. Finally, we define a Riemann sum of f relating P, denoted by R_f^P, by

$$R_f^P = \sum_{i=1}^{n} f(t_i) \Delta x_i,$$

where t_i is any point such that

$$t_i \in [x_{i-1}, x_i], \ \ \forall i \in \{1, \dots, n\}$$

Remark 7.2.3 Observe that, concerning the last definition, we have,

$$s_f^P \le R_f^P \le S_f^P$$

for all partition P of $[a, b]$ and all Riemann sum R_f^P.

Theorem 7.2.4 *Let* $f : [a, b] \to \mathbb{R}$ *be such that* $|f(x)| < K, \ \forall x \in [a, b]$ *for some* $K > 0$. *Let* $P = \{x_0 = a, x_1, x_2, \dots, x_n = b\}$ *be a partition of* $[a, b]$. *Let* $t \in [a, b]$ *be such that* $t \notin P$.

Define $Q = P \cup \{t\}$.

Under such hypotheses, we have

$$s_f^P \le s_f^Q \le S_f^Q \le S_f^P.$$

Proof Observe that there exists $k \in \{1, \dots, n\}$ such that $t \in (x_{k-1}, x_k)$.

Thus,

$$S_f^P - S_f^Q = M_k(x_k - x_{k-1}) - (M_a(t - x_{k-1}) + M_b(x_k - t)),$$

where

$$M_k = \sup\{f(x) \ : \ x \in [x_{k-1}, x_k]\},$$

$$M_a = \sup\{f(x) \ : \ x \in [x_{k-1}, t]\} \le M_k,$$

and

$$M_b = \sup\{f(x) \ : \ x \in [t, x_k]\} \le M_k,$$

Thus,

$$\begin{aligned}
S_f^P - S_f^Q &= M_k(x_k - x_{k-1}) - (M_a(t - x_{k-1}) + M_b(x_k - t)) \\
&= M_k(x_k - t + t - x_{k-1}) - (M_a(t - x_{k-1}) + M_b(x_k - t)) \\
&= (M_k - M_a)(t - x_{k-1}) + (M_k - M_b)(x_k - t) \\
&\ge 0.
\end{aligned} \tag{7.1}$$

Hence,

$$S_f^P \ge S_f^Q.$$

Similarly, we may obtain

$$s_f^Q \ge s_f^P,$$

so that

$$s_f^P \le s_f^Q \le S_f^Q \le S_f^P.$$

The proof is complete.

Remark 7.2.5 Reasoning inductively, having a partition $P = \{x_0 = a, x_1, x_2, \ldots, x_n = b\}$, if we add the points $t_1, \ldots, t_k \in [a, b]$ to P, obtaining $Q = P \cup \{t_1, \ldots, t_k\}$, for each point added we will have an increase (in fact a non-decrease) in the lower sum and a decrease (in fact a non-increase) in the upper sum, so that if $P \subset Q$, we have

$$s_f^P \le s_f^Q \le S_f^Q \le S_f^P.$$

In particular, given two partitions P, Q of $[a, b]$, since $P \subset P \cup Q$, and $Q \subset P \cup Q$, we obtain

$$s_f^P \le s_f^{P \cup Q} \le S_f^{P \cup Q} \le S_f^Q,$$

that is,

$$s_f^P \leq S_f^Q, \quad \forall P, Q, \text{ partitions of } [a, b].$$

Therefore,

$$\sup\{s_f^P \ : \ P \text{ is a partition of } [a, b]\} \leq \inf\{S_f^Q \ : \ Q \text{ is a partition of } [a, b]\}.$$

Definition 7.2.6 Let $f : [a, b] \to \mathbb{R}$ be such that $|f(x)| < K, \ \forall x \in [a, b]$ for some $K > 0$.
    Then we define the lower integral of f on $[a, b]$, denoted by $\underline{I}$, by

$$\underline{I} = \sup\{s_f^P \ : \ P \text{ is a partition of } [a, b]\},$$

and the upper integral of f on $[a, b]$, denoted by $\overline{I}$, by

$$\overline{I} = \inf\{S_f^P \ : \ P \text{ is a partition of } [a, b]\}.$$

Finally, we say that f is Riemann integrable on $[a, b]$ if

$$\underline{I} = \overline{I},$$

and we denote

$$\underline{I} = \overline{I} = I = \int_a^b f(x) \, dx.$$

In such a case $I = \int_a^b f(x) \, dx$ is said to be Riemann integral of f on $[a, b]$.

Theorem 7.2.7 *Let $f : [a, b] \to \mathbb{R}$ be such that $|f(x)| < K, \ \forall x \in [a, b]$ for some $K > 0$. Under such hypotheses, f is Riemann integrable if, and only if, for each $\varepsilon > 0$ there exists a partition P of $[a, b]$ such that*

$$S_f^P - s_f^P < \varepsilon.$$

Proof First, we shall prove the condition sufficiency. Suppose the condition is valid.
    Let $\varepsilon > 0$. Thus, from the condition there exists a partition P of $[a, b]$ such that

$$S_f^P - s_f^P < \varepsilon.$$

Observe that

$$s_f^P \leq \underline{I} \leq \overline{I} \leq S_f^P.$$

Hence,

$$\overline{I} - \underline{I} \leq S_f^P - s_f^P < \varepsilon.$$

Since $\varepsilon > 0$ is arbitrary, we get

$$\overline{I} = \underline{I}.$$

Reciprocally, suppose that

$$\overline{I} = \underline{I} \equiv I.$$

Suppose given a new and not relabeled $\varepsilon > 0$. Thus there exists partitions P_1 and P_2 of $[a, b]$ such that

$$s_f^{P_1} > I - \frac{\varepsilon}{2},$$

and

$$S_f^{P_2} < I + \frac{\varepsilon}{2},$$

Thus,

$$-s_f^{P_1} < -I + \frac{\varepsilon}{2},$$

so that

$$S_f^{P_2} - s_f^{P_1} < I - I + \frac{\varepsilon}{2} + \frac{\varepsilon}{2} = \varepsilon.$$

Observe that

$$s_f^{P_1} \leq s_f^{P_1 \cup P_2} \leq S_f^{P_1 \cup P_2} \leq S_f^{P_2},$$

so that

$$S_f^{P_1 \cup P_2} - s_f^{P_1 \cup P_2} \leq S_f^{P_2} - s_f^{P_1} < \varepsilon.$$

Hence, denoting $P = P_1 \cup P_2$, we obtain

$$S_f^P - s_f^P < \varepsilon.$$

Therefore, the condition is necessary.

The proof is complete.

7.3   Riemann Integral Properties

Theorem 7.3.1 *Let $f_1 : [a, b] \to \mathbb{R}$ and $f_2 : [a, b] \to \mathbb{R}$ be Riemann integrable functions. Under such hypotheses,*

$$f \equiv f_1 + f_2$$

is also Riemann integrable and,

$$\int_a^b (f_1(x) + f_2(x))\, dx = \int_a^b f_1(x)\, dx + \int_a^b f_2(x)\, dx.$$

Proof Let $\varepsilon > 0$ be given.

Denoting,

$$I_1 = \int_a^b f_1(x)\, dx,$$

$$I_2 = \int_a^b f_2(x)\, dx,$$

$$\underline{I} = \sup\{s_f^P \ : \ P \text{ is a partition of } [a, b]\},$$

and

$$\overline{I} = \inf\{s_f^P \ : \ P \text{ is a partition of } [a, b]\},$$

there exist partitions P_1 and P_2 of $[a, b]$ such that

$$I_1 - \frac{\varepsilon}{4} < s_{f_1}^{P_1} \le S_{f_1}^{P_1} < I_1 + \frac{\varepsilon}{4},$$

and

$$I_2 - \frac{\varepsilon}{4} < s_{f_2}^{P_2} \le S_{f_2}^{P_2} < I_2 + \frac{\varepsilon}{4}.$$

Define $Q = P_1 \cup P_2$. Hence,

$$I_1 + I_2 - \frac{\varepsilon}{2}$$

$$< s_{f_1}^{P_1} + s_{f_2}^{P_2}$$

$$\le s_{f_1}^{Q} + s_{f_2}^{Q}$$

$$\leq S^Q_{f_1+f_2}$$

$$\leq \underline{I} \leq \overline{I}$$

$$\leq S^Q_{f_1+f_2}$$

$$\leq S^Q_{f_1} + S^Q_{f_2}$$

$$\leq S^{P_1}_{f_1} + S^{P_2}_{F_2}$$

$$< I_1 + I_2 + \frac{\varepsilon}{2}, \tag{7.2}$$

so that

$$\overline{I} - \underline{I} < \varepsilon.$$

Since $\varepsilon > 0$ is arbitrary, we may conclude that

$$\overline{I} = \underline{I} \equiv I = \int_a^b f(x)\, dx.$$

Moreover, from this and (7.2), we obtain

$$I_1 + I_2 - \frac{\varepsilon}{2} < I < I_1 + I_2 + \frac{\varepsilon}{2},$$

that is,

$$|I - (I_1 + I_2)| < \frac{\varepsilon}{2},$$

Since $\varepsilon > 0$ is arbitrary we have,

$$I = I_1 + I_2,$$

that is

$$\int_a^b f(x)\, dx = \int_a^b f_1(x)\, dx + \int_a^b f_2(x)\, dx.$$

The proof is complete.

Theorem 7.3.2 *Let $f : [a, b] \to \mathbb{R}$ be a Riemann integrable function. Then $-f$ is Riemann integrable and*

$$\int_a^b (-f(x))\, dx = -\int_a^b f(x)\, dx.$$

Proof Suppose given $\varepsilon > 0$. Thus there exists a partition P of $[a, b]$ such that

$$I - \frac{\varepsilon}{2} < s_f^P \le S_f^P < I + \frac{\varepsilon}{2}.$$

Hence,

$$-I - \frac{\varepsilon}{2} < -S_f^P \le -s_f^P < -I + \frac{\varepsilon}{2},$$

so that, denoting

$$\underline{I_1} = \sup\{s_{(-f)}^P \; : \; P \text{ is a partition of } [a, b]\},$$

and

$$\overline{I_1} = \inf\{S_{(-f)}^P \; : \; P \text{ is a partition of } [a, b]\},$$

we have

$$-I - \frac{\varepsilon}{2} < s_{(-f)}^P \le \underline{I_1} \le \overline{I_1} \le S_{(-f)}^P < -I + \frac{\varepsilon}{2}, \tag{7.3}$$

that is,

$$\overline{I_1} - \underline{I_1} < \varepsilon.$$

Since $\varepsilon > 0$ is arbitrary, we have

$$\overline{I_1} = \underline{I_1} \equiv I_1 = \int_a^b (-f(x)) \, dx.$$

From this and (7.3) we obtain,

$$-I - \frac{\varepsilon}{2} < I_1 < -I + \frac{\varepsilon}{2},$$

so that

$$|I_1 - (-I)| < \frac{\varepsilon}{2}.$$

Since $\varepsilon > 0$ is arbitrary, we finally obtain,

$$I_1 = -I,$$

that is,

$$\int_a^b (-f(x)) \, dx = -\int_a^b f(x) \, dx.$$

The proof is complete.

Theorem 7.3.3 *Let* $f : [a, b] \to \mathbb{R}$ *be Riemann integrable and let* $c \in \mathbb{R}$. *Under such hypotheses* cf *is Riemann integrable and*

$$\int_a^b cf(x) \, dx = c \int_a^b f(x) \, dx.$$

Proof Assume first $c > 0$.

The case $c = 0$ is immediate and the case $c < 0$ will be dealt at the end of this proof. Let $\varepsilon > 0$ be given. Thus, there exists a partition P of $[a, b]$ such that

$$I - \frac{\varepsilon}{2c} < s_f^P \leq S_f^P < I + \frac{\varepsilon}{2c}.$$

Hence,

$$cI - \frac{\varepsilon}{2} < cs_f^P \leq cS_f^P < cI + \frac{\varepsilon}{2}.$$

Therefore, denoting,

$$\underline{I_1} = \sup\{s_{(cf)}^P : P \text{ is a partition of } [a, b]\},$$

and

$$\overline{I_1} = \inf\{S_{(cf)}^P : P \text{ is a partition of } [a, b]\},$$

we have

$$cI - \frac{\varepsilon}{2} < s_{(cf)}^P \leq \underline{I_1} \leq \overline{I_1} \leq S_{(cf)}^P < cI + \frac{\varepsilon}{2}. \tag{7.4}$$

so that

$$\overline{I_1} - \underline{I_1} < \varepsilon.$$

Since $\varepsilon > 0$ is arbitrary, we obtain,

$$\overline{I_1} = \underline{I_1} \equiv I_1 = \int_a^b cf(x) \, dx.$$

From this and (7.4), we have,

$$cI - \frac{\varepsilon}{2} < I_1 < cI + \frac{\varepsilon}{2},$$

that is,

$$|I_1 - cI| < \frac{\varepsilon}{2}.$$

Finally, since $\varepsilon > 0$ is arbitrary, we obtain

$$I_1 = cI,$$

that is,

$$\int_a^b cf(x)\,dx = c\int_a^b f(x)\,dx.$$

Now suppose $c < 0$. From this last result and the last theorem, we have

$$-\int_a^b cf(x)\,dx = \int_a^b (-c)f(x)\,dx = -c\int_a^b f(x)\,dx,$$

so that

$$\int_a^b cf(x)\,dx = c\int_a^b f(x)\,dx.$$

The proof is complete.

Theorem 7.3.4 *Let $f : [a, b] \to \mathbb{R}$ be a Riemann integrable function such that*

$$m \le f(x) \le M, \quad \forall x \in [a, b],$$

for some $m, M \in \mathbb{R}$.
  Let $g : [m, M] \to \mathbb{R}$ be a continuous function on $[m, M]$.
  Under such hypotheses, $(g \circ f) : [a, b] \to \mathbb{R}$ is Riemann integrable on $[a, b]$, where

$$(g \circ f)(x) = g(f(x)), \quad \forall x \in [a, b].$$

Proof Let $\varepsilon > 0$ be given. Observe that since $[m, M]$ is compact, g is uniformly continuous on $[m, M]$. Choose $K > 0$ such that

$$|g(t)| < K, \quad \forall t \in [m, M].$$

Thus, there exists $0 < \delta < \frac{\varepsilon}{4K}$ such that if $s, t \in [m, M]$ and $|s - t| < \delta$, then

$$|g(s) - g(t)| < \frac{\varepsilon}{2(b - a)}.$$

Also, since f is integrable, there exists a partition $P = \{x_0 = a, x_1, x_2, \ldots, x_n = b\}$ of $[a, b]$ such that

$$I - \frac{\delta^2}{2} < s_f^P \le S_f^P < I - \frac{\delta^2}{2},$$

where

$$I = \int_a^b f(x)\,dx.$$

Therefore,

$$S_f^P - s_f^P = \sum_{i=1}^n (M_i - m_i)\Delta x_i < \delta^2,$$

where we denote,

$$m_i = \inf\{f(x) \;:\; x \in [x_{i-1}, x_i]\},$$

$$M_i = \sup\{f(x) \;:\; x \in [x_{i-1}, x_i]\},$$

$$m_i^* = \inf\{g(f(x)) \;:\; x \in [x_{i-1}, x_i]\},$$

$$M_i^* = \sup\{g(f(x)) \;:\; x \in [x_{i-1}, x_i]\}.$$

Denote by α the set of indices $i \in \{1, \ldots, n\}$ such that $M_i - m_i < \delta$. Denote by β the set of indices $i \in \{1, \ldots, n\}$ such that $M_i - m_i \geq \delta$.

Observe that

$$\delta \sum_{i \in \beta} \Delta x_i \leq \sum_{i \in \beta} (M_i - m_i)\Delta x_i < \delta^2,$$

and hence,

$$\sum_{i \in \beta} \Delta x_i < \delta < \frac{\varepsilon}{4K}.$$

Observe that if $i \in \alpha$, then $M_i - m_i < \delta$ and thus

$$M_i^* - m_i^* \leq \frac{\varepsilon}{2(b-a)}.$$

Therefore,

$$S_{(g \circ f)}^P - s_{(g \circ f)}^P = \sum_{i=1}^n (M_i^* - m_i^*)\Delta x_i$$

$$= \sum_{i \in \alpha} (M_i^* - m_i^*)\Delta x_i + \sum_{i \in \beta} (M_i^* - m_i^*)\Delta x_i$$

$$< \frac{\varepsilon}{2(b-a)} \sum_{i \in \alpha} \Delta x_i + 2K \sum_{i \in \beta} \Delta x_i$$

$$< \frac{\varepsilon(b-a)}{2(b-a)} + \frac{2K\varepsilon}{4K}$$

$$= \frac{\varepsilon}{2} + \frac{\varepsilon}{2}$$

$$= \varepsilon. \tag{7.5}$$

Summarizing,

$$S^P_{(g \circ f)} - S^P_{(g \circ f)} < \varepsilon.$$

Since $\varepsilon > 0$ is arbitrary, from Theorem 7.2.7 we may conclude that $(g \circ f)$ is Riemann integrable.

The proof is complete.

Proposition 7.3.5 *Let* $f_1, f_2 : [a, b] \to \mathbb{R}$ *be Riemann integrable functionals.*
Under such hypotheses:

1. $f_1 \cdot f_2$ *is Riemann integrable.*
2. $|f_1|$ *is Riemann integrable and*

$$\left| \int_a^b f_1(x) \, dx \right| \leq \int_a^b |f_1(x)| \, dx.$$

Proof

1. With $g(t) = t^2$, from the last theorem and previous results, we have that

$$f_1 \cdot f_2 = \frac{(f_1 + f_2)^2 - (f_1 - f_2)^2}{4}$$

   is Riemann integrable.
2. With $g(t) = |t|$ from the last Theorem $|f_1| = (g \circ f_1)$ is Riemann integrable.
   Moreover since

$$\pm f_1(x) \leq |f_1(x)|, \forall x \in [a, b],$$

   we obtain

$$\pm \int_a^b f_1(x) \, dx \leq \int_a^b |f_1(x)| \, dx$$

so that

$$\left| \int_a^b f_1(x) \, dx \right| \le \int_a^b |f_1(x)| \, dx.$$

The proof is complete.

Theorem 7.3.6 *Let* $f : [a, b] \to \mathbb{R}$ *be a continuous function on* $[a, b]$.
Under such hypotheses f *is Riemann integrable on* $[a, b]$.

Proof Let $\varepsilon > 0$. Since f is continuous on $[a, b]$ and $[a, b]$ is compact, we have
that f is uniformly continuous on $[a, b]$. Thus, there exists $\delta > 0$ such that if $x, y \in$
$[a, b]$ and $|x - y| < \delta$, then

$$|f(y) - f(x)| < \frac{\varepsilon}{2(b - a)}.$$

Let P be a partition of $[a, b]$ such that $0 < |P| < \delta$. Hence,

$$x_i - x_{i-1} < \delta, \quad \forall i \in \{1, \ldots, n\}.$$

Thus,

$$M_i - m_i \le \frac{\varepsilon}{2(b - a)}, \quad \forall i \in \{1, \ldots, n\},$$

so that

$$\begin{aligned}
S_f^P - s_f^P &= \sum_{i=1}^n (M_i - m_i) \Delta x_i \\
&\le \sum_{i=1}^n \frac{\varepsilon}{2(b - a)} \Delta x_i \\
&= \frac{\varepsilon (b - a)}{2(b - a)} \\
&= \frac{\varepsilon}{2} \\
&< \varepsilon.
\end{aligned} \tag{7.6}$$

Since $\varepsilon > 0$ is arbitrary, from this and Theorem 7.2.7, f is integrable on $[a, b]$.

Theorem 7.3.7 *Let* $f : [a, b] \to \mathbb{R}$ *be such that* $|f(x)| < K$, $\forall x \in [a, b]$ *for some*
$K > 0$. *Assume* f *is integrable on* $[a, b]$. *Let* $x_0 \in (a, b)$. *Under such hypotheses,*
f *is integrable on* $[a, x_0]$ *and* $[x_0, b]$ *and*

$$\int_a^b f(x) \, dx = \int_0^{x_0} f(x) \, dx + \int_{x_0}^b f(x) \, dx.$$

Proof We denote,

$$\underline{I_1} = \sup\{s_f^{P_1} \; : \; P_1 \text{ is a partition of } [a, x_0]\},$$

$$\overline{I_1} = \inf\{S_f^{P_1} \; : \; P_1 \text{ is a partition of } [a, x_0]\},$$

$$\underline{I_2} = \sup\{s_f^{P_2} \; : \; P_2 \text{ is a partition of } [x_0, b]\},$$

$$\overline{I_2} = \inf\{S_f^{P_2} \; : \; P_2 \text{ is a partition of } [x_0, b]\}.$$

Let $\varepsilon > 0$. Since f is integrable, there exists a partition P of $[a, b]$ such that

$$I - \frac{\varepsilon}{2} < s_f^P \le S_f^P < I + \frac{\varepsilon}{2}.$$

Define $Q = P \cup \{x_0\}$. Thus we may write

$$Q = P_1 \cup P_2,$$

where P_1 is an appropriate partition of $[a, x_0]$ and P_2 is a partition of $[x_0, b]$. Observe that

$$s_f^Q = s_f^{P_1} + s_f^{P_2},$$

and

$$S_f^Q = S_f^{P_1} + S_f^{P_2},$$

so that

$$\begin{aligned}
I &- \frac{\varepsilon}{2} \\
&< s_f^P \le s_f^Q \\
&= s_f^{P_1} + s_f^{P_2} \\
&\le \underline{I_1} + \underline{I_2} \\
&\le \overline{I_1} + \overline{I_2} \\
&\le S_f^{P_2} + S_f^{P_2} \\
&= S_f^Q \le S_f^P \\
&< I + \frac{\varepsilon}{2}.
\end{aligned} \tag{7.7}$$

From this, we obtain,

$$\overline{I_1} + \overline{I_2} - (\underline{I_1} + \underline{I_2}) < \varepsilon,$$

that is,

$$(\overline{I_1} - \underline{I_1}) + (\overline{I_2} - \underline{I_2}) < \varepsilon,$$

and since, $\overline{I_1} - \underline{I_1} \geq 0$ and $\overline{I_2} - \underline{I_2} \geq 0$ we have,

$$\overline{I_1} - \underline{I_1} < \varepsilon,$$

and

$$\overline{I_2} - \underline{I_2} < \varepsilon.$$

Since $\varepsilon > 0$ is arbitrary, we obtain

$$\overline{I_1} = \underline{I_1} \equiv I_1 = \int_a^{x_0} f(x)\, dx,$$

and

$$\overline{I_2} = \underline{I_2} \equiv I_2 = \int_{x_0}^b f(x)\, dx.$$

Finally, from this and (7.7) we also obtain

$$I - \frac{\varepsilon}{2} < I_1 + I_2 < I + \frac{\varepsilon}{2},$$

that is

$$|I_1 + I_2 - I| < \frac{\varepsilon}{2}.$$

Since $\varepsilon > 0$ is arbitrary, we obtain

$$I_1 + I_2 = I,$$

that is,

$$\int_a^b f(x)\, dx = \int_0^{x_0} f(x)\, dx + \int_{x_0}^b f(x)\, dx.$$

The proof is complete.

Theorem 7.3.8 (The Mean Value Theorem for Integrals) *Let* $f : [a, b] \to \mathbb{R}$ *be a continuous function on* $[a, b]$.

Under such hypotheses, there exists $x_0 \in [a, b]$ *such that*

$$\int_a^b f(x)\, dx = f(x_0)(b - a).$$

Proof Denote

$$m = \min\{ f(x) \; : \; x \in [a, b]\}$$

and

$$M = \max\{ f(x) \; : \; x \in [a, b]\}.$$

Thus,

$$m \le f(x) \le M, \forall x \in [a, b].$$

Observe that

$$m(b - a) \le \int_a^b f(x)\, dx \le M(b - a). \tag{7.8}$$

Define $g : [a, b] \to \mathbb{R}$ by

$$g(x) = f(x)(b - a).$$

Therefore,

$$\min\{g(x) \; : \; x \in [a, b]\} = m(b - a)$$

$$\le \int_a^b f(x)\, dx$$

$$\le M(b - a)$$

$$= \max\{g(x) \; : \; x \in [a, b]\}. \tag{7.9}$$

Since g continuous, from this and the intermediate value theorem, we have that there exists $x_0 \in [a, b]$ such that

$$g(x_0) = f(x_0)(b - a) = \int_a^b f(x)\, dx.$$

The proof is complete.

Theorem 7.3.9 (Fundamental Theorem of Calculus, First Part) *Let* f : $[a, b] \to \mathbb{R}$ *be such that* $|f(x)| < K, \; \forall x \in [a, b]$ *for some* $K > 0$. *Assume* f *is continuous on* $[a, b]$ *and define*

$$F(x) = \int_a^x f(t) \, dt, \forall x \in [a, b]$$

Under such hypotheses

$$F'(x) = f(x), \quad \forall x \in [a, b].$$

Proof Let $x \in (a, b)$. Let $\varepsilon > 0$. Since f is continuous at x, there exists $\delta > 0$ such that if $|t - x| < \delta$, then $|f(x) - f(t)| < \varepsilon$, that is

$$f(x) - \varepsilon < f(t) < f(x) + \varepsilon.$$

Let $0 < h < \delta$.
Thus,

$$\frac{F(x + h) - F(x)}{h} = \frac{\int_a^{x+h} f(t) \, dt - \int_a^x f(t) \, dt}{h}$$

$$= \frac{\int_x^{x+h} f(t) \, dt}{h}, \tag{7.10}$$

so that

$$\frac{(f(x) - \varepsilon) \int_x^{x+h} dt}{h} \leq \frac{F(x + h) - F(x)}{h}$$

$$= \frac{\int_x^{x+h} f(t) \, dt}{h}$$

$$\leq \frac{(f(x) + \varepsilon) \int_x^{x+h} dt}{h}, \tag{7.11}$$

and hence

$$\frac{(f(x) - \varepsilon)h}{h} < \frac{F(x + h) - F(x)}{h} < \frac{(f(x) + \varepsilon)h}{h}, \tag{7.12}$$

that is,

$$f(x) - \varepsilon < \frac{F(x + h) - F(x)}{h} < f(x) + \varepsilon, \forall 0 < h < \delta,$$

so that

$$f(x)-\varepsilon \le \liminf_{h\to 0}\left(\frac{F(x+h)-F(x)}{h}\right) \le \limsup_{h\to 0}\left(\frac{F(x+h)-F(x)}{h}\right) \le f(x)+\varepsilon.$$

Since $\varepsilon > 0$ is arbitrary, we may infer that

$$f(x) = \liminf_{h\to 0}\left(\frac{F(x+h)-F(x)}{h}\right) = \limsup_{h\to 0}\left(\frac{F(x+h)-F(x)}{h}\right),$$

that is

$$f(x) = \lim_{h\to 0}\left(\frac{F(x+h)-F(x)}{h}\right) = F'(x).$$

The case in which either $x = a$ or $x = b$ is dealt similarly with appropriate one-sided limits.

This completes the proof.

Theorem 7.3.10 (Fundamental Theorem of Calculus, Second Part) *Let $G :$ $[a, b] \to \mathbb{R}$ be such that G' is continuous on $[a, b]$.*

Under such hypotheses

$$G(y) - G(x) = \int_x^y G'(t)\, dt, \quad \forall x, y \in [a, b].$$

Proof Define

$$H(x) = \int_a^x G'(t)\, dt, \quad \forall x \in [a, b].$$

From the first part,

$$H'(x) = G'(x), \quad \forall x \in [a, b].$$

Hence $G(x) = H(x) + c, \forall x \in [a, b]$ for some $c \in \mathbb{R}$. In particular

$$G(a) = H(a) + c = 0 + c = c,$$

so that, fixing $x \in [a, b]$, we have,

$$G(x) - G(a) = G(x) - c = H(x) = \int_a^x G'(t)\, dt.$$

Now let $y \in [a, b]$, $y \ge x$.

From the last equation

$$G(y) - G(a) = \int_a^y G'(t)\,dt.$$

Thus,

$$
\begin{aligned}
G(y) - G(x) &= (G(y) - G(a)) - (G(x) - G(a)) \\
&= \int_a^y G'(t)\,dt - \int_a^x G'(t)\,dt \\
&= \int_a^x G'(t)\,dt + \int_x^y G'(t)\,dt - \int_a^x G'(t)\,dt \\
&= \int_x^y G'(t)\,dt.
\end{aligned}
\tag{7.13}
$$

The proof is complete.

7.4   A Criterion of Riemann Integrability

In this section we present a condition necessary and sufficient for Riemann integrability. We start with the definition of exterior and Lebesgue measures.

Definition 7.4.1 (Exterior Measure) Given an open interval (a, b), we define its length, denoted by $l((a, b))$, by $l((a, b)) = b - a$. Given a set $A \subset \mathbb{R}$, we define its exterior measure, denoted by $m^*(A)$ by

$$m^*(A) = \inf\left\{ \sum_{n=1}^{\infty} l(I_n) \ : \ A \subset \cup_{i=1}^n I_n \right\},$$

where I_n is an open interval $\forall n \in \mathbb{N}$.

Definition 7.4.2 (Measurable Set) A set $E \subset \mathbb{R}$ is said to be Lebesgue measurable, if for each $A \subset \mathbb{R}$ we have

$$m^*(A) = m^*(A \cap E) + m^*(A \cap E^c),$$

where

$$E^c = \mathbb{R} \setminus E = \{x \in \mathbb{R} \ : \ x \notin E\}.$$

Thus, if E is Lebesgue measurable, we shall define $m(E) = m^*(E)$ where $m(E)$ is said to be the Lebesgue measure of E.

At this point we present some results relating the exterior measure.

Proposition 7.4.3 *Let $A \subset B \subset \mathbb{R}$. Under such assumptions, we have*

$$m^*(A) \leq m^*(B).$$

Proof If $m^*(B) = +\infty$ the result follows immediately, since in such a case

$$m^*(A) \leq +\infty = m^*(B).$$

Thus, assume $m^*(B) = \alpha < +\infty$.

Let $\varepsilon > 0$ be given.

Since $\alpha + \varepsilon > \alpha = m^*(B)$, there exists a sequence of open intervals $\{I_m\}$ such that

$$B \subset \cup_{m=1}^{\infty} I_m$$

and

$$\alpha \leq \sum_{m=1}^{\infty} l(I_m) < \alpha + \varepsilon.$$

Therefore,

$$A \subset B \subset \cup_{m=1}^{\infty} I_m,$$

so that

$$m^*(A) \leq \sum_{m=1}^{\infty} l(I_m) < \alpha + \varepsilon.$$

Hence,

$$m^*(A) < \alpha + \varepsilon.$$

Since $\varepsilon > 0$ is arbitrary, we may infer that

$$m^*(A) \leq \alpha = m^*(B).$$

The proof is complete.

Also very important is the next result.

Proposition 7.4.4 *Let $A, B \subset \mathbb{R}$. Under such hypotheses, we have*

$$m^*(A \cup B) \leq m^*(A) + m^*(B).$$

Proof If $m^*(A) = +\infty$ or $m^*(B) = +\infty$, the result follows immediately, since in such a case,

$$m^*(A \cup B) \le +\infty = m^*(A) + m^*(B).$$

Thus, suppose $m^*(A) = \alpha_1 < +\infty$ and $m^*(B) = \alpha_2 < +\infty$.
Let $\varepsilon > 0$.
Since $\alpha_1 + \varepsilon/2 > \alpha_1$ and $\alpha_2 + \varepsilon/2 > \alpha_2$ there exist sequences $\{I_m\}$ and $\{J_m\}$ of open real intervals such that

$$A \subset \cup_{m=1}^{\infty} I_m \text{ and } B \subset \cup_{m=1}^{\infty} J_m,$$

and also,

$$\alpha_1 \le \sum_{m=1}^{\infty} l(I_m) < \alpha_1 + \varepsilon/2 \text{ and } \alpha_2 \le \sum_{m=1}^{\infty} l(J_m) < \alpha_2 + \varepsilon/2.$$

Observe that

$$A \cup B \subset [\cup_{m=1}^{\infty} I_m] \cup [\cup_{m=1}^{\infty} J_m],$$

so that

$$m^*(A \cup B) \le \sum_{m=1}^{\infty} l(I_m) + \sum_{m=1}^{\infty} l(J_m)$$
$$< \alpha_1 + \varepsilon/2 + \alpha_2 + \varepsilon/2$$
$$= m^*(A) + m^*(B) + \varepsilon. \qquad (7.14)$$

Since $\varepsilon > 0$ is arbitrary, we may infer that

$$m^*(A \cup B) \le m^*(A) + m^*(B).$$

This completes the proof.

Proposition 7.4.5 *Let $\{E_j\} \subset \mathbb{R}$ be a sequence of real sets. Under such assumptions,*

$$m^*(\cup_{j=1}^{\infty} E_j) \le \sum_{j=1}^{\infty} m^*(E_j).$$

Proof If $m^*(E_j) = +\infty$ for some $j \in \mathbb{N}$, then the result follows immediately, since in such a case,

$$m^*(\cup_{j=1}^{\infty} E_j) \le +\infty = \sum_{j=1}^{\infty} m^*(E_j).$$

Thus, assume

$$m^*(E_j) = \alpha_j < +\infty, \ \forall j \in \mathbb{N}.$$

Since $\alpha_j + \frac{\varepsilon}{2^j} > \alpha_j$, there exists a sequence $\{I_m^j\}_{m \in \mathbb{N}}$ of open intervals such that

$$E_j \subset \cup_{m=1}^\infty I_m^j,$$

and

$$\alpha_j \leq \sum_{m=1}^\infty l(I_m^j) < \alpha_j + \frac{\varepsilon}{2^j},$$

$\forall j \in \mathbb{N}$.
  Therefore,

$$\cup_{j=1}^\infty E_j \subset \cup_{j=1}^\infty [\cup_{m=1}^\infty I_m^j],$$

and

$$m^*(\cup_{j=1}^\infty E_j) \leq \sum_{j=1}^\infty \left(\sum_{m=1}^\infty l(I_m^j) \right)$$

$$< \sum_{j=1}^\infty \left(\alpha_j + \frac{\varepsilon}{2^j} \right)$$

$$= \sum_{j=1}^\infty \alpha_j + \sum_{j=1}^\infty \frac{\varepsilon}{2^j}$$

$$= \sum_{j=1}^\infty \alpha_j + \varepsilon. \tag{7.15}$$

Hence,

$$m^*(\cup_{j=1}^\infty E_j) < \sum_{j=1}^\infty m^*(E_j) + \varepsilon.$$

Since $\varepsilon > 0$ is arbitrary, we may infer that

$$m^*(\cup_{j=1}^\infty E_j) \leq \sum_{j=1}^\infty m^*(E_j).$$

The proof is complete.

Remark 7.4.6 We shall show that if $m^*(E) = 0$, then E is Lebesgue measurable. Indeed, let $E \subset \mathbb{R}$ be such that $m^*(E) = 0$.

Let $A \subset \mathbb{R}$. Since $A \cap E \subset E$ we obtain $m^*(A \cap E) = 0$.

Thus

$$m^*(A \cap E) + m^*(A \cap E^c) = m^*(A \cap E^c) \leq m^*(A). \tag{7.16}$$

Since $m^*(B \cup C) \leq m^*(B) + m^*(C), \forall B, C \subset \mathbb{R}$ we obtain

$$m^*(A) = m^*((A \cap E) \cup (A \cap E^c)) \leq m^*(A \cap E) + m^*(A \cap E^c),$$

so that from this and (7.16) we get

$$m^*(A) = m^*(A \cap E) + m^*(A \cap E^c), \quad \forall A \subset \mathbb{R},$$

and thus E is Lebesgue measurable whenever $m^*(E) = 0$.

Also we recall that in this case

$$0 = m^*(E) = \inf \left\{ \sum_{n=1}^{\infty} l(I_n) \ : \ A \subset \cup_{i=1}^{\infty} I_n \right\},$$

so that, for each $\varepsilon > 0$ we may find a sequence $\{I_n\}$ of real open intervals such that

$$E \subset \cup_{n=1}^{\infty} I_n$$

and

$$\sum_{n=1}^{\infty} l(I_n) < \varepsilon.$$

Moreover, given an open interval (a, b), it is not difficult to prove that (a, b) is Lebesgue measurable and in this case $m((a, b)) = l((a, b)) = (b - a)$. Finally, we may also obtain for a closed interval $m([a, b]) = (b - a)$.

Theorem 7.4.7 *Let $f : [a, b] \to \mathbb{R}$ be such that $|f(x)| < K$, $\forall x \in [a, b]$ for some $K > 0$. Let $A \subset [a, b]$ be the set on which f is not continuous. Assume that $m(A) = 0$ (the Lebesgue measure of A is zero). Under such hypotheses, f is Riemann integrable.*

Proof From the hypotheses $m(A) = 0$. Let $\varepsilon > 0$. Thus, there exists a sequence $\{I_n\}$ of open intervals such that $A \subset \cup_{n=1}^{\infty} I_n$ and

$$\sum_{n=1}^{\infty} m(I_n) < \frac{\varepsilon}{8K}.$$

Let $B = [a, b] \setminus A$. Hence f is continuous on B. Thus, for each $x \in B$ there exists $\delta_x > 0$ such that if $y \in [a, b]$ and $|y - x| < \delta_x$, then

$$|f(y) - f(x)| < \frac{\varepsilon}{4|b - a|}.$$

Thus, denoting $V_{\delta_x} = (x - \delta_x/2, x + \delta_x/2)$ we have that if $y \in V_{\delta_x} \cap [a, b]$, then

$$|f(y) - f(x)| < \frac{\varepsilon}{4|b - a|}. \tag{7.17}$$

Observe that

$$[a, b] \subset \left(\cup_{x \in B} V_{\delta_x}\right) \cup \left(\cup_{n=1}^{\infty} I_n\right),$$

and since $[a, b]$ is compact, there exists $x_1, \ldots, x_n \in B$ and $n_1, \ldots, n_k \in \mathbb{N}$ such that

$$[a, b] \subset \left(\cup_{i=1}^{n} V_{\delta_{x_i}}\right) \cup \left(\cup_{l=1}^{k} I_{n_l}\right).$$

Denote

$$C = \min\{\delta_{x_i}/2, \ i \in \{1, \ldots, n\}\},$$

and

$$D = \min\{m(I_{n_l})/2, \ l \in \{1, \ldots, k\}\},$$

and $\delta = \min\{C, D\}$.

Let $P = \{t_0 = a, t_1, \ldots, t_p = b\}$ be a partition of $[a, b]$ such that $0 < |P| < \delta$.

Denote by α the set of indices $j \in \{1, \ldots, p\}$ such that the interval $P_j = [t_{j-1}, t_j]$ of P intersects $\cup_{l=1}^{k} I_{n_l}$.

Also, denote

$$\beta = \{1, \ldots, p\} \setminus \alpha = \{j \in \{1, \ldots, p\} : j \notin \alpha\}.$$

Thus, if $j \in \beta$, $P_j = [t_{j-1}, t_j]$ does not intersect $\cup_{l=1}^{k} I_{n_l}$.

Let $j \in \alpha$.

Hence $P_j \cap I_{n_l} \neq \emptyset$ for some $l \in \{1, \ldots, k\}$. So, denoting $I_{n_l} = [c_l, d_l]$ since $\delta = \min\{C, D\}$, we have that

$$P_j \subset I_l \equiv [c_l - \delta, d_l + \delta]$$

where

$$m(I_l) = m(I_{n_l}) + 2\delta \leq 2m(I_{n_l}).$$

Therefore, we may conclude that

$$\cup_{j\in\alpha} P_j \subset \cup_{l=1}^{n} I_l$$

so that

$$\sum_{j\in\alpha} m(P_j) = m(\cup_{j\in\alpha} P_j) \leq m(\cup_{l=1}^{n}(I_l)) \leq \sum_{l=1}^{n} m(I_l) \leq 2\sum_{l=1}^{k} m(I_{n_l}) < \frac{\varepsilon}{4K}.$$

Now suppose $j \in \beta$. Thus, $P_j = [t_{j-1}, t_j]$ is such that $t_{j-1} \in V_{\delta_{x_{n_0}}}$ for some $n_0 \in \{1, \ldots, n\}$. Hence,

$$|x_{n_0} - t_{j-1}| < \delta_{x_{n_0}}/2$$

and since

$$|t_j - t_{j-1}| \leq \min\{C, D\} \leq \delta_{x_{n_0}}/2,$$

we obtain

$$\begin{aligned}
|t_j - x_{n_0}| &= |t_j - t_{j-1} + t_{j-1} - x_{n_0}| \\
&\leq |t_j - t_{j-1}| + |t_{j-1} - x_{n_0}| \\
&\leq \delta_{x_{n_0}}/2 + \delta_{x_{n_0}}/2 \\
&= \delta_{x_{n_0}}. \hspace{4cm} (7.18)
\end{aligned}$$

Therefore

$$P_j = [t_{j-1}, t_j] \subset (x_{n_0} - \delta_{x_{n_0}}, x_{n_0} + \delta_{x_{n_0}}).$$

Thus if $t, s \in P_j$, then

$$|f(t) - f(x_{n_0})| < \frac{\varepsilon}{4|b - a|}$$

and

$$|f(s) - f(x_{n_0})| < \frac{\varepsilon}{4|b - a|}$$

so that

$$|f(t) - f(s)| \leq |f(t) - f(x_{n_0})| + |f(x_{n_0}) - f(s)| \leq \frac{\varepsilon}{2|b - a|}, \forall t, s \in P_j$$

and thus, denoting, as usual,

$$M_j = \sup\{f(t) \, : \, t \in P_j\},$$

and

$$m_j = \inf\{f(t), \, : \, t \in P_j\},$$

we have,

$$M_j - m_j \leq \frac{\varepsilon}{2|b-a|},$$

whenever $j \in \beta$.

On the other hand, we have $M_j - m_j \leq 2K$ if $j \in \alpha$.

Thus,

$$
\begin{aligned}
S_f^P - s_f^P &= \sum_{j=1}^{P}(M_j - m_j)m(P_j) \\
&= \sum_{j \in \alpha}(M_j - m_j)m(P_j) + \sum_{j \in \beta}(M_j - m_j)m(P_j) \\
&\leq \sum_{j \in \alpha}2Km(P_j) + \sum_{j \in \beta}\frac{\varepsilon}{2|b-a|}m(P_j) \\
&< 2K\frac{\varepsilon}{4K} + |b-a|\frac{\varepsilon}{2|b-a|} \\
&= \frac{\varepsilon}{2} + \frac{\varepsilon}{2} \\
&= \varepsilon.
\end{aligned}
\tag{7.19}
$$

Summarizing,

$$S_f^P - s_f^P < \varepsilon.$$

Since $\varepsilon > 0$ is arbitrary, from the Theorem 7.2.7, f is Riemann integrable. The proof is complete.

7.4.1   Oscillation

Definition 7.4.8 (Oscillation) Let $f : [a, b] \rightarrow \mathbb{R}$ be such that $|f(x)| < K, \quad \forall x \in [a, b]$, for some $K > 0$.

Let $x \in [a, b]$.

We define the oscillation of f at x, denoted by $\omega_f(x)$ by

$$\omega_f(x) = \lim_{\delta \to 0^+} \sup_{y \in V_\delta(x)} \{|f(y) - f(x)|\},$$

where

$$V_\delta(x) = [a, b] \cap (x - \delta, x + \delta).$$

Remark 7.4.9 Observe that defining

$$g_x(\delta) = \sup_{y \in V_\delta(x)} \{|f(y) - f(x)|\},$$

we have that $g_x(\delta)$ is nonincreasing, so that

$$\omega_f(x) = \lim_{\delta \to 0^+} g_x(\delta)$$

is well defined as the one-sided limit as $\delta \to 0^+$ of a monotone function.

Theorem 7.4.10 *Let $f : [a, b] \to \mathbb{R}$ be such that $|f(x)| < K, \quad \forall x \in [a, b]$ for some $K > 0$. Let $x \in [a, b]$. Under such hypotheses, f is continuous at x if, and only if,*

$$\omega_f(x) = 0.$$

Proof Suppose f is continuous at x. Let $\varepsilon > 0$ be given. Thus there exists $\delta_0 > 0$ such that if $y \in [a, b]$ and $|y - x| < \delta_0$, then

$$|f(y) - f(x)| < \varepsilon.$$

Thus,

$$g_x(\delta) = \sup_{y \in V_\delta(x)} \{|f(y) - f(x)|\} < \varepsilon, \text{ if } 0 < \delta < \delta_0,$$

so that

$$\omega_f(x) = \lim_{\delta \to 0^+} g_x(\delta) \leq \varepsilon.$$

Since $\varepsilon > 0$ is arbitrary, we may conclude that,

$$\omega_f(x) = 0.$$

Reciprocally, suppose $\omega_f(x) = 0$. Let a new and not relabeled $\varepsilon > 0$ be given.

Thus there exists $\delta_0 > 0$ such that if $0 < \delta < \delta_0$, then $g_x(\delta) < \varepsilon$. Thus, for $\delta = \delta_0/2$ we have that $|f(y) - f(x)| < \varepsilon, \forall y \in V_\delta(x)$.

We may conclude that

$$\lim_{y \to x} f(y) = f(x),$$

so that f is continuous at x.

The proof is complete.

Theorem 7.4.11 *Let $f : [a, b] \to \mathbb{R}$ be such that $|f(x)| < K$, $\forall x \in [a, b]$ for some $K > 0$. Suppose f is Riemann integrable. Let $A \subset [a, b]$ be the subset in which f is not continuous. Under such hypotheses, $m(A) = 0$.*

Proof Observe that $x \in A$ if, and only if, f is not continuous at x. In such a case, from the last theorem, $\omega_f(x) > \frac{1}{n}$ for some $n \in \mathbb{N}$. Define

$$B_n = \left\{ x \in [a, b] \ : \ \omega_f(x) > \frac{1}{n} \right\}, \forall n \in \mathbb{N}.$$

Then

$$A = \cup_{n=1}^\infty B_n,$$

Let $n \in \mathbb{N}$. We shall prove that $m^*(B_n) = 0$.

Indeed, suppose given $\varepsilon > 0$. Since f is integrable, there exists a partition $P = \{x_0 = a, x_1, \ldots, x_k = b\}$ of $[a, b]$ such that

$$S_f^P - s_f^P < \varepsilon.$$

Denote by α the set of indices $j \in \{1, \ldots, k\}$ such that the interval $P_j = [x_{j-1}, x_j]$ intersects B_n.

Observe that if $j \in \alpha$, then $M_j - m_j > \frac{1}{n}$.

Moreover, $\cup_{j \in \alpha} P_j \supset B_n$ so that

$$\sum_{j \in \alpha} m^*(P_j) \geq m^*(\cup_{j \in \alpha} P_j) \geq m^*(B_n),$$

where we recall that

$$m^*(P_j) = m(P_j) = \Delta x_j, \forall j \in \{1, \ldots, k\}.$$

Hence

$$\varepsilon > S_f^P - s_f^P$$

$$\geq \sum_{j \in \alpha} (M_j - m_j) m^*(P_j)$$

$$> \frac{1}{n} \sum_{j \in \alpha} m^*(P_j)$$

$$\geq \frac{1}{n} m^*(B_n). \tag{7.20}$$

Summarizing,

$$m^*(B_n) < n\varepsilon.$$

Since $\varepsilon > 0$ is arbitrary, we may infer that $m^*(B_n) = 0$, $\forall n \in \mathbb{N}$. Hence,

$$m^*(A) = m^*(\cup_{n=1}^{\infty} B_n) \leq \sum_{n=1}^{\infty} m^*(B_n) = 0,$$

so that $m^*(A) = 0$, and thus $m(A) = 0$.

The proof is complete.

Definition 7.4.12 Let $f : [a, b] \to \mathbb{R}$ be such that $|f(x)| < K$, $\forall x \in [a, b]$ for some $K > 0$. We say that I is the limit of R_f^P as $|P| \to 0$ as for each $\varepsilon > 0$ there exists $\delta > 0$ such that if $0 < |P| < \delta$ then

$$|R_f^P - I| < \varepsilon,$$

for any Riemann sum R_f^P relating P.

In this case, we denote,

$$\lim_{|P| \to 0} R_f^P = I.$$

Theorem 7.4.13 *Let $f : [a, b] \to \mathbb{R}$ be such that $|f(x)| < K$, $\forall x \in [a, b]$ for some $K > 0$. Suppose f is Riemann integrable.*

Under such hypotheses, we have

$$\lim_{|P| \to 0} R_f^P = I.$$

Reciprocally, suppose there exists $I \in \mathbb{R}$ such that

$$\lim_{|P| \to 0} R_f^P = I.$$

Under such hypotheses, f is integrable and

$$I = \int_a^b f(x)\, dx.$$

Proof Assume f is Riemann integrable.

Suppose given $\varepsilon > 0$. Thus, there exists a partition $P = \{x_0 = a, x_1, x_2, \ldots, x_n\}$ of $[a, b]$ such that

$$I - \frac{\varepsilon}{2} < s_f^P \leq S_f^P < I + \frac{\varepsilon}{2}.$$

Choose $\delta \in \mathbb{R}$ such that

$$0 < \delta < \min\left\{\frac{\varepsilon}{4K[2(n+1)]}, \frac{|P|}{2}\right\}.$$

Let $P_1 = \{t_0 = a, t_1, \ldots, t_k = b\}$ be a partition of $[a, b]$ such that

$$|P_1| < \delta.$$

Denote by α the set of indices $j \in \{1, \ldots, k\}$ such that $P_{1_j} = [t_{j-1}, t_j]$ intersects the partition P.

Observe that each $x_i \in P$ intersects at most two intervals $P_{1_j} = [t_{j-1}, t_j]$ so that α has at most $2(n+1)$ elements, where, as above indicated, $n+1$ is the number of elements of P.

Therefore, if $j \in \{1, \ldots, k\}$ and $j \notin \alpha$, then $P_{1_j} = [t_{j-1}, t_j]$ is contained in the interior of some interval $P_i = [x_{i-1}, x_i]$ relating P.

Thus, we define

$$\beta = \{1, \ldots, k\} \setminus \alpha = \{j \in \{1, \ldots, k\} \ : \ j \notin \alpha\}.$$

Also, define

$$\beta_i = \{j \in \beta \ : \ P_{1_j} \subset P_i\}, \quad \forall i \in \{1, \ldots, n\}.$$

Observe that

$$S_f^P - s_f^P = \sum_{i=1}^n (M_i - m_i) \Delta x_i$$

$$\geq \sum_{i=1}^n \left(\sum_{j \in \beta_i} (M_j^* - m_j^*) \Delta t_j \right)$$

$$= \sum_{j \in \beta} (M_j^* - m_j^*) \Delta t_j. \tag{7.21}$$

where, as usual, we have denoted

$$m_i = \inf\{f(x) \ : \ x \in [x_{i-1}, x_i]\},$$

$$M_i = \sup\{f(x) \ : \ x \in [x_{i-1}, x_i]\},$$

$$m_j^* = \inf\{f(x) \ : \ x \in [t_{j-1}, t_j]\},$$

$$M_j^* = \sup\{f(x) \ : \ x \in [t_{j-1}, t_j]\}.$$

Therefore, we have,

$$
\begin{aligned}
S_f^{P_1} - s_f^{P_1} &= \sum_{j \in \alpha} (M_j^* - m_j^*)\Delta t_j + \sum_{j \in \beta}(M_j^* - m_j^*)\Delta t_j \\
&\leq 2K[2(n+1)]\delta + S_f^P - s_f^P \\
&< \frac{2K\varepsilon}{4K} + \frac{\varepsilon}{2} \\
&= \frac{\varepsilon}{2} + \frac{\varepsilon}{2} \\
&= \varepsilon.
\end{aligned}
\tag{7.22}
$$

Summarizing

$$S_f^{P_1} - s_f^{P_1} < \varepsilon. \tag{7.23}$$

Observe that we have also,

$$s_f^{P_1} \leq R_f^{P_1} \leq S_f^{P_1},$$

and

$$-S_f^{P_1} \leq -I \leq -s_f^{P_1},$$

so that, from this and (7.23), we obtain,

$$-\varepsilon < -S_f^{P_1} + s_f^{P_1} \leq R_f^{P_1} - I \leq S_f^{P_1} - s_f^{P_1} < \varepsilon,$$

that is,

$$-\varepsilon < R_f^{P_1} - I < \varepsilon.$$

Thus, we have got,

$$|R_f^{P_1} - I| < \varepsilon,$$

whenever $0 < |P_1| < \delta$, that is,

$$\lim_{|P|\to 0} R_f^P = I.$$

Conversely, suppose

$$\lim_{|P|\to 0} R_f^P = I.$$

Suppose given a not relabeled $\varepsilon > 0$. Thus there exists $\delta > 0$ such that $|P| < \delta$ then

$$I - \frac{\varepsilon}{2} < R_f^P < I + \frac{\varepsilon}{2},$$

for each Riemann sum relating P.

Choosing a particular P such that $0 < |P| < \delta$, since s_f^P and S_f^P are also arbitrarily close to Riemann sums, we have,

$$I - \frac{\varepsilon}{2} \le s_f^P \le \underline{I_1} \le \overline{I_1} \le S_f^P \le I + \frac{\varepsilon}{2},$$

where

$$\underline{I_1} = \sup\{s_f^P \; : \; P \text{ is partition of } [a, b]\},$$

and

$$\overline{I_1} = \inf\{S_f^P \; : \; P \text{ is partition of } [a, b]\},$$

Thus we have got,

$$\overline{I_1} - \underline{I_1} \le \varepsilon,$$

and since $\varepsilon > 0$ is arbitrary, we obtain

$$\underline{I_1} = \overline{I_1} \equiv I_1 = \int_a^b f(x)\, dx,$$

that is, f is Riemann integrable.

From this and above,

$$I - \frac{\varepsilon}{2} \le I_1 \le I + \frac{\varepsilon}{2},$$

that is,

$$|I - I_1| \le \frac{\varepsilon}{2}.$$

Since $\varepsilon > 0$ is arbitrary, we have got,

$$I = I_1 = \int_a^b f(x)\, dx.$$

This completes the proof.

7.5 Sequences and Series of Functions

We start with the following definition:

Definition 7.5.1 (Continuity in a Metric Space) Let (X, d_X) and (Y, d_Y) be metric spaces. A function $f : X \to Y$ is said to be continuous at $x_0 \in X$, if for each $\varepsilon > 0$ there exists $\delta > 0$ such that if

$$d_X(x, x_0) < \delta,$$

then

$$d_Y(f(x), f(x_0)) < \varepsilon.$$

Theorem 7.5.2 *Let X be a metric space and let $E \subset X$. Let $\{f_n : E \to \mathbb{R}\}$ and $f : E \to \mathbb{R}$, be such that:*
  f_n is continuous, $\forall n \in \mathbb{N}$, and

$$f_n \to f$$

uniformly on E.
  Under such hypotheses, f is continuous on E.

Proof Let $\varepsilon > 0$ be given.
  From the hypotheses, there exists $n_0 \in \mathbb{N}$ such that

$$|f_n(t) - f(t)| < \frac{\varepsilon}{3}, \ \forall n > n_0, \ t \in E.$$

Select $n_1 > n_0$.
  Let $x \in E \cap E'$.
  Since $f_{n_1} : E \to \mathbb{R}$ is continuous, there exists $\delta > 0$ such that if $t \in E$ and $d(x, t) < \delta$, then

$$|f_{n_1}(x) - f_{n_1}(t)| \leq \frac{\varepsilon}{3}.$$

Hence, if $t \in E$ and $d(x, t) < \delta$, we have,

$$
\begin{aligned}
|f(x) - f(t)| &= |f(x) - f_{n_1}(x) + f_{n_1}(x) - f_{n_1}(t) + f_{n_1}(t) - f(t)| \\
&\leq |f(x) - f_{n_1}(x)| + |f_{n_1}(x) - f_{n_1}(t)| + |f_{n_1}(t) - f(t)| \\
&\leq \frac{\varepsilon}{3} + \frac{\varepsilon}{3} + \frac{\varepsilon}{3} \\
&= \varepsilon.
\end{aligned}
\tag{7.24}
$$

Therefore,

$$
\lim_{t \to x} f(t) = f(x),
$$

$\forall x \in E \cap E'$, so that f is continuous on E.

Theorem 7.5.3 *Let* $\{f_n : [a, b] \to \mathbb{R}\}$ *and* $f : [a, b] \to \mathbb{R}$, *be such that:* f_n *is Riemann integrable,* $\forall n \in \mathbb{N}$, *and*

$$
f_n \to f
$$

uniformly on $[a, b]$.

Under such hypotheses, f *is Riemann integrable and*

$$
\lim_{n \to \infty} \int_a^b f_n \, dx = \int_a^b f \, dx.
$$

Proof Let $\varepsilon > 0$. From the hypotheses, there exists $n_0 \in \mathbb{N}$ such that if $n > n_0$, then

$$
|f_n(t) - f(t)| < \frac{\varepsilon}{2(b - a)}, \quad \forall t \in [a, b].
$$

Thus, selecting $n > n_0$ we have

$$
f_n(t) - \frac{\varepsilon}{2(b - a)} < f(t) < f_n(t) + \frac{\varepsilon}{2(b - a)}, \forall t \in [a, b],
$$

so that

$$
\int_a^b f_n(t) \, dt - \frac{\varepsilon}{2} \leq \underline{I} \leq \overline{I} \leq \int_a^b f_n(t) \, dt + \frac{\varepsilon}{2},
\tag{7.25}
$$

where,

$$
\underline{I} = \sup\{s_f^P \; : \; P \text{ is a partition of } [a, b]\},
$$

and

$$\overline{I} = \sup\{s_f^P \ : \ P \text{ is a partition of } [a, b]\}.$$

From the last chain of inequalities, we have,

$$\overline{I} - \underline{I} < \varepsilon.$$

Since $\varepsilon > 0$ is arbitrary, we obtain,

$$\overline{I} = \underline{I} = I = \int_a^b f \, dx.$$

From this and above, if $n > n_0$ we obtain,

$$\int_a^b f_n(t) \, dt - \frac{\varepsilon}{2} \le \underline{I} = \overline{I} = I \le \int_a^b f_n(t) \, dt + \frac{\varepsilon}{2}, \ \forall n > n_0. \qquad (7.26)$$

Therefore,

$$\left| I - \int_a^b f_n(t) \, dt \right| < \varepsilon, \ \forall n > n_0,$$

so that

$$I = \lim_{n \to \infty} \int_a^b f_n \, dx.$$

The proof is complete.

Theorem 7.5.4 (Stone–Weierstrass) *Let $f : [0, 1] \to \mathbb{R}$, be a real continuous function such that $f(0) = f(1) = 0$. Under such hypotheses, there exists a sequence of polynomials $\{P_n\}$ such that*

$$P_n \to f$$

uniformly on $[0, 1]$.

Proof Define

$$f(x) = 0, \forall x \in \mathbb{R} \setminus [0, 1].$$

Thus, f is uniformly continuous on $\mathbb{R}$.
Define $R_n : [-1, 1] \to \mathbb{R}$ by

$$R_n(x) = c_n(1 - x^2)^n, \forall n \in \mathbb{N},$$

where $\{c_n\}$ is such that

$$\int_{-1}^{1} R_n(x)\, dx = 1, \ \forall n \in \mathbb{N}.$$

Observe that,

$$
\begin{aligned}
\int_{-1}^{1} (1 - x^2)^n\, dx &= 2 \int_{0}^{1} (1 - x^2)^n\, dx \\
&\geq 2 \int_{0}^{1/\sqrt{n}} (1 - x^2)^n\, dx \\
&\geq 2 \int_{0}^{1/\sqrt{n}} (1 - n x^2)\, dx \\
&= 2 \left[x - n\frac{x^3}{3} \right]_{0}^{1/\sqrt{n}} \\
&= 2 \left(\frac{1}{\sqrt{n}} - n\frac{(1/\sqrt{n})^3}{3} \right) \\
&= 2 \left(\frac{1}{\sqrt{n}} - \frac{1}{3\sqrt{n}} \right) \\
&= \frac{4}{3\sqrt{n}} \\
&> \frac{1}{\sqrt{n}},
\end{aligned}
\tag{7.27}
$$

so that from

$$\int_{-1}^{1} R_n(x)\, dx = 1, \ \forall n \in \mathbb{N},$$

we obtain

$$c_n < \sqrt{n}, \ \forall n \in \mathbb{N}.$$

Observe also that, for each $0 < \delta < 1$, we have

$$
\begin{aligned}
R_n(x) &\leq c_n (1 - \delta^2)^n \\
&\leq \sqrt{n}(1 - \delta^2)^n,
\end{aligned}
\tag{7.28}
$$

$\forall x$ such that $\delta \leq x \leq 1$.

Observe that,

$$\lim_{n \to +\infty} \sqrt{n}(1 - \delta^2)^n = \lim_{x \to +\infty} \sqrt{x}(1 - \delta^2)^x$$

$$= \lim_{x \to +\infty} \frac{\sqrt{x}}{(1 - \delta^2)^{(-x)}}$$

$$= \lim_{x \to +\infty} \frac{(\sqrt{x})'}{\left((1 - \delta^2)^{(-x)}\right)'}$$

$$= \lim_{x \to +\infty} \frac{1/2x^{-1/2}}{-(1 - \delta^2)^{(-x)} \ln(1 - \delta^2)}$$

$$= \lim_{x \to +\infty} \frac{(1/2)(1 - \delta^2)^x}{- \ln(1 - \delta^2)x^{1/2}}$$

$$= 0, \tag{7.29}$$

so that

$$R_n(x) \to 0$$

uniformly on

$$A_\delta = \{x \in [-1, 1] \ : \ \delta \le |x| \le 1\}.$$

Now define

$$P_n(x) = \int_{-1}^{1} f(x + t)R_n(t)\, dt, \forall x \in [0, 1], \ n \in \mathbb{N}.$$

Observe that, since $\operatorname{supp} f = [0, 1]$, we have (for $0 \le x + t \le 1$, that is, $-x \le t \le 1 - x$),

$$P_n(x) = \int_{-x}^{1-x} f(x + t)R_n(t)\, dt.$$

Also, defining,

$$w = x + t,$$

we obtain,

$$P_n(x) = \int_0^1 f(w)R_n(w - x)\, dw,$$

and from such an expression is clear that P_n is a polynomial in x.

Let $\varepsilon > 0$. Since f is uniformly continuous on $\mathbb{R}$, at this point we may fix $0 < \delta < 1$ such that if

$$|y - x| < 2\delta,$$

then

$$|f(y) - f(x)| < \frac{\varepsilon}{2}.$$

Define,

$$M = \sup\{|f(x)| : x \in \mathbb{R}\}.$$

From $R_n(x) \geq 0$ we get for $0 \leq x \leq 1$,

$$|P_n(x) - f(x)| = \left| \int_{-1}^{1} (f(x + t) - f(x)) R_n(t) \, dt \right|$$

$$\leq 2M \int_{-1}^{-\delta} R_n(t) \, dt + \frac{\varepsilon}{2} \int_{-\delta}^{\delta} R_n(t) \, dt + 2M \int_{\delta}^{1} R_n(t) \, dt$$

$$\leq 2M \sqrt{n}(1 - \delta^2)^n + \frac{\varepsilon}{2} + 2M \sqrt{n}(1 - \delta^2)^n. \qquad (7.30)$$

Observe that, since

$$\lim_{n \to +\infty} \sqrt{n}(1 - \delta^2)^n = 0,$$

there exists $n_0 \in \mathbb{N}$ such that if $n > n_0$, then

$$\sqrt{n}(1 - \delta^2)^n \leq \frac{\varepsilon}{8M}.$$

From this and (7.30), if $n > n_0$ we obtain,

$$|P_n(x) - f(x)| < \frac{\varepsilon}{4} + \frac{\varepsilon}{2} + \frac{\varepsilon}{4} = \varepsilon,$$

$\forall x \in [0, 1]$.

The proof is complete.

Remark 7.5.5 The inequality

$$(1 - x^2)^n \geq 1 - nx^2$$

which is used in the above proof may be obtained from the fact that $h : [0, 1] \to \mathbb{R}$ given by

$$h(x) = (1 - x^2)^n - 1 + nx^2,$$

is such that $h(0) = 0$ and

$$h'(x) = n(1 - x^2)^{n-1}2x + 2nx > 0, \ \forall x \in (0, 1].$$

Remark 7.5.6 In fact a version of this last theorem is valid for any continuous function $f : [0, 1] \to \mathbb{R}$. Indeed, from the last theorem, the result is valid for $g : [0, 1] \to \mathbb{R}$ given by

$$g(x) = f(x) - f(0) - x[f(1) - f(0)], \ \forall x \in [0, 1],$$

from which it is possible to infer that the result may be extended to a general $f : [0, 1] \to \mathbb{R}$.

Finally, the result may be also extended to a general continuous function $f : [a, b] \to \mathbb{R}$, considering its applicability to $g : [0, 1] \to \mathbb{R}$ given by

$$g(x) = f(a + x(b - a)), \ \forall x \in [0, 1].$$

7.6   The Arzela–Ascoli Theorem

In this section we present a classical result in analysis, namely the Arzela–Ascoli theorem.

Definition 7.6.1 (Equi-Continuity) Let $\mathscr{F}$ be a collection of complex functions defined on a metric space (U, d). We say that $\mathscr{F}$ is equicontinuous if for each $\varepsilon > 0$, there exists $\delta > 0$ such that if $u, v \in U$ and $d(u, v) < \delta$ then

$$|f(u) - f(v)| < \varepsilon, \forall f \in \mathscr{F}.$$

Furthermore, we say that $\mathscr{F}$ is point-wise bounded if for each $u \in U$ there exists $M(u) \in \mathbb{R}$ such that

$$|f(u)| < M(u), \forall f \in \mathscr{F}.$$

Theorem 7.6.2 (Arzela–Ascoli) *Suppose $\mathscr{F}$ is a point-wise bounded equicontinuous collection of complex functions defined on a metric space (U, d). Also suppose that U has a countable dense subset E. Thus, each sequence $\{f_n\} \subset \mathscr{F}$ has a subsequence that converges uniformly on every compact subset of U.*

Proof Let $\{u_n\}$ be a countable dense set in (U, d). By hypothesis, $\{f_n(u_1)\}$ is a bounded sequence; therefore it has a convergent subsequence, which is denoted by $\{f_{n_k}(u_1)\}$. Let us denote

$$f_{n_k}(u_1) = \tilde{f}_{1,k}(u_1), \forall k \in \mathbb{N}.$$

Thus there exists $g_1 \in \mathbb{C}$ such that

$$\tilde{f}_{1,k}(u_1) \to g_1, \text{ as } k \to \infty.$$

Observe that $\{f_{n_k}(u_2)\}$ is also bounded and also it has a convergent subsequence, which similarly as above we will denote by $\{\tilde{f}_{2,k}(u_2)\}$. Again there exists $g_2 \in \mathbb{C}$ such that

$$\tilde{f}_{2,k}(u_1) \to g_1, \text{ as } k \to \infty.$$

$$\tilde{f}_{2,k}(u_2) \to g_2, \text{ as } k \to \infty.$$

Proceeding in this fashion for each $m \in \mathbb{N}$ we may obtain $\{\tilde{f}_{m,k}\}$ such that

$$\tilde{f}_{m,k}(u_j) \to g_j, \text{ as } k \to \infty, \forall j \in \{1, \ldots, m\},$$

where the set $\{g_1, g_2, \ldots, g_m\}$ is obtained as above. Consider the diagonal sequence

$$\{\tilde{f}_{k,k}\},$$

and observe that the sequence

$$\{\tilde{f}_{k,k}(u_m)\}_{k>m}$$

is such that

$$\tilde{f}_{k,k}(u_m) \to g_m \in \mathbb{C}, \text{ as } k \to \infty, \forall m \in \mathbb{N}.$$

Therefore we may conclude that from $\{f_n\}$ we may extract a subsequence also denoted by

$$\{f_{n_k}\} = \{\tilde{f}_{k,k}\}$$

which is convergent in

$$E = \{u_n\}_{n \in \mathbb{N}}.$$

Now suppose $K \subset U$, being K compact. Suppose given $\varepsilon > 0$. From the equi-continuity hypothesis there exists $\delta > 0$ such that if $u, v \in U$ and $d(u, v) < \delta$ we have

$$|f_{n_k}(u) - f_{n_k}(v)| < \frac{\varepsilon}{3}, \forall k \in \mathbb{N}.$$

Observe that

$$K \subset \cup_{u \in K} B_{\frac{\delta}{2}}(u),$$

and being K compact we may find $\{\tilde{u}_1, \ldots, \tilde{u}_M\}$ such that

$$K \subset \cup_{j=1}^{M} B_{\frac{\delta}{2}}(\tilde{u}_j).$$

Since E is dense in U, there exists

$$v_j \in B_{\frac{\delta}{2}}(\tilde{u}_j) \cap E, \forall j \in \{1, \ldots, M\}.$$

Fixing $j \in \{1, \ldots, M\}$, from $v_j \in E$ we obtain that

$$\lim_{k \to \infty} f_{n_k}(v_j)$$

exists as $k \to \infty$. Hence there exists $K_{0_j} \in \mathbb{N}$ such that if $k, l > K_{0_j}$, then

$$|f_{n_k}(v_j) - f_{n_l}(v_j)| < \frac{\varepsilon}{3}.$$

Pick $u \in K$, thus

$$u \in B_{\frac{\delta}{2}}(\tilde{u}_{\hat{j}})$$

for some $\hat{j} \in \{1, \ldots, M\}$, so that

$$d(u, v_{\hat{j}}) < \delta.$$

Therefore if

$$k, l > \max\{K_{0_1}, \ldots, K_{0_M}\},$$

then

$$|f_{n_k}(u) - f_{n_l}(u)| \leq |f_{n_k}(u) - f_{n_k}(v_{\hat{j}})| + |f_{n_k}(v_{\hat{j}}) - f_{n_l}(v_{\hat{j}})|$$
$$+ |f_{n_l}(v_{\hat{j}}) - f_{n_l}(u)|$$
$$\leq \frac{\varepsilon}{3} + \frac{\varepsilon}{3} + \frac{\varepsilon}{3} = \varepsilon. \tag{7.31}$$

Since $u \in K$ is arbitrary, we conclude that $\{f_{n_k}\}$ is uniformly Cauchy on K.
    The proof is complete.

7.7   A Note on Fourier Analysis

Consider a sequence of functions $\{D_n\}$, where $D_n : \mathbb{R} \to \mathbb{C}$ is defined by

$$D_n(x) = \sum_{j=-n}^{n} e^{j i x}, \forall n \in \mathbb{N} \cup \{0\}.$$

Denoting $z = e^{i x}$ we have

$$D_n(x) = \sum_{j=-n}^{n} z^j,$$

so that for x such that $z \neq 1$ we have,

$$D_n(x) = \sum_{j=1}^{n} z^j + \sum_{j=0}^{n} z^{-j}$$

$$= \frac{z(1 - z^n)}{1 - z} + \frac{(1 - z^{-n-1})}{1 - 1/z}$$

$$= \frac{z - z^{n+1}}{1 - z} - \frac{z - z^{-n}}{1 - z}$$

$$= \frac{z^{-n} - z^{n+1}}{1 - z}. \tag{7.32}$$

At this point we define $F_N : \mathbb{R} \to \mathbb{C}$ by

$$F_N(x) = \frac{\sum_{n=0}^{N-1} D_n(x)}{N}, \quad \forall N \in \mathbb{N}.$$

From this and (7.32) we obtain,

$$F_N(x) = \sum_{n=0}^{N-1} \frac{z^{-n} - z^{n+1}}{1 - z}$$

$$= \frac{1}{N(1 - z)} \frac{1 - z^{-N}}{1 - z^{-1}} - \frac{1}{N(1 - z)} \frac{1 - z^N}{1 - z}$$

$$= \frac{-z(1 - z^{-N}) - z + z^{N+1}}{N(1 - z)^2}$$

$$= \frac{-2z + z^{-N+1} + z^{N+1}}{N(1 - z)^2}. \tag{7.33}$$

On the other hand,

$$
\begin{aligned}
\left(\frac{\sin(Nx/2)}{\sin(x/2)}\right)^2 &= \frac{(z^{N/2} - z^{-N/2})^2}{(z^{1/2} - z^{-1/2})^2} \\
&= \frac{z^N + z^{-N} - 2}{z + z^{-1} - 2} \\
&= \frac{z}{z}\left(\frac{z^N + z^{-N} - 2}{z + z^{-1} - 2}\right) \\
&= \frac{z^{N+1} + z^{-N+1} - 2z}{z^2 + 1 - 2z} \\
&= \frac{z^{N+1} + z^{-N+1} - 2z}{(1 - z)^2} \\
&= F_N(x)\, N.
\end{aligned}
\tag{7.34}
$$

Summarizing,

$$
F_N(x) = \left(\frac{\sin(Nx/2)}{\sin(x/2)}\right)^2 /N, \forall N \in \mathbb{N},
$$

$\forall x \in [-\pi, \pi]$ such that $x \neq 0$.

Proposition 7.7.1 *Let $F_N : \mathbb{R} \to \mathbb{C}$ be given by*

$$
F_N(x) = \sum_{n=0}^{N-1} D_n(x), \ \forall N \in \mathbb{N},
$$

where

$$
D_n(x) = \sum_{j=-n}^{n} e^{j\,i\,x}.
$$

Under such assumptions and statements, we have,

1. $\int_{-\pi}^{\pi} F_N(x)\, dx = \int_{-\pi}^{\pi} |F_N(x)|\, dx = 2\pi, \ \forall N \in \mathbb{N}.$
2. *Given $\varepsilon > 0$ and $0 < \delta < \pi$, the exists $n_0 \in \mathbb{N}$ such that*

$$
|F_N(x)| < \varepsilon,
$$

$\forall N > n_0$, and $x \in [-\pi, \pi]$, such that $|x| \geq \delta$.

Proof Observe that

$$\int_{-\pi}^{\pi} F_N(x)\, dx = \int_{-\pi}^{\pi} \left(\sum_{n=0}^{N-1} D_n(x) \right) / N,$$

and,

$$\int_{-\pi}^{\pi} D_n(x)\, dx = \sum_{j=-n}^{n} \int_{-\pi}^{\pi} e^{j\,i\,x}\, dx$$

$$= 2\pi + \sum_{j=-n,\ j\neq 0}^{N} \frac{e^{j\,i\,x}\big|_{x=-\pi}^{x=\pi}}{ji}$$

$$= 2\pi + 0$$

$$= 2\pi. \tag{7.35}$$

Hence, since $F_N(x) \geq 0,\ \forall x \in \mathbb{R}$ we obtain

$$\int_{-\pi}^{\pi} F_N(x)\, dx = \int_{-\pi}^{\pi} |F_N(x)|\, dx = (2\pi)N/N = 2\pi,\ \forall N \in \mathbb{N}$$

This proves (1).

To prove (2), let $\varepsilon > 0$ and $0 < \delta < \pi$ be given.

Hence if $x \in [-\pi, \pi]$ and $|x| \geq \delta$, we have

$$|F_N(x)| = \left(\frac{\sin(Nx/2)}{\sin(x/2)} \right)^2 / N$$

$$\leq \frac{1}{N \sin^2(\delta/2)} \to 0,\ \text{as } N \to \infty. \tag{7.36}$$

Hence,

$$\lim_{N \to \infty} F_N(x) = 0,$$

so that there exists $n_0 \in \mathbb{N}$ such that if $N > n_0$, then $0 \leq F_N(x) < \varepsilon$.

This completes the proof.

Theorem 7.7.2 *Let $f : [-\pi, \pi] \to \mathbb{R}$ be a bounded, integrable function which is continuous in $x \in [-\pi, \pi]$.*

Under such assumptions,

$$\lim_{N \to \infty} \frac{1}{2\pi} \int_{-\pi}^{\pi} F_N(x - y) f(y)\, dy = f(x),$$

where the sequence of functions $\{F_N\}$ is the same as in the last proposition.

Proof Suppose given $\varepsilon > 0$.
  Define

$$M = \sup_{x \in [-\pi, \pi]} |f(x)|.$$

Since f is continuous at x, there exists $0 < \delta < \pi$ such that if $y \in [-\pi, \pi]$ and $|y - x| < \delta$, then

$$|f(y) - f(x)| < \frac{\varepsilon}{4\pi}. \tag{7.37}$$

For such $\delta > 0$ and $\varepsilon > 0$, from the last proposition, there exists $n_0 \in \mathbb{N}$ such that if $N > n_0$, $z \in [-\pi, \pi]$ and $|z| \geq \delta$, then

$$\frac{|F(z)|}{2\pi} < \frac{\varepsilon}{4M}. \tag{7.38}$$

Observe that,

$$\left| \frac{1}{2\pi} \int_{-\pi}^{\pi} F_N(x - y) f(y) \, dy - f(x) \right|$$

$$= \left| \frac{1}{2\pi} \int_{-\pi}^{\pi} F_N(x - y)(f(y) - f(x)) \, dy \right|$$

$$\leq \frac{1}{2\pi} \int_{-\pi}^{\pi} |F_N(x - y)| |f(y) - f(x)| \, dy$$

$$= \frac{1}{2\pi} \int_{\Omega_1} |F_N(x - y)| |f(y) - f(x))| \, dy$$

$$+ \frac{1}{2\pi} \int_{\Omega_2} |F_N(x - y)| |f(y) - f(x))| \, dy, \tag{7.39}$$

where

$$\Omega_1 = \{ y \in [-\pi, \pi] : |x - y| < \delta \},$$

and

$$\Omega_2 = \{ y \in [-\pi, \pi] : |x - y| \geq \delta \}.$$

Thus, from (7.37), (7.38), and (7.39), if $N > n_0$, then

$$\left| \frac{1}{2\pi} \int_{-\pi}^{\pi} F_N(x - y) f(y) \, dy - f(x) \right|$$

$$\leq \frac{\varepsilon}{4\pi} \int_{\Omega_1} |F_N(x - y)| \, dy$$

$$+ \frac{1}{2\pi} \int_{\Omega_2} F_N(x - y) \, dy 2M$$

$$\leq \frac{\varepsilon}{4\pi} \int_{\Omega_1} |F_N(z)| \, dz + \frac{\varepsilon}{4\pi} 2\pi$$

$$\leq \frac{\varepsilon}{2} + \frac{\varepsilon}{2}$$

$$= \varepsilon, \tag{7.40}$$

so that,

$$\lim_{N \to \infty} \frac{1}{2\pi} \int_{-\pi}^{\pi} F_N(x - y) f(y) \, dy = f(x).$$

The proof is complete.

Corollary 7.7.3 *For a continuous function* $h : [-\pi, \pi] \to \mathbb{R}$, *we denote generically*

$$\hat{h}(n) = \int_{-\pi}^{\pi} h(x) e^{-inx} \, dx, \ \forall n \in \mathbb{Z}$$

Let $f, g : [-\pi, \pi] \to \mathbb{R}$ *be continuous functions.*
Suppose,

$$\hat{f}(n) = \hat{g}(n), \ \forall n \in \mathbb{Z}.$$

Under such statements and hypotheses, we have,

$$f(x) = g(x), \forall x \in [-\pi, \pi].$$

Proof Observe that

$$F_{N+1}(x) = \sum_{n=0}^{N-1} D_n(x)$$

$$= \sum_{n=-N}^{N} \frac{(N + 1 - |n|) e^{-nxi}}{N + 1}, \tag{7.41}$$

so that, for $x \in [-\pi, \pi]$, we have,

$$\frac{1}{2\pi} \int_{-\pi}^{\pi} F_{N+1}(x - y) f(y) \, dy$$

$$= \frac{1}{2\pi} \int_{-\pi}^{\pi} \sum_{n=-N}^{N} \frac{(N + 1 - |n|) e^{in(x-y)} f(y)}{N + 1} \, dy$$

$$= \sum_{n=-N}^{N} \frac{(N+1-|n|)}{N+1} \int_{-\pi}^{\pi} e^{-iny} \, dy e^{inx}$$

$$= \sum_{n=-N}^{N} \frac{(N+1-|n|)}{N+1} \hat{f}(n) e^{inx}$$

$$= \sum_{n=-N}^{N} \frac{(N+1-|n|)}{N+1} \hat{g}(n) e^{inx}. \tag{7.42}$$

From this and the last theorem, we obtain,

$$f(x) = \lim_{N \to \infty} \sum_{n=-N}^{N} \frac{(N+1-|n|)}{N+1} \hat{f}(n) e^{inx}$$

$$= \lim_{N \to \infty} \sum_{n=-N}^{N} \frac{(N+1-|n|)}{N+1} \hat{g}(n) e^{inx}$$

$$= g(x). \tag{7.43}$$

Since $x \in [-\pi, \pi]$ is arbitrary, the proof is complete.

Theorem 7.7.4 *Let $f : [-\pi, \pi] \to \mathbb{R}$ be a C^2 class function. Define*

$$a_n = \frac{1}{2\pi} \int_{-\pi}^{\pi} f(x) e^{-inx} \, dx, \ \forall n \in \mathbb{Z}.$$

Under such hypotheses,

$$\sum_{n=1}^{\infty} |a_n|$$

is a convergent series.

Proof Observe that

$$2\pi a_n = \int_{-\pi}^{\pi} f(x) e^{-inx} \, dx$$

$$= -\int_{-\pi}^{\pi} f'(x) \frac{e^{-inx}}{-in} \, dx + \left[f(x) \frac{e^{-inx}}{-in} \right]_{x=-\pi}^{x=\pi}$$

$$= -\int_{-\pi}^{\pi} f'(x) \frac{e^{-inx}}{-in} \, dx$$

$$= \int_{-\pi}^{\pi} f''(x)\frac{e^{-inx}}{-n^2}\, dx - \left[f'(x)\frac{e^{-inx}}{i^2 n^2} \right]_{x=-\pi}^{x=\pi}$$

$$= \int_{-\pi}^{\pi} f''(x)\frac{e^{-inx}}{-n^2}\, dx. \tag{7.44}$$

Hence,

$$2\pi |a_n| \leq \frac{1}{n^2} \int_{-\pi}^{\pi} |f''(x)|\, dx$$

$$\leq \frac{2\pi M}{n^2}, \tag{7.45}$$

where

$$M = \sup_{x \in [-\pi, \pi]} |f''(x)|.$$

Therefore,

$$|a_n| \leq \frac{M}{n^2}, \forall n \in \mathbb{Z}.$$

Since the series

$$\sum_{n=1}^{\infty} \frac{1}{n^2}$$

is convergent, from the comparison criterion we may infer that

$$\sum_{n=-\infty}^{\infty} |a_n|$$

is also convergent.

The proof is complete.

Theorem 7.7.5 *Let $f : [-\pi, \pi] \to \mathbb{R}$ be a C^3 class function.*
Define

$$a_n = \hat{f}(n) = \frac{1}{2\pi} \int_{-\pi}^{\pi} f(x)e^{-inx}\, dx, \forall n \in \mathbb{Z}.$$

Under such hypotheses,

$$f(x) = \lim_{N \to \infty} \sum_{n=-N}^{N} \hat{f}_n e^{inx}, \forall x \in [-\pi, \pi]$$

Proof Similar to the proof of the last theorem, we may prove that

$$|a_n| \leq \frac{M_1}{n^3}, \ \forall n \in \mathbb{Z},$$

where here,

$$M_1 = \sup_{x \in [-\pi, \pi]} |f'''(x)|.$$

For each $N \in \mathbb{N}$ define $g_N : [-\pi, \pi] \to \mathbb{C}$ by

$$g_N(x) = \sum_{n=-N}^{N} a_n e^{nxi}.$$

We claim that $g = \lim_{N \to \infty} g_N$ is Lipschitz continuous in $[-\pi, \pi]$.
Pick $x, y \in [-\pi, \pi]$.
Thus,

$$|g_N(x) - g_N(y)| = |\sum_{n=-N}^{N} a_n(e^{inx} - e^{iny})|$$

$$\leq \sum_{n=-N}^{N} |a_n||e^{inx} - e^{iny}|$$

$$\leq \sum_{n=-N}^{N} |a_n|n|x - y|$$

$$\leq \sum_{n=-N}^{N} \frac{M_1}{n^3} n|x - y|$$

$$\leq \left(2 \sum_{n=1}^{\infty} \frac{M_1}{n^2}\right)|x - y|$$

$$\leq C|x - y|, \tag{7.46}$$

where

$$C = 2 \sum_{n=1}^{\infty} \frac{M_1}{n^2}.$$

From this, letting $N \to \infty$, we obtain

$$|g(x) - g(y)| \le C|x - y|, \forall x, y \in [-\pi, \pi],$$

so that the claim holds.

Observe that,

$$
\begin{aligned}
\hat{g}(n) &= \frac{1}{2\pi} \int_{-\pi}^{\pi} g(x) e^{-inx} \, dx \\
&= \frac{a_n}{2\pi} \int_{-\pi}^{\pi} e^{inx} e^{-inx} \, dx \\
&= \frac{a_n}{2\pi} \int_{-\pi}^{\pi} e^0 \, dx \\
&= a_n \\
&= \hat{f}(n), \ \forall n \in \mathbb{Z}.
\end{aligned}
\tag{7.47}
$$

Hence, $\hat{f}(n) = \hat{g}(n), \ \forall n \in \mathbb{Z}$ so that from this and Corollary 7.7.3, we obtain,

$$f(x) = g(x) = \lim_{N \to \infty} \sum_{n=-N}^{N} \hat{f}(n) e^{inx}, \ \forall x \in [-\pi, \pi].$$

The proof is complete.

For the next theorem we recall that $f \in L^2([-\pi, \pi])$ if f is measurable and

$$\int_{-\pi}^{\pi} |f(x)|^2 \, dx < +\infty.$$

About measurable functions and the result of denseness of the set $C^\infty([-\pi, \pi])$ in $L^2([-\pi, \pi])$, we would refer to [3, 4] for more details.

Theorem 7.7.6 *Let $f : [-\pi, \pi] \to \mathbb{R}$ be such that $f \in L^2([-\pi, \pi])$.*

Let $\varepsilon > 0$. Under such hypotheses there exists a sequence $\{a_n\}_{n \in \mathbb{Z}} \subset \mathbb{C}$ and $N_0 \in \mathbb{N}$ such that if $N > N_0$, then

$$\left\| f(x) - \sum_{n=-N}^{N} a_n e^{inx} \right\|_2 < \varepsilon.$$

Remark 7.1 Here observe that both the sequence $\{a_n\}_{n \in \mathbb{Z}}$ and N_0 depends on ε.

Proof Since $C^\infty([-\pi, \pi])$ is dense in $L^2([-\pi, \pi])$, we may find $g \in C^\infty([-\pi, \pi])$ such that

$$\|g - f\|_2 < \frac{\varepsilon}{2}.
\tag{7.48}$$

Observe that $g \in C^3([-\pi, \pi])$, so that from the proof of Corollary 7.7.3, we have that $\{g_N\}_{N \in \mathbb{N}}$ such that

$$g_N : [-\pi, \pi] \to \mathbb{R}$$

is given by

$$g_N(x) = \sum_{n=-N}^{N} \hat{g}(n) e^{inx},$$

is also such that

$$\lim_{N \to \infty} g_N(x) = g(x), \ \forall x \in [-\pi, \pi]. \tag{7.49}$$

Observe, also from the proof of Theorem 7.7.5,

$$a_n = \hat{g}(n) \le \frac{M_1}{n^3}, \ \forall n \in \mathbb{Z} \setminus \{0\}.$$

From this and (7.49), we obtain,

$$\|g - g_N\|_2^2 = \| \sum_{n=-N-1}^{-\infty} a_n e^{inx} + \sum_{n=N}^{\infty} a_n e^{inx} \|_2^2$$

$$= \sum_{n=-N-1}^{-\infty} |a_n|^2 + \sum_{n=N}^{\infty} |a_n|^2$$

$$\to 0, \ \text{as } N \to \infty. \tag{7.50}$$

Summarizing,

$$\|g_N - g\|_2 \to 0, \ \text{as } N \to \infty.$$

Hence, there exists $N_0 \in \mathbb{N}$ such that if $N > N_0$, then

$$\|g - g_N\|_2 < \frac{\varepsilon}{2}.$$

Thus, if $N > N_0$,

$$\|f - g_N\|_2 = \|f - g + g - g_N\|_2$$

$$\le \|f - g\|_2 + \|g - g_N\|_2$$

$$\le \frac{\varepsilon}{2} + \frac{\varepsilon}{2}$$

$$= \varepsilon. \tag{7.51}$$

The proof is complete.

Proposition 7.7.7 *Let* $f : [-\pi, \pi] \to \mathbb{R}$ *be such that* $f \in L^2([-\pi, \pi])$.
Let $N \in \mathbb{N}$.
Define $J : \mathbb{C}^{2N+1} \to \mathbb{R}$ *by*

$$J(\{a_n\}_{n=-N}^N) = \frac{1}{2} \int_{-\pi}^{\pi} \left| f(x) - \sum_{n=-N}^N a_n e^{inx} \right|^2 dx.$$

Under such hypotheses,

$$J(\{\hat{f}(n)\}_{n=-N}^N) = \min_{\{a_n\} \in \mathbb{C}^{2N+1}} J(\{a_n\}_{n=N}^N).$$

Proof Observe that J is a quadratic positive definite functional, so that its global minimum is attained through the solution of equations,

$$\frac{\partial J(\{a_n\})}{\partial a_n} = 0, \forall -N \le n \le N.$$

Such equations stand for,

$$\int_{-\pi}^{\pi} \left(f(x) - \sum_{j=-N}^N a_n e^{inx} \right) e^{-inx} dx,$$

that is,

$$\int_{-\pi}^{\pi} f(x) e^{-inx} dx - a_n(2\pi) = 0,$$

so that,

$$a_n = \frac{1}{2\pi} \int_{-\pi}^{\pi} f(x) e^{-inx} dx = \hat{f}(n), \forall -N \le n \le N.$$

This completes the proof.

Theorem 7.7.8 *Let* $f : [-\pi, \pi] \to \mathbb{R}$ *be such that* $f \in L^2([-\pi, \pi])$.
Define $\{f_N\}$ *by* $f_N : [-\pi, \pi] \to \mathbb{R}$ *where*

$$f_N(x) = \sum_{n=-N}^N \hat{f}(n) e^{inx}, \forall N \in \mathbb{N}.$$

Under such hypotheses,

$$\|f - f_N\|_2 \to 0, \quad as \ N \to \infty,$$

so that, for a subsequence $\{N_k\}$ of $\mathbb{N}$ we have,

$$\lim_{k \to +\infty} f_{N_k} = f(x), \quad a.e. \text{ in } [-\pi, \pi].$$

Moreover, if $f \in C^2([-\pi, \pi])$, then

$$\lim_{N \to \infty} f_N(x) = f(x), \quad a.e. \text{ in } [-\pi, \pi].$$

Hence, in any case, in an appropriate sense, we may denote,

$$f(x) = \sum_{n=-\infty}^{\infty} \hat{f}(n)e^{inx}, \, a.e. \text{ in } [-\pi, \pi].$$

Proof Let $\varepsilon > 0$ be given. From Theorem 7.7.6 there exists a sequence $\{a_n\}_{n \in \mathbb{Z}} \subset \mathbb{C}$ and $N_0 \in \mathbb{N}$ such that if $N > N_0$, then

$$\| f(x) - \sum_{n=-N}^{N} a_n e^{inx} \|_2 < \varepsilon.$$

From this and Proposition 7.7.7, we obtain,

$$\| f(x) - \sum_{n=-N}^{N} \hat{f}(n)e^{inx} \|_2$$

$$\leq \| f(x) - \sum_{n=-N}^{N} a_n e^{inx} \|_2$$

$$< \varepsilon, \text{ if } N > N_0. \tag{7.52}$$

Hence,

$$\lim_{N \to \infty} \| f(x) - \sum_{n=-N}^{N} \hat{f}(n)e^{inx} \|_2 = 0,$$

that is,

$$\lim_{N \to \infty} \| f - f_N \|_2 = 0,$$

From this there exists a subsequence N_k such that

$$\lim_{k \to \infty} f_{N_k} = f(x), \quad a.e. \text{ in } [-\pi, \pi],$$

so that in such an appropriate sense,

$$f(x) = \sum_{n=-\infty}^{\infty} \hat{f}(n)e^{inx}, \text{ a.e. in } [-\pi, \pi].$$

Now assume $f \in C^2[-\pi, \pi]$.
In such a case, as in the proof of Theorem 7.7.4,

$$|\hat{f}(n)| \leq \frac{M}{n^2}, \ \forall n \in \mathbb{N},$$

where

$$M = \sup_{x \in [-\pi, \pi]} |f''(x)|.$$

Hence

$$|\sum_{n=-\infty}^{\infty} \hat{f}(n)e^{inx}| \leq \sum_{n=-\infty}^{\infty} |\hat{f}(n)| \in \mathbb{R},$$

so that the series $\sum_{n=-\infty}^{\infty} \hat{f}(n)e^{inx}$ is absolutely convergent.
Hence, $g : [-\pi, \pi] \to \mathbb{R}$ given point-wisely by

$$g(x) = \lim_{N \to \infty} f_N(x),$$

is well defined.
Observe that,

$$\|g - f_{N_k}\|_2 = \|\sum_{n=-N_k}^{-\infty} \hat{f}(n)e^{inx} + \sum_{n=N_k}^{\infty} \hat{f}(n)e^{inx}\|_2$$

$$\leq \sum_{-N_k}^{-\infty} |a_n| + \sum_{N_k}^{\infty} |a_n|$$

$$\to 0, \text{ as } k \to \infty. \tag{7.53}$$

From this, up to a subsequence of $\{N_k\}$ here also denoted by $\{N_k\}$, we have

$$f(x) = \lim_{k \to \infty} f_{N_k}(x) = g(x), \text{ a.e. in } [-\pi, \pi].$$

Thus,

$$f(x) = g(x)$$

$$= \lim_{N \to \infty} f_N(x)$$

$$= \sum_{n=-\infty}^{\infty} \hat{f}(n)e^{nxi}, \text{ a.e. in } [-\pi, \pi].$$ (7.54)

Lemma 7.1 (Riemann–Lebesgue Lemma) *Let* $f : [-\pi, \pi] \to \mathbb{R}$ *be a bounded Riemann integrable function. Under such hypotheses,*

$$\lim_{n \to \infty} \int_{-\pi}^{\pi} f(x) \sin(nx) \, dx = 0,$$

and,

$$\lim_{n \to \infty} \int_{-\pi}^{\pi} f(x) \cos(nx) \, dx = 0.$$

Proof Since f is bounded and integrable, so is f^2. Define $a_n = \frac{1}{2\pi} \int_{-\pi}^{\pi} f(x)e^{-inx} dx$.
Thus,

$$0 \leq \frac{1}{2\pi} \int_{-\pi}^{\pi} |f(x) - \sum_{n=-N}^{N} a_n e^{-inx}|^2 \, dx$$

$$= \frac{1}{2\pi} \int_{-\pi}^{\pi} f(x)^2 \, dx - \sum_{n=-N}^{N} \frac{1}{2\pi} \int_{-\pi}^{\pi} f(x)e^{-inx} \, dx a_n$$

$$- \sum_{n=-N}^{N} \frac{1}{2\pi} \int_{-\pi}^{\pi} f(x)e^{inx} \, dx \overline{a_n} - \sum_{n=-N}^{N} a_n \overline{a_n}$$

$$= \frac{1}{2\pi} \int_{-\pi}^{\pi} f(x)^2 \, dx - \sum_{n=-N}^{N} |a_n|^2.$$ (7.55)

From this we obtain,

$$\sum_{n=-\infty}^{\infty} |a_n|^2 \leq \frac{1}{2\pi} \int_{-\pi}^{\pi} f(x)^2 \, dx,$$

so that

$$\lim_{n \to \infty} |a_n| = 0,$$

that is,

$$\lim_{n \to \infty} a_n = 0.$$

Hence,

$$\lim_{n \to \infty} \frac{1}{2\pi} \int_{-\pi}^{\pi} f(x) \cos(nx) \, dx = \lim_{n \to \infty} Re(a_n)$$

$$= 0, \tag{7.56}$$

and,

$$\lim_{n \to \infty} \frac{1}{2\pi} \int_{-\pi}^{\pi} f(x) \sin(nx) \, dx = \lim_{n \to \infty} Im(a_n)$$

$$= 0. \tag{7.57}$$

This completes the proof.

Theorem 7.7.9 *Let $f : \mathbb{R} \to \mathbb{R}$ be such that*

$$f(x + 2\pi) = f(x), \ \forall x \in \mathbb{R}.$$

Assume f is bounded and integrable in $[-2\pi, 2\pi]$.
Let $x \in (-\pi, \pi)$ be such that there exists $M > 0$ and $0 < \delta < \pi$ such that if $|z| < \delta$, then

$$|f(x - z) - f(x)| \le M|z|.$$

Under such hypotheses,

$$\lim_{N \to \infty} f_N(x) = f(x),$$

where,

$$f_N(x) = \sum_{n=-N}^{N} \hat{f}(n) e^{inx}, \ \forall x \in \mathbb{R}$$

and

$$\hat{f}(n) = \frac{1}{2\pi} \int_{-\pi}^{\pi} f(x) e^{-inx} \, dx, \ \forall n \in \mathbb{Z}.$$

Proof Define $g : \mathbb{R} \to \mathbb{R}$ by

$$g(z) = \frac{f(x - z) - f(z)}{z}, \ \text{if } z \neq 0,$$

and

$$g(0) = \limsup_{z \to 0} \frac{f(x-z) - f(z)}{z}.$$

Observe that

$$f_N(x) = \sum_{n=-N}^{N} \hat{f}(n) e^{inx}$$

$$= \sum_{n=-N}^{N} \frac{1}{2\pi} \int_{-\pi}^{\pi} f(y) e^{-iny} \, dy e^{inx}$$

$$= \sum_{n=-N}^{N} \frac{1}{2\pi} \int_{-\pi}^{\pi} f(y) e^{in(x-y)} \, dy$$

$$= \frac{1}{2\pi} \int_{-\pi}^{\pi} f(y) D_N(x-y) \, dy$$

$$= \frac{1}{2\pi} \int_{-\pi}^{\pi} f(x-z) D_N(z) \, dz, \qquad (7.58)$$

where

$$D_N(z) = \sum_{j=-n}^{N} e^{-jzi}.$$

Observe that,

$$D_N(z) = \sum_{j=1}^{N} e^{jzi} + \sum_{j=0}^{N} e^{-jzi}.$$

Denoting $w = e^{zi}$ for $z \in [-\pi, \pi]$, $z \neq 0$ (which means $w \neq 1$), we have

$$D_N(z) = \sum_{j=1}^{N} w^j + \sum_{j=0}^{N} w^{-j}$$

$$= \frac{w(1-w^N)}{1-w} + \frac{1-w^{-N-1}}{1-w^{-1}}$$

$$= \frac{w - w^{N+1}}{1-w} - \frac{w(1-w^{-N-1})}{1-w}$$

$$= \frac{w^{-N} - w^{N+1}}{1-w}$$

$$= \frac{w^{-1/2}}{w^{-1/2}} \frac{w^{-N} - w^{N+1}}{1 - w}$$

$$= \frac{w^{-(N+1/2)} - w^{N+1/2}}{w^{-1/2} - w^{1/2}}. \tag{7.59}$$

At this point we recall that,

$$w + e^{iz},$$

so that

$$\frac{w^{-1/2} - w^{1/2}}{2} = \frac{e^{-iz/2} - e^{iz/2}}{2}$$

$$= -i \sin(z/2). \tag{7.60}$$

Also,

$$\frac{w^{-(N+1/2)} - w^{N+1/2}}{2} = \frac{e^{-i(N+1/2)z} - e^{i(N+1/2)z}}{2}$$

$$= -i \sin((N + 1/2)z). \tag{7.61}$$

From these last results and (7.59), we get,

$$D_N(z) = \frac{w^{-(N+1/2)} - w^{N+1/2}}{w^{-1/2} - w^{1/2}}$$

$$= \frac{\sin[(N + 1/2)z])}{\sin(z/2)}, \quad \text{if } z \in [-\pi, \pi], \text{ and } z \neq 0. \tag{7.62}$$

From this and (7.58), considering that

$$\frac{1}{2\pi} \int_{-\pi}^{\pi} D_N(z) \, dz = 1,$$

we have,

$$f_N(x) - f(x) = \frac{1}{2\pi} \int_{-\pi}^{\pi} f(x - z) D_N(z) \, dz - f(x)$$

$$= \frac{1}{2\pi} \int_{-\pi}^{\pi} (f(x - z) - f(x)) D_N(z) \, dz$$

$$= \frac{1}{2\pi} \int_{-\pi}^{\pi} h(z) \, dz, \tag{7.63}$$

where, if $z \in [-\pi, \pi]$ and $z \neq 0$, we have,

$$h(z) = (f(x - z) - f(x))D_N(z)$$

$$= (f(x - z) - f(x))\frac{\sin[(N + 1/2)z]}{\sin(z/2)}$$

$$= \frac{f(x - z) - f(x)}{z}\frac{z}{\sin(z/2)}\sin[(N + 1/2)z]$$

$$= g(z)\frac{z}{\sin(z/2)}\sin[(N + 1/2)z]. \tag{7.64}$$

Hence

$$h(z) = g(z)\frac{z}{\sin(z/2)}\sin[(N + 1/2)z]$$

$$= g(z)\frac{z}{\sin(z/2)}(\sin(Nz)\cos(z/2) + \cos(Nz)\sin(z/2))$$

$$= h_1(z)\sin(Nz) + h_2(z)\cos(Nz), \tag{7.65}$$

where,

$$h_1(z) = g(z)\frac{z}{\sin(z/2)}\cos(z/2)$$

and

$$h_2(z) = g(z)\frac{z}{\sin(z/2)}\sin(z/2).$$

Therefore, h_1 and h_2 are Riemann integrable in $\Omega_{\delta_1} = [-\pi, \delta_1] \cup [\delta_1, \pi]$, for all $0 < \delta_1 < \pi$, and uniformly bounded in Ω_{δ_1}, $\forall 0 < \delta_1 < \pi$, so that h_1 and h_2 are bounded and integrable in $[-\pi, \pi] \setminus \{0\}$.

From this and the Riemann–Lebesgue lemma, we obtain,

$$f_N(x) - f(x) = \frac{1}{2\pi}\int_{-\pi}^{\pi} h_1(z)\sin(Nz)\,dz$$

$$+ \frac{1}{2\pi}\int_{-\pi}^{\pi} h_2(z)\cos(Nz)\,dz$$

$$\to 0, \text{ as } N \to \infty. \tag{7.66}$$

Exercises 7.7.10

1. Let $f : [a, b] \to \mathbb{R}$ be a Riemann integrable function. Prove that the three statements below indicated are equivalent:

   (a) $\int_a^b |f(x)|\,dx = 0$,
   (b) If f is continuous at $c \in [a, b]$, then $f(c) = 0$.
   (c) The set $A = \{x \in [a, b] : f(x) \neq 0\}$ has empty interior.

2. Let $f : [a, b] \to \mathbb{R}$ be a continuous function. Prove that if there exists $c \in [a, b]$ such that $f(c) \neq 0$, then

$$\int_a^b |f(x)| \, dx > 0.$$

3. Let $f : [0, 1] \to \mathbb{R}$ be such that

$$f(x) = \begin{cases} 1, & \text{if } x \in A \\ 1/2, & \text{if } x \in [0, 1] \setminus A. \end{cases} \qquad (7.67)$$

where

$$A = \{1/n : n \in \mathbb{N}\}.$$

Prove that f is Riemann integrable and calculate

$$I = \int_0^1 f(x) \, dx.$$

4. Let $f, g : [a, b] \to \mathbb{R}$ be Riemann integrable functions. Suppose the set

$$A = \{x \in [a, b] \; : \; f(x) \neq g(x)\}$$

has zero Lebesgue measure.
    Show that

$$\int_a^b f(x) \, dx = \int_a^b g(x) \, dx.$$

5. Let $f : [a, b] \to \mathbb{R}$ be a Lipschitzian function. Suppose that

$$A \subset [a, b]$$

has zero Lebesgue measure. Prove that $f(A)$ has also zero Lebesgue measure.

Part II
Multi-Variable Advanced Calculus

Chapter 8
Differential Analysis in $\mathbb{R}^n$

8.1 Introduction

This chapter starts with the basic definitions and results related to scalar functions in $\mathbb{R}^n$. In the first sections we address concepts such as limits, continuity, and differentiability. A study on optimality conditions for critical points is also developed. In the subsequent sections we address these same concepts of limits, continuity, and differentiability for vectorial functions in $\mathbb{R}^n$. In the final sections we develop detailed proofs of the implicit (scalar and vectorial cases) and inverse function theorems. Moreover, results concerning Lagrange Multipliers are also presented through an application of the implicit function theorem for the vectorial case. We finish the chapter with an introduction to differential geometry.

The main references for this chapter are [1, 2, 8, 10, 12].

Specifically about the implicit function theorem for the vectorial case, we would also cite [6, 7, 11].

We start by defining the space $\mathbb{R}^n$.

8.2 The Space $\mathbb{R}^n$

Definition 8.2.1 ($\mathbb{R}^n$) We define the space $\mathbb{R}^n$ as the set of all points $\mathbf{x} = (x_1, \ldots, x_n)$ such that $x_k \in \mathbb{R}, \ \forall k \in \{1, \ldots, n\}$.

In particular, for the $\mathbb{R}^3$, we represent

$$\mathbf{x} = (x_1, x_2, x_3) \in \mathbb{R}^3.$$

© Springer International Publishing AG, part of Springer Nature 2018
F. S. Botelho, *Real Analysis and Applications*,
https://doi.org/10.1007/978-3-319-78631-5_8

The Euclidean norm of a vector $\mathbf{x} \in \mathbb{R}^n$, denoted by $|\mathbf{x}|$, is defined by

$$|\mathbf{x}| = \sqrt{x_1^2 + \cdots + x_n^2}$$

where $\mathbf{x} = (x_1, \ldots, x_n) \in \mathbb{R}^n$.

We also define a sum for $\mathbb{R}^n$, denoted by $(+) : \mathbb{R}^n \times \mathbb{R}^n \to \mathbb{R}^n$, as

$$\mathbf{x} + \mathbf{y} = (x_1 + y_1, \ldots, x_n + y_n),$$

$\forall \mathbf{x} = (x_1, \ldots, x_n) \in \mathbb{R}^n$ and $\mathbf{y} = (y_1, \ldots, y_n) \in \mathbb{R}^n$.

Finally, we define the scalar multiplication, denoted by $(\cdot) : \mathbb{R} \times \mathbb{R}^n \to \mathbb{R}^n$, as

$$\alpha \cdot \mathbf{x} = (\alpha x_1, \ldots, \alpha x_n),$$

$\forall \alpha \in \mathbb{R}, \; \mathbf{x} = (x_1, \ldots, x_n) \in \mathbb{R}^n$.

With such operations the space $\mathbb{R}^n$ is a vectorial one.

8.3   Topology for $\mathbb{R}^n$

Definition 8.3.1 (Open Ball in $\mathbb{R}^n$) For $\mathbf{x} \in \mathbb{R}^n$ and $r > 0$, we define the open ball of center $\mathbf{x}$ and radius r, denoted by $B_r(\mathbf{x})$, as

$$B_r(\mathbf{x}) = \{\mathbf{y} \in \mathbb{R}^n \; : \; |\mathbf{y} - \mathbf{x}| < r\}.$$

For $(x_0, y_0) \in \mathbb{R}^2$ and $r > 0$, we may write,

$$B_r(x_0, y_0) = \left\{ (x, y) \in \mathbb{R}^2 \; : \; \sqrt{(x - x_0)^2 + (y - y_0)^2} < r \right\}.$$

Similarly, the closed ball of center $\mathbf{x} \in \mathbb{R}^n$ and radius $r > 0$, denoted by $\overline{B}_r(\mathbf{x})$, is defined by

$$\overline{B}_r(\mathbf{x}) = \{\mathbf{y} \in \mathbb{R}^n \; : \; |\mathbf{y} - \mathbf{x}| \le r\}.$$

Definition 8.3.2 (Limit Point) Let $E \subset \mathbb{R}^n$. We say that $\mathbf{x} \in \mathbb{R}^n$ is a limit point of E, if for each $r > 0$ there exists $\mathbf{y} \in B_r(\mathbf{x}) \cap E$ such that $\mathbf{y} \neq \mathbf{x}$.

Definition 8.3.3 (Isolated Point) Let $E \subset \mathbb{R}^n$. We say that $\mathbf{x} \in E$ is an isolated point of E, if there exists $r > 0$ such that

$$B_r(\mathbf{x}) \cap E = \{\mathbf{x}\}.$$

For example, let

$$E = B_1(\mathbf{0}) \cup \{(3, 3)\} \subset \mathbb{R}^2.$$

For $r = 1/2$ we have,

$$B_{1/2}(3, 3) \cap E = \{(3, 3)\},$$

so that $P = (3, 3)$ is an isolated point of E.

Definition 8.3.4 (Interior Point) Let $E \subset \mathbb{R}^n$. We say that $\mathbf{x} \in E$ is an interior point of E if there exists $r > 0$ such that

$$B_r(\mathbf{x}) \subset E.$$

For example, consider $B_1(\mathbf{0}) \subset \mathbb{R}^2$. Thus $P = (1/2, 1/2)$ is an interior point of $B_1(\mathbf{0})$. Indeed,

$$B_{1/4}(1/2, 1/2) \subset B_1(\mathbf{0}).$$

Definition 8.3.5 (Open Set) A set $E \subset \mathbb{R}^n$ is said to be open if all its points are interior.

Definition 8.3.6 (Closed Set) A set $E \subset \mathbb{R}^n$ is said to be closed if E^c is open, where

$$E^c = \{\mathbf{x} \in \mathbb{R}^n \ : \ \mathbf{x} \notin E\}.$$

We also denote $A, B \subset \mathbb{R}^n$

$$A \setminus B = \{\mathbf{x} \in A \ : \ \mathbf{x} \notin B\}.$$

Exercise 8.3.7 Let $\mathbf{x} \in \mathbb{R}^n$ and $r > 0$.

Prove that $B_r(\mathbf{x})$ is open.

Solution: We must show that all point of $B_r(\mathbf{x})$ is interior.

Let $\mathbf{y} \in B_r(\mathbf{x})$. Define $d = |\mathbf{y} - \mathbf{x}|$. Thus $d < r$. Define also $r_1 = r - d$.

We are going to show that $B_{r_1}(\mathbf{y}) \subset B_r(\mathbf{x})$.

Indeed, let $\mathbf{z} \in B_{r_1}(\mathbf{y})$.

Thus,

$$\begin{aligned}
|\mathbf{z} - \mathbf{x}| &= |\mathbf{z} - \mathbf{y} + \mathbf{y} - \mathbf{x}| \\
&\leq |\mathbf{z} - \mathbf{y}| + |\mathbf{y} - \mathbf{x}| \\
&< r_1 + d = r.
\end{aligned} \qquad (8.1)$$

Therefore,

$$\mathbf{z} \in B_r(\mathbf{x}), \ \forall \mathbf{z} \in B_{r_1}(\mathbf{y}),$$

that is, $B_{r_1}(\mathbf{y}) \subset B_r(\mathbf{x})$ and therefore $\mathbf{y}$ is an interior point, $\forall \mathbf{y} \in B_r(\mathbf{x})$.

From this we may conclude that $B_r(\mathbf{x})$ is open.

Exercise 8.3.8 Let $E \subset \mathbb{R}^n$. Prove that E is closed if, and only if, $E \supset E'$, where E' is the set of limit points of E.

Solution: Suppose that $E \supset E'$. We are going to show that E is closed.

Let $\mathbf{x} \in E^c$. Thus, $\mathbf{x} \notin E$ and $\mathbf{x} \notin E'$. Hence, there exists $r > 0$ such that $B_r(\mathbf{x}) \cap E = \emptyset$, that is, $B_r(\mathbf{x}) \subset E^c$. Therefore $\mathbf{x}$ is an interior point, $\forall \mathbf{x} \in E^c$, that is, E^c is open, so that E is closed.

Now suppose E is closed. Thus E^c is open.

Let $\mathbf{x} \in E'$.

Let $r > 0$. Thus there exists $\mathbf{y} \in B_r(\mathbf{x}) \cap E$ such that $\mathbf{y} \neq \mathbf{x}$.

Hence $B_r(\mathbf{x}) \nsubseteq E^c$, $\forall r > 0$ and thus $\mathbf{x}$ is not an interior point of E^c, and since E^c is open, we may conclude that $\mathbf{x} \notin E^c$, that is

$$\mathbf{x} \in E, \ \forall \mathbf{x} \in E'.$$

Therefore, $E' \subset E$.

The solution is complete.

Definition 8.3.9 (Bounded Set) Let $E \subset \mathbb{R}^n$. We say that E is bounded if there exists $M > 0$ such that

$$|\mathbf{x}| < M, \ \forall \mathbf{x} \in E.$$

Definition 8.3.10 (Dense Set) Let $A \subset \mathbb{R}^n$. We say that $E \subset \mathbb{R}^n$ is dense in A if $A \subset E \cup E'$.

In particular E is dense in $\mathbb{R}^n$ if $\mathbb{R}^n = E \cup E'$.

For example $\mathbb{Q}^n$ is dense in $\mathbb{R}^n$, where $\mathbb{Q}$ denotes the set of rational numbers.

Definition 8.3.11 (Inner Product in $\mathbb{R}^n$) Let $\mathbf{x}, \mathbf{y} \in \mathbb{R}^n$. We define an inner product for $\mathbf{x} \in \mathbb{R}^n$ and $\mathbf{y} \in \mathbb{R}^n$, denoted by $\mathbf{x} \cdot \mathbf{y}$, by

$$\mathbf{x} \cdot \mathbf{y} = \sum_{k=1}^{n} x_k y_k = x_1 y_1 + \cdots + x_n y_n,$$

where $\mathbf{x} = (x_1, \ldots, x_n)$ and $\mathbf{y} = (y_1, \ldots, y_n) \in \mathbb{R}^n$.

Inner product properties:

1. $\mathbf{x} \cdot \mathbf{x} \geq 0, \forall \mathbf{x} \in \mathbb{R}^n$ and $\mathbf{x} \cdot \mathbf{x} = 0$ if, and only if, $\mathbf{x} = \mathbf{0}$.
2. $\mathbf{x} \cdot \mathbf{x} = x_1^2 + \cdots + x_n^2 = |\mathbf{x}|^2$.
3. $\mathbf{x} \cdot \mathbf{y} = \mathbf{y} \cdot \mathbf{x}, \ \forall \mathbf{x}, \mathbf{y} \in \mathbb{R}^n$.
4. $(\alpha \mathbf{x}) \cdot \mathbf{y} = \alpha(\mathbf{x} \cdot \mathbf{y}) = \mathbf{x} \cdot (\alpha \mathbf{y}), \forall \alpha \in \mathbb{R}, \ \mathbf{x}, \mathbf{y} \in \mathbb{R}^n$.
5. $\mathbf{x} \cdot (\mathbf{y} + \mathbf{z}) = \mathbf{x} \cdot \mathbf{y} + \mathbf{x} \cdot \mathbf{z}, \ \forall \mathbf{x}, \mathbf{y}, \mathbf{z} \in \mathbb{R}^n$.
6. $|\alpha \mathbf{x}| = |\alpha| |\mathbf{x}|, \ \forall \alpha \in \mathbb{R}, \ \mathbf{x} \in \mathbb{R}^n$.

8.3.1 Cauchy–Schwartz Inequality

Theorem 8.3.12 *Let* $\mathbf{x}, \mathbf{y} \in \mathbb{R}^n$. *Under such hypotheses,*

$$|\mathbf{x} \cdot \mathbf{y}| \leq |\mathbf{x}||\mathbf{y}|.$$

Moreover,

$$|\mathbf{x} \cdot \mathbf{y}| = |\mathbf{x}||\mathbf{y}|$$

if, and only if, ($\mathbf{y} = \mathbf{0}$, *or if* $\mathbf{y} \neq \mathbf{0}$, *there exists* $\alpha \in \mathbb{R}$ *such that* $\mathbf{x} = \alpha\mathbf{y}$.)

Proof Assume $\mathbf{y} \neq \mathbf{0}$ (the case $\mathbf{y} = \mathbf{0}$ is immediate).
  Let $\alpha \in \mathbb{R}$.
  Thus

$$(\mathbf{x} - \alpha\mathbf{y}) \cdot (\mathbf{x} - \alpha\mathbf{y}) \geq 0,$$

and hence,

$$\mathbf{x} \cdot \mathbf{x} - 2\alpha\mathbf{x} \cdot \mathbf{y} + \alpha^2\mathbf{y} \cdot \mathbf{y} \geq 0, \ \forall \alpha \in \mathbb{R}.$$

Denoting $a = \mathbf{y} \cdot \mathbf{y}$, $b = -2\mathbf{x} \cdot \mathbf{y}$ and $c = \mathbf{x} \cdot \mathbf{x}$, we must have

$$b^2 - 4ac \leq 0.$$

Therefore,

$$4|\mathbf{x} \cdot \mathbf{y}|^2 - 4(\mathbf{y} \cdot \mathbf{y})(\mathbf{x} \cdot \mathbf{x}) \leq 0,$$

that is,

$$|\mathbf{x} \cdot \mathbf{y}|^2 \leq (\mathbf{y} \cdot \mathbf{y})(\mathbf{x} \cdot \mathbf{x}) = |\mathbf{y}|^2|\mathbf{x}|^2,$$

and thus,

$$|\mathbf{x} \cdot \mathbf{y}| \leq |\mathbf{x}||\mathbf{y}|.$$

Now suppose there exists $\alpha_0 \in \mathbb{R}$ such that $\mathbf{x} = \alpha_0\mathbf{y}$.
  Therefore,

$$\begin{aligned}|\mathbf{x} \cdot \mathbf{y}| &= |(\alpha_0\mathbf{y}) \cdot \mathbf{y}| \\ &= |\alpha_0||\mathbf{y}|^2 \\ &= |\alpha_0\mathbf{y}||\mathbf{y}| = |\mathbf{x}||\mathbf{y}|.\end{aligned} \tag{8.2}$$

Reciprocally, assume that for $\mathbf{y} \neq \mathbf{0}$ we have

$$|\mathbf{x} \cdot \mathbf{y}| = |\mathbf{x}||\mathbf{y}|.$$

Considering the above calculation, this corresponds to

$$b^2 - 4ac = 4|\mathbf{x} \cdot \mathbf{y}|^2 - 4(\mathbf{y} \cdot \mathbf{y})(\mathbf{x} \cdot \mathbf{x}) = 0.$$

Hence, there exists $\alpha_0 \in \mathbb{R}$ such that

$$F(\alpha_0) = 0$$

where

$$F(\alpha) = \mathbf{x} \cdot \mathbf{x} - 2\alpha \mathbf{x} \cdot \mathbf{y} + \alpha^2 \mathbf{y} \cdot \mathbf{y}.$$

Therefore

$$F(\alpha_0) = (\mathbf{x} - \alpha_0 \mathbf{y}) \cdot (\mathbf{x} - \alpha_0 \mathbf{y}) = 0,$$

so that,

$$\mathbf{x} - \alpha_0 \mathbf{y} = \mathbf{0}$$

and thus

$$\mathbf{x} = \alpha_0 \mathbf{y}.$$

This completes the proof.

8.3.2 Triangular Inequality

Proposition 8.3.13 *Let $\mathbf{x}, \mathbf{y} \in \mathbb{R}^n$. Under such hypotheses,*

$$|\mathbf{x} + \mathbf{y}| \leq |\mathbf{x}| + |\mathbf{y}|.$$

Proof Observe that,

$$\begin{aligned}
|\mathbf{x} + \mathbf{y}|^2 &= (\mathbf{x} + \mathbf{y}) \cdot (\mathbf{x} + \mathbf{y}) \\
&= \mathbf{x} \cdot \mathbf{x} + 2\mathbf{x} \cdot \mathbf{y} + \mathbf{y} \cdot \mathbf{y} \\
&= |\mathbf{x}|^2 + 2\mathbf{x} \cdot \mathbf{y} + |\mathbf{y}|^2 \\
&\leq |\mathbf{x}|^2 + 2|\mathbf{x} \cdot \mathbf{y}| + |\mathbf{y}|^2 \\
&\leq |\mathbf{x}|^2 + 2|\mathbf{x}||\mathbf{y}| + |\mathbf{y}|^2 \\
&= (|\mathbf{x}| + |\mathbf{y}|)^2.
\end{aligned} \tag{8.3}$$

Summarizing,

$$|\mathbf{x} + \mathbf{y}|^2 \le (|\mathbf{x}| + |\mathbf{y}|)^2,$$

and therefore,

$$|\mathbf{x} + \mathbf{y}| \le |\mathbf{x}| + |\mathbf{y}|.$$

Exercise: Let $\mathbf{x}, \mathbf{y} \in \mathbb{R}^n$. Prove that

$$||\mathbf{x}| - |\mathbf{y}|| \le |\mathbf{x} - \mathbf{y}|.$$

8.4   Scalar Functions of Several Variables

Definition 8.4.1 Let $D \subset \mathbb{R}^n$ be a set. We say that a binary relation $f : D \to \mathbb{R}$ is a scalar function of several variables (in this case n variables), if for each $\mathbf{x} \in D$ there exists a unique $z \in \mathbb{R}$ such that

$$(\mathbf{x}, z) \in f.$$

In such a case we denote

$$z = f(\mathbf{x}).$$

Moreover, for $f : D \subset \mathbb{R}^n \to \mathbb{R}$, the set D is said to be the domain of f and the set $\mathbb{R}$ is its co-domain.
The range of f, denoted by $R(f)$, is defined by,

$$R(f) = \{f(\mathbf{x}) \ : \ \mathbf{x} \in D\}.$$

Example 8.4.2 Let $f : \mathbb{R}^2 \to \mathbb{R}$ where

$$f(x, y) = x^2 + y^2, \ \forall (x, y) \in \mathbb{R}^2.$$

The domain D of f is $\mathbb{R}^2$ and its co-domain is $\mathbb{R}$. The range of f is $\mathbb{R}^+ = [0, +\infty)$.

8.4.1   Natural Domain

Given an analytical expression $z = f(\mathbf{x})$, where $\mathbf{x} \in \mathbb{R}^n$, we define the natural domain of f (or simply the domain of f), denoted by $D(f)$, as the "greatest" subset of $\mathbb{R}^n$ in which f may represent a function.

Example 8.4.3 For $f(x, y) = \ln(xy - 1)$, let us find its domain.

$$D(f) = \{(x, y) \in \mathbb{R}^2 \ : \ xy - 1 > 0\},$$

Observe that from

$$xy > 1$$

for $x > 0$, we obtain

$$y > \frac{1}{x}.$$

For $x < 0$, we have

$$y < \frac{1}{x}.$$

Thus, we may write,

$$D(f) = D_1 \cup D_2,$$

where

$$D_1 = \left\{ (x, y) \in \mathbb{R}^2 \ : \ x > 0 \text{ and } y > \frac{1}{x} \right\}$$

and

$$D_2 = \left\{ (x, y) \in \mathbb{R}^2 \ : \ x < 0 \text{ and } y < \frac{1}{x} \right\}.$$

8.5 Limits

In this section we develop the limit definition and concerning examples.

We start with the formal definition.

Definition 8.5.1 (Limits) Let $D \subset \mathbb{R}^n$ be a nonempty set and let $\mathbf{x}_0 \in D'$. Let $f : D \to \mathbb{R}$ be a function.

We say that $L \in \mathbb{R}$ is the limit of f as $\mathbf{x}$ approaches $\mathbf{x}_0$, if for each $\varepsilon > 0$ there exists $\delta > 0$ such that if $\mathbf{x} \in D$ and

$$0 < |\mathbf{x} - \mathbf{x}_0| < \delta,$$

then

$$|f(\mathbf{x}) - L| < \varepsilon.$$

In such a case we denote,

$$\lim_{\mathbf{x} \to \mathbf{x}_0} f(\mathbf{x}) = L.$$

Example 8.5.2 Let $f : \mathbb{R}^2 \to \mathbb{R}$ be such that

$$f(x, y) = \begin{cases} 2x + y, & \text{if } (x, y) \neq (1, 1) \\ 8, & \text{if } (x, y) = (1, 1). \end{cases}$$

We are going to show that

$$\lim_{(x,y) \to (1,1)} f(x, y) = 3.$$

Let $\varepsilon > 0$.
Observe that

$$|2x + y - 3| = |2(x - 1) + (y - 1)|$$
$$\leq 2|x - 1| + |y - 1|. \tag{8.4}$$

Let $\delta > 0$. Suppose that

$$\sqrt{(x - 1)^2 + (y - 1)^2} < \delta.$$

Thus,

$$|x - 1| \leq \sqrt{(x - 1)^2 + (y - 1)^2} < \delta,$$

and

$$|y - 1| \leq \sqrt{(x - 1)^2 + (y - 1)^2} < \delta.$$

From this and (8.4), we obtain

$$|2x + y - 3| \leq 2|x - 1| + |y - 1|$$
$$< 2\delta + \delta$$
$$= 3\delta (\equiv \varepsilon). \tag{8.5}$$

Thus, for each $\varepsilon > 0$ there exists $\delta = \varepsilon/3 > 0$ such that if

$$0 < \sqrt{(x - 1)^2 + (y - 1)^2} < \delta,$$

then

$$|2x + y - 3| < 3\delta = \varepsilon.$$

Therefore,

$$\lim_{(x,y)\to(1,1)} f(x, y) = 3.$$

Exercise 8.5.3 Prove formally that

$$\lim_{(x,y)\to(-1,2)} 3x^2 + 2y = 7.$$

Solution: Observe that,

$$
\begin{aligned}
|3x^2 + 2y - 7| &= |3(x^2 - (-1)^2) + 2(y - 2)| \\
&= |3(x + 1)(x - 1) + 2(y - 2)| \\
&\leq 3|x + 1||x - 1| + 2|y - 2|.
\end{aligned}
\tag{8.6}
$$

Let $0 < \delta \leq 1$.
Suppose that

$$0 < \sqrt{(x - (-1))^2 + (y - 2)^2} < \delta.$$

Thus

$$|x + 1| \leq \sqrt{(x + 1)^2 + (y - 2)^2} < \delta,$$

and

$$|y - 2| \leq \sqrt{(x + 1)^2 + (y - 2)^2} < \delta.$$

Therefore,

$$|x + 1| < \delta \leq 1.$$

Thus,

$$
\begin{aligned}
|x + 1| < 1 &\Rightarrow -1 < x + 1 < 1 \\
&\Rightarrow -2 < x < 0 \\
&\Rightarrow -3 < x - 1 < -1 \\
&\Rightarrow |x - 1| < 3.
\end{aligned}
\tag{8.7}
$$

Hence, from this and (8.6), if

$$0 < \delta \le 1$$

and

$$0 < \sqrt{(x+1)^2 + (y-2)^2} < \delta,$$

then

$$|3x^2 + 2y - 7| \le 3|x+1||x-1| + 2|y-2|$$
$$< 3\delta(3) + 2\delta$$
$$= 11\delta(\equiv \varepsilon). \qquad (8.8)$$

Therefore, for each $\varepsilon > 0$ there exists

$$\delta = \min\left\{1, \frac{\varepsilon}{11}\right\}$$

such that if

$$0 < \sqrt{(x+1)^2 + (y-2)^2} < \delta,$$

then

$$|3x^2 + 2y - 7| < 11\delta \le \varepsilon.$$

Thus,

$$\lim_{(x,y)\to(-1,2)} 3x^2 + 2y = 7.$$

Exercise 8.5.4 Prove formally that

$$\lim_{(x,y)\to(-2,3)} 2x^2 - y^2 - 3x + 3y + 1 = 15.$$

Exercise 8.5.5 Prove formally that

$$\lim_{(x,y)\to(2,1)} x^3 + 2y^2 + x - y + 3 = 14.$$

Solution: Let $\varepsilon > 0$.
   Observe that

$$|x^3 + 2y^2 + x - y + 3 - 14|$$
$$= |x^3 + 2y^2 + x - y - 11|$$
$$= |(x^3 - 2^3) + 2(y^2 - 1^2) + (x - 2) - (y - 1)|. \qquad (8.9)$$

However, by a long division, we obtain,

$$x^3 - 2^3 = x^3 - 8 = (x - 2)(x^2 + 2x + 4).$$

Thus,

$$
\begin{aligned}
&|x^3 + 2y^2 + x - y + 3 - 14| \\
&= |(x^3 - 2^3) + 2(y^2 - 1^2) + (x - 2) - (y - 1)| \\
&= |(x - 2)(x^2 + 2x + 4) + 2(y + 1)(y - 1) + (x - 2) - (y - 1)| \\
&\leq |x - 2|(|x|^2 + 2|x| + 4) + 2|y - 1||y + 1| + |x - 2| + |y - 1| \\
&\leq |x - 2|(|x|^2 + 2|x| + 4) + 2|y - 1|(|y| + 1) + |x - 2| + |y - 1|. \quad (8.10)
\end{aligned}
$$

Let $0 < \delta \leq 1$.

Suppose that

$$0 < \sqrt{(x - 2)^2 + (y - 1)^2} < \delta.$$

Thus,

$$|x - 2| \leq \sqrt{(x - 2)^2 + (y - 1)^2} < \delta,$$

and

$$|y - 1| \leq \sqrt{(x - 2)^2 + (y - 1)^2} < \delta,$$

so that,

$$|x - 2| < \delta \leq 1.$$

Hence,

$$
\begin{aligned}
|x - 2| < 1 &\Rightarrow -1 < x - 2 < 1 \\
&\Rightarrow 1 < x < 3 \\
&\Rightarrow |x| < 3. \quad\quad\quad\quad\quad (8.11)
\end{aligned}
$$

Also,

$$|y - 1| < \delta \leq 1.$$

so that,

$$
\begin{aligned}
|y - 1| < 1 &\Rightarrow -1 < y - 1 < 1 \\
&\Rightarrow 0 < y < 2 \\
&\Rightarrow |y| < 2. \quad\quad\quad\quad\quad (8.12)
\end{aligned}
$$

Hence, from this and (8.6), if

$$0 < \delta \leq 1$$

and

$$0 < \sqrt{(x-2)^2 + (y-1)^2} < \delta,$$

then

$$|x^3 + 2y^2 + x - y + 3 - 14|$$
$$\leq |x - 2|(|x|^2 + 2|x| + 4) + 2|y - 1|(|y| + 1) + |x - 2| + |y - 1|$$
$$< \delta(3^2 + 2(3) + 4) + 2\delta(2 + 1) + \delta + \delta$$
$$= 19\delta + 8\delta$$
$$= 27\delta(\equiv \varepsilon). \tag{8.13}$$

Therefore, for each $\varepsilon > 0$ there exists

$$\delta = \min\left\{1, \frac{\varepsilon}{27}\right\}$$

such that if

$$0 < \sqrt{(x-2)^2 + (y-1)^2} < \delta,$$

then

$$|x^3 + 2y^2 + x - y + 3 - 14| < 27\delta \leq \varepsilon.$$

Thus,

$$\lim_{(x,y)\to(2,1)} x^3 + 2y^2 + x - y + 3 = 14.$$

Exercise 8.5.6 Prove formally that

$$\lim_{(x,y)\to(2,3)} x^5 y^4 = (2)^5 (3)^4.$$

Solution: Let $\varepsilon > 0$. Observe that

$$|x^5 y^4 - 2^5 3^4|$$
$$= |x^5 y^4 - 2^5 y^4 + 2^5 y^4 - (2)^5 (3)^4|$$
$$= |(x^5 - 2^5)y^4 + 2^5(y^4 - 3^4)|$$
$$\leq |x^5 - 2^5||y|^4 + 2^5|y^4 - 3^4|. \tag{8.14}$$

However, by long divisions, we obtain

$$x^5 - 2^5 = (x - 2)(x^4 + 2x^3 + 4x^2 + 8x + 16),$$

and

$$y^4 - 3^4 = (y - 3)(y^3 + 3y^2 + 9y + 27).$$

Thus,

$$
\begin{aligned}
&|x^5 y^4 - (2)^5 (3)^4| \\
&\leq |x^5 - 2^5||y|^4 + 2^5|y^4 - 3^4| \\
&\leq |x - 2||x^4 + 2x^3 + 4x^2 + 8x + 16||y|^4 \\
&\quad + 2^5|y - 3||y^3 + 3y^2 + 9y + 27| \\
&\leq |x - 2|(|x|^4 + 2|x|^3 + 4|x|^2 + 8|x| + 16)|y|^4 \\
&\quad + 2^5|y - 3|(|y|^3 + 3|y|^2 + 9|y| + 27). \hspace{2cm} (8.15)
\end{aligned}
$$

Let $0 < \delta \leq 1$.

Suppose

$$0 < \sqrt{(x - 2)^2 + (y - 3)^2} < \delta.$$

Hence,

$$|x - 2| \leq \sqrt{(x - 2)^2 + (y - 3)^2} < \delta,$$

and

$$|y - 3| \leq \sqrt{(x - 2)^2 + (y - 3)^2} < \delta,$$

so that,

$$|x - 2| < \delta \leq 1.$$

Therefore,

$$
\begin{aligned}
|x - 2| < 1 &\Rightarrow -1 < x - 2 < 1 \\
&\Rightarrow 1 < x < 3 \\
&\Rightarrow |x| < 3. \hspace{2cm} (8.16)
\end{aligned}
$$

Also,

$$|y - 3| < \delta \leq 1.$$

so that,

$$|y - 3| < 1 \Rightarrow -1 < y - 3 < 1$$
$$\Rightarrow 2 < y < 4$$
$$\Rightarrow |y| < 4. \tag{8.17}$$

Thus, from this and (8.6), if

$$0 < \delta \le 1$$

and

$$0 < \sqrt{(x - 2)^2 + (y - 3)^2} < \delta,$$

then

$$|x^5 y^4 - (2)^5 (3)^4|$$
$$\le |x - 2|(|x|^4 + 2|x|^3 + 4|x|^2 + 8|x| + 16)|y|^4$$
$$+ 2^5 |y - 3|(|y|^3 + 3|y|^2 + 9|y| + 27)$$
$$\le |x - 2|(3^4 + 2(3)^3 + 4(3)^2 + 8(3) + 16)(4)^4$$
$$+ |y - 3| 2^5 ((4)^3 + 3(4)^2 + 9(4) + 27)$$
$$= \alpha |x - 2| + \beta |y - 3|$$
$$< \alpha \delta + \beta \delta$$
$$= (\alpha + \beta)\delta (\equiv \varepsilon), \tag{8.18}$$

where,

$$\alpha = (3^4 + 2(3)^3 + 4(3)^2 + 8(3) + 16)(4)^4$$

and

$$\beta = 2^5 ((4)^3 + 3(4)^2 + 9(4) + 27).$$

Therefore, for each $\varepsilon > 0$ there exists

$$\delta = \min \left\{ 1, \frac{\varepsilon}{\alpha + \beta} \right\}$$

such that if

$$0 < \sqrt{(x - 2)^2 + (y - 3)^2} < \delta,$$

then

$$|x^5 y^4 - (2)^5 (3)^4| < (\alpha + \beta)\delta \leq \varepsilon.$$

Hence,

$$\lim_{(x,y)\to(2,3)} x^5 y^4 = (2)^5 (3)^4.$$

8.5.1 Limit Uniqueness

We start with the following theorem, which establish the limit uniqueness.

Theorem 8.5.7 *Let $D \subset \mathbb{R}^n$ be a nonempty set and let $f : D \to \mathbb{R}$ be a function. Let $\mathbf{x}_0 \in D'$ and assume*

$$\lim_{\mathbf{x}\to\mathbf{x}_0} f(\mathbf{x}) = L_1 \in \mathbb{R}$$

and

$$\lim_{\mathbf{x}\to\mathbf{x}_0} f(\mathbf{x}) = L_2 \in \mathbb{R}.$$

Under such hypotheses,

$$L_1 = L_2,$$

that is, if the limit exists, it is unique.

Proof Let $\varepsilon > 0$.
  From

$$\lim_{\mathbf{x}\to\mathbf{x}_0} f(\mathbf{x}) = L_1,$$

there exists $\delta_1 > 0$ such that if $\mathbf{x} \in D$ and

$$0 < |\mathbf{x} - \mathbf{x}_0| < \delta_1,$$

then

$$|f(\mathbf{x}) - L_1| < \frac{\varepsilon}{2}. \tag{8.19}$$

  From

$$\lim_{\mathbf{x}\to\mathbf{x}_0} f(\mathbf{x}) = L_2,$$

there exists $\delta_2 > 0$ such that if $\mathbf{x} \in D$ and

$$0 < |\mathbf{x} - \mathbf{x}_0| < \delta_2,$$

then

$$|f(\mathbf{x}) - L_2| < \frac{\varepsilon}{2}. \tag{8.20}$$

Define $\delta = \min\{\delta_1, \delta_2\}$ and choose $\mathbf{x}_1 \in D$ such that

$$0 < |\mathbf{x}_1 - \mathbf{x}_0| < \delta.$$

From this, (8.19) and (8.20), we obtain

$$\begin{aligned}
|L_1 - L_2| &= |L_1 - f(\mathbf{x}_1) + f(\mathbf{x}_1) - L_2| \\
&\le |f(\mathbf{x}_1) - L_1| + |f(\mathbf{x}_1) - L_2| \\
&< \frac{\varepsilon}{2} + \frac{\varepsilon}{2} \\
&= \varepsilon.
\end{aligned} \tag{8.21}$$

Therefore,

$$|L_1 - L_2| < \varepsilon, \forall \varepsilon > 0.$$

Letting $\varepsilon \to 0^+$, we obtain

$$|L_1 - L_2| = 0,$$

that is

$$L_1 = L_2.$$

The proof is complete.

Corollary 8.5.8 *Let $f : D \subset \mathbb{R}^2 \to \mathbb{R}$ be a function and let $(x_0, y_0) \in D'$.
Suppose there exist $r > 0$ and*

$$h_1, h_2 : B_r(x_0) \to \mathbb{R}$$

such that

$$\lim_{x \to x_0} h_1(x) = y_0,$$

$$\lim_{x \to x_0} h_2(x) = y_0$$

and

$$\lim_{x \to x_0} f(x, h_1(x)) \neq \lim_{x \to x_0} f(x, h_2(x)).$$

Under such hypotheses, $\lim_{(x,y) \to (x_0,y_0)} f(x, y)$ *does not exist.*

Proof The proof results directly from the last theorem. We leave the development of its details to the reader.

Example 8.5.9 Let $f : \mathbb{R}^2 \setminus \{(0, 0)\} \to \mathbb{R}$ be defined by

$$f(x, y) = \frac{xy}{x^2 + y^2}.$$

Show that

$$\lim_{(x,y) \to (0,0)} f(x, y) \text{ does not exist.}$$

Let us calculate the limit through the trajectory $y = mx$, where $m \in \mathbb{R}$. Thus,

$$\begin{aligned}
\lim_{x \to 0} f(x, mx) &= \lim_{x \to 0} \frac{x(mx)}{x^2 + m^2 x^2} \\
&= \lim_{x \to 0} \frac{mx^2}{x^2 + m^2 x^2} \\
&= \lim_{x \to 0} \frac{m}{1 + m^2} \frac{x^2}{x^2} \\
&= \lim_{x \to 0} \frac{m}{1 + m^2} \\
&= \frac{m}{1 + m^2}.
\end{aligned} \tag{8.22}$$

Thus, such a limit depends on m. From the last theorem and corollary $\lim_{(x,y) \to (0,0)} f(x, y)$ does not exist.

Example 8.5.10 Let $f : \mathbb{R}^2 \setminus \{(0, 0)\} \to \mathbb{R}$ be defined by

$$f(x, y) = \frac{x^4 y^4}{(x^2 + y^4)^3}.$$

Show that

$$\lim_{(x,y) \to (0,0)} f(x, y) \text{ does not exist.}$$

Let us calculate the limit through the trajectory $x = my^2$, where $m \in \mathbb{R}$. Thus,

$$\lim_{y \to 0} f(my^2, y) = \lim_{x \to 0} \frac{(my^2)^4 y^4}{((my^2)^2 + y^4)^3}$$

$$= \lim_{y \to 0} \frac{m^4 y^{12}}{(m^2 y^4 + y^4)^3}$$

$$= \lim_{y \to 0} \frac{m^4}{(m^2 + 1)^3} \frac{y^{12}}{y^{12}}$$

$$= \lim_{y \to 0} \frac{m^4}{(m^2 + 1)^3}$$

$$= \frac{m^4}{(m^2 + 1)^3}. \tag{8.23}$$

Hence, such a limit depends on m. From the last theorem and corollary $\lim_{(x,y) \to (0,0)} f(x, y)$ does not exist.

Consider the following auxiliary results in exercise form:

Exercise 8.5.11 Let $D \subset \mathbb{R}^n$ be a nonempty set and let $f : D \to \mathbb{R}$ be a function. Let $\mathbf{x}_0 \in D'$ and suppose that

$$\lim_{\mathbf{x} \to \mathbf{x}_0} f(\mathbf{x}) = L \in \mathbb{R}.$$

Show that

$$\lim_{\mathbf{x} \to \mathbf{x}_0} |f(\mathbf{x})| = |L|.$$

Exercise 8.5.12 Let $D \subset \mathbb{R}^n$ be a nonempty set and let $f : D \to \mathbb{R}$ be a function. Let $\mathbf{x}_0 \in D'$ and suppose that

$$\lim_{\mathbf{x} \to \mathbf{x}_0} |f(\mathbf{x})| = 0.$$

Show that

$$\lim_{\mathbf{x} \to \mathbf{x}_0} f(\mathbf{x}) = 0.$$

8.5.2   The Sandwich Theorem

Theorem 8.5.13 (Sandwich Theorem) *Let $D \subset \mathbb{R}^n$ be a nonempty set and let $f, g, h : D \to \mathbb{R}$ be functions.*
Let $\mathbf{x}_0 \in D'$ and assume there exists $\delta_0 > 0$ such that if $\mathbf{x} \in D$ and

$$0 < |\mathbf{x} - \mathbf{x}_0| < \delta_0,$$

then

$$f(\mathbf{x}) \le g(\mathbf{x}) \le h(\mathbf{x}). \tag{8.24}$$

Suppose also that,

$$\lim_{\mathbf{x} \to \mathbf{x}_0} f(\mathbf{x}) = L \in \mathbb{R}$$

and

$$\lim_{\mathbf{x} \to \mathbf{x}_0} h(\mathbf{x}) = L \in \mathbb{R}.$$

Under such hypotheses,

$$\lim_{\mathbf{x} \to \mathbf{x}_0} g(\mathbf{x}) = L.$$

Proof Let $\varepsilon > 0$. From

$$\lim_{\mathbf{x} \to \mathbf{x}_0} f(\mathbf{x}) = L,$$

there exists $\delta_1 > 0$ such that if $\mathbf{x} \in D$ and

$$0 < |\mathbf{x} - \mathbf{x}_0| < \delta_1,$$

then

$$|f(\mathbf{x}) - L| < \varepsilon$$

that is,

$$L - \varepsilon < f(\mathbf{x}) < L + \varepsilon. \tag{8.25}$$

Similarly, from

$$\lim_{\mathbf{x} \to \mathbf{x}_0} h(\mathbf{x}) = L,$$

there exists $\delta_2 > 0$ such that if $\mathbf{x} \in D$ and

$$0 < |\mathbf{x} - \mathbf{x}_0| < \delta_2,$$

then

$$|h(\mathbf{x}) - L| < \varepsilon$$

that is,

$$L - \varepsilon < h(\mathbf{x}) < L + \varepsilon. \tag{8.26}$$

Define $\delta = \min\{\delta_0, \delta_1, \delta_2\}$.

Thus, if $\mathbf{x} \in D$ and

$$0 < |\mathbf{x} - \mathbf{x}_0| < \delta,$$

from (8.24), (8.25), and (8.26), we obtain

$$L - \varepsilon < f(\mathbf{x}) \le g(\mathbf{x}) \le h(\mathbf{x}) < L + \varepsilon,$$

that is,

$$|g(\mathbf{x}) - L| < \varepsilon,$$

if $\mathbf{x} \in D$ and $0 < |\mathbf{x} - \mathbf{x}_0| < \delta$.

Therefore,

$$\lim_{\mathbf{x} \to \mathbf{x}_0} g(\mathbf{x}) = L.$$

The proof is complete.

Example 8.5.14 Let $f : \mathbb{R}^2 \setminus \{(0,0)\} \to \mathbb{R}$ where

$$f(x, y) = \frac{x^2 y}{x^2 + y^2}.$$

Observe that if $(x, y) \ne (0, 0)$, then

$$0 \le \left| \frac{x^2 y}{x^2 + y^2} \right| = \frac{x^2 |y|}{x^2 + y^2} \le |y|. \tag{8.27}$$

Observe also that

$$\lim_{(x,y)\to(0,0)} |y| = 0$$

From this, (8.27) and the Sandwich theorem, we obtain

$$\lim_{(x,y)\to(0,0)} \left| \frac{x^2 y}{x^2 + y^2} \right| = 0,$$

and thus,

$$\lim_{(x,y)\to(0,0)} \frac{x^2 y}{x^2 + y^2} = 0.$$

8.5.3 Properties of Limits

We start with the following preliminary results.

Theorem 8.5.15 *Let $D \subset \mathbb{R}^n$ be a nonempty set. Let $\mathbf{x}_0 \in D'$ and let $f : D \to \mathbb{R}$ be a function.*
   Suppose that

$$\lim_{\mathbf{x}\to\mathbf{x}_0} f(\mathbf{x}) = L \in \mathbb{R}$$

   Under such hypotheses, there exist $\delta > 0$ and $A > 0$ such that if $\mathbf{x} \in D$ and $0 < |\mathbf{x} - \mathbf{x}_0| < \delta$, then

$$|f(\mathbf{x})| < A.$$

Proof Let $\varepsilon = 1$. From

$$\lim_{\mathbf{x}\to\mathbf{x}_0} f(\mathbf{x}) = L,$$

there exists $\delta > 0$ such that if $\mathbf{x} \in D$ and $0 < |\mathbf{x} - \mathbf{x}_0| < \delta$, then

$$|f(\mathbf{x}) - L| < \varepsilon = 1. \tag{8.28}$$

Thus,

$$|f(\mathbf{x})| - |L| \le |f(\mathbf{x}) - L| < 1,$$

that is,

$$|f(\mathbf{x})| < |L| + 1 \equiv A,$$

if $\mathbf{x} \in D$ and $0 < |\mathbf{x} - \mathbf{x}_0| < \delta$.

The proof is complete.

Exercise 8.5.16 Let $D \subset \mathbb{R}^n$ be a nonempty and let $f, g : D \to \mathbb{R}$ be functions. Let $\mathbf{x}_0 \in D'$ and assume there exist $\delta_0 > 0$ and $K > 0$ such that

$$|g(\mathbf{x})| \le K, \forall \mathbf{x} \in D \text{ such that } 0 < |\mathbf{x} - \mathbf{x}_0| < \delta_0.$$

Suppose also that

$$\lim_{\mathbf{x} \to \mathbf{x}_0} f(\mathbf{x}) = 0.$$

Under such hypotheses, show that

$$\lim_{\mathbf{x} \to \mathbf{x}_0} f(\mathbf{x}) g(\mathbf{x}) = 0.$$

Solution: Let $\varepsilon > 0$. From

$$\lim_{\mathbf{x} \to \mathbf{x}_0} f(\mathbf{x}) = 0,$$

there exists $\delta_1 > 0$ such that if $\mathbf{x} \in D$ and $0 < |\mathbf{x} - \mathbf{x}_0| < \delta_1$, then

$$|f(\mathbf{x})| < \frac{\varepsilon}{K}.$$

Define $\delta = \min\{\delta_0, \delta_1\}$.

Hence, if $\mathbf{x} \in D$ and $0 < |\mathbf{x} - \mathbf{x}_0| < \delta$, then

$$|f(\mathbf{x}) g(\mathbf{x})| = |f(\mathbf{x})||g(\mathbf{x})| < \frac{\varepsilon}{K} K = \varepsilon,$$

that is

$$|f(\mathbf{x}) g(\mathbf{x})| < \varepsilon.$$

We may conclude that

$$\lim_{\mathbf{x}\to\mathbf{x}_0} f(\mathbf{x})g(\mathbf{x}) = 0.$$

The solution is complete.

Exercise 8.5.17 Let $f : \mathbb{R}^2 \to \mathbb{R}$ be such that

$$f(x, y) = \begin{cases} (x + y) \sin\left(\frac{1}{x^2+y^2}\right), & \text{if } (x, y) \neq (0, 0) \\ 5, & \text{if } (x, y) = (0, 0). \end{cases}$$

show that

$$\lim_{(x,y)\to(0,0)} f(x, y) = 0.$$

Solution: Observe that

$$\left| \sin\left(\frac{1}{x^2 + y^2}\right) \right| \leq 1, \ \forall (x, y) \neq (0, 0).$$

Moreover

$$\lim_{(x,y)\to(0,0)} (x + y) = 0.$$

Therefore, from these last three lines and the last exercise, we obtain

$$\lim_{(x,y)\to(0,0)} f(x, y) = 0.$$

Theorem 8.5.18 *Let $D \subset \mathbb{R}^n$ be a nonempty set. Let $\mathbf{x}_0 \in D'$ and let $f, g : D \to \mathbb{R}$ be functions.*
Suppose that

$$\lim_{\mathbf{x}\to\mathbf{x}_0} f(\mathbf{x}) = L \in \mathbb{R}$$

and

$$\lim_{\mathbf{x}\to\mathbf{x}_0} g(\mathbf{x}) = M \in \mathbb{R}.$$

Under such hypotheses, we have

1.

$$\lim_{\mathbf{x}\to\mathbf{x}_0} \alpha f(\mathbf{x}) = \alpha L, \forall \alpha \in \mathbb{R},$$

2.

$$\lim_{\mathbf{x} \to \mathbf{x}_0} [f(\mathbf{x}) + g(\mathbf{x})] = L + M,$$

3.

$$\lim_{\mathbf{x} \to \mathbf{x}_0} f(\mathbf{x})g(\mathbf{x}) = LM,$$

4.

$$\lim_{\mathbf{x} \to \mathbf{x}_0} \frac{f(\mathbf{x})}{g(\mathbf{x})} = \frac{L}{M}, \ if \ M \neq 0.$$

Proof We are going to prove just the items 3 and 4. We leave the proof of the remaining items as exercises.

To prove 3, observe that

$$|f(\mathbf{x})g(\mathbf{x}) - LM| = |f(\mathbf{x})g(\mathbf{x}) - f(\mathbf{x})M + f(\mathbf{x})M - LM|$$

$$= |f(\mathbf{x})(g(\mathbf{x}) - M) + M(f(\mathbf{x}) - L)|$$

$$\leq |f(\mathbf{x})||g(\mathbf{x}) - M| + |M||f(\mathbf{x}) - L|. \qquad (8.29)$$

On the other hand, from

$$\lim_{\mathbf{x} \to \mathbf{x}_0} f(\mathbf{x}) = L$$

and from Theorem 8.5.15 there exist $A > 0$ and $\delta_0 > 0$ such that if $\mathbf{x} \in D$ and $0 < |\mathbf{x} - \mathbf{x}_0| < \delta_0$, then

$$|f(\mathbf{x})| < A. \qquad (8.30)$$

Let $\varepsilon > 0$.
From

$$\lim_{\mathbf{x} \to \mathbf{x}_0} f(\mathbf{x}) = L,$$

there exists $\delta_1 > 0$ such that if $\mathbf{x} \in D$ and $0 < |\mathbf{x} - \mathbf{x}_0| < \delta_1$, then

$$|f(\mathbf{x}) - L| < \frac{\varepsilon}{2(|M| + 1)}. \qquad (8.31)$$

Moreover, from

$$\lim_{\mathbf{x} \to \mathbf{x}_0} g(\mathbf{x}) = M,$$

there exists $\delta_2 > 0$ such that if $\mathbf{x} \in D$ and $0 < |\mathbf{x} - \mathbf{x}_0| < \delta_2$, then

$$|g(\mathbf{x}) - M| < \frac{\varepsilon}{2A}. \tag{8.32}$$

Define $\delta = \min\{\delta_0, \delta_1, \delta_2\}$. Thus, if $\mathbf{x} \in D$ and $0 < |\mathbf{x} - \mathbf{x}_0| < \delta$, from (8.29), (8.30), (8.31), and (8.32), we obtain

$$|f(\mathbf{x})g(\mathbf{x}) - LM| \leq |f(\mathbf{x})||g(\mathbf{x}) - M| + |M||f(\mathbf{x}) - L|$$
$$< A\frac{\varepsilon}{2A} + |M|\frac{\varepsilon}{2(|M| + 1)}$$
$$< \varepsilon. \tag{8.33}$$

Therefore,

$$\lim_{\mathbf{x} \to \mathbf{x}_0} f(\mathbf{x})g(\mathbf{x}) = LM.$$

To prove 4, for $M \neq 0$, we are going to prove firstly that

$$\lim_{\mathbf{x} \to \mathbf{x}_0} \frac{1}{g(\mathbf{x})} = \frac{1}{M}.$$

Observe that,

$$\left|\frac{1}{g(\mathbf{x})} - \frac{1}{M}\right| = \frac{|g(\mathbf{x}) - M|}{|g(\mathbf{x})||M|}. \tag{8.34}$$

Let

$$\varepsilon_1 = \frac{|M|}{2} > 0.$$

From

$$\lim_{\mathbf{x} \to \mathbf{x}_0} g(\mathbf{x}) = M,$$

there exists $\delta_3 > 0$ such that if $\mathbf{x} \in D$ and $0 < |\mathbf{x} - \mathbf{x}_0| < \delta_3$, then

$$|g(\mathbf{x}) - M| < \varepsilon_1 = \frac{|M|}{2}. \tag{8.35}$$

Therefore,

$$||g(\mathbf{x})| - |M|| \leq |g(\mathbf{x}) - M| < \frac{|M|}{2},$$

that is

$$-\frac{|M|}{2} < |g(\mathbf{x})| - |M| < \frac{|M|}{2},$$

so that

$$|g(\mathbf{x})| > \frac{|M|}{2} > 0$$

and thus,

$$\frac{1}{|g(\mathbf{x})|} < \frac{2}{|M|}. \tag{8.36}$$

also, from

$$\lim_{\mathbf{x} \to \mathbf{x}_0} g(\mathbf{x}) = M,$$

there exists $\delta_4 > 0$ such that if $\mathbf{x} \in D$ and $0 < |\mathbf{x} - \mathbf{x}_0| < \delta_4$, then

$$|g(\mathbf{x}) - M| < \frac{\varepsilon |M|^2}{2}. \tag{8.37}$$

Define $\delta_5 = \min\{\delta_3, \delta_4\}$. Therefore, if $\mathbf{x} \in D$ and $0 < |\mathbf{x} - \mathbf{x}_0| < \delta_5$, from (8.34), (8.36), and (8.37), we obtain

$$\left| \frac{1}{g(\mathbf{x})} - \frac{1}{M} \right| = \frac{|g(\mathbf{x}) - M|}{|g(\mathbf{x})||M|}$$

$$< \frac{\varepsilon |M|^2}{2} \frac{2}{|M||M|}$$

$$= \varepsilon \tag{8.38}$$

Hence,

$$\lim_{\mathbf{x} \to \mathbf{x}_0} \frac{1}{g(\mathbf{x})} = \frac{1}{M}.$$

Finally, from this and item 4, we have

$$\lim_{\mathbf{x} \to \mathbf{x}_0} \frac{f(\mathbf{x})}{g(\mathbf{x})} = \lim_{\mathbf{x} \to \mathbf{x}_0} f(\mathbf{x}) \cdot \lim_{\mathbf{x} \to \mathbf{x}_0} \frac{1}{g(\mathbf{x})} = \frac{L}{M}.$$

The proof is complete.

8.5.4 Limits for Composed Functions

We start with the main result.

Theorem 8.5.19 *Let $D \subset \mathbb{R}^n$ and $E \subset \mathbb{R}$ be nonempty sets. Let $f : D \to \mathbb{R}$ and $g : E \to \mathbb{R}$ be functions such that $f(D) \subset E$.*
   Let $\mathbf{x}_0 \in D'$ and $y_0 \in E \cap E'$.
   Suppose that

$$\lim_{\mathbf{x} \to \mathbf{x}_0} f(\mathbf{x}) = y_0$$

and

$$\lim_{y \to y_0} g(y) = g(y_0).$$

   Under such hypotheses,

$$\lim_{\mathbf{x} \to \mathbf{x}_0} g(f(\mathbf{x})) = g(y_0).$$

Proof Let $\varepsilon > 0$. From

$$\lim_{y \to y_0} g(y) = g(y_0)$$

there exists $\eta > 0$ such that if $y \in E$ and $|y - y_0| < \eta$, then

$$|g(y) - g(y_0)| < \varepsilon. \tag{8.39}$$

   From

$$\lim_{\mathbf{x} \to \mathbf{x}_0} f(\mathbf{x}) = y_0,$$

there exists $\delta > 0$ such that if $\mathbf{x} \in D$ and $0 < |\mathbf{x} - \mathbf{x}_0| < \delta$, then

$$|f(\mathbf{x}) - y_0| < \eta.$$

   From this and (8.39), we obtain

$$|g(f(\mathbf{x})) - g(y_0)| < \varepsilon,$$

if $\mathbf{x} \in D$ and $0 < |\mathbf{x} - \mathbf{x}_0| < \delta$.

Since $\varepsilon > 0$ is arbitrary, we may conclude that

$$\lim_{\mathbf{x} \to \mathbf{x}_0} g(f(\mathbf{x})) = g(y_0).$$

8.6 Continuous Functions

Definition 8.6.1 Let $D \subset \mathbb{R}^n$ be a nonempty set and let $f : D \to \mathbb{R}$ be a function. For $\mathbf{x}_0 \in D$, we say that f is continuous at $\mathbf{x}_0$, if for each $\varepsilon > 0$ there exists $\delta > 0$ such that if $\mathbf{x} \in D$ and $|\mathbf{x} - \mathbf{x}_0| < \delta$, then

$$|f(\mathbf{x}) - f(\mathbf{x}_0)| < \varepsilon.$$

Remark 8.6.2 In particular, if $\mathbf{x}_0 \in D \cap D'$, f will be continuous at $\mathbf{x}_0$ if, and only if,

$$\lim_{\mathbf{x} \to \mathbf{x}_0} f(\mathbf{x}) = f(\mathbf{x}_0).$$

Also, if $\mathbf{x}_0 \in D$ is an isolated point of D, then all function $f : D \to \mathbb{R}$ is continuous at $\mathbf{x}_0$.

Indeed, in such a case, there exists $r > 0$ such that

$$B_r(\mathbf{x}_0) \cap D = \{\mathbf{x}_0\}.$$

Hence, it suffices to choose $0 < \delta < r$ to satisfy the continuity definition, $\forall \varepsilon > 0$.

Theorem 8.6.3 *Let $D \subset \mathbb{R}^n$ be a nonempty set and let $f, g : D \to \mathbb{R}$ be continuous functions at $\mathbf{x}_0 \in D \cap D'$.*

Under such hypotheses,

1. *αf is continuous at $\mathbf{x}_0$, $\forall \alpha \in \mathbb{R}$.*
2. *$f + g$ is continuous at $\mathbf{x}_0$.*
3. *$f \cdot g$ is continuous at $\mathbf{x}_0$.*
4.

$$\frac{f}{g} \text{ is continuous at } \mathbf{x}_0, \text{ if } g(\mathbf{x}_0) \neq 0.$$

Proof The proof results directly from the limit properties.

For example, for the item 2, from the continuity of f and g at $\mathbf{x}_0$, we obtain

$$\lim_{\mathbf{x} \to \mathbf{x}_0} f(\mathbf{x}) = f(\mathbf{x}_0)$$

and

$$\lim_{\mathbf{x} \to \mathbf{x}_0} g(\mathbf{x}) = g(\mathbf{x}_0)$$

so that from the limit properties, we obtain

$$\lim_{\mathbf{x} \to \mathbf{x}_0} f(\mathbf{x}) + g(\mathbf{x}) = f(\mathbf{x}_0) + g(\mathbf{x}_0).$$

The remaining items may be proven similarly.

Exercise 8.6.4 Verify if $f : \mathbb{R}^2 \to \mathbb{R}$ is continuous at $(0, 0)$, where

$$f(x, y) = \begin{cases} \sin\left(\frac{x^2 + 3xy}{\sqrt{x^2 + y^2}}\right), & \text{if } (x, y) \neq (0, 0) \\ 1, & \text{if } (x, y) = (0, 0). \end{cases}$$

Solution: Observe that if $(x, y) \neq (0, 0)$, then

$$0 \leq \left| \frac{x^2 + 3xy}{\sqrt{x^2 + y^2}} \right|$$

$$\leq \frac{x^2 + 3|xy|}{\sqrt{x^2 + y^2}}$$

$$= \frac{|x||x| + 3|x||y|}{\sqrt{x^2 + y^2}}$$

$$= \frac{|x||x|}{\sqrt{x^2 + y^2}} + \frac{3|x||y|}{\sqrt{x^2 + y^2}}$$

$$\leq |x| + 3|x|$$

$$= 4|x|. \tag{8.40}$$

Since

$$\lim_{(x,y) \to (0,0)} 4|x| = 0,$$

from this, (8.40) and the Sandwich Theorem, we obtain

$$\lim_{(x,y) \to (0,0)} \left| \frac{x^2 + 3xy}{\sqrt{x^2 + y^2}} \right| = 0,$$

so that,

$$\lim_{(x,y)\to(0,0)} \frac{x^2 + 3xy}{\sqrt{x^2 + y^2}} = 0.$$

From this, since

$$\lim_{y\to 0} \operatorname{sen}(y) = \operatorname{sen}(0) = 0,$$

from Theorem 8.5.19 (limit for composed functions) we have

$$\lim_{(x,y)\to(0,0)} f(x, y) = \lim_{(x,y)\to(0,0)} \sin\left(\frac{x^2 + 3xy}{\sqrt{x^2 + y^2}}\right)$$

$$= \sin\left(\lim_{(x,y)\to(0,0)} \frac{x^2 + 3xy}{\sqrt{x^2 + y^2}}\right)$$

$$= \sin(0)$$

$$= 0. \tag{8.41}$$

Thus,

$$\lim_{(x,y)\to(0,0)} f(x, y) = 0 \neq 1 = f(0, 0).$$

We may conclude that f is not continuous at $(0, 0)$.
The solution is complete.

Exercise 8.6.5 Let $(x_0, y_0) \in \mathbb{R}^2$.
Prove by induction that

$$\lim_{(x,y)\to(x_0,y_0)} x^n = x_0^n, \ \forall n \in \mathbb{N}.$$

Similarly, conclude that

$$\lim_{(x,y)\to(x_0,y_0)} y^m = y_0^m, \ \forall m \in \mathbb{N}.$$

Exercise 8.6.6 Let $f : \mathbb{R}^2 \to \mathbb{R}$ be such

$$f(x, y) = \sum_{n=0}^{N} \sum_{m=0}^{M} a_{mn} x^m y^n.$$

Let $(x_0, y_0) \in \mathbb{R}^2$. Through the limit properties and the last exercise, show that

$$\lim_{(x,y)\to(x_0,y_0)} f(x, y) = f(x_0, y_0).$$

Exercise 8.6.7 Let $f : D \to \mathbb{R}$ be such that

$$f(x, y) = \frac{\sum_{n=0}^{N} \sum_{m=0}^{M} a_{mn} x^m y^n}{\sum_{n=0}^{N_1} \sum_{m=0}^{M_1} b_{mn} x^m y^n},$$

where

$$D = \left\{ (x, y) \in \mathbb{R}^2 : \sum_{n=0}^{N_1} \sum_{m=0}^{M_1} b_{mn} x^m y^n \neq 0 \right\}.$$

Let $(x_0, y_0) \in D$, that is, $(x_0, y_0) \in \mathbb{R}^2$ such that

$$\sum_{n=0}^{N_1} \sum_{m=0}^{M_1} b_{mn} x_0^m y_0^n \neq 0.$$

Through the limit properties and the last exercise, show that

$$\lim_{(x,y)\to(x_0,y_0)} f(x, y) = f(x_0, y_0).$$

8.6.1   Types of Discontinuities

Definition 8.6.8 Let $D \subset \mathbb{R}^n$ and let $f : D \to \mathbb{R}$ be a function. Let $\mathbf{x}_0 \in D \cap D'$. Suppose that f is discontinuous (not continuous) at $\mathbf{x}_0$.

We say that such discontinuity is removable if it exists

$$\lim_{\mathbf{x}\to\mathbf{x}_0} f(\mathbf{x}).$$

Observe that in such a case, we have,

$$\lim_{\mathbf{x}\to\mathbf{x}_0} f(\mathbf{x}) \neq f(\mathbf{x}_0).$$

Thus, we may "remove" such a discontinuity redefining f at $\mathbf{x}_0$, now denoting it by $\tilde{f} : D \to \mathbb{R}$, where

$$\tilde{f}(\mathbf{x}) = \begin{cases} f(\mathbf{x}), & \text{if } \mathbf{x} \neq \mathbf{x}_0 \\ \lim_{\mathbf{y} \to \mathbf{x}_0} f(\mathbf{y}), & \text{if } \mathbf{x} = \mathbf{x}_0. \end{cases}$$

Hence, the new function $\tilde{f} : D \to \mathbb{R}$ is continuous at $\mathbf{x}_0$ (the discontinuity has been removed).

Finally, if a discontinuity at $\mathbf{x}_0$ is not removable, it is said to be essential. In such case it does not exist $\lim_{\mathbf{x} \to \mathbf{x}_0} f(\mathbf{x})$.

Exercise 8.6.9 Let $f : \mathbb{R}^2 \to \mathbb{R}$ where,

$$f(x, y) = \begin{cases} \frac{xy}{|x|+|y|}, & \text{if } (x, y) \neq (0, 0) \\ 1, & \text{if } (x, y) = (0, 0). \end{cases}$$

Find in which points of $\mathbb{R}^2$ f is discontinuous and classify such discontinuities in removable or essential.

Solution: Observe that

$$\lim_{(x,y) \to (x_0,y_0)} f(x, y) = \lim_{(x,y) \to (x_0,y_0)} \frac{xy}{|x| + |y|} = \frac{x_0 y_0}{|x_0| + |y_0|} = f(x_0, y_0),$$

$\forall (x_0, y_0) \neq (0, 0)$.

Thus, the only point where possibly f is discontinuous is $(0, 0)$.

Observe that if $(x, y) \neq (0, 0)$, then

$$0 \leq \left| \frac{xy}{|x| + |y|} \right| = \frac{|x||y|}{|x| + |y|} \leq |x|. \tag{8.42}$$

Since

$$\lim_{(x,y) \to (0,0)} |x| = 0,$$

from this, (8.42) and the sandwich theorem, we obtain

$$\lim_{(x,y) \to (0,0)} \left| \frac{xy}{|x| + |y|} \right| = 0,$$

so that

$$\lim_{(x,y) \to (0,0)} \frac{xy}{|x| + |y|} = 0,$$

Thus,

$$\lim_{(x,y) \to (0,0)} f(x, y) = 0 \neq 1 = f(0, 0),$$

that is, f is discontinuous at $(0, 0)$ and such a discontinuity is removable.

The solution is complete.

8.7   Partial Derivatives

Definition 8.7.1 Let $D \subset \mathbb{R}^n$ be a nonempty set and let $f : D \to \mathbb{R}$ be a function. Let $\mathbf{x} \in D^0$, where D^0 denotes the interior of D. We define the partial derivative of f related to x_j at the point $\mathbf{x} = (x_1, \ldots, x_n) \in D^0$, denoted by

$$\frac{\partial f(\mathbf{x})}{\partial x_j},$$

as

$$\frac{\partial f(\mathbf{x})}{\partial x_j} = \lim_{h \to 0} \frac{f(x_1, \ldots, x_j + h, \ldots, x_n) - f(x_1, \ldots, x_j, \ldots, x_n)}{h},$$

if such a limit exists.

Remark 8.7.2 In particular for $D \subset \mathbb{R}^2$ and $\mathbf{x} = (x, y) \in D^0$, we denote,

$$\frac{\partial f(x, y)}{\partial x} = \lim_{\Delta x \to 0} \frac{f(x + \Delta x, y) - f(x, y)}{\Delta x},$$

and

$$\frac{\partial f(x, y)}{\partial y} = \lim_{\Delta y \to 0} \frac{f(x, y + \Delta y) - f(x, y)}{\Delta y}.$$

Example 8.7.3 Let $f : D \to \mathbb{R}$ where

$$D = \{(x, y) \in \mathbb{R}^2 \; : \; 2x + 3y \neq 0\}$$

and

$$f(x, y) = \frac{1}{2x + 3y}.$$

Through the definition of partial derivative, we are going to obtain

$$\frac{\partial f(x, y)}{\partial x}$$

at

$$\mathbf{x} = (x, y) \in D^0 = D.$$

$$\frac{\partial f(x, y)}{\partial x} = \lim_{\Delta x \to 0} \frac{f(x + \Delta x, y) - f(x, y)}{\Delta x}$$

$$= \lim_{\Delta x \to 0} \frac{\frac{1}{2(x + \Delta x) + 3y} - \frac{1}{2x + 3y}}{\Delta x}$$

$$= \lim_{\Delta x \to 0} \frac{1}{\Delta x} \left(\frac{2x + 3y - (2(x + \Delta x) + 3y)}{(2(x + \Delta x) + 3y)(2x + 3y)} \right)$$

$$= \lim_{\Delta x \to 0} \frac{-2\Delta x}{\Delta x[2(x + \Delta x) + 3y](2x + 3y)}$$

$$= \lim_{\Delta x \to 0} \frac{-2}{[2(x + \Delta x) + 3y](2x + 3y)}$$

$$= \frac{-2}{(2x + 3y)^2}. \tag{8.43}$$

Therefore,

$$\frac{\partial f(x, y)}{\partial x} = \frac{-2}{(2x + 3y)^2}.$$

Definition 8.7.4 (Directional Derivative) Let $D \subset \mathbb{R}^n$ be a nonempty set and let $f : D \to \mathbb{R}$ be a function.

Let $\mathbf{x} \in D^0$ and $\mathbf{v} \in \mathbb{R}^n$. We define the derivative of f related to $\mathbf{v}$ at $\mathbf{x}$, denoted by

$$\frac{\partial f(\mathbf{x})}{\partial \mathbf{v}},$$

as

$$\frac{\partial f(\mathbf{x})}{\partial \mathbf{v}} = \lim_{h \to 0} \frac{f(\mathbf{x} + h\mathbf{v}) - f(\mathbf{x})}{h},$$

if such a limit exists.

Example 8.7.5 Let $f : \mathbb{R}^2 \to \mathbb{R}$ where

$$f(x, y) = x^2 + y^2.$$

Let $\mathbf{x} = (x, y) \in \mathbb{R}^2$ and

$$\mathbf{v} = (\cos(\theta), \operatorname{sen}(\theta)) \in \mathbb{R}^2.$$

We are going to calculate

$$\frac{\partial f(x, y)}{\partial \mathbf{v}}.$$

Observe that,

$$
\begin{aligned}
\frac{\partial f(x, y)}{\partial \mathbf{v}} &= \lim_{h \to 0} \frac{f(\mathbf{x} + h\mathbf{v}) - f(\mathbf{x})}{h} \\
&= \lim_{h \to 0} \frac{f(x + h\cos(\theta), y + h\sin(\theta)) - f(x, y)}{h} \\
&= \lim_{h \to 0} \frac{(x + h\cos(\theta))^2 + (y + h\sin(\theta))^2 - (x^2 + y^2)}{h} \\
&= \lim_{h \to 0} \frac{x^2 + 2xh\cos(\theta) + h^2\cos^2(\theta) + y^2 + 2yh\sin(\theta) + h^2\sin^2(\theta) - x^2 - y^2}{h} \\
&= \lim_{h \to 0} \frac{2xh\cos(\theta) + h^2\cos^2(\theta) + 2yh\sin(\theta) + h^2\sin^2(\theta)}{h} \\
&= \lim_{h \to 0} [2x\cos(\theta) + h\cos^2(\theta) + 2y\sin(\theta) + h\sin^2(\theta)] \\
&= 2x\cos(\theta) + 2y\sin(\theta).
\end{aligned}
\tag{8.44}
$$

Theorem 8.7.6 (Mean Value Theorem for $\mathbb{R}^n$) *Let $D \subset \mathbb{R}^n$ be a nonempty set and $f : D \to \mathbb{R}$ be a continuous function.*
Assume $\mathbf{x} \in D^0$ and $\mathbf{v} \in \mathbb{R}^n$ are such that

$$A \subset D^0$$

where,

$$A = \{\mathbf{x} + t\mathbf{v} \; : \; t \in [0, 1]\}.$$

Suppose

$$\frac{\partial f(\mathbf{y})}{\partial \mathbf{v}}$$

exists $\forall \mathbf{y} \in A$.

Under such hypotheses, there exists $t_0 \in (0, 1)$ such that

$$f(\mathbf{x} + \mathbf{v}) - f(\mathbf{x}) = \frac{\partial f(\mathbf{x} + t_0 \mathbf{v})}{\partial \mathbf{v}}.$$

Proof Define $g : [0, 1] \to \mathbb{R}$ where

$$g(t) = f(\mathbf{x} + t\mathbf{v}), \ \forall t \in [0, 1].$$

Observe that

$$g(0) = f(\mathbf{x}),$$

$$g(1) = f(\mathbf{x} + \mathbf{v})$$

and

$$\begin{aligned} g'(t) &= \lim_{h \to 0} \frac{g(t + h) - g(t)}{h} \\ &= \lim_{h \to 0} \frac{f(\mathbf{x} + (t + h)\mathbf{v}) - f(\mathbf{x} + t\mathbf{v})}{h} \\ &= \frac{\partial f(\mathbf{x} + t\mathbf{v})}{\partial \mathbf{v}}, \end{aligned} \tag{8.45}$$

and such a derivative is well defined $\forall t \in (0, 1)$.

Therefore, from the mean value theorem for $\mathbb{R}^1$, there exists $t_0 \in (0, 1)$ such that

$$g(1) - g(0) = g'(t_0)(1 - 0) = g'(t_0),$$

that is,

$$f(\mathbf{x} + \mathbf{v}) - f(\mathbf{x}) = \frac{\partial f(\mathbf{x} + t_0 \mathbf{v})}{\partial \mathbf{v}}.$$

The proof is complete.

8.8   Differentiability in $\mathbb{R}^n$

We start with the main definition.

Definition 8.8.1 (Differentiability) Let $D \subset \mathbb{R}^n$ be a nonempty set and let $f : D \to \mathbb{R}$ be a function and $\mathbf{x} \in D^\circ$.

We say that f is differentiable at $\mathbf{x}$, if there exists a vector $\mathbf{a} = (a_1, \ldots, a_n) \in \mathbb{R}^n$ such that the function $r : B_{r_0}(\mathbf{0}) \to \mathbb{R}$ defined through the relation

$$f(\mathbf{x} + \mathbf{v}) = f(\mathbf{x}) + \mathbf{a} \cdot \mathbf{v} + r(\mathbf{v}), \; \forall \mathbf{v} \in B_{r_0}(\mathbf{0}),$$

for some $r_0 > 0$, that is,

$$r(\mathbf{v}) = f(\mathbf{x} + \mathbf{v}) - f(\mathbf{x}) - \mathbf{a} \cdot \mathbf{v},$$

is also such that

$$\lim_{\mathbf{v} \to \mathbf{0}} \frac{r(\mathbf{v})}{|\mathbf{v}|} = 0.$$

Remark 8.8.2 Observe that in the context of this last definition, in particular for $\mathbf{v} = t\mathbf{e}_j$, we have

$$f(\mathbf{x} + t\mathbf{e}_j) - f(\mathbf{x}) = \mathbf{a} \cdot (t\mathbf{e}_j) + r(t\mathbf{e}_j)$$

and thus,

$$\begin{aligned}
\frac{f(\mathbf{x} + t\mathbf{e}_j) - f(\mathbf{x})}{t} &= \mathbf{a} \cdot (\mathbf{e}_j) + \frac{r(t\mathbf{e}_j)}{t} \\
&= a_j + \frac{r(t\mathbf{e}_j)}{t} \\
&\to a_j, \text{ as } t \to 0.
\end{aligned} \tag{8.46}$$

Therefore

$$\begin{aligned}
a_j &= \lim_{t \to 0} \frac{f(\mathbf{x} + t\mathbf{e}_j) - f(\mathbf{x})}{t} \\
&= \frac{\partial f(\mathbf{x})}{\partial \mathbf{e}_j} \\
&= \frac{\partial f(\mathbf{x})}{\partial x_j}.
\end{aligned} \tag{8.47}$$

Hence, if f is differentiable at $\mathbf{x}$, then necessarily

$$\mathbf{a} = \left(\frac{\partial f(\mathbf{x})}{\partial x_1}, \ldots, \frac{\partial f(\mathbf{x})}{\partial x_n} \right).$$

Defining the gradient of f at $\mathbf{x}$ by

$$\nabla f(\mathbf{x}) = \left(\frac{\partial f(\mathbf{x})}{\partial x_1}, \dots, \frac{\partial f(\mathbf{x})}{\partial x_n} \right),$$

we have

$$f(\mathbf{x} + \mathbf{v}) - f(\mathbf{x}) = \sum_{j=1}^{n} a_j v_j + r(\mathbf{v})$$

$$= \sum_{j=1}^{n} \frac{\partial f(\mathbf{x})}{\partial x_j} v_j + r(\mathbf{v})$$

$$= \nabla f(\mathbf{x}) \cdot \mathbf{v} + r(\mathbf{v}), \tag{8.48}$$

and, in particular

$$f(\mathbf{x} + t\mathbf{v}) - f(\mathbf{x}) = \nabla f(\mathbf{x}) \cdot (t\mathbf{v}) + r(t\mathbf{v}), \tag{8.49}$$

so that

$$\frac{f(\mathbf{x} + t\mathbf{v}) - f(\mathbf{x})}{t} = \nabla f(\mathbf{x}) \cdot (\mathbf{v}) + \frac{r(t\mathbf{v})}{t}$$

$$\rightarrow \nabla f(\mathbf{x}) \cdot (\mathbf{v}), \text{ as } t \rightarrow 0. \tag{8.50}$$

Therefore,

$$\frac{\partial f(\mathbf{x})}{\partial \mathbf{v}} = \lim_{t \to 0} \frac{f(\mathbf{x} + t\mathbf{v}) - f(\mathbf{x})}{t} = \nabla f(\mathbf{x}) \cdot (\mathbf{v}). \tag{8.51}$$

Proposition 8.8.3 *Let $D \subset \mathbb{R}^n$ be a nonempty set. Let $f : D \to \mathbb{R}$ be a function. Assume that f is differentiable at $\mathbf{x} \in D^\circ$.*

Under such hypotheses, f is continuous at $\mathbf{x}$.

Proof From the hypotheses, there exists $r_0 > 0$ such that $r : B_{r_0}(\mathbf{0}) \to \mathbb{R}$ defined through the relation

$$f(\mathbf{x} + \mathbf{v}) = f(\mathbf{x}) + \nabla f(\mathbf{x}) \cdot \mathbf{v} + r(\mathbf{v}), \forall \mathbf{v} \in B_{r_0}(\mathbf{0}),$$

is also such that

$$\frac{r(\mathbf{v})}{|\mathbf{v}|} \rightarrow 0, \text{ when } \mathbf{v} \rightarrow \mathbf{0}.$$

Let $\mathbf{y} \in B_{r_0}(\mathbf{x})$ and let $\mathbf{v} \in B_{r_0}(\mathbf{0})$ be such that $\mathbf{y} = \mathbf{x} + \mathbf{v}$. Therefore, $\mathbf{y} \rightarrow \mathbf{x}$ if, and only if, $\mathbf{v} \rightarrow \mathbf{0}$.

Thus,

$$f(\mathbf{y}) = f(\mathbf{x}) + \nabla f(\mathbf{x}) \cdot (\mathbf{y} - \mathbf{x}) + r(\mathbf{y} - \mathbf{x}),$$

where

$$r(\mathbf{y} - \mathbf{x}) \to 0, \quad \text{as } \mathbf{y} \to \mathbf{x}.$$

Hence,

$$\lim_{\mathbf{y} \to \mathbf{x}} f(\mathbf{y}) = f(\mathbf{x}),$$

that is, f is continuous at $\mathbf{x}$.

The proof is complete.

8.9 More Details About the Differentiability in $\mathbb{R}^2$

Let us see specifically for the case $\mathbb{R}^2$, more details about the differentiability definition.

Definition 8.9.1 Let $f : D \subset \mathbb{R}^2 \to \mathbb{R}$ be a function and let $(x_0, y_0) \in D^0$. Let $r_0 > 0$ be such that

$$B_{r_0}(x_0, y_0) \subset D.$$

Assume that

$$\frac{\partial f(x_0, y_0)}{\partial x} \text{ and } \frac{\partial f(x_0, y_0)}{\partial y}$$

exist.

In such a case, we say that f is differentiable at (x_0, y_0), if $r(\Delta x, \Delta y)$ defined by

$$r(\Delta x, \Delta y) = f(x_0 + \Delta x, y_0 + \Delta y) - f(x_0, y_0) - \frac{\partial f(x_0, y_0)}{\partial x} \Delta x - \frac{\partial f(x_0, y_0)}{\partial y} \Delta y,$$

$\forall (\Delta x, \Delta y) \in B_{r_0}(0, 0)$, is also such that

$$\lim_{(\Delta x, \Delta y) \to (0,0)} \frac{r(\Delta x, \Delta y)}{\sqrt{\Delta x^2 + \Delta_y^2}} = 0.$$

Example 8.9.2 Let $f : \mathbb{R}^2 \to \mathbb{R}$ defined by

$$f(x, y) = 3x^2 + 2y.$$

Let us show that f is differentiable in $\mathbb{R}^2$.
Let $(x, y) \in \mathbb{R}^2$.
Thus,

$$
\begin{aligned}
f(x + \Delta x, y + \Delta y) - f(x, y) &= 3(x + \Delta x)^2 + (y + \Delta y) - (3x^2 + y) \\
&= 3(x^2 + 2x\Delta x + \Delta x^2) + (y + \Delta y) - (3x^2 + y) \\
&= 6x\Delta x + \Delta y + 3\Delta x^2.
\end{aligned} \tag{8.52}
$$

Also,

$$\frac{\partial f(x, y)}{\partial x}\Delta x + \frac{\partial f(x, y)}{\partial y}\Delta y = 6x\Delta x + \Delta y.$$

Hence, if $(\Delta x, \Delta y) \neq (0, 0)$, then

$$
\begin{aligned}
\left| \frac{r(\Delta x, \Delta y)}{\sqrt{\Delta x^2 + \Delta y^2}} \right| &= \left| \frac{f(x + \Delta x, y + \Delta y) - f(x, y) - \frac{\partial f(x,y)}{\partial x}\Delta x - \frac{\partial f(x,y)}{\partial y}\Delta y}{\sqrt{\Delta x^2 + \Delta y^2}} \right| \\
&= \frac{3\Delta x^2}{\sqrt{\Delta x^2 + \Delta y^2}} \\
&= \frac{|\Delta x||\Delta x|}{\sqrt{\Delta x^2 + \Delta y^2}} \\
&\leq |\Delta x|
\end{aligned} \tag{8.53}
$$

Since

$$\lim_{(\Delta x, \Delta y) \to (0,0)} |\Delta x| = 0,$$

from this, (8.53) and from the sandwich theorem, we obtain

$$\lim_{(\Delta x, \Delta y) \to (0,0)} \frac{r(\Delta x, \Delta y)}{\sqrt{\Delta x^2 + \Delta y^2}} = 0.$$

We may conclude that f is differentiable at (x, y), $\forall (x, y) \in \mathbb{R}^2$.

Example 8.9.3 Consider the example in which $f : \mathbb{R}^2 \to \mathbb{R}$ is defined by

$$f(x, y) = \frac{y}{x^2 + 1}.$$

We are going to show that f is differentiable in $\mathbb{R}^2$.

Let $(x, y) \in \mathbb{R}^2$.

Since the partial derivatives of f are well defined, in order that such a function be differentiable at (x, y), it suffices the function $r(\Delta x, \Delta y)$, defined through the relation

$$f(x + \Delta x, y + \Delta y) = f(x, y) + \nabla f(x, y) \cdot (\Delta x, \Delta y) + r(\Delta x, \Delta y),$$

that is,

$$r(\Delta x, \Delta y) = f(x + \Delta x, y + \Delta y) - f(x, y) - \frac{\partial f(x, y)}{\partial x} \Delta x - \frac{\partial f(x, y)}{\partial y} \Delta y,$$

be such that

$$\lim_{(\Delta x, \Delta y) \to (0,0)} \frac{r(\Delta x, \Delta y)}{\sqrt{\Delta x^2 + \Delta y^2}} = 0.$$

Observe that

$$\frac{\partial f(x, y)}{\partial x} = -\frac{2yx}{(1 + x^2)^2},$$

and

$$\frac{\partial f(x, y)}{\partial y} = \frac{1}{1 + x^2},$$

and thus, for $(\Delta x, \Delta y) \neq (0, 0)$, we have that,

$$r(\Delta x, \Delta y) = \frac{y + \Delta y}{1 + (x + \Delta x)^2} - \frac{y}{1 + x^2}$$

$$+ \frac{2yx \Delta x}{(1 + x^2)^2} - \frac{\Delta y}{1 + x^2}$$

$$= \frac{(y + \Delta y)(1 + x^2) - (1 + (x + \Delta x)^2)y}{(1 + x^2)(1 + (x + \Delta x)^2)}$$

$$+ \frac{2yx \Delta x - (1 + x^2)\Delta y}{(1 + x^2)^2}$$

$$= \frac{\Delta y(1+x)^2 - 2xy\Delta x - \Delta x^2 y}{(1+x^2)(1+(x+\Delta x)^2)}$$

$$+ \frac{2yx\Delta x - (1+x^2)\Delta y}{(1+x^2)^2}$$

$$= -2xy\left(\frac{1}{(1+x^2)(1+(x+\Delta x)^2)} - \frac{1}{(1+x^2)^2}\right)\Delta x$$

$$+ \left(\frac{1}{1+(x+\Delta x)^2} - \frac{1}{1+x^2}\right)\Delta y$$

$$- \frac{y\Delta x^2}{(1+x^2)(1+(x+\Delta x)^2)}. \tag{8.54}$$

Thus,

$$\left|\frac{r(\Delta x, \Delta y)}{\sqrt{\Delta x^2 + \Delta y^2}}\right| \le \left|-2xy\left(\frac{1}{(1+x^2)(1+(x+\Delta x)^2)} - \frac{1}{(1+x^2)^2}\right)\right| \cdot \left|\frac{\Delta x}{\sqrt{\Delta x^2 + \Delta y^2}}\right|$$

$$+ \left|\left(\frac{1}{1+(x+\Delta x)^2} - \frac{1}{1+x^2}\right)\right|\left|\frac{\Delta y}{\sqrt{\Delta x^2 + \Delta y^2}}\right|$$

$$\left|-\frac{y}{(1+x^2)(1+(x+\Delta x)^2)}\right| \cdot \frac{|\Delta x|^2}{\sqrt{\Delta x^2 + \Delta y^2}}. \tag{8.55}$$

so that

$$\left|\frac{r(\Delta x, \Delta y)}{\sqrt{\Delta x^2 + \Delta y^2}}\right| \le \left|-2xy\left(\frac{1}{(1+x^2)(1+(x+\Delta x)^2)} - \frac{1}{(1+x^2)^2}\right)\right|$$

$$+ \left|\left(\frac{1}{1+(x+\Delta x)^2} - \frac{1}{1+x^2}\right)\right|$$

$$\left|-\frac{y}{(1+x^2)(1+(x+\Delta x)^2)}\right| \cdot |\Delta x|. \tag{8.56}$$

Observe that

$$\lim_{(\Delta x, \Delta y) \to (0,0)} \left|-2xy\left(\frac{1}{(1+x^2)(1+(x+\Delta x)^2)} - \frac{1}{(1+x^2)^2}\right)\right|$$

$$+ \lim_{(\Delta x, \Delta y) \to (0,0)} \left|\left(\frac{1}{1+(x+\Delta x)^2} - \frac{1}{1+x^2}\right)\right|$$

$$+ \lim_{(\Delta x, \Delta y) \to (0,0)} \left| -\frac{y}{(1+x^2)(1+(x+\Delta x)^2)} \right| \cdot \lim_{(\Delta x, \Delta y) \to (0,0)} |\Delta x|$$
$$= 0. \tag{8.57}$$

From this, (8.56) and from the sandwich theorem, we may conclude that

$$\lim_{(\Delta x, \Delta y) \to (0,0)} \frac{r(\Delta x, \Delta y)}{\sqrt{\Delta x^2 + \Delta y^2}} = 0.$$

Thus f is differentiable at (x, y), $\forall (x, y) \in \mathbb{R}^2$.

8.9.1   Interpretation of Differentiability

Particularly if $f : D \subset \mathbb{R}^2$ is differentiable at $(x, y) \in D^0$ we have that

$$r(\Delta x, \Delta y) = f(x + \Delta x, y + \Delta y) - f(x, y) - \frac{\partial f(x, y)}{\partial x} \Delta x - \frac{\partial f(x, y)}{\partial y} \Delta y,$$

is such that

$$\lim_{(\Delta x, \Delta y) \to (0,0)} \frac{r(\Delta x, \Delta y)}{\sqrt{\Delta x^2 + \Delta y^2}} = 0.$$

Thus, denoting

$$\eta(\Delta x, \Delta y) = \frac{r(\Delta x, \Delta y)}{\sqrt{\Delta x^2 + \Delta y^2}}$$

we have,

$$f(x + \Delta x, y + \Delta y) - f(x, y) = \frac{\partial f(x, y)}{\partial x} \Delta x + \frac{\partial f(x, y)}{\partial y} \Delta y + r(\Delta x, \Delta y)$$

$$= \frac{\partial f(x, y)}{\partial x} \Delta x + \frac{\partial f(x, y)}{\partial y} \Delta y$$

$$+ \eta(\Delta x, \Delta y) \sqrt{\Delta x^2 + \Delta y^2}, \tag{8.58}$$

pagebreak
where

$$\lim_{(\Delta x, \Delta y) \to (0,0)} \eta(\Delta x, \Delta y) = \lim_{(\Delta x, \Delta y) \to (0,0)} \frac{r(\Delta x, \Delta y)}{\sqrt{\Delta x^2 + \Delta y^2}} = 0.$$

Therefore, if $(\Delta x, \Delta y) \approx (0, 0)$, we may write

$$\Delta f(x, y, \Delta x, \Delta y) \approx \frac{\partial f(x, y)}{\partial x} \Delta x + \frac{\partial f(x, y)}{\partial y} \Delta y,$$

where

$$\Delta f(x, y, \Delta x, \Delta y) = f(x + \Delta x, y + \Delta y) - f(x, y).$$

Hence, at this point of this text, we are going to define the differential of a function at a point.

Definition 8.9.4 (Differential) Let $f : D \subset \mathbb{R}^2 \to \mathbb{R}$ be a differentiable function at $(x, y) \in D^0$.

We define the differential of f at (x, y), denoted by $df(x, y)$, as

$$df(x, y) = \frac{\partial f(x, y)}{\partial x} dx + \frac{\partial f(x, y)}{\partial y} dy.$$

Observe that the definition of differential is an abstraction. Its meaning is that if f is differentiable at (x, y) and $\Delta x, \Delta y$ are "small" in absolute value, then

$$\Delta f(x, y, \Delta x, \Delta y) = f(x + \Delta x, y + \Delta y) - f(x, y) \approx \frac{\partial f(x, y)}{\partial x} \Delta x + \frac{\partial f(x, y)}{\partial y} \Delta y,$$

or, more precisely,

$$\Delta f(x, y, \Delta x, \Delta y) = \frac{\partial f(x, y)}{\partial x} \Delta x + \frac{\partial f(x, y)}{\partial y} \Delta y + \eta(\Delta x, \Delta y) \sqrt{\Delta x^2 + \Delta y^2},$$

where

$$\lim_{(\Delta x, \Delta y) \to (0,0)} \eta(\Delta x, \Delta y) = 0.$$

8.9.2 Sufficient Conditions for Differentiability

In this section we are going to establish sufficient conditions for a function in $\mathbb{R}^2$ be differentiable at a point. Indeed, such a result is more general and may be easily extended for $\mathbb{R}^n$.

Theorem 8.9.5 Let $D \subset \mathbb{R}^2$ be a nonempty set and $f : D \to \mathbb{R}$ a function. Let $(x_0, y_0) \in D^\circ$ be such that there exists $r_0 > 0$ such that $f_x(x, y)$ and $f_y(x, y)$ exist on $B_{r_0}(x_0, y_0)$ and are both continuous at (x_0, y_0), where

$$B_{r_0}(x_0, y_0) = \left\{ (x, y) \in \mathbb{R}^2 : \sqrt{(x - x_0)^2 + (y - y_0)^2} < r_0 \right\}.$$

Under such hypotheses, f is differentiable at (x_0, y_0).

Proof For $(\Delta x, \Delta y) \in B_{r_0}(\mathbf{0}) \setminus \{(0, 0)\}$, define

$$r(\Delta x, \Delta y) = \Delta f(x_0, y_0, \Delta x, \Delta y) - f_x(x_0, y_0)\Delta x - f_y(x_0, y_0)\Delta y,$$

where from the mean value theorem for $\mathbb{R}^1$,

$$\begin{aligned}
\Delta f(x_0, y_0, \Delta x, \Delta y) &= f(x_0 + \Delta x, y_0 + \Delta y) - f(x_0, y_0) \\
&= f(x_0 + \Delta x, y_0 + \Delta y) - f(x_0, y_0 + \Delta y) \\
&\quad + f(x_0, y_0 + \Delta y) - f(x_0, y_0) \\
&= f_x(x_0 + \varepsilon_1 \Delta x, y_0 + \Delta y)\Delta x \\
&\quad + f_y(x_0, y_0 + \varepsilon_2 \Delta y)\Delta y,
\end{aligned} \tag{8.59}$$

where the functions $\varepsilon_1(\Delta x, \Delta y)$ and $\varepsilon_2(\Delta x, \Delta y)$ are such that

$$0 < \varepsilon_1 < 1 \text{ and } 0 < \varepsilon_2 < 1.$$

Therefore,

$$\begin{aligned}
r(\Delta x, \Delta y) &= (f_x(x_0 + \varepsilon_1 \Delta x, y_0 + \Delta y) - f_x(x_0, y_0))\Delta x \\
&\quad + (f_y(x_0, y_0 + \varepsilon_2 \Delta y) - f_y(x_0, y_0))\Delta y,
\end{aligned} \tag{8.60}$$

and thus, for $(\Delta x, \Delta y) \neq (0, 0)$, we have

$$\begin{aligned}
\left| \frac{r(\Delta x, \Delta y)}{\sqrt{\Delta x^2 + \Delta y^2}} \right| &\leq |f_x(x_0 + \varepsilon_1 \Delta x, y_0 + \Delta y) - f_x(x_0, y_0)| \left| \frac{\Delta x}{\sqrt{\Delta x^2 + \Delta y^2}} \right| \\
&\quad + |f_y(x_0, y_0 + \varepsilon_2 \Delta y) - f_y(x_0, y_0)| \left| \frac{\Delta y}{\sqrt{\Delta x^2 + \Delta y^2}} \right| \\
&\leq |f_x(x_0 + \varepsilon_1 \Delta x, y_0 + \Delta y) - f_x(x_0, y_0)| \\
&\quad + |f_y(x_0, y_0 + \varepsilon_2 \Delta y) - f_y(x_0, y_0)| \\
&\to 0, \text{ when } (\Delta x, \Delta y) \to (0, 0).
\end{aligned} \tag{8.61}$$

From this we may conclude that f is differentiable at (x_0, y_0).
The proof is complete.

Exercise 8.9.6 Let $f : \mathbb{R}^2 \to \mathbb{R}$ where

$$f(x, y) = \begin{cases} \frac{x^2 y^2}{x^2 + y^2}, & \text{if } (x, y) \neq (0, 0) \\ 0, & \text{if } (x, y) = (0, 0). \end{cases}$$

Through the last theorem 8.9.5, show that f is differentiable in $\mathbb{R}^2$.

8.10 The Chain Rule in $\mathbb{R}^n$

Theorem 8.10.1 (The Chain Rule) *Let $D \subset \mathbb{R}^m$ and $V \subset \mathbb{R}^n$ be open sets. Let $\mathbf{f}$: $D \to \mathbb{R}^n$, that is, $\mathbf{f} = (f_1, \ldots, f_n)$ be such that $\mathbf{f}(D) \subset V$, where f_k is differentiable at $\mathbf{x}_0 \in D \ \forall k \in \{1, \ldots, n\}$.*
Assume also $g : V \to \mathbb{R}$ is differentiable at $\mathbf{y}_0 = f(\mathbf{x}_0) \in V$.
Under such hypotheses,

$$\frac{\partial (g \circ \mathbf{f})(\mathbf{x}_0)}{\partial x_j} = \sum_{k=1}^{n} \frac{\partial g(\mathbf{y}_0)}{\partial y_k} \frac{\partial f_k(\mathbf{x}_0)}{\partial x_j},$$

$\forall j \in \{1, \ldots, m\}$.
Moreover, $(g \circ \mathbf{f}) : D \to \mathbb{R}$ is differentiable at $\mathbf{x}_0$.

Proof Fix $j \in \{1, \ldots, m\}$.

Since g is differentiable at $\mathbf{y}_0 = \mathbf{f}(\mathbf{x}_0)$, we have that $r(\mathbf{y})$ defined through the relation,

$$g(\mathbf{f}(\mathbf{x}_0) + \mathbf{y}) - g(\mathbf{f}(\mathbf{x}_0)) = \nabla g(\mathbf{f}(\mathbf{x}_0)) \cdot \mathbf{y} + r(\mathbf{y}), \tag{8.62}$$

that is,

$$r(\mathbf{y}) = g(\mathbf{f}(\mathbf{x}_0) + \mathbf{y}) - g(\mathbf{f}(\mathbf{x}_0)) - \nabla g(\mathbf{f}(\mathbf{x}_0)) \cdot \mathbf{y}$$

is such that

$$\lim_{\mathbf{y} \to 0} \frac{r(\mathbf{y})}{|\mathbf{y}|} = 0.$$

Hence, in particular for

$$\mathbf{y} = \mathbf{f}(\mathbf{x}_0 + t\mathbf{e}_j) - \mathbf{f}(\mathbf{x}_0) \equiv \Delta \mathbf{f}(\mathbf{x}_0, t),$$

there exists $r_0 > 0$ such that if $0 < |t| < r_0$, then

$$g(\mathbf{f}(\mathbf{x}_0 + t\mathbf{e}_j)) - g(\mathbf{f}(\mathbf{x}_0)) = \nabla g(\mathbf{f}(\mathbf{x}_0)) \cdot (\mathbf{f}(\mathbf{x}_0 + t\mathbf{e}_j) - \mathbf{f}(\mathbf{x}_0)) + r(\Delta \mathbf{f}(\mathbf{x}_0, t))$$

$$= \nabla g(\mathbf{f}(\mathbf{x}_0)) \cdot (\mathbf{f}(\mathbf{x}_0 + t\mathbf{e}_j) - \mathbf{f}(\mathbf{x}_0))$$

$$+ w(\Delta \mathbf{f}(\mathbf{x}_0, t)) |\Delta \mathbf{f}(\mathbf{x}_0, t)|, \tag{8.63}$$

where

$$w(\Delta \mathbf{f}(\mathbf{x}_0, t)) = \frac{r(\Delta \mathbf{f}(\mathbf{x}_0, t))}{|\Delta \mathbf{f}(\mathbf{x}_0, t)|}, \text{ if } \Delta \mathbf{f}(\mathbf{x}_0, t) \neq \mathbf{0},$$

and

$$w(\Delta \mathbf{f}(\mathbf{x}_0, t)) = 0, \text{ if } \Delta \mathbf{f}(\mathbf{x}_0, t) = \mathbf{0}.$$

Observe that,

$$\Delta \mathbf{f}(\mathbf{x}_0, t) = \mathbf{f}(\mathbf{x}_0 + t\mathbf{e}_j) - \mathbf{f}(\mathbf{x}_0) \to \mathbf{0},$$

as $t \to 0$.

From this and the hypotheses of differentiability, we have

$$w(\Delta \mathbf{f}(\mathbf{x}_0, t)) \to 0, \text{ as } t \to 0,$$

so that from this and (8.63), we obtain

$$
\begin{aligned}
\frac{g(\mathbf{f}(\mathbf{x}_0 + t\mathbf{e}_j)) - g(\mathbf{f}(\mathbf{x}_0))}{t} &= \nabla g(\mathbf{f}(\mathbf{x}_0)) \cdot \frac{(\mathbf{f}(\mathbf{x}_0 + t\mathbf{e}_j) - \mathbf{f}(\mathbf{x}_0))}{t} \\
&\quad + w(\Delta \mathbf{f}(x_0, t)) \frac{|\Delta \mathbf{f}(\mathbf{x}_0, t)|}{t} \\
&\to \nabla g(\mathbf{y}_0) \cdot \frac{\partial \mathbf{f}(\mathbf{x}_0)}{\partial \mathbf{e}_j}, \text{ as } t \to 0 \\
&= \sum_{k=1}^{n} \frac{\partial g(\mathbf{y}_0)}{\partial y_k} \frac{\partial f_k(\mathbf{x}_0)}{\partial x_j}.
\end{aligned}
\tag{8.64}
$$

Indeed,

$$
\begin{aligned}
\frac{\Delta \mathbf{f}(\mathbf{x}_0, t)}{t} &= \frac{(\mathbf{f}(\mathbf{x}_0 + t\mathbf{e}_j) - \mathbf{f}(\mathbf{x}_0))}{t} \\
&\to \frac{\partial \mathbf{f}(\mathbf{x}_0)}{\partial \mathbf{e}_j}, \text{ as } t \to 0 \\
&= \frac{\partial \mathbf{f}(\mathbf{x}_0)}{\partial x_j} \\
&= \sum_{k=1}^{n} \frac{\partial f_k(\mathbf{x}_0)}{\partial x_j} \mathbf{e}_k.
\end{aligned}
\tag{8.65}
$$

Therefore,

$$w(\Delta \mathbf{f}(\mathbf{x}_0, t)) \frac{|\Delta \mathbf{f}(\mathbf{x}_0, t)|}{t} \to 0, \text{ as } t \to 0.$$

Summarizing, we have obtained

$$\frac{\partial (g \circ \mathbf{f})(\mathbf{x}_0)}{\partial x_j} = \lim_{t \to 0} \frac{g(\mathbf{f}(\mathbf{x}_0 + t\mathbf{e}_j)) - g(\mathbf{f}(\mathbf{x}_0))}{t}$$

$$= \sum_{k=1}^{n} \frac{\partial g(\mathbf{y}_0)}{\partial y_k} \frac{\partial f_k(\mathbf{x}_0)}{\partial x_j}. \tag{8.66}$$

Finally, to show that $(g \circ \mathbf{f})$ is differentiable at $\mathbf{x}_0$, observe that

$$g(\mathbf{f}(\mathbf{x}_0 + \mathbf{v})) - g(\mathbf{f}(\mathbf{x}_0)) = \nabla g(\mathbf{f}(\mathbf{x}_0)) \cdot (\mathbf{f}(\mathbf{x}_0 + \mathbf{v}) - \mathbf{f}(\mathbf{x}_0))$$

$$+ w(\Delta \mathbf{f}(\mathbf{x}_0, \mathbf{v})) |\Delta \mathbf{f}(\mathbf{x}_0, \mathbf{v})|, \tag{8.67}$$

where

$$\Delta \mathbf{f}(\mathbf{x}_0, \mathbf{v}) = \mathbf{f}(\mathbf{x}_0 + \mathbf{v}) - \mathbf{f}(\mathbf{x}_0) \to \mathbf{0},$$

when $\mathbf{v} \to \mathbf{0}$ so that

$$w(\Delta \mathbf{f}(\mathbf{x}_0, \mathbf{v})) \to 0, \text{ as } \mathbf{v} \to \mathbf{0}.$$

Thus,

$$g(\mathbf{f}(\mathbf{x}_0 + \mathbf{v})) - g(\mathbf{f}(\mathbf{x}_0)) = \nabla g(\mathbf{f}(\mathbf{x}_0)) \cdot \left(\sum_{k=1}^{n} \left(\frac{\partial f_k(\mathbf{x}_0)}{\partial \mathbf{v}} + r_k(\mathbf{v}) \right) \mathbf{e}_k \right)$$

$$+ w(\Delta \mathbf{f}(x_0, \mathbf{v})) \left| \sum_{k=1}^{n} \left(\frac{\partial f_k(\mathbf{x}_0)}{\partial \mathbf{v}} + r_k(\mathbf{v}) \right) \mathbf{e}_k \right|$$

$$= \nabla g(\mathbf{f}(\mathbf{x}_0)) \cdot \left(\sum_{k=1}^{n} \left(\frac{\partial f_k(\mathbf{x}_0)}{\partial \mathbf{v}} \right) \mathbf{e}_k \right) + \tilde{r}(\mathbf{v}), \tag{8.68}$$

where

$$r_k(\mathbf{v}) = f_k(\mathbf{x}_0 + \mathbf{v}) - f_k(\mathbf{x}_0) - \frac{\partial f_k(\mathbf{x}_0)}{\partial \mathbf{v}}$$

and such that

$$\frac{r_k(\mathbf{v})}{|\mathbf{v}|} \to 0, \text{ as } \mathbf{v} \to \mathbf{0}, \ \forall k \in \{1, \ldots, n\}.$$

Moreover,

$$\tilde{r}(\mathbf{v}) = \nabla g(\mathbf{f}(\mathbf{x}_0)) \cdot \left(\sum_{k=1}^{n} r_k(\mathbf{v})\mathbf{e}_k \right)$$

$$+ w(\Delta \mathbf{f}(x_0, \mathbf{v})) \left| \sum_{k=1}^{n} \left(\frac{\partial f_k(\mathbf{x}_0)}{\partial \mathbf{v}} + r_k(\mathbf{v}) \right) \mathbf{e}_k \right|. \qquad (8.69)$$

Therefore,

$$\left| \frac{\tilde{r}(\mathbf{v})}{|\mathbf{v}|} \right| \leq \left| \nabla g(\mathbf{f}(\mathbf{x}_0)) \cdot \left(\sum_{k=1}^{n} \frac{r_k(\mathbf{v})}{|\mathbf{v}|}\mathbf{e}_k \right) \right|$$

$$+ |w(\Delta \mathbf{f}(x_0, \mathbf{v}))| \left| \sum_{k=1}^{n} \left(\nabla f_k(\mathbf{x}_0) \cdot \frac{\mathbf{v}}{|\mathbf{v}|} + \frac{r_k(\mathbf{v})}{|\mathbf{v}|} \right) \mathbf{e}_k \right|$$

$$\to 0, \text{ as } \mathbf{v} \to \mathbf{0}. \qquad (8.70)$$

From this we may conclude $(g \circ \mathbf{f})$ is differentiable at $\mathbf{x}_0$.
The proof is complete.

At this point we present a simplified chain rule version for the case $\mathbb{R}^2$

Theorem 8.10.2 (Chain Rule for $\mathbb{R}^2$) *Let $U : D \subset \mathbb{R}^2$ and $\mathbf{f} : V \subset \mathbb{R}^2 \to \mathbb{R}^2$ be functions such that $\mathbf{f}(V) \subset D$ where D, V are open sets. Here we point-wisely denote $\mathbf{f}(r, s) = (X(r, s), Y(r, s))$.*

Assume $\mathbf{f}$ is continuous and its partial derivatives X_r, Y_r, X_s, Y_s to exist at $(r_0, s_0) \in V$. Moreover suppose U is differentiable at $(x_0, y_0) = (X(r_0, s_0), Y(r_0, y_0))$.

Under such hypotheses,

$$\frac{\partial U(\mathbf{f}(r_0, s_0))}{\partial r} = \frac{\partial U(x_0, y_0)}{\partial x}\frac{\partial X(r_0, s_0)}{\partial r} + \frac{\partial U(x_0, y_0)}{\partial y}\frac{\partial Y(r_0, s_0)}{\partial r},$$

and

$$\frac{\partial U(\mathbf{f}(r_0, s_0))}{\partial s} = \frac{\partial U(x_0, y_0)}{\partial x}\frac{\partial X(r_0, s_0)}{\partial s} + \frac{\partial U(x_0, y_0)}{\partial y}\frac{\partial Y(r_0, s_0)}{\partial s}.$$

Proof Let $R > 0$ be such that $B_R(r_0, s_0) \subset V$. Let $\Delta r \in \mathbb{R}$ be such that $0 < |\Delta r| < R$.

Denote

$$\Delta X = X(r_0 + \Delta r, s_0) - X(r_0, s_0)$$

and

$$\Delta Y = Y(r_0 + \Delta r, s_0) - Y(r_0, s_0).$$

Observe that

$$
\begin{aligned}
\mathbf{f}(r_0 + \Delta r, s_0) - \mathbf{f}(r_0, s_0) &= (X(r_0 + \Delta r, s_0), Y(r_0 + \Delta r, s_0)) - (X(r_0, s_0), Y(r_0, s_0)) \\
&= (X(r_0 + \Delta r, s_0) - X(r_0, s_0), Y(r_0 + \Delta r, s_0) - Y(r_0, s_0)) \\
&= (\Delta X, \Delta Y).
\end{aligned}
\tag{8.71}
$$

From this and the differentiability of U at $(x_0, y_0) = (X(r_0, s_0), Y(r_0, s_0))$, we may write

$$
\begin{aligned}
U(\mathbf{f}(r_0 + \Delta r, s_0)) - U(\mathbf{f}(r_0, s_0)) &= U(\mathbf{f}(r_0, s_0) + (\Delta X, \Delta Y)) - U(\mathbf{f}(r_0, s_0)) \\
&= \nabla U(x_0, y_0) \cdot (\Delta X, \Delta Y) \\
&\quad + \eta(\Delta X, \Delta Y)\sqrt{\Delta X^2 + \Delta Y^2} \\
&= \frac{\partial U(x_0, y_0)}{\partial x}\Delta X + \frac{\partial U(x_0, y_0)}{\partial y}\Delta Y \\
&\quad + \eta(\Delta X, \Delta Y)\sqrt{\Delta X^2 + \Delta Y^2},
\end{aligned}
\tag{8.72}
$$

where

$$\Delta X = X(r_0 + \Delta r, s_0) - X(r_0, s_0) \to 0, \text{ as } \Delta r \to 0,$$

$$\Delta Y = Y(r_0 + \Delta r, s_0) - Y(r_0, s_0) \to 0, \text{ as } \Delta r \to 0,$$

and in this case,

$$\eta(\Delta X, \Delta Y) \to 0, \text{ as } \Delta r \to 0.$$

Moreover,

$$
\begin{aligned}
\frac{\sqrt{\Delta X^2 + \Delta Y^2}}{|\Delta r|} &= \sqrt{\left(\frac{\Delta X}{\Delta r}\right)^2 + \left(\frac{\Delta Y}{\Delta r}\right)^2} \\
&\to \sqrt{\left(\frac{\partial X(r_0, s_0)}{\partial r}\right)^2 + \left(\frac{\partial Y(r_0, s_0)}{\partial r}\right)^2},
\end{aligned}
$$

so that,

$$\eta(\Delta X, \Delta Y)\frac{\sqrt{\Delta X^2 + \Delta Y^2}}{\Delta r} \to 0, \text{ as } \Delta r \to 0.$$

From such results and (8.73), we obtain

$$\frac{U(\mathbf{f}(r_0+\Delta r, s_0))-U(\mathbf{f}(r_0, s_0))}{\Delta r} = \frac{\partial U(x_0, y_0)}{\partial x}\frac{\Delta X}{\Delta r} + \frac{\partial U(x_0, y_0)}{\partial y}\frac{\Delta Y}{\Delta r}$$

$$+\eta(\Delta X, \Delta Y)\frac{\sqrt{\Delta X^2 + \Delta Y^2}}{\Delta r}$$

$$\to \frac{\partial U(x_0, y_0)}{\partial x}\frac{\partial X(r_0, s_0)}{\partial r} + \frac{\partial U(x_0, y_0)}{\partial y}\frac{\partial Y(r_0, s_0)}{\partial r},$$

as $\Delta r \to 0$.

Thus,

$$\frac{\partial U(\mathbf{f}(r_0, s_0))}{\partial r} = \lim_{\Delta r \to 0}\frac{U(\mathbf{f}(r_0 + \Delta r, s_0)) - U(\mathbf{f}(r_0, s_0))}{\Delta r}$$

$$= \frac{\partial U(x_0, y_0)}{\partial x}\frac{\partial X(r_0, s_0)}{\partial r} + \frac{\partial U(x_0, y_0)}{\partial y}\frac{\partial Y(r_0, s_0)}{\partial r}.$$

Similarly we may calculate

$$\frac{\partial U(\mathbf{f}(r_0, s_0))}{\partial s}.$$

The proof is complete.

8.10.1   Some Exercises About Partial Derivatives and the Chain Rule

Exercises 8.10.3

1. Let $f : \mathbb{R}^3 \setminus \{(0, 0, 0)\} \to \mathbb{R}$ be such that

$$f(x, y, z) = \frac{1}{\sqrt{x^2 + y^2 + z^2}}.$$

Let

$$x(r, s, t) = e^{-r+s+t},$$

$$y(r, s, t) = e^{r-s+t},$$

and

$$z(r, s, t) = e^{r+s-t}.$$

Calculate f_r, f_s, f_t and show that

$$f_r + f_s + f_t = -f.$$

2. Let $\phi : \mathbb{R} \to \mathbb{R}$ be a differentiable function.
   Let $f : D \subset \mathbb{R}^2 \to \mathbb{R}$ where

$$f(x, y) = (x + y)\phi\left(\frac{x}{y}\right).$$

Find the domain D of f and for $(x, y) \in D^0$, calculate f_x, f_y and show that

$$x f_x(x, y) + y f_y(x, y) = f(x, y).$$

3. Let $\phi : \mathbb{R} \to \mathbb{R}$ be a differentiable function.
   Let $f : \mathbb{R}^2 \to \mathbb{R}$ where

$$f(x, y) = e^{x+y}\phi(x - y).$$

Calculate f_x, f_y and show that

$$f_x(x, y) + f_y(x, y) = 2f(x, y).$$

4. Let $\phi : \mathbb{R} \to \mathbb{R}$ be a differentiable function.
   Let

$$f(x, y) = e^{x/y}\phi\left(\frac{x}{y}\right).$$

Show that

$$x f_x(x, y) + y f_y(x, y) = 0.$$

5. Let

$$f(x, y, z) = e^{\frac{x+y}{z}}.$$

Show that

$$x f_x(x, y, z) + y f_y(x, y, z) + z f_z(x, y, z) = 0,$$

where such derivatives too exist.

6. Let $f : \mathbb{R}^3 \to \mathbb{R}$ be such that Let

$$f(x, y, z) = \int_0^{\cos^2(x^3 + y^2 + z)} \sqrt{1 + t^4}\, dt.$$

Find

$$f_x(x, y, z), \quad f_y(x, y, z) \text{ and } f_z(x, y, z).$$

7. Let

$$f(x, y) = \operatorname{sen}\left(e^{x/y}\right),$$

where

$$x = \operatorname{sen}(u + v)$$

and

$$y = \cos(u + v).$$

Calculate f_u and f_v and show that

$$\cos^2(u + v) f_u = \cot\left(e^{x/y}\right) e^{x/y} f.$$

8. Let $f(x, y)$ be a differentiable function in $\mathbb{R}^2$. For

$$x = g(u - v)$$

and

$$y = h(u - v),$$

where $g, h : \mathbb{R} \to \mathbb{R}$ are differentiable functions.
   Show that

$$f_u + f_v = 0.$$

8.10.2 Revision Exercises

Exercises 8.10.4

1. Sketch the domain of $f(x, y)$ where

   (a)
   $$f(x, y) = \sqrt{x - y} + \ln(x + y + 3).$$

   (b)
   $$f(x, y) = \frac{\sqrt{y - x^2}}{\sqrt{2x + y - 3}}.$$

   (c)
   $$f(x, y) = \frac{x - y}{\cos(x) - \cos(y)}.$$

2. Let $f(x, y)$ be the functions below specified. Find their domains and sketch them. Sketch also the respective graphics.

   (a)
   $$f(x, y) = \sqrt{5 - x^2 - y^2},$$

   (b)
   $$z = f(x, y),$$

   where
   $$x^2 + 4y^2 + z^2 = 5, \quad \text{with } z \geq 0.$$

   (c)
   $$f(x, y) = \frac{1}{5 - x^2 - y^2}.$$

   (d)
   $$f(x, y) = \sqrt{5 - |x| - |y|}.$$

3. Let $f : \mathbb{R}^2 \to \mathbb{R}$ where

$$f(x, y) = \begin{cases} \frac{x^4 - 2y^3}{x^2 + y^2}, & \text{if } (x, y) \neq (0, 0) \\ 0, & \text{if } (x, y) = (0, 0). \end{cases}$$

Calculate f_x and f_y in $\mathbb{R}^2$.

8.11   Higher Order Partial Derivatives

Definition 8.11.1 Let $D \subset \mathbb{R}^2$ be a nonempty set. Let $f : D \to \mathbb{R}$ be a function and $(x, y) \in D^\circ$ such that $f_x(x, y)$ and $f_y(x, y)$ exist in $B_r(x, y)$, some $r > 0$.
   We may define the second order partial derivatives

$$\frac{\partial^2 f(x, y)}{\partial x^2}, \quad \frac{\partial^2 f(x, y)}{\partial x \partial y}, \quad \frac{\partial^2 f(x, y)}{\partial y \partial x} \quad \text{and} \quad \frac{\partial^2 f(x, y)}{\partial y^2},$$

by

$$\frac{\partial^2 f(x, y)}{\partial x^2} = \lim_{\Delta x \to 0} \frac{f_x(x + \Delta x, y) - f_x(x, y)}{\Delta x},$$

$$\frac{\partial^2 f(x, y)}{\partial y^2} = \lim_{\Delta y \to 0} \frac{f_y(x, y + \Delta y) - f_y(x, y)}{\Delta y},$$

$$\frac{\partial^2 f(x, y)}{\partial x \partial y} = \lim_{\Delta x \to 0} \frac{f_y(x + \Delta x, y) - f_y(x, y)}{\Delta x},$$

$$\frac{\partial^2 f(x, y)}{\partial y \partial x} = \lim_{\Delta y \to 0} \frac{f_x(x, y + \Delta y) - f_x(x, y)}{\Delta y},$$

and similarly,

$$\frac{\partial^3 f(x, y)}{\partial x^3} = \lim_{\Delta x \to 0} \frac{f_{xx}(x + \Delta x, y) - f_{xx}(x, y)}{\Delta x},$$

$$\frac{\partial^3 f(x, y)}{\partial y^3} = \lim_{\Delta y \to 0} \frac{f_{yy}(x, y + \Delta y) - f_{yy}(x, y)}{\Delta y},$$

and so on, if such limits too exist.

8.11.1 *About the Equality of the Second Mixed Partial Derivatives* f_{xy} *and* f_{yx}

We start with the exercise.

Exercise 8.11.2 Let $f : \mathbb{R}^2 \to \mathbb{R}$ be such that

$$f(x, y) = \begin{cases} \frac{x^3 y + 3xy^3}{x^2 + y^2}, & \text{if } (x, y) \neq (0, 0) \\ 0, & \text{if } (x, y) = (0, 0). \end{cases}$$

Calculate $f_{xy}(0, 0)$ and $f_{yx}(0, 0)$.

Solution: Let $(x, y) \neq (0, 0)$.

Thus,

$$\begin{aligned}
f_x(x, y) &= \left(\frac{x^3 y + 3xy^3}{x^2 + y^2} \right)_x \\
&= \frac{(x^3 y + 3xy^3)_x (x^2 + y^2) - (x^3 y + 3xy^3)(x^2 + y^2)_x}{(x^2 + y^2)^2} \\
&= \frac{(3x^2 y + 3y^3)(x^2 + y^2) - (x^3 y + 3xy^3)(2x)}{(x^2 + y^2)^2} \\
&= \frac{3x^4 y + 3x^2 y^3 + 3y^3 x^2 + 3y^5 - 2x^4 y - 6x^2 y^3}{(x^2 + y^2)^2} \\
&= \frac{x^4 y + 3y^5}{(x^2 + y^2)^2}.
\end{aligned} \tag{8.73}$$

Also,

$$\begin{aligned}
f_x(0, 0) &= \lim_{\Delta x \to 0} \frac{f(\Delta x, 0) - f(0, 0)}{\Delta x} \\
&= \lim_{\Delta x \to 0} \frac{\Delta x^3 (0)/\Delta x^2 - 0}{\Delta x} \\
&= \lim_{\Delta x \to 0} \frac{0}{\Delta x} = 0.
\end{aligned} \tag{8.74}$$

Hence,

$$\begin{aligned}
f_{xy}(0, 0) &= \lim_{\Delta y \to 0} \frac{f_x(0, \Delta y) - f_x(0, 0)}{\Delta x} \\
&= \lim_{\Delta y \to 0} \frac{3\Delta y^5 / \Delta y^4 - 0}{\Delta y}
\end{aligned}$$

$$= \lim_{\Delta y \to 0} \frac{3\Delta y^5}{\Delta y^5}$$

$$= \lim_{\Delta y \to 0} 3$$

$$= 3. \tag{8.75}$$

Similarly, for $(x, y) \neq (0, 0)$ we obtain

$$f_y(x, y) = \left(\frac{x^3 y + 3xy^3}{x^2 + y^2} \right)_y$$

$$= \frac{(x^3 y + 3xy^3)_y (x^2 + y^2) - (x^3 y + 3xy^3)(x^2 + y^2)_y}{(x^2 + y^2)^2}$$

$$= \frac{(x^3 + 9xy^2)(x^2 + y^2) - (x^3 y + 3xy^3)(2y)}{(x^2 + y^2)^2}$$

$$= \frac{x^5 + x^3 y^2 + 9x^3 y^2 + 9xy^4 - 2x^3 y^2 - 6xy^4}{(x^2 + y^2)^2}$$

$$= \frac{x^5 + 8x^3 y^2 + 3xy^4}{(x^2 + y^2)^2}. \tag{8.76}$$

Also,

$$f_y(0, 0) = \lim_{\Delta y \to 0} \frac{f(0, \Delta y) - f(0, 0)}{\Delta y}$$

$$= \lim_{\Delta y \to 0} \frac{3\Delta y^3 (0)/\Delta y^2 - 0}{\Delta y}$$

$$= \lim_{\Delta y \to 0} \frac{0}{\Delta y} = 0. \tag{8.77}$$

Thus,

$$f_{yx}(0, 0) = \lim_{\Delta x \to 0} \frac{f_y(\Delta x, 0) - f_y(0, 0)}{\Delta x}$$

$$= \lim_{\Delta x \to 0} \frac{\Delta x^5 / \Delta x^4 - 0}{\Delta x}$$

$$= \lim_{\Delta x \to 0} \frac{\Delta x^5}{\Delta x^5}$$

$$= \lim_{\Delta x \to 0} 1$$

$$= 1. \tag{8.78}$$

We have got

$$f_{xy}(0, 0) = 3 \neq 1 = f_{yx}(0, 0).$$

The next theorem establishes sufficient conditions in order that f_{xy} and f_{yx} be equal.

Theorem 8.11.3 *Let $D \subset \mathbb{R}^2$ be a nonempty set. Let $f : D \to \mathbb{R}$ be a function and let $(x_0, y_0) \in D^\circ$ be such that there exists $r > 0$ such that f_x, f_y, f_{xy}, and f_{yx} exist in $B_r(x_0, y_0)$. Assume also f_x and f_y are continuous at $B_r(x_0, y_0)$. Finally, suppose f_{xy} and f_{yx} are continuous at (x_0, y_0).*
Under such hypotheses,

$$f_{xy}(x_0, y_0) = f_{yx}(x_0, y_0).$$

Proof Let $\varepsilon > 0$ be such that

$$(x_0 - \varepsilon, x_0 + \varepsilon) \times (y_0 - \varepsilon, y_0 + \varepsilon) \subset B_r(x_0, y_0).$$

For $0 < |h| < \varepsilon$ define

$$\Delta = f(x_0 + h, y_0 + h) - f(x_0 + h, y_0) - f(x_0, y_0 + h) + f(x_0, y_0).$$

Define also the function G by

$$G(x) = f(x, y_0 + h) - f(x, y_0).$$

Hence,

$$\Delta = G(x_0 + h) - G(x_0),$$

and

$$G'(x) = f_x(x, y_0 + h) - f_x(x, y_0).$$

From the mean value theorem for $\mathbb{R}^1$, there exist $c_1 \in (x_0, x_0 + h)$ and $d_1 \in (y_0, y_0 + h)$ such that

$$
\begin{aligned}
G(x_0 + h) - G(x_0) &= G'(c_1)h \\
&= [f_x(c_1, y_0 + h) - f_x(c_1, y_0)]h \\
&= f_{xy}(c_1, d_1)h^2. \tag{8.79}
\end{aligned}
$$

Thus, we have got,

$$\Delta = f_{xy}(c_1, d_1)h^2.$$

Define now

$$H(y) = f(x_0 + h, y) - f(x_0, y).$$

Therefore

$$\Delta = H(y_0 + h) - H(y_0),$$

and hence

$$H'(y) = f_y(x_0 + h, y) - f_y(x_0, y).$$

From the mean value theorem for $\mathbb{R}^1$, there exist $c_2 \in (x_0, x_0 + h)$ and $d_2 \in (y_0, y_0 + h)$ such that

$$\begin{aligned} H(y_0 + h) - H(y_0) &= H'(d_2)h \\ &= [f_y(x_0 + h, d_2) - f_y(x_0, d_2)]h \\ &= f_{yx}(c_2, d_2)h^2. \end{aligned} \tag{8.80}$$

We have obtained

$$\Delta = f_{yx}(c_2, d_2)h^2 = f_{xy}(c_1, d_1)h^2,$$

and thus,

$$f_{xy}(c_1, d_1) = f_{yx}(c_2, d_2), \forall 0 < |h| < \varepsilon. \tag{8.81}$$

Letting $h \to 0$, we obtain

$$c_1, c_2 \to x_0,$$

and

$$d_1, d_2 \to y_0,$$

and therefore

$$f_{xy}(c_1, d_1) \to f_{xy}(x_0, y_0),$$

and

$$f_{yx}(c_2, d_2) \to f_{yx}(x_0, y_0).$$

From such results and (8.81), we have

$$f_{xy}(x_0, y_0) = f_{yx}(x_0, y_0).$$

The proof is complete.

Exercises 8.11.4

1. Let $f : \mathbb{R}^2 \to \mathbb{R}$ where

$$f(x, y) = \begin{cases} \frac{x^3 + y^2 e^y x}{x^2 + y^2}, & \text{if } (x, y) \neq (0, 0) \\ 0, & \text{if } (x, y) = (0, 0). \end{cases}$$

Calculate $f_{xy}(0, 0)$.

2. Let $f : \mathbb{R}^2 \to \mathbb{R}$ where

$$f(x, y) = \begin{cases} \frac{x^3 + y^2 \cos(y) x}{x^2 + y^2}, & \text{if } (x, y) \neq (0, 0) \\ 0, & \text{if } (x, y) = (0, 0). \end{cases}$$

Calculate $f_{xy}(0, 0)$.

3. Let $f : \mathbb{R}^2 \to \mathbb{R}$ where

$$f(x, y) = \begin{cases} \frac{x^4 + y^2 \operatorname{sen}(y) x}{x^2 + y^2}, & \text{if } (x, y) \neq (0, 0) \\ 0, & \text{if } (x, y) = (0, 0). \end{cases}$$

Calculate $f_{xy}(0, 0)$.

4. Let $f, g : \mathbb{R} \to \mathbb{R}$ be functions of C^2 class. Let $a \in \mathbb{R}$. Show that

$$u(x, t) = f(x + at) + g(x - at),$$

is such that

$$u_{tt}(x, t) = a^2 u_{xx}(x, t).$$

5. Let $f : \mathbb{R}^2 \to \mathbb{R}$ be such that

$$f(x, y) = \int_0^{x^3 - y^5} \left(\int_0^u \ln(1 + t^2) \, dt \right) du.$$

Calculate $f_{xy}(x, y)$.

8.12   The Taylor Theorem for $\mathbb{R}^n$

In this section we develop the Taylor formula for $\mathbb{R}^n$.

Theorem 8.12.1 (Taylor Theorem) *Let $D \subset \mathbb{R}^n$ be an open set, let $f : D \to \mathbb{R}$ be a function and $\mathbf{x}_0 \in D$.*

Suppose there exists $r > 0$ such that all the partial derivatives of f of order up to $m - 1$ exist and are continuous in $B_r(\mathbf{x}_0)$. Moreover, assume the partial derivatives of f of order m too exist in $B_r(\mathbf{x}_0)$.

Let $\mathbf{x} \in B_r(\mathbf{x}_0)$ be such that $\mathbf{x} \neq \mathbf{x}_0$. Under such assumptions and denoting $\mathbf{v} = \mathbf{x} - \mathbf{x}_0$, so that $\mathbf{x} = \mathbf{x}_0 + \mathbf{v}$, we have that there exists $\tilde{t} \in (0, 1)$ such that

$$f(\mathbf{x}_0 + \mathbf{v}) = f(\mathbf{x}_0) + \frac{df(\mathbf{x}_0) \cdot \mathbf{v}}{1!}$$
$$+ \frac{d^2 f(\mathbf{x}_0) \cdot \mathbf{v}^2}{2!} + \cdots + \frac{d^{m-1} f(\mathbf{x}_0) \cdot \mathbf{v}^{m-1}}{(m-1)!}$$
$$+ \frac{d^m f(\mathbf{x}_0 + \tilde{t}\mathbf{v}) \cdot \mathbf{v}^m}{m!}, \tag{8.82}$$

where,

$$df(\mathbf{x}_0) \cdot \mathbf{v} = \sum_{j=1}^{n} \frac{\partial f(\mathbf{x}_0)}{\partial x_j} v_j = \nabla f(\mathbf{x}_0) \cdot \mathbf{v},$$

$$d^2 f(\mathbf{x}_0) \cdot \mathbf{v}^2 = \sum_{i,j=1}^{n} \frac{\partial^2 f(\mathbf{x}_0)}{\partial x_i \partial x_j} v_i v_j,$$

$$d^3 f(\mathbf{x}_0) \cdot \mathbf{v}^3 = \sum_{i,j,k=1}^{n} \frac{\partial^3 f(\mathbf{x}_0)}{\partial x_i \partial x_j \partial x_k} v_i v_j v_k,$$

and so on, up to

$$d^m f(\mathbf{x}_0) \cdot \mathbf{v}^m = \sum_{j_1, j_2, \ldots, j_m = 1}^{n} \frac{\partial^m f(\mathbf{x}_0)}{\partial x_{j_1} \partial x_{j_2} \cdots \partial x_{j_m}} v_{j_1} v_{j_2} \cdots v_{j_m}.$$

Proof Define

$$g(t) = f(\mathbf{x}_0 + t\mathbf{v}), \ \forall t \in [0, 1].$$

Thus

$$g(0) = f(\mathbf{x_0}),$$

$$g(1) = f(\mathbf{x_0} + \mathbf{v}),$$

so that $g^{(1)}$, $g^{(2)}, \ldots, g^{(m-1)}$ exist and are continuous in $[0, 1]$. Moreover $g^{(m)}$ exists in $(0, 1)$.

Also,

$$g'(t) = df(\mathbf{x_0} + t\mathbf{v}) \cdot \mathbf{v},$$

$$g''(t) = d^2 f(\mathbf{x_0} + t\mathbf{v}) \cdot \mathbf{v}^2,$$

$$g^{(3)}(t) = d^3 f(\mathbf{x_0} + t\mathbf{v}) \cdot \mathbf{v}^3,$$

and so on, up to

$$g^{(m)}(t) = d^m f(\mathbf{x_0} + t\mathbf{v}) \cdot \mathbf{v}^m.$$

From the Taylor theorem for $\mathbb{R}^1$, there exists $\tilde{t} \in (0, 1)$ such that

$$g(1) = g(0) + \sum_{j=1}^{m-1} \frac{g^{(j)}(0)}{j!} + \frac{1}{m!} g^{(m)}(\tilde{t}),$$

that is,

$$f(\mathbf{x_0} + \mathbf{v}) = f(\mathbf{x_0}) + \frac{df(\mathbf{x_0}) \cdot \mathbf{v}}{1!}$$

$$+ \frac{d^2 f(\mathbf{x_0}) \cdot \mathbf{v}^2}{2!} + \cdots + \frac{d^{m-1} f(\mathbf{x_0}) \cdot \mathbf{v}^{m-1}}{(m-1)!}$$

$$+ \frac{d^m f(\mathbf{x_0} + \tilde{t}\mathbf{v}) \cdot \mathbf{v}^m}{m!}. \tag{8.83}$$

The proof is complete.

8.12.1   Taylor Formula for $\mathbb{R}^2$ with Lagrange Remainder of Second Order

Although the proof of the next theorem results directly from the last one, for pedagogical reasons we present its proof in detail.

Theorem 8.12.2 *Let $D \subset \mathbb{R}^2$ be an open set and $f : D \to \mathbb{R}$ a C^2 class function.*
Let $(x_0, y_0) \in D$.
   Under such hypotheses, there exists $r > 0$ such that for each $(\Delta x, \Delta y) \in B_r(0, 0)$, there exists $\tilde{t} \in (0, 1)$ such that

$$f(x_0 + \Delta x, y_0 + \Delta y) = f(x_0, y_0) + f_x(x_0, y_0)\Delta x + f_y(x_0, y_0)\Delta y$$

$$+ \frac{1}{2}\left(f_{xx}(\overline{x}, \overline{y})\Delta x^2 + 2f_{xy}(\overline{x}, \overline{y})\Delta x \Delta y + f_{yy}(\overline{x}, \overline{y})\Delta y^2 \right),$$

$$(8.84)$$

where

$$\overline{x} = x_0 + \tilde{t}\Delta x \text{ and } \overline{y} = y_0 + \tilde{t}\Delta y.$$

Proof Since $(x_0, y_0) \in D$ and D is open, there exists $r > 0$ such that $B_r(x_0, y_0) \subset D$.

Let

$$(\Delta x, \Delta y) \in B_r(0, 0) \setminus \{(0, 0)\}.$$

Thus

$$(x_0 + \Delta x, y_0 + \Delta y) \in B_r(x_0, y_0) \subset D.$$

Define $g : [0, 1] \to \mathbb{R}$ by

$$g(t) = f(x_0 + t\Delta x, y_0 + t\Delta y).$$

Therefore

$$g(1) = f(x_0 + \Delta x, y_0 + \Delta y), \qquad (8.85)$$

and

$$g(0) = f(x_0, y_0). \qquad (8.86)$$

Moreover,

$$g'(t) = f_x(x_0 + t\Delta x, y_0 + t\Delta y)\Delta x + f_y(x_0 + t\Delta x, y_0 + t\Delta y)\Delta y, \qquad (8.87)$$

and

$$g''(t) = f_{xx}(x_0 + t\Delta x, y_0 + t\Delta y)\Delta x^2 + 2f_{xy}(x_0 + t\Delta x, y_0 + t\Delta y)\Delta x \Delta y$$

$$+ f_{yy}(x_0 + t\Delta x, y_0 + t\Delta y)\Delta y^2. \qquad (8.88)$$

From the Taylor Formula for $\mathbb{R}^1$, there exists $\tilde{t} \in (0, 1)$ such that,

$$g(1) = g(0) + \frac{1}{1!}g'(0)(1 - 0) + \frac{1}{2!}g''(\tilde{t})(1 - 0)^2,$$

and thus, from this, (8.85), (8.86), (8.87), and (8.88), we obtain,

$$f(x_0+\Delta x, y_0+\Delta y) = f(x_0, y_0) + f_x(x_0, y_0)\Delta x + f_y(x_0, y_0)\Delta y$$
$$+\frac{1}{2}\left(f_{xx}(\overline{x}, \overline{y})\Delta x^2 + 2f_{xy}(\overline{x}, \overline{y})\Delta x\Delta y + f_{yy}(\overline{x}, \overline{y})\Delta y^2\right),$$

$$(8.89)$$

where

$$\overline{x} = x_0 + \tilde{t}\Delta x \text{ and } \overline{y} = y_0 + \tilde{t}\Delta y.$$

The proof is complete.

Exercise 8.12.3 Considering the statement of the last theorem and its hypotheses, denote

$$E(\Delta x, \Delta y) = \frac{1}{2}(f_{xx}(\overline{x}, \overline{y})\Delta x^2 + 2f_{xy}(\overline{x}, \overline{y})\Delta x\Delta y$$
$$+ f_{yy}(\overline{x}, \overline{y})\Delta y^2)$$
$$= f(x_0 + \Delta x, y_0 + \Delta y) - f(x_0, y_0)$$
$$+ f_x(x_0, y_0)\Delta x + f_y(x_0, y_0)\Delta y. \quad (8.90)$$

Prove that there exist $0 < r_1 < r$ and $K > 0$ such that

$$|E(\Delta x, \Delta y)| < K(\Delta x^2 + \Delta y^2), \forall(\Delta x \Delta y) \in B_{r_1}(0.0),$$

Solution: Denote

$$E(\Delta x, \Delta y) = a\Delta x^2 + b\Delta x\Delta y + c\Delta y^2$$

where

$$a = \frac{1}{2}f_{xx}(\overline{x}, \overline{y}), \ b = f_{xy}(\overline{x}, \overline{y}), \text{ and } c = \frac{1}{2}f_{yy}(\overline{x}, \overline{y}).$$

From the continuity of $f_{xx}(x, y)$, $f_{xy}(x, y)$ and $f_{yy}(x, y)$ in D, there exist $r_1 > 0$ and $K > 0$ such that $0 < r_1 < r$ and

$$|f_{xx}(x, y)| < K, |f_{xy}(x, y)| < K, \text{ and } |f_{yy}(x, y)| < K, \ \forall(x, y) \in B_{r_1}(x_0, y_0).$$

Thus

$$|a| < K/2, |b| < K \text{ and } |c| < K/2, \quad \forall (\Delta x, \Delta y) \in B_{r_1}(0, 0).$$

Therefore,

$$|E(\Delta x, \Delta y)| \leq |a|\Delta x^2 + |b||\Delta x||\Delta y| + |c|\Delta y^2$$

$$< \frac{K}{2}\Delta x^2 + K|\Delta x||\Delta y| + \frac{K}{2}\Delta y^2, \tag{8.91}$$

On the other hand,

$$(|\Delta x| - |\Delta y|)^2 = \Delta x^2 - 2|\Delta x||\Delta y| + \Delta y^2 \geq 0, \forall (\Delta x, \Delta y) \in \mathbb{R}^2,$$

so that

$$|\Delta x||\Delta y| \leq \frac{\Delta x^2 + \Delta y^2}{2}, \forall (\Delta x, \Delta y) \in \mathbb{R}^2.$$

From this and (8.91), we obtain

$$|E(\Delta x, \Delta y)| < \frac{K}{2}\Delta x^2 + K\frac{\Delta x^2 + \Delta y^2}{2} + \frac{K}{2}\Delta y^2$$

$$= K(\Delta x^2 + \Delta y^2), \tag{8.92}$$

$\forall (\Delta x, \Delta y) \in B_{r_1}(0, 0).$
The solution is complete.

Exercise 8.12.4 Let $D = [0, \pi] \times [0, \pi] \subset \mathbb{R}^2$ and $f : D \to \mathbb{R}$ where

$$f(x, y) = \cos^2(x^2 + y^2).$$

Let (x_0, y_0) and $(x, y) \in D^0$.
Show that

$$|f(x, y) - f(x_0, y_0)| \leq 4\sqrt{2}\pi \sqrt{(x - x_0)^2 + (y - y_0)^2},$$

Exercise 8.12.5 Let $D = [0, \pi] \times [0, \pi] \subset \mathbb{R}^2$ and $f : D \to \mathbb{R}$ be such that

$$f(x, y) = \text{sen}(x^2 + y^2).$$

Let (x_0, y_0) and $(x, y) \in D^0$.

Show that

$$|f(x, y) - f(x_0, y_0) - f_x(x_0, y_0)(x - x_0) - f_y(x_0, y_0)(y - y_0)|$$
$$\leq (4\pi^2 + 2)[(x - x_0)^2 + (y - y_0)^2]. \tag{8.93}$$

8.13   Local and Global Extremal Points for Functions in $\mathbb{R}^n$

In this section we develop the theory concerning local and global extremals for functions in $\mathbb{R}^n$.

We start with the basic definitions.

Definition 8.13.1 Let $D \subset \mathbb{R}^n$ be a nonempty set. We say that $x_0 \in D$ is a point of local minimum for a function $f : D \to \mathbb{R}$ in D, if there exists $r > 0$ such that,

$$f(\mathbf{x}) \geq f(\mathbf{x}_0), \ \forall \mathbf{x} \in B_r(\mathbf{x}_0) \cap D.$$

We say that $\mathbf{x}_0 \in D$ is a point of local maximum for f in D, if there exists $r > 0$ such that,

$$f(\mathbf{x}) \leq f(\mathbf{x}_0), \ \forall \mathbf{x} \in B_r(\mathbf{x}_0) \cap D.$$

We say that $\mathbf{x}_0 \in D$ is a point of global minimum for f in D, if

$$f(\mathbf{x}) \geq f(\mathbf{x}_0), \ \forall \mathbf{x} \in D.$$

We say that $\mathbf{x}_0 \in D$ is a point of global maximum for f in D, if

$$f(\mathbf{x}) \leq f(\mathbf{x}_0), \ \forall \mathbf{x} \in D.$$

Theorem 8.13.2 *Let $D \subset \mathbb{R}^n$ be an open set. Suppose a function $f : D \to \mathbb{R}$ has a point of local minimum at $\mathbf{x}_0 \in D$. Assume also the first order partial derivatives of f to exist at $\mathbf{x}_0$. Under such hypotheses,*

$$\frac{\partial f(\mathbf{x}_0)}{\partial x_j} = 0, \ \forall j \in \{1, \ldots, n\}.$$

Proof Choose $j \in \{1, \ldots, n\}$. From the hypotheses, there exists $r > 0$ such that

$$f(\mathbf{x}) \geq f(\mathbf{x}_0), \ \forall \mathbf{x} \in B_r(\mathbf{x}_0).$$

In particular, for $0 < h < r$ we have

$$f(\mathbf{x}_0 + h\mathbf{e}_j) \geq f(\mathbf{x}_0),$$

that is,

$$\frac{f(\mathbf{x}_0 + h\mathbf{e}_j) - f(\mathbf{x}_0)}{h} \geq 0.$$

Letting $h \to 0^+$, we obtain

$$\frac{\partial f(\mathbf{x}_0)}{\partial x_j} \geq 0. \tag{8.94}$$

On the other hand, for $-r < h < 0$, we have

$$f(\mathbf{x}_0 + h\mathbf{e}_j) - f(\mathbf{x}_0) \geq 0,$$

so that

$$\frac{f(\mathbf{x}_0 + h\mathbf{e}_j) - f(\mathbf{x}_0)}{h} \leq 0.$$

Letting $h \to 0^-$, we obtain

$$\frac{\partial f(\mathbf{x}_0)}{\partial x_j} \leq 0. \tag{8.95}$$

From (8.94) and (8.95) we have,

$$\frac{\partial f(\mathbf{x}_0)}{\partial x_j} = 0. \tag{8.96}$$

Since $j \in \{1, \ldots, n\}$ is arbitrary, the proof is complete.

8.14   Critical Points

Definition 8.14.1 Let $D \subset \mathbb{R}^n$ be an open set and let $f : D \to \mathbb{R}$ be a function. We say that $\mathbf{x}_0 \in D$ is a critical point of f, if

$$\frac{\partial f(x_0)}{\partial x_j} = 0, \ \forall j \in \{1, \ldots, m\}.$$

Remark 8.14.2 Consider the functions $f, g : \mathbb{R}^2 \to \mathbb{R}$ where

$$f(x, y) = x^2 + y^2,$$

and

$$g(x, y) = x^2 - y^2, \ \forall \mathbf{x} = (x, y) \in \mathbb{R}^2.$$

Thus, f has a point of global minimum at $\mathbf{x}_0 = (0, 0)$, and indeed, from $f_x(x, y) = 2x$ and $f_y(x, y) = 2y$ we get

$$f_x(0, 0) = 0,$$

and

$$f_y(0, 0) = 0.$$

Therefore $\mathbf{x}_0 = (0, 0)$ is a critical point of f and also a critical point of g.

However $(0, 0)$ is neither a local minimum nor a local maximum for g, in fact, it is a saddle point, that is, a minimum relating the variable x and a maximum relating the variable y.

Thus, to qualitatively classify the critical points, we need the so-called test of second partial derivatives.

Theorem 8.14.3 (Test of Second Partial Derivatives) *Let $D \subset \mathbb{R}^2$ be an open set. Let $f : D \to \mathbb{R}$ be a function and suppose $(x_0, y_0) \in D$ is such that*

$$f_x(x_0, y_0) = 0,$$

and

$$f_y(x_0, y_0) = 0.$$

Suppose there exists $r > 0$ such that the first and second partial derivatives of f too exist and be continuous on $B_r(x_0, y_0)$.

Under such hypotheses, denoting

$$\Delta(x_0, y_0) = f_{xx}(x_0, y_0) f_{yy}(x_0, y_0) - f_{xy}(x_0, y_0)^2,$$

we have:

1. *if $\Delta(x_0, y_0) > 0$ and $f_{xx}(x_0, y_0) > 0$, then (x_0, y_0) is a point of local minimum for f.*
2. *if $\Delta(x_0, y_0) > 0$ and $f_{xx}(x_0, y_0) < 0$, then (x_0, y_0) is a point of local maximum for f.*
3. *if $\Delta(x_0, y_0) < 0$, (x_0, y_0) is not a point of local extremal for f.*
4. *if $\Delta(x_0, y_0) = 0$, we cannot get any conclusion about (x_0, y_0).*

Proof We shall prove only the item 1, leaving the proof of the remaining items as exercises.

Assume $\Delta(x_0, y_0) > 0$ and $f_{xx}(x_0, y_0) > 0$.

From the continuity of second order partial derivatives, there exists $r_1 > 0$ such that if $(x, y) \in B_{r_1}(x_0, y_0)$, then

$$\Delta(x, y) > 0 \text{ and } f_{xx}(x, y) > 0.$$

Let $(\Delta x, \Delta y) \in \mathbb{R}^2 \setminus \{(0, 0)\}$ be such that

$$\sqrt{\Delta x^2 + \Delta y^2} < r_1,$$

that is,

$$(x_0 + \Delta x, y_0 + \Delta y) \in B_{r_1}(x_0, y_0).$$

Define

$$g(t) = f(x_0 + t\Delta x, y_0 + t\Delta y), \quad \forall t \in [0, 1].$$

Thus,

$$g(0) = f(x_0, y_0)$$

and

$$g(1) = f(x_0 + \Delta x, y_0 + \Delta y).$$

From Taylor's Theorem for one variable, there exists $\tilde{t} \in (0, 1)$ such that

$$g(1) = g(0) + \frac{1}{1!}g'(0) + \frac{1}{2!}g''(\tilde{t}), \qquad (8.97)$$

where

$$g'(t) = f_x(x_0 + t\Delta x, y_0 + t\Delta y)\Delta x + f_y(x_0 + t\Delta x, y_0 + t\Delta y)\Delta y,$$

and thus

$$g'(0) = f_x(x_0, y_0)\Delta x + f_y(x_0, y_0)\Delta y = 0.$$

Moreover

$$g''(\tilde{t}) = f_{xx}(x_0 + \tilde{t}\Delta x, y_0 + \tilde{t}\Delta y)\Delta x^2$$
$$+ 2f_{xy}(x_0 + \tilde{t}\Delta x, y_0 + \tilde{t}\Delta y)\Delta x\Delta y$$

$$+f_{yy}(x_0 + \tilde{t}\Delta x, y_0 + \tilde{t}\Delta y)\Delta y^2$$

$$= f_{xx}(\tilde{x})\Delta x^2 + 2f_{xy}(\tilde{x})\Delta x \Delta y + f_{yy}(\tilde{x})\Delta y^2$$

$$= a\Delta x^2 + 2b\Delta x \Delta y + c\Delta y^2$$

$$\equiv H(\Delta x, \Delta y), \tag{8.98}$$

where we have denoted

$$\tilde{x} = (x_0 + \tilde{t}\Delta x, y_0 + \tilde{t}\Delta y),$$

$$a = f_{xx}(\tilde{x}), \quad b = f_{xy}(\tilde{x}) \text{ and } c = f_{yy}(\tilde{x}).$$

Hence, from these last results and (8.97) we obtain

$$f(x_0 + \Delta x, y_0 + \Delta y) - f(x_0, y_0) = H(\Delta x, \Delta y). \tag{8.99}$$

Define,

$$H_1(\Delta\tilde{x}, \Delta\tilde{y}) = a\Delta\tilde{x}^2 + 2b\Delta\tilde{x}\Delta\tilde{y} + c\Delta\tilde{y}^2.$$

Observe that, from the hypotheses,

$$ac - b^2 > 0 \text{ and } a > 0.$$

Fix $\Delta\tilde{y} \in \mathbb{R}$.

Consider the quadratic form in $\Delta\tilde{x}$ with $\Delta\tilde{y}$ fixed, relating H_1, that is

$$H_1(\Delta\tilde{x}, \Delta\tilde{y}) = \tilde{a}\Delta\tilde{x}^2 + \tilde{b}\Delta\tilde{x} + \tilde{c},$$

where,

$$\tilde{a} = a,$$

$$\tilde{b} = 2b\Delta\tilde{y},$$

and

$$\tilde{c} = c\Delta\tilde{y}^2.$$

Observe that,

$$\tilde{b}^2 - 4\tilde{a}\tilde{c} = 4b^2\Delta\tilde{y}^2 - 4ac\Delta\tilde{y}^2$$

$$= 4\Delta\tilde{y}^2(b^2 - ac) \leq 0, \quad \forall\Delta\tilde{y} \in \mathbb{R}. \tag{8.100}$$

From this and $a > 0$, we may conclude that H_1 is a positive definite quadratic form, that is,

$$H_1(\Delta \tilde{x}, \Delta \tilde{y}) \geq 0, \ \forall (\Delta \tilde{x}, \Delta \tilde{y}) \in \mathbb{R}^2.$$

In particular, we obtain,

$$H(\Delta x, \Delta y) \geq 0,$$

so that

$$f(x_0 + \Delta x, y_0 + \Delta y) - f(x_0, y_0) = H(\Delta x, \Delta y) \geq 0,$$

that is,

$$f(x_0 + \Delta x, y_0 + \Delta y) \geq f(x_0, y_0),$$

$\forall (\Delta x, \Delta y) \in \mathbb{R}^2$, such that $\sqrt{\Delta x^2 + \Delta y^2} < r_1$.
Therefore, (x_0, y_0) is a point of local minimum for f.
The proof is complete.

Example 8.14.4 Consider the example in which $f : \mathbb{R}^2 \to \mathbb{R}$ is given by

$$f(x, y) = x^2 - 4xy + y^3 + 4y, \ \forall (x, y) \in \mathbb{R}^2.$$

We are going to obtain and classify the critical points of f.
Observe that,

$$f_x(x, y) = 2x - 4y,$$

and

$$f_y(x, y) = -4x + 3y^2 + 4.$$

The critical points are defined by the equations,

$$f_x(x, y) = 2x - 4y = 0 \tag{8.101}$$

$$f_y(x, y) = -4x + 3y^2 + 4 = 0. \tag{8.102}$$

From (8.101) we obtain

$$x = 2y. \tag{8.103}$$

Replacing such a result into Eq. (8.102), we get,

$$-8y + 3y^2 + 4 = 0,$$

so that $y_1 = 2$ and $y_2 = 2/3$ are the solutions of this last equation.

For $y_1 = 2$, from (8.103) we obtain the corresponding $x_1 = 2y_1 = 2(2) = 4$. For $y_1 = 2/3$, from (8.103) we obtain the corresponding $x_2 = 2y_2 = 2(2/3) = 4/3$.

Therefore, $(x_1, y_1) = (4, 2)$ and $(x_2, y_2) = (4/3, 2/3)$ are the critical points of f.

On the other hand, for the second derivatives, we have,

$$f_{xx}(x, y) = 2,$$

$$f_{yy}(x, y) = 6y,$$

$$f_{xy}(x, y) = -4.$$

Thus,

$$\Delta(x, y) = f_{xx}(x, y) f_{yy}(x, y) - f_{xy}(x, y)^2 = 12y - 16.$$

Hence,

$$\Delta(x_1, y_1) = \Delta(4, 2) = 12(2) - 16 = 24 - 16 = 8 > 0.$$

Since $f_{xx}(x_1, y_1) = 2 > 0$, we may infer that $(x_1, y_1) = (4, 2)$ is a point of local minimum for f.

Also,

$$\Delta(x_2, y_2) = \Delta(4/3, 2/3) = 12(2/3) - 16 = 8 - 16 = -8 < 0,$$

so that $(x_2, y_2) = (4/3, 2/3)$ is not a point of local extremal for f.

8.15 The Implicit Function Theorem

8.15.1 Introduction

Consider a function $f : D \subset \mathbb{R}^2 \to \mathbb{R}$ of C^1 class and let $(x_0, y_0) \in D^0$.

Assume $f_y(x_0, y_0) \neq 0$. Under such hypotheses, the implicit function theorem, to be stated and proved in the next lines, guarantees the existence of r, $r_1 > 0$ and a function $y : B_r(x_0) = (x_0 - r, x_0 + r) \to B_{r_1}(y_0) = (y_0 - r_1, y_0 + r_1)$, defined implicitly by the equation

$$f(x, y) = f(x_0, y_0) \equiv c,$$

that is,

$$f(x, y(x)) = c, \forall x \in B_r(x_0).$$

Moreover, also from such a theorem, $y'(x)$ exists in $B_r(x_0)$, so that, from the chain rule, we may obtain

$$\frac{df(x, y(x))}{dx} = \frac{dc}{dx} = 0,$$

that is,

$$f_x(x, y(x)) + f_y(x, y(x))y'(x) = 0,$$

so that

$$y'(x) = -\frac{f_x(x, y(x))}{f_y(x, y(x))}.$$

Similarly, for the case in which $f : D \subset \mathbb{R}^3 \to \mathbb{R}$ is of C^1 class and $(x_0, y_0, z_0) \in D^0$, if $f_z(x_0, y_0, z_0) \neq 0$, from this same theorem, we may obtain $r, r_1 > 0$ and a function $z : B_r(x_0, y_0) \to B_{r_1}(z_0)$ defined implicitly by

$$f(x, y, z) = f(x_0, y_0, z_0) \equiv c,$$

that is,

$$f(x, y, z(x, y)) = c, \ \forall (x, y) \in B_r(x_0, y_0)$$

Moreover, also as a result of the implicit function theorem, z_x and z_y exist in $B_r(x_0, y_0)$, so that, from the chain rule, we have,

$$\frac{df(x, y, z(x, y))}{dx} = \frac{dc}{dx} = 0,$$

that is,

$$f_x(x, y, z(x, y)) + f_z(x, y, z(x, y))z_x(x, y) = 0,$$

so that

$$z_x(x, y) = -\frac{f_x(x, y, z(x, y))}{f_z(x, y, z(x, y))}.$$

Similarly,

$$\frac{df(x, y, z(x, y))}{dy} = \frac{dc}{dy} = 0,$$

so that

$$f_y(x, y, z(x, y)) + f_z(x, y, z(x, y))z_y(x, y) = 0,$$

that is,

$$z_y(x, y) = -\frac{f_y(x, y, z(x, y))}{f_z(x, y, z(x, y))}.$$

So, with such a discussion in mind, at this point we introduce the implicit function theorem for the scalar case.

8.16 Implicit Function Theorem, Scalar Case

Theorem 8.16.1 *Let $D \to \mathbb{R}$ be an open set and let $f : D \to \mathbb{R}$ be a function. Assume f is such that for $(\mathbf{x}_0, y_0) \in D \subset \mathbb{R}^n$, where $\mathbf{x}_0 \in \mathbb{R}^{n-1}$ and $y_0 \in \mathbb{R}$, we have:*

$$f(\mathbf{x}_0, y_0) = c \in \mathbb{R}$$

and

$$f_y(\mathbf{x}_0, y_0) \neq 0.$$

Suppose also f and its first partial derivatives are continuous on $B_r(\mathbf{x}_0, y_0)$, for some $r > 0$. Under such hypotheses, there exist $\delta > 0$ and $\varepsilon > 0$ such that, for

$$B_\delta(\mathbf{x}_0) \times J,$$

where $J = (y_0 - \varepsilon, y_0 + \varepsilon)$, we have that there exists a unique continuous function

$$\xi : B_\delta(\mathbf{x}_0) \to J,$$

such that for each $\mathbf{x} \in B_\delta(\mathbf{x}_0)$, we have

$$f(\mathbf{x}, \xi(\mathbf{x})) = c,$$

and

$$\frac{\partial \xi(\mathbf{x})}{\partial x_j} = \frac{-f_{x_j}(\mathbf{x}, \xi(\mathbf{x}))}{f_y(\mathbf{x}, \xi(\mathbf{x}))},$$

$\forall j \in \{1, \ldots, n-1\}$.

The function $y = \xi(\mathbf{x})$ is said to be implicitly defined by the equation $f(\mathbf{x}, y) = c$ in a neighborhood of $(\mathbf{x}_0, y_0)$.

Proof With no loss in generality assume $f_y(\mathbf{x}_0, y_0) > 0$.

Since f_y is continuous on D, there exists $\delta_1 > 0$ and $\varepsilon > 0$ such that,

$$f_y(\mathbf{x}, y) > 0, \ \forall (\mathbf{x}, y) \in B_{\delta_1}(\mathbf{x}_0) \times \overline{J} \subset D.$$

Hence the function $f(\mathbf{x}_0, y)$ is strictly increasing in $\overline{J}$ and therefore:

$$f(\mathbf{x}_0, y_0 - \varepsilon) < c = f(\mathbf{x}_0, y_0) < f(\mathbf{x}_0, y_0 + \varepsilon).$$

From this, since f is continuous, there exists $\delta \in \mathbb{R}$ such that $0 < \delta < \delta_1$ and also such that

$$f(\mathbf{x}, y_0 - \varepsilon) < c < f(\mathbf{x}, y_0 + \varepsilon),$$

$\forall \mathbf{x} \in B_\delta(\mathbf{x}_0)$.

From the intermediate value theorem, there exists $y \in (y_0 - \varepsilon, y_0 + \varepsilon) = J$ such that

$$f(\mathbf{x}, y) = c.$$

We claim that such a $y \in J$ is unique.

Suppose, to obtain contradiction, that there exists $y_1 \in J$ such that $y_1 \neq y$ and

$$f(\mathbf{x}, y_1) = c$$

Hence

$$f(\mathbf{x}, y) = f(\mathbf{x}, y_1) = c.$$

From this and the mean value theorem, there exists y_2 between y and y_1 such that

$$f_y(\mathbf{x}, y_2) = 0.$$

This contradicts $f_y(\mathbf{x}, y) > 0$ in $B_\delta(\mathbf{x}_0) \times \overline{J}$.

Hence, the y in question is unique, and we shall denote it by

$$y = \xi(\mathbf{x}).$$

Thus,

$$f(\mathbf{x}, \xi(\mathbf{x})) = c, \ \forall \mathbf{x} \in B_\delta(\mathbf{x}_0).$$

Now, we will prove that the function $\xi : B_\delta(\mathbf{x}_0) \to J$ is continuous.
Let $\mathbf{x} \in B_\delta(\mathbf{x}_0)$.
Let $\{\mathbf{x}_n\} \subset B_\delta(\mathbf{x}_0)$ be such that

$$\lim_{n \to \infty} \mathbf{x}_n = \mathbf{x}.$$

It suffices to show that

$$\lim_{n \to \infty} \xi(\mathbf{x}_n) = \xi(\mathbf{x}) = y.$$

Suppose, to obtain contradiction, we do not have

$$\lim_{n \to \infty} \xi(\mathbf{x}_n) = y.$$

Thus, there exists $\varepsilon_0 > 0$ such that for each $k \in \mathbb{N}$ there exists $n_k \in \mathbb{N}$ such that $n_k > k$ and

$$|\xi(\mathbf{x}_{n_k}) - y| \geq \varepsilon_0.$$

Observe that $\{\xi(\mathbf{x}_{n_k})\} \subset J$ and such a set is bounded.
Thus, there exists a subsequence of $\{\xi(\mathbf{x}_{n_k})\}$ which we shall also denote by $\{\xi(\mathbf{x}_{n_k})\}$ and $y_1 \in \overline{J}$ such that $y_1 \neq y$ and

$$\lim_{k \to \infty} \xi(\mathbf{x}_{n_k}) = y_1,$$

Hence, since f is continuous, we obtain:

$$f(x, y_1) = \lim_{k \to \infty} f(\mathbf{x}_{n_k}, \xi(\mathbf{x}_{n_k})) = \lim_{k \to \infty} c = c.$$

Therefore,

$$f(\mathbf{x}, y) = f(\mathbf{x}, y_1) = c.$$

From this and the mean value theorem, there exists y_2 between y and y_1 such that

$$f_y(\mathbf{x}, y_2) = 0.$$

This contradicts $f_y(\mathbf{x}, y) > 0$ in $B_\delta(\mathbf{x}_0) \times \overline{J}$.
Therefore

$$\lim_{n \to \infty} \xi(\mathbf{x}_n) = y = \xi(\mathbf{x}).$$

Since $\{\mathbf{x}_n\} \subset B_\delta(\mathbf{x}_0)$ such that

$$\lim_{n \to \infty} \mathbf{x}_n = \mathbf{x},$$

is arbitrary, we may conclude that ξ is continuous at $\mathbf{x}$, $\forall \mathbf{x} \in B_\delta(\mathbf{x}_0)$.
For the final part, choose $j \in \{1, \ldots, n-1\}$.
Let $\mathbf{x} \in B_\delta(\mathbf{x}_0)$ and $h \in \mathbb{R}$ be such that

$$\mathbf{x} + h\mathbf{e}_j \in B_\delta(\mathbf{x}_0).$$

Denote

$$v = \xi(\mathbf{x} + h\mathbf{e}_j) - \xi(\mathbf{x}),$$

that is,

$$\xi(\mathbf{x} + h\mathbf{e}_j) = \xi(\mathbf{x}) + v.$$

From the continuity of ξ we have:

$$v \to 0, \text{ as } h \to 0.$$

Observe that

$$f(\mathbf{x}, \xi(\mathbf{x})) = c,$$

and

$$f(\mathbf{x} + h\mathbf{e}_j, \xi(\mathbf{x} + h\mathbf{e}_j)) = f(\mathbf{x} + h\mathbf{e}_j, \xi(\mathbf{x}) + v) = c,$$

so that

$$f(\mathbf{x} + h\mathbf{e}_j, \xi(\mathbf{x}) + v) - f(\mathbf{x}, \xi(\mathbf{x})) = c - c = 0,$$

and therefore from the mean value theorem there exists $\tilde{t} \in (0, 1)$ such that

$$0 = f(\mathbf{x} + h\mathbf{e}_j, \xi(\mathbf{x}) + v) - f(\mathbf{x}, \xi(\mathbf{x}))$$

$$= f_{x_j}(\mathbf{x} + \tilde{t}h\mathbf{e}_j, \xi(\mathbf{x}) + \tilde{t}v)h + f_y(\mathbf{x} + \tilde{t}h\mathbf{e}_j, \xi(\mathbf{x}) + \tilde{t}v)v, \qquad (8.104)$$

so that,

$$\frac{\xi(\mathbf{x} + h\mathbf{e}_j) - \xi(\mathbf{x})}{h} = \frac{v}{h}$$

$$= \frac{-f_{x_j}(\mathbf{x} + \tilde{t}h\mathbf{e}_j, \xi(\mathbf{x}) + \tilde{t}v)}{f_y(\mathbf{x} + \tilde{t}h\mathbf{e}_j, \xi(\mathbf{x}) + \tilde{t}v)}$$

$$\rightarrow \frac{-f_{x_j}(\mathbf{x}, \xi(\mathbf{x}))}{f_y(\mathbf{x}, \xi(\mathbf{x}))}, \quad \text{as } h \rightarrow 0. \qquad (8.105)$$

Summarizing the result obtained,

$$\frac{\partial \xi(\mathbf{x})}{\partial x_j} = \lim_{h \rightarrow 0} \frac{\xi(\mathbf{x} + h\mathbf{e}_j) - \xi(\mathbf{x})}{h}$$

$$= \frac{-f_{x_j}(\mathbf{x}, \xi(\mathbf{x}))}{f_y(\mathbf{x}, \xi(\mathbf{x}))}. \qquad (8.106)$$

The proof is complete.

Example 8.16.2 Consider the example in which $F : \mathbb{R}^2 \rightarrow \mathbb{R}$ is given by

$$F(x, y) = x^2y^2 - x\cos(y) - 1, \quad \forall (x, y) \in \mathbb{R}^2$$

Observe that

$$F(1, \pi/2) = \frac{\pi^2}{4} - 1$$

and

$$F_y(x, y) = 2x^2y + x\sin(y),$$

so that

$$F_y(1, \pi/2) = \pi + 1 > 0.$$

Hence, from the implicity function theorem, the equation

$$F(x, y) = F(1, \pi/2) = \frac{\pi^2}{4} - 1,$$

defines implicitly a function $y(x)$ in a neighborhood of $x = 1$.

Moreover, from this same theorem, the derivative of y in such a neighborhood is given by,

$$\frac{dy(x)}{dx} = -\frac{F_x(x, y(x))}{F_y(x, y(x))}.$$

Thus, informally,

$$\frac{dy(x)}{dx} = \frac{-2xy^2 + \cos(y)}{2x^2y + x\sin(y)},$$

where it is understood the dependence $y(x)$, so that

$$\frac{dy(1)}{dx} = \frac{-2(1)(\pi/2)^2 + \cos(\pi/2)}{2(1)(\pi/2) + 1\sin(\pi/2)} = \frac{-\pi^2}{2(\pi + 1)}.$$

Example 8.16.3 Let $f, g : \mathbb{R}^n \to \mathbb{R}$ be such that

$$g(\mathbf{x}) = f(\mathbf{x})[1 + f(\mathbf{x})^6 + 5f(\mathbf{x})^8], \ \forall \mathbf{x} \in \mathbb{R}^n.$$

Assume g is of C^1 class. We are going to show that f is of C^1 class as well. Define $F : \mathbb{R}^n \times \mathbb{R} \to \mathbb{R}$ by

$$F(\mathbf{x}, y) = g(\mathbf{x}) - y(1 + y^6 + 5y^8).$$

Let $\mathbf{x} \in \mathbb{R}^n$. Thus

$$F(\mathbf{x}, f(\mathbf{x})) = 0.$$

Observe that

$$F_y(\mathbf{x}, y) = -1 - 7y^6 - 45y^8 < 0, \ \forall y \in \mathbb{R}.$$

From the implicity function theorem, the equation

$$F(\mathbf{x}, y) = 0,$$

defines a unique function $y(\mathbf{x})$ in an neighborhood of $\mathbf{x}$. Since

$$F(\mathbf{x}, f(\mathbf{x})) = 0, \ \forall \mathbf{x} \in \mathbb{R}^n,$$

necessarily we must have

$$y(\mathbf{x}) = f(\mathbf{x})$$

in the concerning neighborhood of $\mathbf{x}$.

Also from the implicit function theorem,

$$\frac{\partial y(\mathbf{x})}{\partial x_j} = \frac{\partial f(\mathbf{x})}{\partial x_j} = -\frac{F_{x_j}(\mathbf{x}, f(\mathbf{x}))}{F_y(\mathbf{x}, f(\mathbf{x}))},$$

so that,

$$\frac{\partial f(\mathbf{x})}{\partial x_j} = \frac{-g_{x_j}(\mathbf{x})}{-1 - 7f(\mathbf{x})^6 - 45f(\mathbf{x})^8},$$

Since $\mathbf{x} \in \mathbb{R}^n$ and $j \in \{1, \ldots, n\}$ are arbitrary, we may infer that f is of C^1 class.

8.17 Vectorial Functions in $\mathbb{R}^n$

In this section we develop some fundamental results concerning vectorial functions.

Definition 8.17.1 (Vectorial Functions) Let $D \subset \mathbb{R}^n$ be a set. A binary relation $\mathbf{f} : D \to \mathbb{R}^m$ where $m \geq 2$ is said to be a vectorial function, if for each $\mathbf{x} \in D$ there exists only one $\mathbf{z} \in \mathbb{R}^m$ such that $(\mathbf{x}, \mathbf{z}) \in \mathbf{f}$.

In such a case we denote,

$$\mathbf{z} = f(\mathbf{x}).$$

Example 8.17.2 Let $\mathbf{f} : \mathbb{R}^2 \to \mathbb{R}^3$ be such that

$$\mathbf{f}(x, y) = x^2 \mathbf{e}_1 + (y + x)\mathbf{e}_2 + \sin(x + y)\mathbf{e}_3, \ \forall (x, y) \in \mathbb{R}^2.$$

we may also denote

$$\mathbf{f}(\mathbf{x}) = \begin{pmatrix} f_1(x, y) \\ f_2(x, y) \\ f_3(x, y) \end{pmatrix} \tag{8.107}$$

where

$$f_1(x, y) = x^2, \quad f_2(x, y) = x + y, \quad f_3(x, y) = \sin(x + y).$$

Definition 8.17.3 (Limits for Vectorial Functions) Let $D \subset \mathbb{R}^n$ be a nonempty set. Let $\mathbf{f} : D \to \mathbb{R}^m$ be a vectorial function and let $\mathbf{x}_0 \in D'$.

We say that $\mathbf{L} \in \mathbb{R}^m$ is the limit of $\mathbf{f}$ as $\mathbf{x}$ approaches $\mathbf{x}_0$, as for each $\varepsilon > 0$, there exists $\delta > 0$ such that if $\mathbf{x} \in D$ and $0 < |\mathbf{x} - \mathbf{x}_0| < \delta$, then

$$|\mathbf{f}(\mathbf{x}) - \mathbf{L}| < \varepsilon.$$

In such a case, we denote,

$$\lim_{\mathbf{x} \to \mathbf{x}_0} \mathbf{f}(\mathbf{x}) = \mathbf{L} \in \mathbb{R}^m.$$

Theorem 8.17.4 *Let $D \subset \mathbb{R}^n$ be a nonempty set. Let $\mathbf{f} : D \to \mathbb{R}^m$ be a vectorial function and let $\mathbf{x}_0 \in D'$. Denoting*

$$\mathbf{f}(\mathbf{x}) = \begin{pmatrix} f_1(\mathbf{x}) \\ f_2(\mathbf{x}) \\ \vdots \\ f_m(\mathbf{x}) \end{pmatrix} \tag{8.108}$$

we have that

$$\lim_{\mathbf{x} \to \mathbf{x}_0} \mathbf{f}(\mathbf{x}) = \mathbf{L} = \sum_{k=1}^{m} l_k \mathbf{e}_k \in \mathbb{R}^m,$$

if and only if,

$$\lim_{\mathbf{x} \to \mathbf{x}_0} f_k(\mathbf{x}) = l_k, \quad \forall k \in \{1, \ldots, m\}.$$

Proof Suppose that

$$\lim_{\mathbf{x} \to \mathbf{x}_0} \mathbf{f}(\mathbf{x}) = \mathbf{L} = \sum_{k=1}^{m} l_k \mathbf{e}_k \in \mathbb{R}^m.$$

Let $\varepsilon > 0$ be given. Thus there exists $\delta > 0$ such that if $\mathbf{x} \in D$ and $0 < |\mathbf{x} - \mathbf{x}_0| < \delta$, then

$$|\mathbf{f}(\mathbf{x}) - \mathbf{L}| < \varepsilon.$$

Select $k \in \{1, \ldots, m\}$.

Therefore,

$$|f_k(\mathbf{x}) - l_k| \leq \sqrt{\sum_{j=1}^{N} |f_j(\mathbf{x}) - l_j|^2}$$

$$= |\mathbf{f}(\mathbf{x}) - \mathbf{L}|$$

$$< \varepsilon, \tag{8.109}$$

if $\mathbf{x} \in D$ and $0 < |\mathbf{x} - \mathbf{x}_0| < \delta$, so that

$$\lim_{\mathbf{x} \to \mathbf{x}_0} f_k(x) = l_k, \ \forall k \in \{1, \dots, m\}.$$

Reciprocally, suppose

$$\lim_{\mathbf{x} \to \mathbf{x}_0} f_k(\mathbf{x}) = l_k, \ \forall k \in \{1, \dots, m\}.$$

Thus, for each $k \in \{1, \dots, m\}$ there exists $\delta_k > 0$ such that if $\mathbf{x} \in D$ and $0 < |\mathbf{x} - \mathbf{x}_0| < \delta_k$, then

$$|f(x_k) - l_k| < \frac{\varepsilon}{\sqrt{m}}.$$

Define $\delta = \min\{\delta_k, \ k \in \{1, \dots, m\}\}$.
Thus, if $\mathbf{x} \in D$ and $0 < |\mathbf{x} - \mathbf{x}_0| < \delta$, then

$$|\mathbf{f}(\mathbf{x}) - \mathbf{L}| = \sqrt{\sum_{k=1}^{m} |f_k(\mathbf{x}) - l_k|^2} < \sqrt{\sum_{k=1}^{m} \frac{\varepsilon^2}{m}} = \varepsilon,$$

so that

$$\lim_{\mathbf{x} \to \mathbf{x}_0} \mathbf{f}(\mathbf{x}) = \mathbf{L}.$$

The proof is complete.

8.17.1 Limit Proprieties

Theorem 8.17.5 *Let $D \subset \mathbb{R}^m$ be a nonempty set. Let $\mathbf{f}, \mathbf{g} : D \to \mathbb{R}^m$ be vectorial functions and let $\mathbf{x}_0 \in D'$.*

Suppose that

$$\lim_{\mathbf{x} \to \mathbf{x}_0} \mathbf{f}(\mathbf{x}) = \mathbf{L} \in \mathbb{R}^m$$

and

$$\lim_{\mathbf{x} \to \mathbf{x}_0} \mathbf{g}(\mathbf{x}) = \mathbf{M} \in \mathbb{R}^m.$$

Upon such assumptions, we have

1.

$$\lim_{\mathbf{x} \to \mathbf{x}_0} \alpha \mathbf{f}(\mathbf{x}) = \alpha \mathbf{L}, \quad \forall \alpha \in \mathbb{R}.$$

2.

$$\lim_{\mathbf{x} \to \mathbf{x}_0} \mathbf{f}(\mathbf{x}) + \mathbf{g}(\mathbf{x}) = \mathbf{L} + \mathbf{M},$$

3.

$$\lim_{\mathbf{x} \to \mathbf{x}_0} \mathbf{f}(\mathbf{x}) \cdot \mathbf{g}(\mathbf{x}) = \mathbf{L} \cdot \mathbf{M}.$$

The proof of such results is similar to those of the case of scalar functions and it is left as an exercise.

Definition 8.17.6 (Continuous Vectorial Function) Let $D \subset \mathbb{R}^n$ be a nonempty set and let $\mathbf{f} : D \to \mathbb{R}^m$ be a vectorial function. Let $\mathbf{x}_0 \in D$. We say that $\mathbf{f}$ is continuous at $\mathbf{x}_0$ as for each $\varepsilon > 0$ there exists $\delta > 0$ such that if $\mathbf{x} \in D$ and $|\mathbf{x} - \mathbf{x}_0| < \delta$, then

$$|\mathbf{f}(\mathbf{x}) - \mathbf{f}(\mathbf{x}_0)| < \varepsilon.$$

Therefore, if $\mathbf{x}_0$ is an isolated point of D, then $\mathbf{f}$ will be automatically continuous at $\mathbf{x}_0$, considering that in such a case, there exists $r > 0$ such that

$$B_r(\mathbf{x}_0) \cap D = \{\mathbf{x}_0\}.$$

thus, for a given $\varepsilon > 0$, it suffices to take $\delta = \frac{r}{2}$, so that if $\mathbf{x} \in D$ and $|\mathbf{x} - \mathbf{x}_0| < \delta$, then necessarily $\mathbf{x} = \mathbf{x}_0$ and hence

$$|\mathbf{f}(\mathbf{x}) - \mathbf{f}(\mathbf{x}_0)| = |\mathbf{f}(\mathbf{x}_0) - \mathbf{f}(\mathbf{x}_0)| = 0 < \varepsilon.$$

On the other hand, if $\mathbf{x}_0 \in D \cap D'$ in such a case the definition of continuity coincide with the one of limit, so that $\mathbf{f}$ will be continuous at $\mathbf{x}_0$, if and only if,

$$\lim_{\mathbf{x} \to \mathbf{x}_0} \mathbf{f}(\mathbf{x}) = \mathbf{f}(\mathbf{x}_0).$$

Finally, if **f** is continuous on all its domain D, we simply say that **f** is a continuous function.

Theorem 8.17.7 *Let $D \subset \mathbb{R}^m$ be a nonempty set. Let $\mathbf{f}, \mathbf{g} : D \to \mathbb{R}^m$ be vectorial continuous functions at $\mathbf{x}_0 \in D \cap D'$.*

Upon such assumptions,

1. $\alpha \mathbf{f}$ is continuous at $\mathbf{x}_0$, $\forall \alpha \in \mathbb{R}$.
2. $\mathbf{f} + \mathbf{g}$ is continuous at $\mathbf{x}_0$.
3. $\mathbf{f} \cdot \mathbf{g}$ is continuous at $\mathbf{x}_0$.

The proof results from the limit properties and it is left as an exercise.

Definition 8.17.8 (Directional Derivative) Let $D \subset \mathbb{R}^n$ be a nonempty set and let $\mathbf{f} : D \to \mathbb{R}^m$ be a function. Suppose $\mathbf{x}_0 \in D°$ and let $\mathbf{v} \in \mathbb{R}^n$. We define the derivative of **f** relating **v** at the point $\mathbf{x}_0$, denoted by

$$\frac{\partial \mathbf{f}(\mathbf{x}_0)}{\partial \mathbf{v}},$$

by

$$\frac{\partial \mathbf{f}(\mathbf{x}_0)}{\partial \mathbf{v}} = \lim_{h \to 0} \frac{\mathbf{f}(\mathbf{x}_0 + h\mathbf{v}) - \mathbf{f}(\mathbf{x}_0)}{h},$$

so that denoting,

$$\mathbf{f}(\mathbf{x}) = \begin{pmatrix} f_1(\mathbf{x}) \\ f_2(\mathbf{x}) \\ \vdots \\ f_m(\mathbf{x}) \end{pmatrix} \tag{8.110}$$

we have

$$\frac{\partial \mathbf{f}(\mathbf{x})}{\partial \mathbf{v}} = \begin{pmatrix} \frac{\partial f_1(\mathbf{x})}{\partial \mathbf{v}} \\ \frac{\partial f_2(\mathbf{x})}{\partial \mathbf{v}} \\ \vdots \\ \frac{\partial f_m(\mathbf{x})}{\partial \mathbf{v}} \end{pmatrix} \tag{8.111}$$

and, in particular,

$$\frac{\partial \mathbf{f}(\mathbf{x})}{\partial x_j} = \begin{pmatrix} \frac{\partial f_1(\mathbf{x})}{\partial x_j} \\ \frac{\partial f_2(\mathbf{x})}{\partial x_j} \\ \vdots \\ \frac{\partial f_m(\mathbf{x})}{\partial x_j} \end{pmatrix} \tag{8.112}$$

where,

$$\frac{\partial \mathbf{f}(\mathbf{x})}{\partial x_j} = \lim_{h \to 0} \frac{\mathbf{f}(\mathbf{x} + h\mathbf{e}_j) - \mathbf{f}(\mathbf{x})}{h}, \ \forall j \in \{1, \ldots, n\}.$$

if such limits exist.

Definition 8.17.9 (Differentiable Vectorial Functions) Let $D \subset \mathbb{R}^n$ be a nonempty set and $\mathbf{f} : D \to \mathbb{R}^m$ a function. Suppose $\mathbf{x}_0 \in D^\circ$ and let $r > 0$ be such that $B_r(\mathbf{x}_0) \subset D$. We say that $\mathbf{f}$ is differentiable at $\mathbf{x}_0$, if there exists a matrix $m \times n$ $M(\mathbf{x}_0)$ such that the function $\mathbf{r}(\mathbf{v})$ defined through the relation

$$\mathbf{f}(\mathbf{x}_0 + \mathbf{v}) = f(\mathbf{x}_0) + M(\mathbf{x}_0) \cdot \mathbf{v} + \mathbf{r}(\mathbf{v}), \ \forall \mathbf{v} \in B_r(\mathbf{0})$$

that is

$$\mathbf{r}(\mathbf{v}) = \mathbf{f}(\mathbf{x}_0 + \mathbf{v}) - f(\mathbf{x}_0) - M(\mathbf{x}_0) \cdot \mathbf{v},$$

is such that

$$\lim_{\mathbf{v} \to 0} \frac{\mathbf{r}(\mathbf{v})}{|\mathbf{v}|} = \mathbf{0}.$$

Remark 8.17.10 In the context of this last definition, for all $0 < |h| < r$, we have

$$f(\mathbf{x}_0 + h\mathbf{e}_j) - f(\mathbf{x}_0) = M(\mathbf{x}_0) \cdot \mathbf{e}_j h + r(h\mathbf{e}_j),$$

so that

$$\frac{f(\mathbf{x}_0 + h\mathbf{e}_j) - f(\mathbf{x}_0)}{h} = M(\mathbf{x}_0) \cdot \mathbf{e}_j + \frac{\mathbf{r}(h\mathbf{e}_j)}{h} \to M(\mathbf{x}_0) \cdot \mathbf{e}_j, \text{ as } h \to 0.$$

Thus,

$$M(\mathbf{x}_0) \cdot \mathbf{e}_j = \frac{\partial \mathbf{f}(\mathbf{x}_0)}{\partial x_j}.$$

Hence, denoting

$$M(\mathbf{x}_0) = \begin{pmatrix} M_{11} & M_{12} & \cdots & M_{1n} \\ M_{21} & M_{22} & \cdots & M_{2n} \\ \vdots & \vdots & \ddots & \vdots \\ M_{m1} & M_{m2} & \cdots & M_{mn} \end{pmatrix} \tag{8.113}$$

we obtain

$$M(\mathbf{x}_0) \cdot \mathbf{e}_j = \begin{pmatrix} M_{1j} \\ M_{2j} \\ \vdots \\ M_{mj} \end{pmatrix} = \begin{pmatrix} \frac{\partial f_1(\mathbf{x})}{\partial x_j} \\ \frac{\partial f_2(\mathbf{x})}{\partial x_j} \\ \vdots \\ \frac{\partial f_m(\mathbf{x})}{\partial x_j} \end{pmatrix} = \frac{\partial \mathbf{f}(\mathbf{x})}{\partial x_j} \qquad (8.114)$$

so that

$$M(\mathbf{x}_0) = \{M_{ij}\}_{m \times n} = \left(\frac{\partial \mathbf{f}(\mathbf{x}_0)}{\partial x_j} \right)_{m \times n}.$$

Therefore, we define

$$\mathbf{f}'(\mathbf{x}_0) = \{M_{ij}\}_{m \times n} = \left(\frac{\partial \mathbf{f}(\mathbf{x}_0)}{\partial x_j} \right)_{m \times n},$$

where $\mathbf{f}'(\mathbf{x}_0)$ is to be the derivative matrix (or Jacobian matrix) of $\mathbf{f}$ at $\mathbf{x}_0$, that is

$$\mathbf{f}'(\mathbf{x}_0) = \begin{pmatrix} M_{11} & M_{12} & \cdots & M_{1n} \\ M_{21} & M_{22} & \cdots & M_{2n} \\ \vdots & \vdots & \ddots & \vdots \\ M_{m1} & M_{m2} & \cdots & M_{mn} \end{pmatrix} = \begin{pmatrix} \frac{\partial f_1(\mathbf{x})}{\partial x_1} & \frac{\partial f_1(\mathbf{x})}{\partial x_2} & \cdots & \frac{\partial f_1(\mathbf{x})}{\partial x_n} \\ \frac{\partial f_2(\mathbf{x})}{\partial x_1} & \frac{\partial f_2(\mathbf{x})}{\partial x_2} & \cdots & \frac{\partial f_2(\mathbf{x})}{\partial x_n} \\ \vdots & \vdots & \ddots & \vdots \\ \frac{\partial f_m(\mathbf{x})}{\partial x_1} & \frac{\partial f_m(\mathbf{x})}{\partial x_2} & \cdots & \frac{\partial f_m(\mathbf{x})}{\partial x_n} \end{pmatrix}. \qquad (8.115)$$

Theorem 8.17.11 *Let* $D \subset \mathbb{R}^m$ *be a nonempty set and* $\mathbf{f} : D \to \mathbb{R}^m$ *a vectorial function, where we denote*

$$\mathbf{f}(\mathbf{x}) = \begin{pmatrix} f_1(\mathbf{x}) \\ f_2(\mathbf{x}) \\ \vdots \\ f_m(\mathbf{x}) \end{pmatrix} \qquad (8.116)$$

for $f_k : D \to \mathbb{R}$, $\forall k \in \{1, \dots, m\}$.

Suppose $\mathbf{x}_0 \in D^\circ$. *Under such hypotheses,* $\mathbf{f}$ *is differentiable at* $\mathbf{x}_0$ *if, and only if,* f_k *is differentiable at* $\mathbf{x}_0$, $\forall k \in \{1, \dots, m\}$.

Proof Suppose **f** is differentiable at $\mathbf{x}_0$.

Let $r_0 > 0$ be such that $B_{r_0}(\mathbf{x}_0) \subset D$. Thus, $\mathbf{r} : B_{r_0}(\mathbf{0}) \to \mathbb{R}^m$ defined through the relation

$$\mathbf{f}(\mathbf{x}_0 + \mathbf{v}) - \mathbf{f}(\mathbf{x}_0) = \mathbf{f}'(\mathbf{x}_0)\mathbf{v} + \mathbf{r}(\mathbf{v}), \ \forall \mathbf{v} \in B_{r_0}(\mathbf{0}),$$

that is,

$$\mathbf{r}(\mathbf{v}) = \mathbf{f}(\mathbf{x}_0 + \mathbf{v}) - \mathbf{f}(\mathbf{x}_0) - \mathbf{f}'(\mathbf{x}_0)\mathbf{v},$$

is such that

$$\frac{\mathbf{r}(\mathbf{v})}{|\mathbf{v}|} \to \mathbf{0}, \ \text{as } \mathbf{v} \to \mathbf{0}.$$

From this, we obtain

$$\begin{pmatrix} f_1(\mathbf{x}_0 + \mathbf{v}) \\ f_2(\mathbf{x}_0 + \mathbf{v}) \\ \vdots \\ f_m(\mathbf{x}_0 + \mathbf{v}) \end{pmatrix} - \begin{pmatrix} f_1(\mathbf{x}_0) \\ f_2(\mathbf{x}_0) \\ \vdots \\ f_m(\mathbf{x}_0) \end{pmatrix} = \left[\frac{\partial f_i(\mathbf{x}_0)}{\partial x_j} \right] \mathbf{v} + \begin{pmatrix} r_1(\mathbf{v}) \\ r_2(\mathbf{v}) \\ \vdots \\ r_m(\mathbf{v}) \end{pmatrix} \quad (8.117)$$

so that from the kth line of this equation, we have

$$f_k(\mathbf{x}_0 + \mathbf{v}) - f_k(\mathbf{x}_0) = \sum_{j=1}^{m} \frac{\partial f_k(\mathbf{x}_0)}{\partial x_j} v_j + r_k(\mathbf{v}),$$

where

$$\frac{|r_k(\mathbf{v})|}{|\mathbf{v}|} \le \left| \frac{\mathbf{r}(\mathbf{v})}{|\mathbf{v}|} \right| \to 0, \ \text{as } \mathbf{v} \to \mathbf{0}.$$

Therefore, f_k is differentiable at $\mathbf{x}_0, \forall k \in \{1, \ldots, m\}$.

Reciprocally, suppose f_k is differentiable at $\mathbf{x}_0, \forall k \in \{1, \ldots, m\}$.

Choose $k \in \{1, \ldots, m\}$. Thus $r_k : B_{r_0}(\mathbf{0}) \to \mathbb{R}$ defined through the relation

$$f_k(\mathbf{x}_0 + \mathbf{v}) - f_k(\mathbf{x}_0) = \sum_{j=1}^{m} \frac{\partial f_k(\mathbf{x}_0)}{\partial x_j} v_j + r_k(\mathbf{v}), \quad (8.118)$$

is such that,

$$\frac{r_k(\mathbf{v})}{|\mathbf{v}|} \to \mathbf{0}, \ \text{as } \mathbf{v} \to \mathbf{0}.$$

Therefore, from (8.118), we obtain

$$
\begin{pmatrix} f_1(\mathbf{x}_0 + \mathbf{v}) \\ f_2(\mathbf{x}_0 + \mathbf{v}) \\ \vdots \\ f_m(\mathbf{x}_0 + \mathbf{v}) \end{pmatrix} - \begin{pmatrix} f_1(\mathbf{x}_0) \\ f_2(\mathbf{x}_0) \\ \vdots \\ f_m(\mathbf{x}_0) \end{pmatrix} = \left[\frac{\partial f_i(\mathbf{x}_0)}{\partial x_j} \right] \mathbf{v} + \begin{pmatrix} r_1(\mathbf{v}) \\ r_2(\mathbf{v}) \\ \vdots \\ r_m(\mathbf{v}) \end{pmatrix}, \qquad (8.119)
$$

that is,

$$
\mathbf{r}(\mathbf{v}) = \mathbf{f}(\mathbf{x}_0 + \mathbf{v}) - \mathbf{f}(\mathbf{x}_0) - \mathbf{f}'(\mathbf{x}_0)\mathbf{v},
$$

where

$$
\mathbf{r}(\mathbf{v}) = \begin{pmatrix} r_1(\mathbf{v}) \\ r_2(\mathbf{v}) \\ \vdots \\ r_m(\mathbf{v}) \end{pmatrix},
$$

is such that

$$
\frac{\mathbf{r}(\mathbf{v})}{|\mathbf{v}|} = \begin{pmatrix} r_1(\mathbf{v})/|\mathbf{v}| \\ r_2(\mathbf{v})/|\mathbf{v}| \\ \vdots \\ r_m(\mathbf{v})/|\mathbf{v}| \end{pmatrix} \rightarrow \begin{pmatrix} 0 \\ 0 \\ \vdots \\ 0 \end{pmatrix} = \mathbf{0} \text{ as } \mathbf{v} \rightarrow \mathbf{0}.
$$

Hence, $\mathbf{f}$ is differentiable at $\mathbf{x}_0$.
The proof is complete.

Remark 8.17.12 Specially for the case $n = 1$ we shall define a new different notation for $\mathbf{f}$.
We denote,

$$
\mathbf{f}(t) = \begin{pmatrix} f_1(t) \\ f_2(t) \\ \vdots \\ f_m(t) \end{pmatrix} = \begin{pmatrix} x_1(t) \\ x_2(t) \\ \vdots \\ x_m(t) \end{pmatrix} \equiv \mathbf{r}(t)^T \qquad (8.120)
$$

so that

$$
\mathbf{r}(t) = (x_1(t), \ldots, x_m(t)), \forall t \in [a, b],
$$

for a one variable vectorial function

$$\mathbf{r} : [a, b] \to \mathbb{R}^m.$$

We are going to define the derivative of $\mathbf{r}$, denoted by $\mathbf{r}'$, by

$$\mathbf{r}'(t) = (x_1'(t), \ldots, x_m'(t)),$$

if the derivatives in question exist at $t \in [a, b]$.

Theorem 8.17.13 (Mean Value Incquality for One Variable Vectorial Functions) *Let $\mathbf{r} : [a, b] \to \mathbb{R}^m$ be a continuous one variable vectorial function, such that $\mathbf{r}'$ is continuous on (a, b), where $a < b$. Under such hypotheses, there exists $\tilde{t} \in (a, b)$ such that*

$$|\mathbf{r}(b) - \mathbf{r}(a)| \le (b - a)|\mathbf{r}'(\tilde{t})|.$$

Proof Define $\mathbf{z} = \mathbf{r}(b) - \mathbf{r}(a)$, and

$$\phi(t) = \mathbf{z} \cdot \mathbf{r}(t), \; \forall t \in [a, b].$$

Thus, ϕ is continuous on $[a, b]$ and it is differentiable on (a, b). From the mean value theorem, there exists $\tilde{t} \in (a, b)$ such that

$$\phi(b) - \phi(a) = (b - a)\phi'(\tilde{t}).$$

Thus,

$$
\begin{aligned}
\mathbf{z} \cdot (\mathbf{r}(b) - \mathbf{r}(a)) = |\mathbf{z}|^2 &= (b - a)(\mathbf{z} \cdot \mathbf{r}'(\tilde{t})) \\
&= (b - a)|\mathbf{z} \cdot \mathbf{r}'(\tilde{t})| \\
&\le (b - a)|\mathbf{z}||\mathbf{r}'(\tilde{t})|, \quad\quad\quad (8.121)
\end{aligned}
$$

so that

$$|\mathbf{r}(b) - \mathbf{r}(a)| = |\mathbf{z}| \le (b - a)|\mathbf{r}'(\tilde{t})|.$$

This completes the proof.

Theorem 8.17.14 (Mean Value Inequality for Vectorial Functions) *Let $D \subset \mathbb{R}^n$ be an open set and let $\mathbf{f} : D \to \mathbb{R}^m$ be a differentiable function at each point of the set*

$$A = \{\mathbf{x}_0 + t\mathbf{v} : t \in (0, 1)\},$$

for a given $\mathbf{v} \in \mathbb{R}^n$.

Assume also **f** *is continuous on*

$$\overline{A} = \{\mathbf{x}_0 + t\mathbf{v} \; : \; t \in [0, 1]\},$$

$$|\mathbf{f}'(\mathbf{x})| \leq M, \; \forall \mathbf{x} \in A,$$

for some $M \in \mathbb{R}^+$.
Under such hypotheses

$$|\mathbf{f}(\mathbf{x}_0 + \mathbf{v}) - \mathbf{f}(\mathbf{x}_0)| \leq M|\mathbf{v}|.$$

Proof Let $\mathbf{r} : [0, 1] \to \mathbb{R}^m$ be defined by

$$\mathbf{r}(t) = \mathbf{f}(\mathbf{x}_0 + t\mathbf{v}), \forall t \in [0, 1].$$

From the hypotheses $\mathbf{r}$ is continuous on $[0, 1]$ and differentiable in $(0, 1)$.
From the mean value inequality for one variable vectorial functions, there exists
$\tilde{t} \in (0, 1)$ such that

$$|\mathbf{r}(1) - \mathbf{r}(0)| \leq |\mathbf{r}'(\tilde{t})|(1 - 0),$$

so that

$$|\mathbf{f}(\mathbf{x}_0 + \mathbf{v}) - \mathbf{f}(\mathbf{x}_0)| \leq |\mathbf{r}'(\tilde{t})|$$
$$= |\mathbf{f}'(\mathbf{x}_0 + \tilde{t}\mathbf{v}) \cdot \mathbf{v}|$$
$$= |\mathbf{f}'(\mathbf{x}_0 + \tilde{t}\mathbf{v})||\mathbf{v}|$$
$$\leq M|\mathbf{v}|. \tag{8.122}$$

The proof is complete.

8.18   Implicit Function Theorem for the Vectorial Case

We start with the following auxiliary result. Indeed its proof is very similar to that
of the one-dimensional case; however we present again the proof for the sake of
completeness.

Theorem 8.18.1 *Let* $0 \leq \lambda < 1$. *Suppose that* $\{\mathbf{x}_n\} \subset \mathbb{R}^n$ *is such that,*

$$|\mathbf{x}_{n+2} - \mathbf{x}_{n+1}| \leq \lambda |\mathbf{x}_{n+1} - \mathbf{x}_n|, \forall n \in \mathbb{N}.$$

Upon such assumptions $\{\mathbf{x}_n\}$ *is a Cauchy sequence so that it is converging.*

Proof Observe that

$$|\mathbf{x}_3 - \mathbf{x}_2| \leq \lambda |\mathbf{x}_2 - \mathbf{x}_1|$$

$$|\mathbf{x}_4 - \mathbf{x}_3| \leq \lambda |\mathbf{x}_3 - \mathbf{x}_2| \leq \lambda^2 |\mathbf{x}_2 - \mathbf{x}_1|$$

$$\cdots \quad \cdots \quad \cdots$$

$$|\mathbf{x}_{n+1} - \mathbf{x}_n| \leq \lambda^{n-1} |\mathbf{x}_2 - \mathbf{x}_1|. \tag{8.123}$$

Thus, for $n, p \in \mathbb{N}$ we have

$$|\mathbf{x}_{n+p} - \mathbf{x}_n| = |\mathbf{x}_{n+p} - \mathbf{x}_{n+p-1} + \mathbf{x}_{n+p-1} - \mathbf{x}_{n+p-2} + \mathbf{x}_{n+p-1}$$

$$+ \cdots - \mathbf{x}_{n+1} + \mathbf{x}_{n+1} - \mathbf{x}_n|$$

$$\leq |\mathbf{x}_{n+p} - \mathbf{x}_{n+p-1}| + |\mathbf{x}_{n+p-1} - \mathbf{x}_{n+p-2}| + \cdots + |\mathbf{x}_{n+1} - \mathbf{x}_n|$$

$$\leq (\lambda^{n+p-2} + \lambda^{n+p-3} + \cdots + \lambda^{n-1})|\mathbf{x}_2 - \mathbf{x}_1|. \tag{8.124}$$

Therefore,

$$|\mathbf{x}_{n+p} - \mathbf{x}_n| \leq \lambda^{n-1}(\lambda^{p-1} + \lambda^{p-2} + \cdots + 1)|\mathbf{x}_2 - \mathbf{x}_1|$$

$$\leq \frac{\lambda^{n-1}(1 - \lambda^p)}{1 - \lambda}|\mathbf{x}_2 - \mathbf{x}_1|$$

$$\leq \frac{\lambda^{n-1}}{1 - \lambda}|\mathbf{x}_2 - \mathbf{x}_1|. \tag{8.125}$$

Observe that

$$\lim_{n \to \infty} \frac{\lambda^{n-1}}{1 - \lambda}|\mathbf{x}_2 - \mathbf{x}_1| = 0.$$

Let $\varepsilon > 0$. Thus there exists $n_0 \in \mathbb{N}$ such that if $n > n_0$, then

$$\frac{\lambda^{n-1}}{1 - \lambda}|\mathbf{x}_2 - \mathbf{x}_1| < \varepsilon.$$

From this and (8.125), we get,

$$|\mathbf{x}_{n+p} - \mathbf{x}_n| < \varepsilon, \text{ if } n > n_0.$$

Hence, for $m = p + n$ we obtain

$$|\mathbf{x}_m - \mathbf{x}_n| < \varepsilon, \text{ if } m > n > n_0.$$

Thus, $\{\mathbf{x}_n\}$ is a Cauchy sequence, therefore it is converging.

At this point we present in detail the Banach contractor function theorem.

Theorem 8.18.2 *Let $D \subset \mathbb{R}^m$ be a nonempty, closed, and convex set. Let $\mathbf{f} : D \to D$ be a continuous vectorial function such that*

$$|\mathbf{f}(\mathbf{x}) - \mathbf{f}(\mathbf{y})| \leq \lambda |\mathbf{x} - \mathbf{y}|, \ \forall \mathbf{x}, \ \mathbf{y} \in D,$$

for some

$$0 \leq \lambda < 1,$$

that is, $\mathbf{f}$ is a contractor function.

Choose $\mathbf{x}_1 \in D$. With such a choice in mind, the sequence defined by

$$\mathbf{x}_{k+1} = \mathbf{f}(\mathbf{x}_k), \ \forall k \in \mathbb{N}$$

is such that there exists $\mathbf{x}_0 \in D$ such that

$$\lim_{k \to \infty} \mathbf{x}_k = \mathbf{x}_0$$

and

$$\mathbf{f}(\mathbf{x}_0) = \mathbf{x}_0,$$

there is, $\mathbf{x}_0$ is a fixed point for $\mathbf{f}$.

Moreover, such a $\mathbf{x}_0 \in D$ is unique.

Proof Observe that

$$|\mathbf{x}_{n+2} - \mathbf{x}_{n+1}| = |\mathbf{f}(\mathbf{x}_{n+1}) - \mathbf{f}(\mathbf{x}_n)| \leq \lambda |\mathbf{x}_{n+1} - \mathbf{x}_n|,$$

$\forall n \in \mathbb{N}$.

From the last theorem and since D is closed, $\{\mathbf{x}_n\}$ converges to some $\mathbf{x}_0 \in D$. Observe that

$$\mathbf{x}_{n+1} = \mathbf{f}(\mathbf{x}_n), \ \forall n \in \mathbb{N}.$$

From the continuity of $\mathbf{f}$ we obtain,

$$\mathbf{x}_0 = \lim_{n \to \infty} \mathbf{x}_{n+1} = \lim_{n \to \infty} \mathbf{f}(\mathbf{x}_n) = \mathbf{f}(\mathbf{x}_0),$$

so that

$$\mathbf{x}_0 = \mathbf{f}(\mathbf{x}_0).$$

Now, suppose $\mathbf{y} \in D$ is such that

$$\mathbf{f}(\mathbf{y}) = \mathbf{y}.$$

Thus,

$$|\mathbf{y} - \mathbf{x}_0| = |\mathbf{f}(\mathbf{y}) - \mathbf{f}(\mathbf{x}_0)| \le \lambda |\mathbf{y} - \mathbf{x}_0|,$$

so that,

$$(1 - \lambda)|\mathbf{y} - \mathbf{x}_0| \le 0.$$

Since $1 - \lambda > 0$ we have got $|\mathbf{y} - \mathbf{x}_0| = 0$, that is,

$$\mathbf{y} = \mathbf{x}_0.$$

From this we may infer that $\mathbf{x}_0$ is unique.
The proof is complete.

Remark 8.18.3 For $D \subset \mathbb{R}^m$ closed and convex, if $\mathbf{f} : D \to D$ is a differentiable vectorial function such that

$$|\mathbf{f}'(\mathbf{x})| < \lambda, \forall \mathbf{x} \in D,$$

for some $0 \le \lambda < 1$, then from mean value inequality, given $\mathbf{x}, \mathbf{y} \in D$, there exists $t \in (0, 1)$ such that

$$|\mathbf{f}(\mathbf{y}) - \mathbf{f}(\mathbf{x})| \le |\mathbf{f}'(\mathbf{x} + t(\mathbf{y} - \mathbf{x}))||\mathbf{y} - \mathbf{x}| \le \lambda |\mathbf{y} - \mathbf{x}|.$$

Therefore, in such a case, $\mathbf{f}$ is a contractor function, so that from the last theorem, there exists one and only one $\mathbf{x}_0 \in D$ such that

$$\mathbf{x}_0 = \mathbf{f}(\mathbf{x}_0).$$

Theorem 8.18.4 (Implicit Function Theorem, Vectorial Case) *Denote* $(\mathbf{x}, \mathbf{y}) \in \mathbb{R}^{n+m}$ *where* $\mathbf{x} = (x_1, \dots, x_n) \in \mathbb{R}^n$ *and* $\mathbf{y} = (y_1, \dots, y_m) \in \mathbb{R}^m$. *Let* $D \subset \mathbb{R}^{n+m}$ *be a nonempty open set and let* $\mathbf{f} : D \to \mathbb{R}^m$ *be a function of* C^1 *class on* D, *that is,* $\mathbf{f}$ *and its first order partial derivatives are continuous on* D, *so that* $\mathbf{f}$ *is differentiable on* D. *Denote*

$$\mathbf{f}'(\mathbf{x}, \mathbf{y}) = \left\{ \frac{\partial f_k(\mathbf{x}, \mathbf{y})}{\partial x_j} |_{j=1}^n , \quad \frac{\partial f_k(\mathbf{x}, \mathbf{y})}{\partial y_j} |_{j=1}^m \right\},$$

where $k \in \{1, \dots, m\}$.

Thus, we may also denote:

$$\mathbf{f}'(\mathbf{x}, \mathbf{y}) = \left\{ \mathbf{f}_x(\mathbf{x}, \mathbf{y}) , \ \mathbf{f}_y(\mathbf{x}, \mathbf{y}) \right\},$$

where,

$$\mathbf{f}_x(\mathbf{x}, \mathbf{y}) = \left\{ \frac{\partial f_k(\mathbf{x}, \mathbf{y})}{\partial x_j} \right\}_{j=1}^{n},$$

$$\mathbf{f}_y(\mathbf{x}, \mathbf{y}) = \left\{ \frac{\partial f_k(\mathbf{x}, \mathbf{y})}{\partial y_j} \right\}_{j=1}^{m},$$

where,

$$\mathbf{f}(\mathbf{x}, \mathbf{y}) = (f_k(\mathbf{x}, \mathbf{y})) = \begin{pmatrix} f_1(\mathbf{x}, \mathbf{y}) \\ f_2(\mathbf{x}, \mathbf{y}) \\ \vdots \\ f_m(\mathbf{x}, \mathbf{y}) \end{pmatrix} \tag{8.126}$$

Assume $\mathbf{f}(\mathbf{x}_0, \mathbf{y}_0) = \mathbf{0}$, *and that*

$$\det(\mathbf{f}_y(\mathbf{x}_0, \mathbf{y}_0)) \neq 0,$$

where

$$(\mathbf{x}_0, \mathbf{y}_0) \in D.$$

Let $\mathbf{v} \in \mathbb{R}^n$. *Under such hypotheses, there exist* $\delta_1 > 0$ *and* $\delta_2 > 0$ *such that for each* $\mathbf{x} \in B_{\delta_1}(\mathbf{x}_0)$ *there exists a unique* $\mathbf{y} \in \overline{B}_{\delta_2}(\mathbf{y}_0)$ *such that*

$$\mathbf{f}(\mathbf{x}, \mathbf{y}) = \mathbf{0}.$$

Moreover, denoting

$$\mathbf{y} = \xi(\mathbf{x})$$

we have

$$\mathbf{f}(\mathbf{x}, \xi(\mathbf{x})) = \mathbf{0}, \forall \mathbf{x} \in B_\delta(\mathbf{x}_0),$$

and also

$$\frac{\partial \xi(\mathbf{x})}{\partial \mathbf{v}} = -[\mathbf{f}_y(\mathbf{x}, \xi(\mathbf{x}))]^{-1} \cdot [\mathbf{f}_x(\mathbf{x}, \xi(\mathbf{x})) \cdot \mathbf{v}].$$

Proof Denote $A = \mathbf{f}_y(\mathbf{x}_0, \mathbf{y}_0)$.

From the hypotheses A^{-1} exists. Define

$$\lambda = \frac{1}{2|A^{-1}|} > 0. \tag{8.127}$$

Since $\mathbf{f}_y$ is continuous on D, there exist $\tilde{\delta}_1 > 0$ and $\delta_2 > 0$ such that if $x \in B_{\tilde{\delta}_1}(\mathbf{x}_0)$ and $\mathbf{y} \in \overline{B}_{\delta_2}(\mathbf{y}_0)$, then

$$\det(\mathbf{f}_y(\mathbf{x}, \mathbf{y})) \neq 0$$

and

$$|\mathbf{f}_y(\mathbf{x}, \mathbf{y}) - A| = |\mathbf{f}_y(\mathbf{x}, \mathbf{y}) - \mathbf{f}_y(\mathbf{x}_0, \mathbf{y}_0)| < \lambda \tag{8.128}$$

Define

$$\varepsilon = \frac{\delta_2}{2|A^{-1}|} > 0.$$

From the continuity of $\mathbf{f}$, there exist $\delta_1 > 0$ such that $\delta_1 < \tilde{\delta}_1$ and also such that if $\mathbf{x} \in B_{\delta_1}(\mathbf{x}_0)$, then

$$|\mathbf{f}(\mathbf{x}, \mathbf{y}_0)| = |\mathbf{f}(\mathbf{x}, \mathbf{y}_0) - \mathbf{f}(\mathbf{x}_0, \mathbf{y}_0)| < \varepsilon.$$

Let $\mathbf{x} \in B_{\delta_1}(\mathbf{x}_0)$.

Define

$$\phi(\mathbf{y}) = \mathbf{y} - A^{-1}(\mathbf{f}(\mathbf{x}, \mathbf{y})).$$

Thus,

$$\begin{aligned}
\phi'(\mathbf{y}) &= I - A^{-1}(\mathbf{f}_y(\mathbf{x}, \mathbf{y})) \\
&= A^{-1}A - A^{-1}(\mathbf{f}_y(\mathbf{x}, \mathbf{y})) \\
&= A^{-1}(A - \mathbf{f}_y(\mathbf{x}, \mathbf{y})),
\end{aligned} \tag{8.129}$$

so that for $\mathbf{y} \in \overline{B}_{\delta_2}(\mathbf{y}_0)$, we obtain

$$\begin{aligned}
|\phi'(\mathbf{y})| &\leq |A^{-1}||A - \mathbf{f}_y(\mathbf{x}, \mathbf{y})| \\
&\leq \frac{1}{2\lambda}\lambda = \frac{1}{2},
\end{aligned} \tag{8.130}$$

so that

$$|\phi'(\mathbf{y})| < \frac{1}{2}, \ \forall \mathbf{y} \in \overline{B}_{\delta_2}(\mathbf{y}_0).$$

From this and the mean value inequality for vectorial functions, we get

$$|\phi(\mathbf{y}_1) - \phi(\mathbf{y}_2)| \leq \frac{1}{2}|\mathbf{y}_1 - \mathbf{y}_2|, \ \forall \mathbf{y}_1, \mathbf{y}_2 \in \overline{B}_{\delta_2}(\mathbf{y}_0). \tag{8.131}$$

We are going to prove that

$$\phi(\overline{B}_{\delta_2}(\mathbf{y}_0)) \subset B_{\delta_2}(\mathbf{y}_0).$$

Let $\mathbf{y} \in B_{\delta_2}(\mathbf{y}_0)$, thus,

$$\begin{aligned}
|\phi(\mathbf{y}) - \mathbf{y}_0| &= |\phi(\mathbf{y}) - \phi(\mathbf{y}_0) + \phi(\mathbf{y}_0) - \mathbf{y}_0| \\
&\leq |\phi(\mathbf{y}) - \phi(\mathbf{y}_0)| + |\phi(\mathbf{y}_0) - \mathbf{y}_0| \\
&\leq \frac{1}{2}|\mathbf{y} - \mathbf{y}_0| + |\phi(\mathbf{y}_0) - \mathbf{y}_0| \\
&\leq \frac{\delta_2}{2} + |\phi(\mathbf{y}_0) - \mathbf{y}_0|.
\end{aligned} \tag{8.132}$$

On the other hand,

$$\phi(\mathbf{y}_0) = \mathbf{y}_0 - A^{-1}(f(\mathbf{x}, \mathbf{y}_0)),$$

that is

$$\begin{aligned}
|\phi(\mathbf{y}_0) - \mathbf{y}_0| &\leq |A^{-1}||\mathbf{f}(\mathbf{x}, \mathbf{y}_0)| \\
&\leq |A^{-1}|\varepsilon \\
&\leq \frac{|A^{-1}|\delta_2}{2|A^{-1}|} \\
&= \frac{\delta_2}{2}.
\end{aligned} \tag{8.133}$$

From this and (8.132) we obtain

$$\begin{aligned}
|\phi(\mathbf{y}) - \mathbf{y}_0| &\leq \frac{\delta_2}{2} + |\phi(\mathbf{y}_0) - \mathbf{y}_0| \\
&< \frac{\delta_2}{2} + \frac{\delta_2}{2} \\
&= \delta_2
\end{aligned} \tag{8.134}$$

Therefore,

$$\phi(\mathbf{y}) \in \overline{B}_{\delta_2}(\mathbf{y}_0), \ \forall \mathbf{y} \in \overline{B}_{\delta_2}(\mathbf{y}_0).$$

From these last results we may infer that ϕ is a contractor mapping on $\overline{B}_{\delta_2}(\mathbf{y}_0)$.

From the contractor mapping theorem, there exists a unique $\mathbf{y} \in \overline{B}_{\delta_2}(\mathbf{y}_0)$ such that

$$\phi(\mathbf{y}) = \mathbf{y},$$

so that

$$\phi(\mathbf{y}) = \mathbf{y} - A^{-1}(\mathbf{f}(\mathbf{x}, \mathbf{y})) = \mathbf{y},$$

so that

$$A^{-1}(\mathbf{f}(\mathbf{x}, \mathbf{y})) = \mathbf{0}$$

and since $\det(A^{-1}) \neq 0$, we obtain

$$\mathbf{f}(\mathbf{x}, \mathbf{y}) = \mathbf{0}.$$

Since, for each $\mathbf{x} \in B_{\delta_1}(\mathbf{x}_0)$ such a $\mathbf{y}$ is unique, we denote

$$\mathbf{y} = \xi(\mathbf{x})$$

so that,

$$\mathbf{f}(\mathbf{x}, \xi(\mathbf{x})) = \mathbf{0}, \ \forall \mathbf{x} \in B_{\delta_1}(\mathbf{x}_0).$$

Next we show that ξ is continuous on $B_{\delta_1}(\mathbf{x}_0)$.

Let $\mathbf{x} \in B_{\delta_1}(\mathbf{x}_0)$.

Let $\{\mathbf{x}_n\} \subset B_{\delta_1}(\mathbf{x}_0)$ be such that

$$\lim_{n \to \infty} \mathbf{x}_n = \mathbf{x}.$$

It suffices to show that

$$\lim_{n \to \infty} \xi(\mathbf{x}_n) = \xi(\mathbf{x}) = \mathbf{y}.$$

Suppose, to obtain contradiction, we do not have

$$\lim_{n \to \infty} \xi(\mathbf{x}_n) = \mathbf{y}.$$

Thus, there exists $\varepsilon_0 > 0$ such that for each $k \in \mathbb{N}$ there exists $n_k \in \mathbb{N}$ such that $n_k > k$ and

$$|\xi(\mathbf{x}_{n_k}) - \mathbf{y}| \geq \varepsilon_0.$$

Observe that $\{\xi(\mathbf{x}_{n_k})\} \subset \overline{B}_{\delta_2}(\mathbf{y}_0)$ and such a set is compact.

Thus, there exists a subsequence of $\{\xi(\mathbf{x}_{n_k})\}$ which we shall also denote by $\{\xi(\mathbf{x}_{n_k})\}$ and $\mathbf{y}_1 \in \overline{B}_{\delta_2}(\mathbf{y}_0)$ such that $\mathbf{y}_1 \neq \mathbf{y}$ and

$$\lim_{k \to \infty} \xi(\mathbf{x}_{n_k}) = \mathbf{y}_1,$$

Hence, since $\mathbf{f}$ is continuous, we obtain:

$$\mathbf{f}(\mathbf{x}, \mathbf{y}_1) = \lim_{k \to \infty} \mathbf{f}(\mathbf{x}_{n_k}, \xi(\mathbf{x}_{n_k})) = \lim_{k \to \infty} \mathbf{0} = \mathbf{0}.$$

Therefore,

$$\mathbf{f}(\mathbf{x}, \mathbf{y}) = \mathbf{f}(\mathbf{x}, \mathbf{y}_1) = \mathbf{0}.$$

so that from $f(\mathbf{x}, \mathbf{y}_1) = \mathbf{0}$ we obtain

$$\phi(\mathbf{y}_1) = \mathbf{y}_1 - A^{-1}(\mathbf{f}(\mathbf{x}, \mathbf{y}_1)) = \mathbf{y}_1,$$

that is

$$\phi(\mathbf{y}_1) = \mathbf{y}_1.$$

Since the solution of this last equation is unique, we would obtain

$$\mathbf{y}_1 = \mathbf{y},$$

which contradicts

$$\mathbf{y}_1 \neq \mathbf{y}.$$

Therefore

$$\lim_{n \to \infty} \xi(\mathbf{x}_n) = \mathbf{y} = \xi(\mathbf{x}).$$

Since $\{\mathbf{x}_n\} \subset B_\delta(\mathbf{x}_0)$ such that

$$\lim_{n \to \infty} \mathbf{x}_n = \mathbf{x},$$

is arbitrary, we may conclude that ξ is continuous at $\mathbf{x}$, $\forall \mathbf{x} \in B_{\delta_1}(\mathbf{x}_0)$.

For the final part, let $\mathbf{x} \in B_{\delta_1}(\mathbf{x}_0)$ and $h \in \mathbb{R}$ be such that

$$\mathbf{x} + h\mathbf{v} \in B_\delta(\mathbf{x}_0).$$

Denote

$$\mathbf{u} = \xi(\mathbf{x} + h\mathbf{v}) - \xi(\mathbf{x}),$$

that is,

$$\xi(\mathbf{x} + h\mathbf{v}) = \xi(\mathbf{x}) + \mathbf{u}.$$

From the continuity of ξ we have:

$$\mathbf{u} \to \mathbf{0}, \text{ as } h \to 0.$$

Observe that

$$\mathbf{f}(\mathbf{x}, \xi(\mathbf{x})) = \mathbf{0},$$

and

$$\mathbf{f}(\mathbf{x} + h\mathbf{v}, \xi(\mathbf{x} + h\mathbf{v})) = \mathbf{f}(\mathbf{x} + h\mathbf{v}, \xi(\mathbf{x}) + \mathbf{u}) = \mathbf{0},$$

so that

$$\mathbf{f}(\mathbf{x} + h\mathbf{v}, \xi(\mathbf{x}) + \mathbf{u}) - \mathbf{f}(\mathbf{x}, \xi(\mathbf{x})) = \mathbf{0} - \mathbf{0} = \mathbf{0},$$

Therefore

$$\begin{aligned}
0 &= \mathbf{f}(\mathbf{x} + h\mathbf{v}, \xi(\mathbf{x}) + \mathbf{u}) - \mathbf{f}(\mathbf{x}, \xi(\mathbf{x})) \\
&= \mathbf{f}_x(\mathbf{x}, \xi(\mathbf{x})) \cdot h\mathbf{v} + \mathbf{f}_y(\mathbf{x}, \xi(\mathbf{x})) \cdot \mathbf{u} \\
&\quad + \mathbf{w}(h, \mathbf{u})(|h||\mathbf{v}| + |\mathbf{u}|),
\end{aligned} \tag{8.135}$$

Observe that

$$\mathbf{u} \to \mathbf{0}, \text{ as } h \to 0,$$

so that from the differentiability definition,

$$\mathbf{w}(h, \mathbf{u}) \to \mathbf{0}, \text{ as } h \to 0.$$

From (8.135), we have,

$$-\mathbf{f}_y(\mathbf{x}, \xi(\mathbf{x})) \cdot \frac{\mathbf{u}}{h} = \mathbf{f}_x(\mathbf{x}, \xi(\mathbf{x})) \cdot \mathbf{v} + \mathbf{w}(h, \mathbf{u}) \left(\frac{|h|}{h} |\mathbf{v}| + \frac{|\mathbf{u}|}{h} \right) \qquad (8.136)$$

so that

$$\frac{\mathbf{u}}{h} = -[\mathbf{f}_y(\mathbf{x}, \xi(\mathbf{x}))]^{-1} \cdot [\mathbf{f}_x(\mathbf{x}, \xi(\mathbf{x})) \cdot \mathbf{v}] +$$

$$-[\mathbf{f}_y(\mathbf{x}, \xi(\mathbf{x}))]^{-1} \cdot \left[\mathbf{w}(h, \mathbf{u}) \left(\frac{|h|}{h} |\mathbf{v}| + \frac{|\mathbf{u}|}{h} \right) \right] \qquad (8.137)$$

At this point we denote

$$B = -[\mathbf{f}_y(\mathbf{x}, \xi(\mathbf{x}))]^{-1},$$

and

$$C = -[\mathbf{f}_y(\mathbf{x}, \xi(\mathbf{x}))]^{-1} \cdot [\mathbf{f}_x(\mathbf{x}, \xi(\mathbf{x})) \cdot \mathbf{v}],$$

so that

$$\left| \frac{\mathbf{u}}{h} \right| \leq |C| + |B| \left[|\mathbf{w}(h, \mathbf{u})| \left(\frac{|h|}{|h|} |\mathbf{v}| + \frac{|\mathbf{u}|}{|h|} \right) \right] \qquad (8.138)$$

and thus, since

$$\mathbf{w}(h, \mathbf{u}) \to \mathbf{0}, \text{ as } h \to 0,$$

we obtain, for all h sufficiently small,

$$\left| \frac{\mathbf{u}}{h} \right| \leq \frac{|C| + |B| \left[|\mathbf{w}(h, \mathbf{u})| |\mathbf{v}| \right]}{(1 - |B| |\mathbf{w}(h, \mathbf{u})|)}$$

$$\to |C|, \text{ as } h \to 0. \qquad (8.139)$$

Thus, there exists $h_0 > 0$ such that if $0 < |h| < h_0$, then

$$\left| \frac{\mathbf{u}}{h} \right| \leq |C| + 1,$$

so that from this we obtain

$$|\mathbf{w}(h, \mathbf{u})| \left(\frac{|\mathbf{u}|}{|h|} \right) \to 0, \text{ as } h \to 0.$$

Hence, from this and (8.138) we have

$$
\frac{\mathbf{u}}{h} = -[\mathbf{f}_y(\mathbf{x}, \xi(\mathbf{x}))]^{-1} \cdot [\mathbf{f}_x(\mathbf{x}, \xi(\mathbf{x})) \cdot \mathbf{v}] +
$$

$$
-[\mathbf{f}_y(\mathbf{x}, \xi(\mathbf{x}))]^{-1} \cdot \left[\mathbf{w}(h, \mathbf{u}) \left(\frac{|h|}{h} |\mathbf{v}| + \frac{|\mathbf{u}|}{h} \right) \right]
$$

$$
\rightarrow -[\mathbf{f}_y(\mathbf{x}, \xi(\mathbf{x}))]^{-1} \cdot [\mathbf{f}_x(\mathbf{x}, \xi(\mathbf{x})) \cdot \mathbf{v}] \tag{8.140}
$$

as $h \to 0$.
Therefore,

$$
\frac{\partial \xi(\mathbf{x})}{\partial \mathbf{v}} = \lim_{h \to 0} \frac{\xi(\mathbf{x} + h\mathbf{v}) - \xi(\mathbf{x})}{h}
$$

$$
= \lim_{h \to 0} \left(\frac{\mathbf{u}}{h} \right)
$$

$$
= -[\mathbf{f}_y(\mathbf{x}, \xi(\mathbf{x}))]^{-1} \cdot [\mathbf{f}_x(\mathbf{x}, \xi(\mathbf{x})) \cdot \mathbf{v}] \tag{8.141}
$$

The proof is complete.

8.19 Lagrange Multipliers

In this section we develop necessary conditions for extremals with equality restrictions by using the implicit function theorem for vectorial functions.

Let $D \subset \mathbb{R}^4$ be an open set. Let $f, g, h : D \to \mathbb{R}$ be functions of C^1 class. Suppose $(x_0, y_0, u_0, v_0) \in D$ is a point of local minimum of f subject to

$$
\begin{cases} g(x, y, u, v) = 0 \\ h(x, y, u, v) = 0 \end{cases} \tag{8.142}
$$

Suppose that

$$
\det \begin{bmatrix} g_u(\mathbf{x}_0) & g_v(\mathbf{x}_0) \\ h_u(\mathbf{x}_0) & h_v(\mathbf{x}_0) \end{bmatrix} \neq 0 \tag{8.143}
$$

Under such hypotheses, since

$$
\begin{cases} g(x_0, y_0, u_0, v_0) = 0 \\ h(x_0, y_0, u_0, v_0) = 0 \end{cases} \tag{8.144}
$$

from the implicit theorem for vectorial functions, the equations

$$\begin{cases} g(x, y, u, v) = 0 \\ h(x, y, u, v) = 0 \end{cases} \qquad (8.145)$$

define implicitly the functions $u(x, y)$, $v(x, y)$ at $B_\delta(x_0, y_0)$ for some $\delta > 0$, so that

$$\begin{cases} g(x, y, u(x, y), v(x, y)) = 0 \\ h(x, y, u(x, y), v(x, y)) = 0, \ \forall(x, y) \in B_\delta(x_0, y_0). \end{cases} \qquad (8.146)$$

Therefore, the original problem of restricted optimization will correspond to the local minimization of the function $F : B_\delta(x_0, y_0) \to \mathbb{R}$ given by

$$F(x, y) = f(x, y, u(x, y), v(x, y)).$$

Hence, the necessary conditions for a local minimum will be:

$$F_x(x_0, y_0) = 0 \text{ and } F_y(x_0, y_0) = 0.$$

Observe that from $F_x(x_0, y_0) = 0$, we obtain:

$$f_x(\mathbf{x}_0) + f_u(\mathbf{x}_0)u_x(x_0, y_0) + f_v(\mathbf{x}_0)v_x(x_0, y_0) = 0, \qquad (8.147)$$

and from $F_y(x_0, y_0) = 0$, we obtain:

$$f_y(\mathbf{x}_0) + f_u(\mathbf{x}_0)u_y(x_0, y_0) + f_v(\mathbf{x}_0)v_y(x_0, y_0) = 0. \qquad (8.148)$$

On the other hand, from

$$\begin{cases} g(x, y, u(x, y), v(x, y)) = 0 \\ h(x, y, u(x, y), v(x, y)) = 0, \ \forall(x, y) \in B_\delta(x_0, y_0). \end{cases} \qquad (8.149)$$

we get

$$\frac{dg(\mathbf{x}_0)}{dx} = g_x(\mathbf{x}_0) + g_u(\mathbf{x}_0)u_x(x_0, y_0) + g_v(\mathbf{x}_0)v_x(x_0, y_0) = 0,$$

and

$$\frac{dh(\mathbf{x}_0)}{dx} = h_x(\mathbf{x}_0) + h_u(\mathbf{x}_0)u_x(x_0, y_0) + h_v(\mathbf{x}_0)v_x(x_0, y_0) = 0,$$

so that

$$\frac{dg(\mathbf{x}_0)}{dy} = g_y(\mathbf{x}_0) + g_u(\mathbf{x}_0)u_y(x_0, y_0) + g_v(\mathbf{x}_0)v_y(x_0, y_0) = 0,$$

and

$$\frac{dh(\mathbf{x}_0)}{dy} = h_y(\mathbf{x}_0) + h_u(\mathbf{x}_0)u_y(x_0, y_0) + ghv(\mathbf{x}_0)v_y(x_0, y_0) = 0,$$

so that

$$\begin{bmatrix} u_x(x_0, y_0) \\ v_x(x_0, y_0) \end{bmatrix} = -\begin{bmatrix} g_u(\mathbf{x}_0) & g_v(\mathbf{x}_0) \\ h_u(\mathbf{x}_0) & h_v(\mathbf{x}_0) \end{bmatrix}^{-1} \begin{bmatrix} g_x(\mathbf{x}_0) \\ h_x(\mathbf{x}_0) \end{bmatrix} \qquad (8.150)$$

and

$$\begin{bmatrix} u_y(x_0, y_0) \\ v_y(x_0, y_0) \end{bmatrix} = -\begin{bmatrix} g_u(\mathbf{x}_0) & g_v(\mathbf{x}_0) \\ h_u(\mathbf{x}_0) & h_v(\mathbf{x}_0) \end{bmatrix}^{-1} \begin{bmatrix} g_y(\mathbf{x}_0) \\ h_y(\mathbf{x}_0) \end{bmatrix}. \qquad (8.151)$$

Thus,

$$f_x(\mathbf{x}_0) + [f_u \quad f_v]\begin{bmatrix} u_x(x_0, y_0) \\ v_x(x_0, y_0) \end{bmatrix} = 0, \qquad (8.152)$$

so that

$$f_x(\mathbf{x}_0) - [f_u \quad f_v]\begin{bmatrix} g_u(\mathbf{x}_0) & g_v(\mathbf{x}_0) \\ h_u(\mathbf{x}_0) & h_v(\mathbf{x}_0) \end{bmatrix}^{-1} \begin{bmatrix} g_x(\mathbf{x}_0) \\ h_x(\mathbf{x}_0) \end{bmatrix} = 0. \qquad (8.153)$$

and thus

$$f_x(\mathbf{x}_0) + \lambda_1 g_x(\mathbf{x}_0) + \lambda_2 h_x(\mathbf{x}_0) = 0.$$

where:

$$[\lambda_1 \quad \lambda_2] = -[f_u \quad f_v]\begin{bmatrix} g_u(\mathbf{x}_0) & g_v(\mathbf{x}_0) \\ h_u(\mathbf{x}_0) & h_v(\mathbf{x}_0) \end{bmatrix}^{-1}. \qquad (8.154)$$

Similarly, we may obtain:

$$f_y(\mathbf{x}_0) + \lambda_1 g_y(\mathbf{x}_0) + \lambda_2 h_y(\mathbf{x}_0) = 0.$$

Finally, observe that

$$f_u(\mathbf{x}_0) + \lambda_1 g_u(\mathbf{x}_0) + \lambda_2 h_u(\mathbf{x}_0)$$

$$= f_u + [\lambda_1 \quad \lambda_2]\begin{bmatrix} g_u(\mathbf{x}_0) \\ h_u(\mathbf{x}_0) \end{bmatrix}$$

$$= f_u - [f_u \quad f_v] \begin{bmatrix} g_u(\mathbf{x}_0) & g_v(\mathbf{x}_0) \\ h_u(\mathbf{x}_0) & h_v(\mathbf{x}_0) \end{bmatrix}^{-1} \begin{bmatrix} g_u(\mathbf{x}_0) \\ h_u(\mathbf{x}_0) \end{bmatrix}$$

$$= f_u - [f_u \quad f_v] \begin{bmatrix} h_v(\mathbf{x}_0) & -g_v(\mathbf{x}_0) \\ -h_u(\mathbf{x}_0) & g_u(\mathbf{x}_0) \end{bmatrix} \begin{bmatrix} g_u(\mathbf{x}_0) \\ h_u(\mathbf{x}_0) \end{bmatrix} / (g_u h_v - h_u g_v)$$

$$= f_u - [f_u \quad f_v] \begin{bmatrix} 1 \\ 0 \end{bmatrix}$$

$$= f_u - f_u = 0, \tag{8.155}$$

Similarly, we obtain:

$$f_v + \lambda_1 g_v + \lambda_2 h_v = 0.$$

We may summarize the set of necessary conditions as indicated below:

$$\begin{cases} f_x + \lambda_1 g_x + \lambda_2 h_x = 0, \\ f_y + \lambda_1 g_y + \lambda_2 h_y = 0, \\ f_u + \lambda_1 g_u + \lambda_2 h_u = 0, \\ f_v + \lambda_1 g_v + \lambda_2 h_v = 0, \\ g = 0 \\ h = 0, \end{cases} \tag{8.156}$$

with all functions in question considered at the point $\mathbf{x}_0 = (x_0, y_0, u_0, v_0)$.

8.20 Lagrange Multipliers: The General Case

Let $D \subset \mathbb{R}^{n+m}$ be a nonempty open set, where we denote

$$(\mathbf{x}, \mathbf{y}) \in D \subset \mathbf{R}^{n+m}$$

and where

$$\mathbf{x} = (x_1, \dots, x_n) \in \mathbb{R}^n$$

and

$$\mathbf{y} = (y_1, \dots, y_m) \in \mathbb{R}^m.$$

Let $f, g_1, \dots, g_m : D \to \mathbb{R}$ be functions of C^1 class.

Suppose that

$$(\mathbf{x}_0, \mathbf{y}_0) \in D$$

is a point of local minimum of f subject to

$$\begin{cases} g_1(\mathbf{x}, \mathbf{y}) = 0, \\ g_2(\mathbf{x}, \mathbf{y}) = 0, \\ \quad \vdots \\ g_m(\mathbf{x}, \mathbf{y}) = 0. \end{cases} \tag{8.157}$$

Thus, there exists $\delta > 0$ such that

$$f(\mathbf{x}, \mathbf{y}) \geq f(\mathbf{x}_0, \mathbf{y}_0),$$

$\forall (\mathbf{x}, \mathbf{y}) \in B_\delta(\mathbf{x}_0, \mathbf{y}_0)$ such that,

$$\begin{cases} g_1(\mathbf{x}, \mathbf{y}) = 0, \\ g_2(\mathbf{x}, \mathbf{y}) = 0, \\ \quad \vdots \\ g_m(\mathbf{x}, \mathbf{y}) = 0. \end{cases} \tag{8.158}$$

We shall define $G : D \to \mathbb{R}^m$ by

$$G(\mathbf{x}, \mathbf{y}) = \begin{bmatrix} g_1(\mathbf{x}, \mathbf{y}) = 0 \\ g_2(\mathbf{x}, \mathbf{y}) = 0 \\ \vdots \\ g_m(\mathbf{x}, \mathbf{y}) = 0 \end{bmatrix} \tag{8.159}$$

Therefore, the restriction in question is equivalent to

$$G(\mathbf{x}, \mathbf{y}) = \mathbf{0} \in \mathbb{R}^m.$$

Denoting

$$G_y(\mathbf{x}_0, \mathbf{y}_0) = \left\{ \frac{\partial g_i(\mathbf{x}_0, \mathbf{y}_0)}{\partial y_j} \right\},$$

where $i \in \{1, \ldots, m\}$ and $j \in \{1, \ldots, m\}$, assume

$$\det(G_y(\mathbf{x}_0, \mathbf{y}_0)) \neq 0.$$

From the implicit theorem, vectorial case, there exist $\delta_1 > 0$ and $\delta_2 > 0$ such that for each $\mathbf{x} \in B_{\delta_1}(\mathbf{x_0})$ there exists a unique $\mathbf{y} \in \overline{B}_{\delta_2}(\mathbf{y_0})$ such that

$$G(\mathbf{x}, \mathbf{y}) = \mathbf{0}.$$

Denoting such an $\mathbf{y}$ by $\mathbf{y} = \mathbf{y}(\mathbf{x})$, we obtain

$$G(\mathbf{x}, \mathbf{y}(\mathbf{x})) = \mathbf{0}, \ \forall \mathbf{x} \in B_{\delta_1}(\mathbf{x_0}).$$

Therefore, the original constrained optimization problem may be seen as the non-constrained local minimization of $F : B_{\delta_1}(\mathbf{x_0})$, where

$$F(\mathbf{x}) = f(\mathbf{x}, \mathbf{y}(\mathbf{x})).$$

Observe that $\mathbf{x_0}$ is a point of local minimum for F so that the first order optimality conditions stand for,

$$F_{x_j}(\mathbf{x_0}) = 0, \forall j \in \{1, \ldots, n\}.$$

Thus,

$$\frac{\partial f(\mathbf{x_0}, \mathbf{y}(\mathbf{x_0}))}{\partial x_j} = 0,$$

that is,

$$\frac{\partial f(\mathbf{x_0}, \mathbf{y_0})}{\partial x_j} + \sum_{k=1}^{m} \frac{\partial f(\mathbf{x_0}, \mathbf{y_0})}{\partial y_k} \frac{\partial y_k(\mathbf{x_0})}{\partial x_j} = 0, \tag{8.160}$$

$\forall j \in \{1, \ldots, n\}$.

On the other hand, from implicit function theorem, vectorial case, we have

$$\left\{ \frac{\partial y_k(\mathbf{x_0})}{\partial x_j} \right\} = -\left[G_y(\mathbf{x_0}, \mathbf{y_0}) \right]^{-1} \left[G_{x_j}(\mathbf{x_0}, \mathbf{y_0}) \right]$$

$$= -\left\{ \frac{\partial g_i(\mathbf{x_0}, \mathbf{y_0})}{\partial y_p} \right\}_{m \times m}^{-1} \begin{bmatrix} (g_1)_{x_j}(\mathbf{x_0}, \mathbf{y_0}) \\ (g_2)_{x_j}(\mathbf{x_0}, \mathbf{y_0}) \\ \vdots \\ (g_m)_{x_j}(\mathbf{x_0}, \mathbf{y_0}) \end{bmatrix}_{m \times 1}, \tag{8.161}$$

where $i \in \{1, \ldots, m\}$, $p \in \{1, \ldots, m\}$, $\forall i \in \{1, \ldots, n\}$. From this and (8.160) we obtain

$$\frac{\partial f(\mathbf{x}_0, \mathbf{y}_0)}{\partial x_j} - \left[\begin{array}{ccc} \frac{\partial f(\mathbf{x}_0, \mathbf{y}_0)}{\partial y_1} & \cdots & \frac{\partial f(\mathbf{x}_0, \mathbf{y}_0)}{\partial y_m} \end{array}\right]_{1 \times m} \begin{bmatrix} (y_1)_{x_j}(\mathbf{x}_0) \\ (y_2)_{x_j}(\mathbf{x}_0) \\ \vdots \\ (y_m)_{x_j}(\mathbf{x}_0) \end{bmatrix}_{m \times 1}$$

$$= \frac{\partial f(\mathbf{x}_0, \mathbf{y}_0)}{\partial x_j}$$

$$- \left[\begin{array}{ccc} \frac{\partial f(\mathbf{x}_0, \mathbf{y}_0)}{\partial y_1} & \cdots & \frac{\partial f(\mathbf{x}_0, \mathbf{y}_0)}{\partial y_m} \end{array}\right]_{1 \times m} \left[\frac{\partial g_i(\mathbf{x}_0, \mathbf{y}_0)}{\partial y_p}\right]^{-1}_{m \times m} \begin{bmatrix} (g_1)_{x_j}(\mathbf{x}_0, \mathbf{y}_0) \\ (g_2)_{x_j}(\mathbf{x}_0, \mathbf{y}_0) \\ \vdots \\ (g_m)_{x_j}(\mathbf{x}_0, \mathbf{y}_0) \end{bmatrix}_{m \times 1}$$

$$= \frac{\partial f(\mathbf{x}_0, \mathbf{y}_0)}{\partial x_j} + \left[\begin{array}{ccc} \lambda_1 & \cdots & \lambda_m \end{array}\right]_{1 \times m} \begin{bmatrix} (g_1)_{x_j}(\mathbf{x}_0, \mathbf{y}_0) \\ (g_2)_{x_j}(\mathbf{x}_0, \mathbf{y}_0) \\ \vdots \\ (g_m)_{x_j}(\mathbf{x}_0, \mathbf{y}_0) \end{bmatrix}_{m \times 1}$$

$$= 0, \tag{8.162}$$

Indeed, this last equation stands for,

$$\frac{\partial f(\mathbf{x}_0, \mathbf{y}_0)}{\partial x_j} + \sum_{k=1}^{m} \lambda_k \frac{\partial g_k(\mathbf{x}_0, \mathbf{y}_0)}{\partial x_j} = 0, \tag{8.163}$$

$\forall j \in \{1, \ldots, n\}$, where,

$$[\lambda_1, \ldots, \lambda_m] = -\left[\begin{array}{ccc} \frac{\partial f(\mathbf{x}_0, \mathbf{y}_0)}{\partial y_1} & \cdots & \frac{\partial f(\mathbf{x}_0, \mathbf{y}_0)}{\partial y_m} \end{array}\right]_{1 \times m} \left[\frac{\partial g_i(\mathbf{x}_0, \mathbf{y}_0)}{\partial y_p}\right]^{-1}_{m \times m}.$$

On the other hand,

$$\frac{\partial f(\mathbf{x}_0, \mathbf{y}_0)}{\partial y_j} + \sum_{k=1}^{m} \lambda_k \frac{\partial g_k(\mathbf{x}_0, \mathbf{y}_0)}{\partial y_j}$$

$$= \frac{\partial f(\mathbf{x}_0, \mathbf{y}_0)}{\partial y_j}$$

$$+ \left[\begin{array}{ccc} \lambda_1 & \cdots & \lambda_m \end{array}\right]_{1 \times m} \begin{bmatrix} (g_1)_{y_j}(\mathbf{x}_0, \mathbf{y}_0) \\ (g_2)_{y_j}(\mathbf{x}_0, \mathbf{y}_0) \\ \vdots \\ (g_m)_{y_j}(\mathbf{x}_0, \mathbf{y}_0) \end{bmatrix}_{m \times 1}$$

$$= \frac{\partial f(\mathbf{x}_0, \mathbf{y}_0)}{\partial x_j}$$

$$- \left[\frac{\partial f(\mathbf{x}_0, \mathbf{y}_0)}{\partial y_1} \cdots \frac{\partial f(\mathbf{x}_0, \mathbf{y}_0)}{\partial y_m} \right]_{1 \times m} \left[\frac{\partial g_i(\mathbf{x}_0, \mathbf{y}_0)}{\partial y_p} \right]^{-1}_{m \times m} \begin{bmatrix} (g_1)_{y_j}(\mathbf{x}_0, \mathbf{y}_0) \\ (g_2)_{y_j}(\mathbf{x}_0, \mathbf{y}_0) \\ \vdots \\ (g_m)_{y_j}(\mathbf{x}_0, \mathbf{y}_0) \end{bmatrix}_{m \times 1} \tag{8.164}$$

At this point define

$$\mathbf{z} = \left[\frac{\partial g_i(\mathbf{x}_0, \mathbf{y}_0)}{\partial y_p} \right]^{-1}_{m \times m} \begin{bmatrix} (g_1)_{y_j}(\mathbf{x}_0, \mathbf{y}_0) \\ (g_2)_{y_j}(\mathbf{x}_0, \mathbf{y}_0) \\ \vdots \\ (g_m)_{y_j}(\mathbf{x}_0, \mathbf{y}_0) \end{bmatrix}_{m \times 1},$$

so that

$$\begin{bmatrix} (g_1)_{y_1}(\mathbf{x}_0, \mathbf{y}_0) & (g_1)_{y_2}(\mathbf{x}_0, \mathbf{y}_0) & \cdots & (g_1)_{y_m}(\mathbf{x}_0, \mathbf{y}_0) \\ (g_2)_{y_1}(\mathbf{x}_0, \mathbf{y}_0) & (g_2)_{y_2}(\mathbf{x}_0, \mathbf{y}_0) & \cdots & (g_2)_{y_m}(\mathbf{x}_0, \mathbf{y}_0) \\ \vdots & \vdots & \ddots & \vdots \\ (g_m)_{y_1}(\mathbf{x}_0, \mathbf{y}_0) & (g_m)_{y_2}(\mathbf{x}_0, \mathbf{y}_0) & \cdots & (g_m)_{y_m}(\mathbf{x}_0, \mathbf{y}_0) \end{bmatrix}_{m \times m} \begin{bmatrix} z_1 \\ z_2 \\ \vdots \\ z_m \end{bmatrix}_{m \times 1}$$

$$= \begin{bmatrix} (g_1)_{y_j}(\mathbf{x}_0, \mathbf{y}_0) \\ (g_2)_{y_j}(\mathbf{x}_0, \mathbf{y}_0) \\ \vdots \\ (g_m)_{y_j}(\mathbf{x}_0, \mathbf{y}_0) \end{bmatrix}_{m \times 1}. \tag{8.165}$$

From this last equation is clear that $z_j = 1$ and $z_k = 0$, if $k \neq j$. Therefore, we have obtained,

$$\frac{\partial f(\mathbf{x}_0, \mathbf{y}_0)}{\partial y_j} + \sum_{k=1}^{m} \lambda_k \frac{\partial g_k(\mathbf{x}_0, \mathbf{y}_0)}{\partial y_j}$$

$$= \frac{\partial f(\mathbf{x}_0, \mathbf{y}_0)}{\partial y_j} - \left[\frac{\partial f(\mathbf{x}_0, \mathbf{y}_0)}{\partial y_1} \cdots \frac{\partial f(\mathbf{x}_0, \mathbf{y}_0)}{\partial y_m} \right]_{1 \times m} \mathbf{z}$$

$$= \frac{\partial f(\mathbf{x}_0, \mathbf{y}_0)}{\partial y_j} - \frac{\partial f(\mathbf{x}_0, \mathbf{y}_0)}{\partial y_j}$$

$$= 0, \forall j \in \{1, \ldots, m\}. \tag{8.166}$$

Summarizing, the first order optimality conditions for the local minimization of f subject to $G(\mathbf{x}, \mathbf{y}) = \mathbf{0}$ will be

$$\frac{\partial f(\mathbf{x}_0, \mathbf{y}_0)}{\partial x_i} + \sum_{k=1}^{m} \lambda_k \frac{\partial g_k(\mathbf{x}_0, \mathbf{y}_0)}{\partial x_i} = 0, \ \forall i \in \{1, \ldots, n\},$$

$$\frac{\partial f(\mathbf{x}_0, \mathbf{y}_0)}{\partial y_j} + \sum_{k=1}^{m} \lambda_k \frac{\partial g_k(\mathbf{x}_0, \mathbf{y}_0)}{\partial y_j} = 0, \ \forall j \in \{1, \ldots, m\},$$

$$g_k(\mathbf{x}_0, \mathbf{y}_0) = 0, \ \forall k \in \{1, \ldots, m\}.$$

We have $n + 2m$ variables and $n + 2m$ equations.

Observe that such a system corresponds to a critical point of the Lagrangian L, where

$$L(\mathbf{x}, \mathbf{y}, \lambda_1, \ldots, \lambda_m) = f(\mathbf{x}, \mathbf{y}) + \sum_{k=1}^{m} \lambda_k g_k(\mathbf{x}, \mathbf{y}).$$

Finally, $\lambda_1, \ldots, \lambda_m$ are said to be the Lagrange multipliers relating the corresponding constraints of the original problem.

8.21 Inverse Function Theorem

Theorem 8.21.1 *Let $D \subset \mathbb{R}^n$ be an open set and let $\mathbf{f} : D \to \mathbb{R}^n$ be a C^1 class function on D.*

Let $\mathbf{x}_0 \in D$ be such that $\det(\mathbf{f}'(\mathbf{x}_0)) \neq 0$.

Under such hypotheses, denoting $\mathbf{y}_0 = \mathbf{f}(\mathbf{x}_0)$ we have

1. *There exist open sets $U, V \subset \mathbb{R}^n$ such that $\mathbf{x}_0 \in U$ and $\mathbf{y}_0 \in V$ and such that $\mathbf{f}(U) = V$. Moreover $\mathbf{f}$ is injective on U.*
2. *Defining the local inverse of $\mathbf{f}$, denoted by*

$$\mathbf{f}^{-1} = \mathbf{g} : V \to U,$$

by

$$\mathbf{g}(\mathbf{y}) = \mathbf{x} \Leftrightarrow \mathbf{y} = \mathbf{f}(\mathbf{x}), \ \forall \mathbf{x} \in U, \ \mathbf{y} \in V$$

we have that

$$\mathbf{g}'(\mathbf{y}) = [\mathbf{f}'(\mathbf{g}(\mathbf{y}))]^{-1}.$$

Moreover, $\mathbf{g}$ is also of C^1 class.

Proof Denote $\mathbf{f}'(\mathbf{x}_0) = A$. From the hypotheses, $\det(A) \neq 0$.
Define $\lambda = \frac{1}{2|A^{-1}|}$.

Since $\mathbf{f}'$ is continuous on D, there exists $\delta_1 > 0$ such that if $\mathbf{x} \in \overline{B}_{\delta_1}(\mathbf{x}_0)$, then

$$\det(\mathbf{f}'(\mathbf{x})) \neq 0,$$

and

$$|\mathbf{f}'(\mathbf{x}) - A| = |\mathbf{f}'(\mathbf{x}) - \mathbf{f}'(\mathbf{x}_0)| < \lambda.$$

Fix $\mathbf{y} \in \mathbb{R}^n$ and define

$$\phi(\mathbf{x}) = \mathbf{x} + A^{-1}(\mathbf{y} - \mathbf{f}(\mathbf{x})).$$

Hence,

$$\begin{aligned}
\phi'(\mathbf{x}) &= I + A^{-1}(-\mathbf{f}'(\mathbf{x})) \\
&= A^{-1}A - A^{-1}(\mathbf{f}'(\mathbf{x})) \\
&= A^{-1}(A - \mathbf{f}'(\mathbf{x})).
\end{aligned} \tag{8.167}$$

Therefore,

$$|\phi'(\mathbf{x})| \leq |A^{-1}||A - \mathbf{f}'(\mathbf{x})| \leq \frac{1}{2\lambda}\lambda = \frac{1}{2}.$$

Thus,

$$|\phi'(\mathbf{x})| \leq \frac{1}{2}, \quad \forall \mathbf{x} \in \overline{B}_{\delta_1}(\mathbf{x}), \ \mathbf{y} \in \mathbb{R}^n.$$

From the mean value inequality we obtain

$$|\phi(\mathbf{x}_2) - \phi(\mathbf{x}_1)| \leq \frac{1}{2}|\mathbf{x}_2 - \mathbf{x}_1|,$$

$\forall \mathbf{x}_1, \mathbf{x}_2 \in \overline{B}_{\delta_1}(\mathbf{x}_0), \ \mathbf{y} \in \mathbb{R}^n$.
Define $\delta_2 = \frac{\delta_1}{2|A^{-1}|}$.
Let $\mathbf{y} \in B_{\delta_2}(\mathbf{y}_0)$.
Observe that for such a specific $\mathbf{y}$, we have

$$\begin{aligned}
|\phi(\mathbf{x}_0) - \mathbf{x}_0| &= |A^{-1}(\mathbf{y} - \mathbf{f}(\mathbf{x}_0))| \\
&\leq |A^{-1}||\mathbf{y} - \mathbf{y}_0|
\end{aligned}$$

$$\leq \frac{|A^{-1}|\delta_1}{2|A^{-1}|}$$

$$= \frac{\delta_1}{2}, \tag{8.168}$$

so that

$$|\phi(\mathbf{x}_0) - \mathbf{x}_0| < \frac{\delta_1}{2}.$$

Thus, for $\mathbf{x} \in \overline{B}_{\delta_1}(\mathbf{x}_0)$, we have that

$$|\phi(\mathbf{x}) - \mathbf{x}_0| = |\phi(\mathbf{x}) - \phi(\mathbf{x}_0) + \phi(\mathbf{x}_0) - \mathbf{x}_0|$$

$$\leq |\phi(\mathbf{x}) - \phi(\mathbf{x}_0)| + \frac{\delta_1}{2}$$

$$\leq \frac{1}{2}|\mathbf{x} - \mathbf{x}_0| + \frac{\delta_1}{2}$$

$$< \frac{\delta_1}{2} + \frac{\delta_1}{2} = \delta_1. \tag{8.169}$$

Therefore, we may infer that

$$|\phi(\mathbf{x}) - \mathbf{x}_0| < \delta_1, \ \forall \mathbf{x} \in \overline{B}_{\delta_1}(\mathbf{x}_0),$$

so that,

$$\phi(\overline{B}_{\delta_1}(\mathbf{x}_0)) \subset \overline{B}_{\delta_1}(\mathbf{x}_0),$$

in fact, from the results exposed above,

$$\phi(\overline{B}_{\delta_1}(\mathbf{x}_0)) \subset B_{\delta_1}(\mathbf{x}_0).$$

We may conclude that ϕ is a contractor mapping on $\overline{B}_{\delta_1}(\mathbf{x}_0)$.

Thus, from the contractor mapping theorem, there exists a unique $\mathbf{x} \in B_\delta(\mathbf{x}_0)$ such that $\phi(\mathbf{x}) = \mathbf{x}$.

From

$$\phi(\mathbf{x}) = \mathbf{x} + A^{-1}(\mathbf{y} - \mathbf{f}(\mathbf{x})),$$

we obtain

$$A^{-1}(\mathbf{y} - \mathbf{f}(\mathbf{x})) = \mathbf{0},$$

and since $\det(A^{-1}) \neq 0$, we get

$$\mathbf{y} - \mathbf{f}(\mathbf{x}) = \mathbf{0},$$

that is,

$$\mathbf{y} = \mathbf{f}(\mathbf{x}),$$

where such a $\mathbf{x} \in B_{\delta_1}(\mathbf{x}_0)$ is unique, $\forall \mathbf{y} \in B_{\delta_2}(\mathbf{y}_0)$.
  Hence, $\mathbf{g}(\mathbf{y}) = \mathbf{x}$.
  Define

$$V = B_{\delta_2}(\mathbf{y}_0).$$

we claim that $\mathbf{g}$ is injective on V. Suppose $\mathbf{y}_1, \mathbf{y}_2 \in V$ are such that

$$\mathbf{g}(\mathbf{y}_1) = \mathbf{g}(\mathbf{y}_2) = \mathbf{x}.$$

  From above

$$\mathbf{y}_1 = \mathbf{y}_2 = \mathbf{f}(\mathbf{x}),$$

so that we may conclude that $\mathbf{g}$ is injective on V.
  Define $U = f^{-1}(V) \cap B_{\delta_1}(\mathbf{x}_0)$. From the injectivity of $\mathbf{g}$ on V we infer that $\mathbf{f}$ is injective on U.
  Observe that U and V are open sets and

$$\mathbf{f}(U) = V.$$

  Finally, $\mathbf{g} : V \to U$ is such that

$$\mathbf{g}(\mathbf{y}) = \mathbf{x} \Leftrightarrow \mathbf{y} = \mathbf{f}(\mathbf{x}).$$

  For the next item, let $\mathbf{y}, \mathbf{v} \in \mathbb{R}^n$ be such that $\mathbf{y}$ and $\mathbf{y} + \mathbf{v} \in V$.
Thus, there exists $\mathbf{x}, \mathbf{x}_1 \in U$ such that

$$\mathbf{y} = \mathbf{f}(\mathbf{x}) \text{ and } \mathbf{y} + \mathbf{v} = \mathbf{f}(\mathbf{x}_1).$$

  Here we denote,

$$\mathbf{h} = \mathbf{x}_1 - \mathbf{x},$$

so that $\mathbf{x}_1 = \mathbf{x} + \mathbf{h}$, and therefore

$$\mathbf{y} + \mathbf{v} = \mathbf{f}(\mathbf{x} + \mathbf{h}),$$

and moreover,

$$g(\mathbf{y}) = \mathbf{x},$$

and

$$g(\mathbf{y} + \mathbf{v}) = \mathbf{x} + \mathbf{h}.$$

Observe that, from the exposed above,

$$|\phi(\mathbf{x} + \mathbf{h}) - \phi(\mathbf{x})| \le \frac{1}{2}|\mathbf{h}|, \tag{8.170}$$

where,

$$\phi(\mathbf{x} + \mathbf{h}) = \mathbf{x} + \mathbf{h} + A^{-1}(\mathbf{y} - \mathbf{f}(\mathbf{x} + \mathbf{h})),$$

$$\phi(\mathbf{x}) = \mathbf{x} + A^{-1}(\mathbf{y} - \mathbf{f}(\mathbf{x})),$$

therefore, from this and (8.170), we obtain,

$$\begin{aligned}
\frac{1}{2}|\mathbf{h}| &\ge |\phi(\mathbf{x} + \mathbf{h}) - \phi(\mathbf{x})| \\
&= |\mathbf{h} + A^{-1}(\mathbf{f}(\mathbf{x}) - \mathbf{f}(\mathbf{x} + \mathbf{h}))| \\
&= |\mathbf{h} + A^{-1}(\mathbf{y} - (\mathbf{y} + \mathbf{v}))| \\
&= |\mathbf{h} - A^{-1}\mathbf{v}| \\
&\ge |\mathbf{h}| - |A^{-1}\mathbf{v}|, \tag{8.171}
\end{aligned}$$

so that

$$\frac{1}{2}|\mathbf{h}| \ge |\mathbf{h}| - |A^{-1}\mathbf{v}|,$$

and thus,

$$|A^{-1}\mathbf{v}| \ge \frac{1}{2}|\mathbf{h}|.$$

From this, we may infer that

$$|\mathbf{h}| \le 2|A^{-1}||\mathbf{v}| \le \lambda^{-1}|\mathbf{v}|, \tag{8.172}$$

so that

$$\frac{1}{|\mathbf{v}|} \le \frac{1}{\lambda|\mathbf{h}|}.$$

Observe that from (8.172), we have, $\mathbf{h} \to \mathbf{0}$ as $\mathbf{v} \to \mathbf{0}$, so that $\mathbf{g}$ is continuous on V.

Observe also that $\det(\mathbf{f}'(\mathbf{x})) \neq 0$ and define

$$T = [\mathbf{f}'(\mathbf{x})]^{-1}.$$

Therefore,

$$\mathbf{g}(\mathbf{y} + \mathbf{v}) - \mathbf{g}(\mathbf{y}) - T\mathbf{v} = \mathbf{h} - T\mathbf{v}$$
$$= TT^{-1}\mathbf{h} - T\mathbf{v}$$
$$= -T(\mathbf{v} - T^{-1}\mathbf{h})$$
$$= -T(\mathbf{f}(\mathbf{x} + \mathbf{h}) - \mathbf{f}(\mathbf{x}) - \mathbf{f}'(\mathbf{x})\mathbf{h}), \qquad (8.173)$$

so that

$$\frac{|\mathbf{g}(\mathbf{y} + \mathbf{v}) - \mathbf{g}(\mathbf{y}) - T\mathbf{v}|}{|\mathbf{v}|} \leq \frac{|-T||\mathbf{f}(\mathbf{x} + \mathbf{h}) - \mathbf{f}(\mathbf{x}) - \mathbf{f}'(\mathbf{x})\mathbf{h}|}{|\mathbf{v}|}$$
$$\leq \frac{|T|}{\lambda} \frac{|\mathbf{f}(\mathbf{x} + \mathbf{h}) - \mathbf{f}(\mathbf{x}) - \mathbf{f}'(\mathbf{x})\mathbf{h}|}{|\mathbf{h}|}$$
$$\to 0 \quad \text{as } \mathbf{v} \to \mathbf{0}. \qquad (8.174)$$

From this we may conclude that g is differentiable at $\mathbf{y}$ and also

$$\mathbf{g}'(\mathbf{y}) = T = [\mathbf{f}'(\mathbf{x})]^{-1} = [\mathbf{f}'(\mathbf{g}(\mathbf{y}))]^{-1}.$$

Moreover, from $\det[\mathbf{f}'(\mathbf{x})]^{-1} \neq 0$ on $B_{\delta_1}(\mathbf{x}_0)$, and being $\mathbf{g}$ continuous on V and $\mathbf{f}'$ continuous on $B_{\delta_1}(\mathbf{x}_0)$, we may conclude that $\mathbf{g}$ is of C^1 class on V.

The proof is complete.

Example 8.21.2 In this example, we define $\mathbf{f} : \mathbb{R}^2 \to \mathbb{R}^2$ by

$$\mathbf{f}(x, y) = \begin{pmatrix} f_1(\mathbf{x}) \\ f_2(\mathbf{x}) \end{pmatrix} \qquad (8.175)$$

where

$$f_1(x, y) = \sin^2[\pi(x + y)] + y,$$

and

$$f_2(x, y) = \ln(x^2 + y^2 + 1).$$

Let $\mathbf{x}_0 = (1, 0) \in \mathbb{R}^2$.

Observe that

$$\mathbf{f}'(x, y) = \begin{pmatrix} (f_1)_x(x, y) & (f_1)_y(x, y) \\ (f_2)_x(x, y) & (f_2)_y(x, y) \end{pmatrix}$$

$$= \begin{bmatrix} 2\sin[\pi(x + y)]\cos[\pi(x + y)]\pi & 2\sin[\pi(x + y)]\cos[\pi(x + y)]\pi + 1 \\ \frac{2x}{x^2+y^2+1} & \frac{2y}{x^2+y^2+1} \end{bmatrix} \quad (8.176)$$

Hence,

$$\mathbf{f}'(1, 0) = \begin{bmatrix} 0 & 1 \\ 1 & 0 \end{bmatrix},$$

so that

$$\det(\mathbf{f}'(1, 0)) = -1 \neq 0.$$

Thus, from the inverse function theorem, the inverse $\mathbf{g} = \mathbf{f}^{-1}$ of f is well defined in a neighborhood of

$$\mathbf{f}(\mathbf{x}_0) = \mathbf{f}(1, 0) = (0, \ln 2)^T.$$

Also from such a theorem,

$$\mathbf{g}'(\mathbf{f}(\mathbf{x}_0)) = \mathbf{g}'(0, \ln 2) = [\mathbf{f}'(\mathbf{x}_0)]^{-1} = [\mathbf{f}'(1, 0)]^{-1},$$

that is,

$$\mathbf{g}'(0, \ln 2) = \begin{bmatrix} 0 & 1 \\ 1 & 0 \end{bmatrix}^{-1} = \begin{bmatrix} 0 & -1 \\ -1 & 0 \end{bmatrix}/(-1) = \begin{bmatrix} 0 & 1 \\ 1 & 0 \end{bmatrix}.$$

Exercises 8.21.3

1. Let $A, B \subset \mathbb{R}^n$ be open sets.
   Prove that $A \cup B$ and $A \cap B$ are open.
2. Let $\{A_\alpha, \ \alpha \in L\} \subset \mathbb{R}^n$ be a collection of sets such that A_α is open $\forall \alpha \in L$.
   Show that $\cup_{\alpha \in L} A_\alpha$ is open.
3. Let $\{A_\alpha, \ \alpha \in L\} \subset \mathbb{R}^n$ be a collection of sets such that A_α is closed $\forall \alpha \in L$.
   Show that $\cap_{\alpha \in L} A_\alpha$ is closed.
4. Let $A, B \subset \mathbb{R}^n$ be closed sets.
   Prove that $A \cup B$ is closed.
5. Through the proof of the Heine–Borel for $\mathbb{R}$, prove a version of such a theorem for $\mathbb{R}^n$.

   That is, prove that for $E \subset \mathbb{R}^n$, the following three properties are equivalent.

(a) E is compact.

(b) E is closed and bounded.

(c) Every infinite set contained in E has a limit point in E.

6. Given $A, B \subset \mathbb{R}^n$, we define the distance between A and B, denoted by $d(A, B)$, as

$$d(A, B) = \inf\{|u - v| \ : \ u \in A \text{ and } v \in B\}.$$

Let $K, F \subset \mathbb{R}^n$ be sets such that K is compact, F is closed, and

$$K \cap F = \emptyset.$$

Prove that $d(K, F) > 0$ and there exist $u_0 \in K$ and $v_0 \in F$ such that

$$d(K, F) = |u_0 - v_0|.$$

7. Let $K, V \subset \mathbb{R}^n$ be sets such that K is compact, V is open, and

$$K \subset V.$$

Prove that there exists an open set $W \subset \mathbb{R}^n$ such that

$$K \subset W \subset \overline{W} \subset V.$$

8. Calculate the limits:

(a)

$$\lim_{(x,y)\to(0,0)} \frac{x - y}{\sqrt{x} - \sqrt{y}},$$

(b)

$$\lim_{(x,y)\to(2,2)} \frac{x + y - 4}{\sqrt{x + y} - 2},$$

9. Prove formally that:

(a) $\lim_{(x,y)\to(1,3)} 3x - 5y + 7 = -5$,

(b) $\lim_{(x,y)\to(-2,1)} -x + 4y + 4 = 10$,

(c) $\lim_{(x,y)\to(1,3)} x^2 + y^2 - 2x + 1 = 9$,

(d) $\lim_{(x,y)\to(-1,2)} 3x^2 - 2y^2 - 2x + 3y + 5 = 8$.

10. For the functions indicated below, show that $\lim_{(x,y)\to(0,0)} f(x,y)$ does not exist, where

(a)

$$f(x,y) = \frac{x}{\sqrt{x^2+y^2}},$$

(b)

$$f(x,y) = \frac{xy}{|xy|},$$

(c)

$$f(x,y) = \frac{x^2-y^2}{x^2+y^2},$$

(d)

$$f(x,y) = \frac{x^4+3x^2y^2+2xy^3}{(x^2+y^2)^2},$$

(e)

$$f(x,y) = \frac{x^9y}{(x^6+y^2)^2}.$$

11. For the functions $f : \mathbb{R}^2 \setminus \{(0,0)\} \to \mathbb{R}$ indicated below, show that $\lim_{(x,y)\to(0,0)} f(x,y)$ exists and calculate its value, where,

(a)

$$f(x,y) = x \cos\left(\frac{1}{x^2+y^2}\right),$$

(b)

$$f(x,y) = \frac{x^2+3xy}{\sqrt{x^2+y^2}},$$

(c)

$$f(x,y) = \frac{x^2y+xy^2}{x^2+y^2},$$

(d)

$$f(x, y) = \cos\left(\frac{x^3 - y^3}{x^2 + y^2}\right).$$

12. For the functions $f : \mathbb{R}^2 \setminus \{(0, 0)\} \to \mathbb{R}$ indicated below, calculate, if they exist, the limits $\lim_{(x,y)\to(0,0)} f(x, y)$ and discuss about the possibility or not of such functions to be continuously extended to $(0, 0)$, by appropriately defining $f(0, 0)$, where

(a)

$$f(x, y) = \ln\left(\frac{3x^4 - x^2 y^2 + 3y^4}{x^2 + y^2} + 2\right),$$

(b)

$$f(x, y) = \ln\left(x^2 \cos^2\left(\frac{1}{x^2 + y^2}\right) + 3\right).$$

13. Let $a, b, c \in \mathbb{R}$, where $a \neq 0$ or $b \neq 0$.
    Prove formally that,

$$\lim_{(x,y)\to(x_0,y_0)} ax + by + c = ax_0 + by_0 + c.$$

14. Let $a, b, c, d, e, f \in \mathbb{R}$ where $a \neq 0, b \neq 0$ or $c \neq 0$.
    Prove formally that,

$$\lim_{(x,y)\to(x_0,y_0)} ax^2 + by^2 + cxy + dx + ey + f = ax_0^2 + by_0^2 + cx_0 y_0 + dx_0 + ey_0 + f.$$

15. Let $D \subset \mathbb{R}^n$ be an open set, let $f, g : D \to \mathbb{R}$ be real functions, and let $\mathbf{x_0} \in D$.
    Suppose there exist $K > 0$ and $\delta > 0$ such that $|g(\mathbf{x})| < K$, if $0 < |\mathbf{x} - \mathbf{x_0}| < \delta$.
    Assume

$$\lim_{\mathbf{x}\to\mathbf{x_0}} f(\mathbf{x}) = 0.$$

Under such hypotheses, prove that

$$\lim_{\mathbf{x}\to\mathbf{x_0}} f(\mathbf{x})g(\mathbf{x}) = 0.$$

16. Use the item 15 to prove that

$$\lim_{(x,y)\to(0,0)} f(x, y) = 0,$$

where $f : \mathbb{R}^2 \to \mathbb{R}$ is defined by

$$f(x, y) = \begin{cases} (x^2 + y^2 + x - y) \sin\left(\frac{1}{x^2+y^2}\right), & \text{if } (x, y) \neq (0, 0) \\ 5, & \text{if } (x, y) = (0, 0). \end{cases}$$

17. Let $D \subset \mathbb{R}^n$ be an open set and let $\mathbf{x_0} \in D$.
     Assume $f, g : D \to \mathbb{R}$ are such that

$$\lim_{\mathbf{x}\to\mathbf{x_0}} f(\mathbf{x}) = L \in \mathbb{R}$$

and

$$\lim_{\mathbf{x}\to\mathbf{x_0}} g(\mathbf{x}) = M \in \mathbb{R},$$

where $L < M$.
     Prove that there exists $delta > 0$ such that if $0 < |\mathbf{x} - \mathbf{x_0}| < \delta$, then

$$f(\mathbf{x}) < \frac{L + M}{2} < g(\mathbf{x}).$$

     Hint: Define $\varepsilon = \frac{M-L}{2}$.

18. Let $D \subset \mathbb{R}^n$ be an open set and let $f : D \to \mathbb{R}$ be a continuous function.
     Assume $A \subset \mathbb{R}$ is open. Prove that $f^{-1}(A)$ is open, where,

$$f^{-1}(A) = \{\mathbf{x} \in D : f(\mathbf{x}) \in A\}.$$

19. Let $f : \mathbb{R}^n \to \mathbb{R}$ be a continuous function and let $c \in \mathbb{R}$.
     Prove that the sets B and C are closed, where,

(a) $B = \{\mathbf{x} \in \mathbb{R}^n : f(\mathbf{x}) \leq c\}$.
(b) $C = \{\mathbf{x} \in \mathbb{R}^n : f(\mathbf{x}) = c\}$.

20. Let $A, F \subset \mathbb{R}^n$ be such that A is open and F is closed. Prove that $A \setminus F$ is open and $F \setminus A$ is closed.

21. Let $D \subset \mathbb{R}^n$ be an open set and let $f : D \to \mathbb{R}$ be a continuous function.
     Assume $F \subset \mathbb{R}$ is closed. Prove that there exists a closed set $F_1 \subset \mathbb{R}^n$ such that $f^{-1}(F) = D \cap F_1$, where

$$f^{-1}(F) = \{\mathbf{x} \in D : f(\mathbf{x}) \in F\}.$$

22. Let $f : \mathbb{R}^2 \to \mathbb{R}$ be such that

$$f(x, y) = \begin{cases} \frac{\sin(x+y)}{x+y}, & \text{if } x + y \neq 0 \\ 1, & \text{if } x + y = 0. \end{cases}$$

Prove that f is continuous on $\mathbb{R}^2$.

23. Let $f : \mathbb{R}^2 \to \mathbb{R}$ be such that

$$f(x, y) = \begin{cases} \frac{xy}{|x|+|y|}, & \text{if } (x, y) \neq (0, 0) \\ 0, & \text{if } (x, y) = (0, 0). \end{cases}$$

Prove that f is continuous on $\mathbb{R}^2$.

24. Through the definition of partial derivative, calculate

$$\frac{\partial f(x, y)}{\partial x} \text{ and } \frac{\partial f(x, y)}{\partial y},$$

where

$$f(x, y) = x^2 - 3y^2 + x.$$

25. Through the definition of partial derivative, for $(x, y) \in \mathbb{R}^2$ such that $x + 2y > 0$, calculate

$$\frac{\partial f(x, y)}{\partial x} \text{ and } \frac{\partial f(x, y)}{\partial y},$$

where

$$f(x, y) = \frac{1}{\sqrt{x + 2y}}.$$

26. Through the definition of partial derivative, for $(x, y) \in \mathbb{R}^2$ such that $x^2 - y \neq 0$, calculate

$$\frac{\partial f(x, y)}{\partial y},$$

where

$$f(x, y) = \frac{x + 2y}{x^2 - y}.$$

27. Through the definition of differentiability, prove that the functions below indicated are differentiable on the respective domains,

(a) $f(x, y) = 3x^2 - 2xy + 5y^2$,
(b) $f(x, y) = 2xy^2 - 3xy$,
(c) $f(x, y) = \frac{x^2}{y}$.

28. Let $f : \mathbb{R}^2 \to \mathbb{R}$ be defined by

$$f(x, y) = \begin{cases} \frac{(x^3 + y^3)}{x^2 + y^2}, & \text{if } (x, y) \neq (0, 0) \\ 0, & \text{if } (x, y) = (0, 0). \end{cases}$$

Calculate $f_x(0, 0)$ and $f_y(0, 0)$.

29. Let $f : \mathbb{R}^2 \to \mathbb{R}$ defined by

$$f(x, y) = \begin{cases} \frac{3x^2 y^2}{x^4 + y^4}, & \text{if } (x, y) \neq (0, 0) \\ 0, & \text{if } (x, y) = (0, 0). \end{cases}$$

Prove that $f_x(0, 0)$ and $f_y(0, 0)$ exist; however f is not differentiable at $(0, 0)$.

30. Let $f : \mathbb{R}^2 \to \mathbb{R}$ be defined by

$$f(x, y) = \begin{cases} \frac{xy(x^2 - y^2)}{x^2 + y^2}, & \text{if } (x, y) \neq (0, 0) \\ 0, & \text{if } (x, y) = (0, 0). \end{cases}$$

Prove that $f_x(0, 0)$ and $f_y(0, 0)$ exist and f is differentiable at $(0, 0)$.

31. Let $f : \mathbb{R}^3 \to \mathbb{R}$ be defined by

$$f(x, y, z) = \begin{cases} \frac{xyz^2}{x^2 + y^2 + z^2}, & \text{if } (x, y, z) \neq (0, 0, 0) \\ 0, & \text{if } (x, y, z) = (0, 0, 0). \end{cases}$$

Prove that f is differentiable at $(0, 0, 0)$.

32. Let $f : \mathbb{R}^2 \to \mathbb{R}$ be defined by

$$f(x, y) = \begin{cases} (x^2 + y^2) \sin\left(\frac{1}{\sqrt{x^2 + y^2}}\right), & \text{if } (x, y) \neq (0, 0) \\ 0, & \text{if } (x, y) = (0, 0). \end{cases}$$

(a) Obtain $\Delta f(0, 0, \Delta x, \Delta y)$.
(b) Calculate $f_x(0, 0)$ and $f_y(0, 0)$.
(c) Through the definition of differentiability, show that f is differentiable at $(0, 0)$.

33. For the functions indicated below, obtain the respective domains and prove that they are differentiable (on the domains in question):

    (a) $f(x, y) = \frac{x+y}{x^2+5y}$
    (b) $f(x, y) = y \ln x - x/y$,
    (c) $f(x, y) = \arctan(x^2 - y) + \frac{1}{\sqrt{x^2-y}}$,

34. Let $D \subset \mathbb{R}^n$ be an open connected set. Suppose all partial derivatives of f are zero on D.

    Prove that f and constant on D.

35. Let $D \subset \mathbb{R}^2$ be an open rectangle and let $f : D \to \mathbb{R}$ be a function. Assume f has partial derivatives well defined on D. Let (x, y) and $(x + u, y + v) \in D$.

    Prove that there exists $\lambda \in (0, 1)$ such that

$$f(x + u, y + v) - f(x, y) = f_x(x + \lambda u, y + v)u + f_y(x, y + \lambda v)v.$$

36. Let $D \subset \mathbb{R}^n$ be an open convex set and let $f : D \to \mathbb{R}$ be a function. Suppose there exists $K > 0$ such that

$$\left| \frac{\partial f(\mathbf{x})}{\partial x_j} \right| \le K, \ \forall \mathbf{x} \in D, \ j \in \{1, \dots, n\}.$$

Prove that

$$|f(\mathbf{x}) - f(\mathbf{y})| \le Kn|\mathbf{x} - \mathbf{y}|, \ \forall \mathbf{x}, \mathbf{y} \in D.$$

37. Let $D \subset \mathbb{R}^n$ be an open set and let $f : D \to \mathbb{R}$ be a differentiable function at $\mathbf{x_0} \in D$. Prove that there exist $\delta > 0$ and $K > 0$ such that if $|\mathbf{h}| < \delta$, then $\mathbf{x_0} + \mathbf{h} \in D$ and

$$|f(\mathbf{x_0} + \mathbf{h}) - f(\mathbf{x_0})| < K|\mathbf{h}|.$$

38. Let $f : \mathbb{R}^n \setminus \{\mathbf{0}\} \to \mathbb{R}$ be defined by $f(\mathbf{x}) = |\mathbf{x}|^c$, where $c \in \mathbb{R}$. Let $\mathbf{x} = (x_1, \dots, x_n)$ and $\mathbf{v} = (v_1, \dots, v_n) \in \mathbb{R}^n$.

    Calculate

$$\nabla f(\mathbf{x}) \cdot \mathbf{v}.$$

39. Let $f : \mathbb{R}^3 \to \mathbb{R}$ be defined by

$$f(x, y, t) = \frac{t^2 + y}{e^t + x^2 + t^2}.$$

Suppose the functions $x : \mathbb{R} \to \mathbb{R}$ and $y : \mathbb{R} \to \mathbb{R}$ be defined by

$$x(t) = \cos^2(t^3),$$

and

$$y(t) = e^{t^2}.$$

Through the chain rule, calculate $g'(t)$ where $g(t) = f(x(t), y(t), t)$, $\forall t \in \mathbb{R}$.

Finally, obtain the equation of the tangent line to the graph of g at the points $t = 0$ and $t = \pi$.

40. Let $g : \mathbb{R}^3 \to \mathbb{R}$ be defined by

$$g(x, y, z) = \frac{x^2 + y^2 + xy}{z^2 + e^x + \cos^2(y)}.$$

Let $z(x, y) = \cos^2(x^2 + y^2)$ and define $h : \mathbb{R}^2 \to \mathbb{R}$ by

$$h(x, y) = g(x, y, z(x, y)).$$

Through the chain rule, calculate $h_x(x, y)$ and $h_y(x, y)$.

Find the equation of the normal line and the equation of the tangent plane, to the graph of h at the point $(1, 0)$.

41. Let $f : \mathbb{R} \to \mathbb{R}$ be a differentiable function. Let $u(x, y) = bx - ay$. Show that $z(x, y) = f(u(x, y))$ satisfies the equation,

$$a\frac{\partial z}{\partial x} + b\frac{\partial z}{\partial y} = 0.$$

42. Let $f : \mathbb{R}^2 \to \mathbb{R}$ be a differentiable function.

Denoting $u(r, \theta) = f(x, y)$, where $x = r \cos \theta$ and $y = r \sin \theta$ show that

$$\frac{\partial u}{\partial x} = \frac{\partial u}{\partial r} \cos \theta - \frac{\partial u}{\partial \theta} \frac{\sin \theta}{r},$$

and

$$\frac{\partial u}{\partial y} = \frac{\partial u}{\partial r} \sin \theta + \frac{\partial u}{\partial \theta} \frac{\cos \theta}{r}.$$

43. Consider the ellipsoid of equation

$$\frac{x^2}{a^2} + \frac{y^2}{b^2} + \frac{z^2}{c^2} = 1,$$

where $a, b, c > 0$.

Find the closest points on such surface to the origin $(0, 0, 0)$.

44. Let A be a matrix $m \times n$. Let $\mathbf{y}_0 \in \mathbb{R}^m$ and let $f : \mathbb{R}^n \to \mathbb{R}$ be defined by

$$f(\mathbf{x}) = \langle (A\mathbf{x}), \mathbf{y}_0 \rangle,$$

where $\langle \cdot, \cdot \rangle : \mathbb{R}^m \times \mathbb{R}^m \to \mathbb{R}$ denotes the usual inner product in $\mathbb{R}^m$. Through the method of Lagrange multipliers, find the points of minimum and maximum of $f(\mathbf{x})$ subject to $|\mathbf{x}| = 1$.

45. A function $f : \mathbb{R}^n \to \mathbb{R}$ is said to be convex if

$$f(\lambda \mathbf{x} + (1 - \lambda \mathbf{y})) \leq \lambda f(\mathbf{x}) + (1 - \lambda) f(\mathbf{y}), \ \forall \mathbf{x}, \mathbf{y} \in \mathbb{R}^n, \ \lambda \in [0, 1].$$

(a) Suppose that $f : \mathbb{R}^n \to \mathbb{R}$ is differentiable. Show that f is convex if, and only if,

$$f(\mathbf{y}) - f(\mathbf{x}) \geq \nabla f(\mathbf{x}) \cdot (\mathbf{y} - \mathbf{x}), \ \forall \mathbf{x}, \mathbf{y} \in \mathbb{R}^n.$$

(b) Prove that if f is convex, differentiable, and $\nabla f(\mathbf{x}) = \mathbf{0}$, then $\mathbf{x} \in \mathbb{R}^n$ is a point of global minimum for f.

46. Let $f : \mathbb{R}^n \to \mathbb{R}$ be a twice differentiable function such that

$$H(\mathbf{x}) = \left\{ \frac{\partial^2 f(\mathbf{x})}{\partial x_i \partial x_j} \right\}$$

is a positive definite matrix. $\forall \mathbf{x} \in \mathbb{R}^n$.

Show that f is convex on $\mathbb{R}^n$.

47. Let $F, G : \mathbb{R}^4 \to \mathbb{R}$ be defined by $F(x, y, u, v) = x^2 + y^3 - u + v^2$ and $G(x, y, u, v) = e^{2x} + e^{3y} + 2uv + 3v^2$. Assuming the hypotheses of the vectorial case of implicit function theorem, consider the functions $u(x, y)$ and $v(x, y)$ implicitly defined on a neighborhood of a point $(x, y, u, v) \in \mathbb{R}^4$ such that

$$F(x, y, u, v) = 0 \quad \text{and} \quad G(x, y, u, v) = 0.$$

Find u_x, u_y, v_x and v_y on such neighborhood.

8.22 Some Topics on Differential Geometry

Definition 8.22.1 (Limit) Let $C \subset \mathbb{R}^3$ be a curve defined by a one variable vectorial function $\mathbf{r} : [a, b] \to \mathbb{R}^3$. Let $t_0 \in (a, b)$. We say that $\mathbf{A}$ is the limit of **r** as t approaches t_0, as for each $\varepsilon > 0$, there exists $\delta > 0$ such that if $t \in [a, b]$ and $0 < |t - t_0| < \delta$, then $|\mathbf{r}(t) - \mathbf{A}| < \varepsilon$.

In such a case, we denote,

$$\lim_{t \to t_0} \mathbf{r}(t) = \mathbf{A}.$$

Theorem 8.22.2 *Let* $\mathbf{r} : [a, b] \to \mathbb{R}^3$ *be point-wisely expressed by*

$$\mathbf{r}(t) = x_1(t)\mathbf{e}_1 + x_2(t)\mathbf{e}_2 + x_3(t)\mathbf{e}_3,$$

where $\{\mathbf{e}_1, \mathbf{e}_2, \mathbf{e}_3\}$ *is the canonical basis for* $\mathbb{R}^3$.

Let $t_0 \in [a, b]$.

Under such hypotheses,

$$\lim_{t \to t_0} \mathbf{r}(t) = A_1\mathbf{e}_1 + A_2\mathbf{e}_2 + A_3\mathbf{e}_3 \equiv \mathbf{A},$$

if, and only if,

$$\lim_{t \to t_0} x_j(t) = A_j, \ \forall j \in \{1, 2, 3\}.$$

Proof Assume first

$$\lim_{t \to t_0} \mathbf{r}(t) = A_1\mathbf{e}_1 + A_2\mathbf{e}_2 + A_3\mathbf{e}_3.$$

Let $\varepsilon > 0$. Thus there exists $\delta > 0$ such that if $t \in [a, b]$ and $0 < |t - t_0| < \delta$, then

$$|\mathbf{r}(t) - \mathbf{A}| < \varepsilon.$$

In particular

$$|x_k(t) - A_k| \leq \sqrt{\sum_{k=1}^{3} |x_k(t) - A_k|^2}$$
$$= |\mathbf{r}(t) - \mathbf{A}| < \varepsilon, \tag{8.177}$$

$\forall t \in [a, b]$, such that $0 < |t - t_0| < \delta$.

Thus,

$$\lim_{t \to t_0} x_k(t) = A_k, \ \forall k \in \{1, 2, 3\}.$$

Reciprocally, suppose that

$$\lim_{t \to t_0} x_k(t) = A_k, \quad \forall k \in \{1, 2, 3\}.$$

Let $\varepsilon > 0$. Let $k \in \{1, 2, 3\}$.

Thus, there exists $\delta_k > 0$ such that if $0 < |t - t_0| < \delta_k$, then

$$|x_k(t) - A_k| < \varepsilon/\sqrt{3}.$$

Therefore, denoting $\delta = \min\{\delta_1, \delta_2, \delta_3\}$, if $t \in [a, b]$ and $0 < |t - t_0| < \delta$, we have,

$$
\begin{aligned}
|\mathbf{r}(t) - \mathbf{A}| &= \sqrt{\sum_{k=1}^{3} |x_k(t) - A_k|^2} \\
&< \sqrt{\sum_{k=1}^{3} \varepsilon^2/3} \\
&= \sqrt{\varepsilon^2} \\
&= \varepsilon,
\end{aligned}
\tag{8.178}
$$

Therefore,

$$\lim_{t \to t_0} \mathbf{r}(t) = \mathbf{A}.$$

The proof is complete.

In the next lines we present the definition of derivative for a one variable vectorial function.

Definition 8.22.3 (Derivative) Let $\mathbf{r} : [a, b] \to \mathbb{R}^3$ be a one variable vectorial function.

Let $t_0 \in (a, b)$. We define the derivative of $\mathbf{r}$ relating t at t_0, denoted by $\mathbf{r}'(t_0)$, by

$$\mathbf{r}'(t_0) = \lim_{h \to 0} \frac{\mathbf{r}(t_0 + h) - \mathbf{r}(t_0)}{h},$$

if such a limit exists.

Similarly, if $t_0 = a$ or $t_0 = b$ we define the derivative of $\mathbf{r}$ at these values for t_0, through one-sided limits.

Remark 8.22.4 Considering this last definition and the last theorem, if

$$\mathbf{r}(t) = \sum_{k=1}^{3} x_k(t)\mathbf{e}_k,$$

on $[a, b]$, for some $t \in [a, b]$, we may infer that $\mathbf{r}'(t)$ exists, if and only if, $x'_k(t)$ exists, $\forall k \in \{1, 2, 3\}$, and in such a case

$$\mathbf{r}'(t) = \sum_{k=1}^{3} x'_k(t)\mathbf{e}_k.$$

8.22.1 Arc Length

Consider a curve $C \subset \mathbb{R}^3$ defined by the one variable vectorial differentiable function $\mathbf{r} : [a, b] \to \mathbb{R}^3$ point-wisely represented by

$$\mathbf{r}(t) = \sum_{k=1}^{3} x_k(t)\mathbf{e}_k.$$

Consider a partition P of $[a, b]$, given by

$$P = \{t_0 = a, t_1, \ldots, t_n = b\}.$$

Thus, a first approximation for the length of C, denoted by L, is given by

$$L \approx \sum_{j=1}^{n} |\Delta \mathbf{r}_j|,$$

where,

$$\Delta \mathbf{r}_j = \mathbf{r}(t_j) - \mathbf{r}(t_{j-1})$$

$$= \frac{d\mathbf{r}(t_j)}{dt} \Delta t_j + \mathcal{O}(\Delta t_j^2), \tag{8.179}$$

where $\Delta t_j = t_j - t_{j-1}$, $\forall j \in \{1, \ldots, n-1\}$.
   If $|P| = \max\{\Delta t_j,\ j \in \{1, \ldots, n-1\}\}$ is small enough, we have

$$L \approx \sum_{j=1}^{n} |\Delta \mathbf{r}_j|$$

$$\approx \sum_{j=1}^{m} \left| \frac{d\mathbf{r}(t_j)}{dt} \right| \Delta t_j$$

$$\equiv S_{\mathbf{r}}^{P}. \tag{8.180}$$

With such an approximation in mind, assuming $\mathbf{r}$ is of C^1 class or at least piecewise continuous, we define the length of C, by

$$L = \lim_{|P| \to 0} S_{\mathbf{r}}^P = \int_a^b \left| \frac{d\mathbf{r}(t)}{dt} \right| \, dt,$$

so that

$$L = \int_a^b \sqrt{\sum_{j=1}^3 (x_j'(t))^2} \, dt,$$

8.22.2 The Arc Length Function

Definition 8.22.5 Consider a C^1 class curve defined by the function $\mathbf{r} : [a, b] \to \mathbb{R}^3$.

We define the relating arc length function to C, denoted by $s : [a, b] \to \mathbb{R}^+$, point-wisely by

$$s(t) = \int_a^t \sqrt{\mathbf{r}'(u) \cdot \mathbf{r}'(u)} \, du, \quad \forall t \in [a, b].$$

Hence $s(t)$ provides a measure of the arc length of C between a and t.

Observe that

$$\frac{ds(t)}{dt} = \sqrt{\mathbf{r}'(t) \cdot \mathbf{r}'(t)},$$

so that

$$ds(t) = \sqrt{\mathbf{r}'(t) \cdot \mathbf{r}'(t)} dt$$
$$= \sqrt{d\mathbf{r}(t) \cdot d\mathbf{r}(t)}, \tag{8.181}$$

and thus, in a differential form, we may denote,

$$ds(t)^2 = d\mathbf{r}(t) \cdot d\mathbf{r}(t).$$

Moreover, we denote,

$$s(t) = \int_a^t \sqrt{\mathbf{r}'(u) \cdot \mathbf{r}'(u)} \, du$$
$$= f(t), \tag{8.182}$$

so that

$$t = f^{-1}(s) \equiv t(s).$$

With such results in mind, we define the parametrization of C by its arc length, denoted point-wisely by $\hat{\mathbf{r}}(s)$, by

$$\hat{\mathbf{r}}(s) = \mathbf{r}(t(s)).$$

When the meaning is clear, we denote

$$\dot{\mathbf{r}} = \frac{d\mathbf{r}(t(s))}{ds},$$

$$\ddot{\mathbf{r}} = \frac{d^2\mathbf{r}(t(s))}{ds^2},$$

$$\mathbf{r}' = \frac{d\mathbf{r}(t)}{dt},$$

$$\mathbf{r}'' = \frac{d^2\mathbf{r}(t)}{dt^2}.$$

Definition 8.22.6 (Unit Tangent Vector) Let C be a C^1 class curve defined by $\mathbf{r} : [a, b] \to \mathbb{R}^3$.

Let $t \in [a, b]$. We define the unit tangent vector to C at t, denoted by $\mathbf{T}(t)$, by

$$\mathbf{T}(t) = \frac{\mathbf{r}'(t)}{|\mathbf{r}'(t)|},$$

where we recall to have assumed through the C^1 class definition, $\mathbf{r}'(t) \neq \mathbf{0}$, for all $t \in [a, b]$.

Observe that

$$\mathbf{T}(t) = \frac{\mathbf{r}'(t)}{|\mathbf{r}'(t)|}$$

$$= \frac{\mathbf{r}'(t)}{s'(t)}$$

$$= \frac{\frac{d\mathbf{r}(t)}{dt}}{\frac{ds(t)}{dt}}$$

$$= \frac{d\mathbf{r}(t(s))}{ds}$$

$$= \hat{\mathbf{T}}(s). \tag{8.183}$$

Therefore

$$\hat{\mathbf{T}} = \dot{\mathbf{r}},$$

and

$$\dot{\hat{\mathbf{T}}} = \ddot{\mathbf{r}}.$$

Definition 8.22.7 Let C be a curve defined by a C^1 class function $\mathbf{r} : [a, b] \to \mathbb{R}^3$. Let $t \in [a, b]$.

We define the curvature vector of C at t, denoted by $\mathbf{K}(t)$, by

$$\mathbf{K}(t) = \frac{d\mathbf{T}(t)}{ds}.$$

Observe that

$$\mathbf{K}(t) = \frac{d\mathbf{T}(t)}{ds}$$

$$= \frac{d\mathbf{T}(t)}{dt} \frac{dt}{ds}$$

$$= \frac{\mathbf{T}'(t)}{\frac{ds(t)}{dt}}$$

$$= \frac{\mathbf{T}'(t)}{|\mathbf{r}'(t)|}. \tag{8.184}$$

We may also define

$$\hat{\mathbf{K}}(s) = \frac{d\hat{\mathbf{T}}(s)}{ds}$$

$$= \dot{\hat{\mathbf{T}}}(s)$$

$$= \frac{d^2 \mathbf{r}(t(s))}{ds^2}. \tag{8.185}$$

Observe that $|\mathbf{T}(t)| = 1$, $\forall t \in [a, b]$, that is,

$$\mathbf{T}(t) \cdot \mathbf{T}(t) = 1,$$

so that

$$\mathbf{T}'(t) \cdot \mathbf{T}(t) = 0.$$

Hence $\mathbf{T}'(t)$ has a direction orthogonal to the one of $\mathbf{T}(t)$.
So, at this point we introduce the next definition.

Definition 8.22.8 Let C be a curve defined by a C^1 class function $\mathbf{r} : [a, b] \to \mathbb{R}^3$.
Let $t \in [a, b]$. We define the normal vector to C at t, denoted by $\mathbf{N}(t)$, by

$$\mathbf{N}(t) = \frac{\mathbf{T}'(t)}{|\mathbf{T}'(t)|}.$$

Also, we define the bi-normal vector to C, denoted by $\mathbf{B}(t)$, by

$$\mathbf{B}(t) = \mathbf{T}(t) \times \mathbf{N}(t).$$

In the context of the last definition, from

$$\mathbf{B} = \mathbf{T} \times \mathbf{N},$$

we obtain

$$\begin{aligned}
\frac{d\mathbf{B}}{ds} &= \frac{d\mathbf{T}}{ds} \times \mathbf{N} + \mathbf{T} \times \frac{d\mathbf{N}}{ds} \\
&= \frac{\mathbf{T}'(t)}{|\mathbf{r}'(t)|} \times \mathbf{N} + \mathbf{T} \times \frac{d\mathbf{N}}{ds} \\
&= \mathbf{T} \times \frac{d\mathbf{N}}{ds}.
\end{aligned} \tag{8.186}$$

Therefore, $\frac{d\mathbf{B}}{ds}$ is orthogonal to $\mathbf{T}$.
Also, from $\mathbf{B} \cdot \mathbf{B} = 1$, we obtain,

$$\frac{d\mathbf{B}}{ds} \cdot \mathbf{B} = 0,$$

and hence

$$\frac{d\mathbf{B}}{ds}$$

is orthogonal to $\mathbf{T}$ and $\mathbf{B}$, so that it has the direction of $\mathbf{N}$. So, we may denote

$$\frac{d\mathbf{B}}{ds} = \tau \mathbf{N},$$

where the scalar τ is said to be the torsion of the curve C at the point corresponding to $t \in [a, b]$.

Moreover, we define the osculating plane to C at t, as the one which has as normal directional the vector $\mathbf{B}(t)$ and contains the point $\mathbf{r}(t)$. The normal plane to C at t is the one which has as normal direction the vector $\mathbf{T}(t)$ and contains the point $\mathbf{r}(t)$. Finally, the rectifying plane to C at t is the one which has $\mathbf{N}(t)$ as the normal direction and it contains $\mathbf{r}(t)$.

Finally, observe also that from

$$\mathbf{N} = \mathbf{B} \times \mathbf{T},$$

we obtain

$$\begin{aligned}
\frac{d\mathbf{N}}{ds} &= \frac{d\mathbf{B}}{ds} \times \mathbf{T} + \mathbf{B} \times \frac{d\mathbf{T}}{ds} \\
&= \tau\mathbf{N} \times \mathbf{T} + \mathbf{B} \times \kappa\mathbf{N} \\
&= -\kappa\mathbf{T} - \tau\mathbf{B}.
\end{aligned} \tag{8.187}$$

Here we have denoted

$$\kappa = |\mathbf{K}(t)| = \left|\frac{d\mathbf{T}}{ds}\right| = \left|\frac{\mathbf{T}'(t)}{|r'(t)|}\right|.$$

Since $\mathbf{T}'$ and $\mathbf{N}$ has the same direction, we may write,

$$\mathbf{K}(t) = \frac{d\mathbf{T}}{ds} = \kappa(t)\mathbf{N}(t).$$

$\kappa(t)$ is said to be the scalar curvature of C at the point corresponding to t.

8.23   Some Notes About the Scalar Curvature in a Surface in $\mathbb{R}^3$

Consider a surface $S \subset \mathbb{R}^3$, where

$$S = \{\mathbf{r}(\mathbf{u}) \; : \; \mathbf{u} \in D\},$$

where $\mathbf{r} : D \to \mathbb{R}^3$ is a C^1 class function and $D \subset \mathbb{R}^2$ is an open, bounded, simply connect set with a C^1 class boundary ∂D. Here $\mathbf{u} = (u_1, u_2) \in \mathbb{R}^2$, and

$$\mathbf{r}(u_1, u_2) = X_1(u_1, u_2)\mathbf{e}_1 + X_2(u_1, u_2)\mathbf{e}_2 + X_3(u_1, u_2)\mathbf{e}_3,$$

and $\{\mathbf{e}_1, \mathbf{e}_2, \mathbf{e}_3\}$ is the canonical basis of $\mathbb{R}^3$.

Generically, we shall denote

$$\mathbf{r}_\alpha = \frac{\partial \mathbf{r}}{\partial u_\alpha},$$

$$\mathbf{r}_{\alpha\beta} = \frac{\partial^2 \mathbf{r}}{\partial u_\alpha \partial u_\beta}.$$

Let $(\mathbf{r} \circ \mathbf{u}) : [a, b] \to S$, be a curve on S, that is,

$$\mathbf{r}(\mathbf{u}(t)) = \mathbf{r}(u_1(t), u_2(t)), \ \forall t \in [a, b].$$

For

$$s(t) = \int_a^t |(\mathbf{r} \circ \mathbf{u})'(v)| \, dv,$$

we shall consider this same curve parameterized by its arc length, where, as the meaning is clear we denote simply

$$\hat{\mathbf{r}}(s) = (\mathbf{r} \circ \mathbf{u})(t(s)) \equiv \mathbf{r},$$

so that

$$\dot{\mathbf{r}} = \frac{\partial \mathbf{r}}{\partial u_1} \frac{\partial u_1}{ds} + \frac{\partial \mathbf{r}}{\partial u_2} \frac{\partial u_2}{ds} = \mathbf{r}_\alpha \dot{u}_\alpha.$$

Thus,

$$\ddot{\mathbf{r}} = \mathbf{r}_{\alpha\beta} \dot{u}_\alpha \dot{u}_\beta + \mathbf{r}_\alpha \ddot{u}_\alpha.$$

Hence, for $\mathbf{n} = \frac{\mathbf{r}_\alpha \times \mathbf{r}_\beta}{|\mathbf{r}_\alpha \times \mathbf{r}_\beta|}$, we get

$$\ddot{\mathbf{r}} \cdot \mathbf{n} = (\mathbf{r}_{\alpha\beta} \cdot \mathbf{n}) \dot{u}_\alpha \dot{u}_\beta$$

$$= b_{\alpha\beta} \dot{u}_\alpha \dot{u}_\beta, \tag{8.188}$$

where

$$\{b_{\alpha\beta}\} = \{\mathbf{r}_{\alpha\beta} \cdot \mathbf{n}\},$$

is the curvature tensor of S at the point corresponding to s, relating C.

Observe that $\mathbf{r}_\alpha \cdot \mathbf{n} = 0$, so that

$$\mathbf{r}_{\alpha\beta} \cdot \mathbf{n} + \mathbf{r}_\alpha \cdot \mathbf{n}_\beta = 0,$$

and thus,

$$\begin{aligned} b_{\alpha\beta} &= \mathbf{r}_{\alpha\beta} \cdot \mathbf{n} \\ &= -\mathbf{r}_\alpha \cdot \mathbf{n}_\beta, \end{aligned} \tag{8.189}$$

where

$$\mathbf{n}_\beta = \frac{\partial \mathbf{n}}{\partial u_\beta}.$$

At this point we recall that:

$$\begin{aligned} \ddot{\mathbf{r}} &= \dot{\hat{\mathbf{T}}}(s) \\ &= \hat{\mathbf{K}}(s) \\ &= \kappa(s)\mathbf{N}(t(s)), \end{aligned} \tag{8.190}$$

Let γ be the angle between

$$\mathbf{n} = \frac{\mathbf{r}_\alpha \times \mathbf{r}_\beta}{|\mathbf{r}_\alpha \times \mathbf{r}_\beta|},$$

and

$$\mathbf{N}(t(s)) \equiv \mathbf{N}.$$

so that

$$\begin{aligned} \cos(\gamma) &= \mathbf{N} \cdot \mathbf{n} \\ &= \frac{\ddot{\mathbf{r}}}{\kappa} \cdot \mathbf{n}, \end{aligned} \tag{8.191}$$

and thus,

$$\ddot{\mathbf{r}} \cdot \mathbf{n} = \kappa \cos(\gamma),$$

and hence

$$\begin{aligned} \ddot{\mathbf{r}} \cdot \mathbf{n} &= (\mathbf{r}_{\alpha\beta} \cdot \mathbf{n})\dot{u}_\alpha \dot{u}_\beta \\ &= b_{\alpha\beta} \dot{u}_\alpha \dot{u}_\beta \\ &= \kappa \cos(\gamma). \end{aligned} \tag{8.192}$$

On the other hand,

$$
\begin{aligned}
ds^2 &= d\mathbf{r} \cdot d\mathbf{r} \\
&= \left(\frac{\partial \mathbf{r}}{\partial u_\alpha} \, du_\alpha \right) \cdot \left(\frac{\partial \mathbf{r}}{\partial u_\beta} \, du_\beta \right) \\
&= \mathbf{r}_\alpha \cdot \mathbf{r}_\beta du_\alpha du_\beta \\
&= g_{\alpha\beta} du_\alpha du_\beta,
\end{aligned}
\tag{8.193}
$$

where,

$$
\mathbf{g}_\alpha = \mathbf{r}_\alpha,
$$

$$
g_{\alpha\beta} = \mathbf{g}_\alpha \cdot \mathbf{g}_\beta,
$$

so that from this and

$$
b_{\alpha\beta} = -\mathbf{r}_\alpha \cdot \mathbf{n}_\beta,
$$

we obtain

$$
\begin{aligned}
b_{\alpha\beta} du_\alpha du_\beta &= -\mathbf{r}_\alpha \cdot \mathbf{n}_\beta du_\alpha du_\beta \\
&= -\left(\frac{\partial \mathbf{r}}{\partial u_\alpha} du_\alpha \right) \cdot \left(\frac{\partial \mathbf{n}}{\partial u_\beta} du_\beta \right) \\
&= -d\mathbf{r} \cdot \mathbf{n}.
\end{aligned}
\tag{8.194}
$$

Observe also that

$$
\begin{aligned}
\kappa \cos(\gamma) &= b_{\alpha\beta} \dot{u}_\alpha \dot{u}_\beta \\
&= b_{\alpha\beta} \frac{\partial u_\alpha}{\partial s} \frac{\partial u_\beta}{\partial s} \\
&= b_{\alpha\beta} \frac{u'_\alpha u'_\beta}{s' s'} \\
&= b_{\alpha\beta} \frac{u'_\alpha u'_\beta}{g_{\alpha\beta} u'_\alpha u'_\beta},
\end{aligned}
\tag{8.195}
$$

so that denoting

$$
\kappa_n = \kappa \cos(\gamma),
$$

we obtain

$$\kappa_n = \frac{b_{\alpha\beta} u'_\alpha u'_\beta}{g_{\alpha\beta} u'_\alpha u'_\beta},$$

so that,

$$(b_{\alpha\beta} - \kappa_n g_{\alpha\beta}) c_\alpha c_\beta = 0, \qquad (8.196)$$

where we have denoted $c_\alpha = u'_\alpha$, for $\alpha \in \{1, 2\}$.

Observe that c_α, κ_n depend on the curve in question, but $\{b_{\alpha\beta}\}$ and $\{g_{\alpha\beta}\}$ do not depend, so that we shall differentiate (8.196) looking for the conditions for which κ_n has extremal values, that is

$$\frac{\partial \kappa_n}{\partial c_\alpha} = 0.$$

So, denoting

$$a_{\alpha\beta} = b_{\alpha\beta} - \kappa_n g_{\alpha\beta},$$

we obtain

$$\frac{\partial}{\partial c_\gamma} [a_{\alpha\beta} c_\alpha c_\beta] = 0,$$

so that

$$
\begin{aligned}
0 = a_{\alpha\beta} &\left(\frac{\partial c_\alpha}{\partial c_\gamma} c_\beta + c_\alpha \frac{\partial c_\beta}{\partial c_\gamma} \right) \\
&= a_{\alpha\beta} (\delta^\alpha_\gamma c_\beta + c_\alpha \delta^\beta_\gamma) \\
&= a_{\gamma\beta} c_\beta + a_{\alpha\gamma} c_\alpha \\
&= (a_{\gamma\alpha} + a_{\alpha\gamma}) c_\alpha, \qquad (8.197)
\end{aligned}
$$

Since $a_{\alpha\gamma} = a_{\gamma\alpha}$, from this and the last above equation, we get,

$$a_{\alpha\gamma} c_\alpha = 0,$$

so that in particular

$$(b_{\alpha\gamma} - k_n g_{\alpha\gamma}) c_\gamma = 0,$$

so that

$$b_\alpha^\beta c_\alpha - k_n c_\beta = 0,$$

that is,

$$\begin{bmatrix} b_1^1 - \kappa_n & b_2^1 \\ b_1^2 & b_2^2 - \kappa_n \end{bmatrix} \begin{bmatrix} c_1 \\ c_2 \end{bmatrix} = \begin{bmatrix} 0 \\ 0 \end{bmatrix},$$

so that for having nonzero solutions, the determinant concerning this last system must be zero, and hence

$$\kappa_n^2 - \kappa_n b_\alpha^\alpha + \det(\{b_\alpha^\beta\}) = 0.$$

Since

$$\det\{b_\alpha^\beta\} = \frac{b}{g},$$

we obtain,

$$\kappa_n^2 - b_{\alpha\beta} g^{\alpha\beta} \kappa_n + b/g = 0,$$

so that denoting by κ_1 and κ_2 the solutions of this last equation, we have,

$$\kappa_1 \kappa_2 = b/g.$$

Finally,

$$K = \kappa_1 \kappa_2 = b/g$$

is said to be the Gaussian curvature of S at the point in question.
   Moreover,

$$H = \frac{1}{2}(\kappa_1 + \kappa_2) = \frac{1}{2} b_{\alpha\beta} g^{\alpha\beta} = \frac{1}{2} b_\alpha^\alpha,$$

is said to be the mean curvature of S at the same point.
   About the notation, we recall that

$$\mathbf{r}_{\alpha\beta} = \frac{\partial^2 \mathbf{r}}{\partial u_\alpha u_\beta},$$

and we shall denote

$$\mathbf{r}_{\alpha\beta} = \Gamma^\gamma_{\alpha\beta}\mathbf{r}_\gamma + a_{\alpha\beta}\mathbf{n}.$$

Thus,

$$b_{\alpha\beta} = \mathbf{r}_{\alpha\beta} \cdot \mathbf{n} = a_{\alpha\beta}\mathbf{n} \cdot \mathbf{n} = a_{\alpha\beta}.$$

At this point, we start to specify the coefficients $\Gamma^\gamma_{\alpha\beta}$. Observe that,

$$\begin{aligned} \mathbf{r}_{\alpha\beta} \cdot \mathbf{r}_\lambda &= \Gamma^\gamma_{\alpha\beta}\mathbf{r}_\gamma \cdot \mathbf{r}_\lambda + b_{\alpha\beta}\mathbf{n} \cdot \mathbf{r}_\lambda \\ &= \Gamma^\gamma_{\alpha\beta}\mathbf{r}_\gamma \cdot \mathbf{r}_\lambda \\ &= \Gamma^\gamma_{\alpha\beta}g_{\gamma\lambda}. \end{aligned} \qquad (8.198)$$

Hence,

$$\Gamma^\rho_{\alpha\beta} = \mathbf{r}_{\alpha\beta} \cdot \mathbf{r}_\lambda g^{\lambda\rho}.$$

Let us denote,

$$\Gamma_{\alpha\beta\gamma} = \mathbf{r}_{\alpha\beta} \cdot \mathbf{r}_\lambda,$$

so that

$$\Gamma_{\alpha\beta\lambda} = g_{\gamma\lambda}\Gamma^\lambda_{\alpha\beta},$$

and also,

$$\Gamma^\rho_{\alpha\beta} = g^{\lambda\rho}\Gamma_{\alpha\beta\lambda}.$$

Observe that from

$$\mathbf{r}_{\alpha\beta} = \mathbf{r}_{\beta\alpha},$$

we get

$$\Gamma_{\alpha\beta\lambda} = \Gamma_{\beta\alpha\lambda},$$

and

$$\Gamma^\rho_{\alpha\beta} = \Gamma^\rho_{\beta\alpha}.$$

Observe also that,

$$g_{\alpha\lambda} = \mathbf{r}_\alpha \cdot \mathbf{r}_\lambda,$$

so that

$$\frac{\partial g_{\alpha\lambda}}{\partial u_\beta} = \mathbf{r}_{\alpha\beta} \cdot \mathbf{r}_\lambda + \mathbf{r}_\alpha \cdot \mathbf{r}_{\lambda\beta},$$

so that

$$\frac{\partial g_{\alpha\lambda}}{\partial u_\beta} = \Gamma_{\alpha\beta\lambda} + \Gamma_{\lambda\beta\alpha}, \tag{8.199}$$

$$\frac{\partial g_{\lambda\beta}}{\partial u_\alpha} = \Gamma_{\lambda\alpha\beta} + \Gamma_{\beta\alpha\lambda}, \tag{8.200}$$

and,

$$\frac{\partial g_{\beta\alpha}}{\partial u_\lambda} = \Gamma_{\beta\lambda\alpha} + \Gamma_{\alpha\lambda\beta}, \tag{8.201}$$

so that from (8.199), (8.200), and (8.201), we obtain

$$\Gamma_{\alpha\beta\lambda} = \frac{1}{2}\left[\frac{\partial g_{\beta\lambda}}{\partial u_\alpha} + \frac{\partial g_{\lambda\alpha}}{\partial u_\beta} - \frac{\partial g_{\alpha\beta}}{\partial u_\lambda}\right].$$

We recall also that,

$$\mathbf{r}_{\alpha\beta} = \Gamma_{\alpha\beta}^\gamma \mathbf{r}_\gamma + b_{\alpha\beta}\mathbf{n}.$$

Observe that for a C^3 class surface, we have

$$\mathbf{r}_{\alpha\beta\lambda} = \mathbf{r}_{\alpha\lambda\beta},$$

Hence,

$$\mathbf{r}_{\alpha\beta\lambda} = \frac{\partial \Gamma_{\alpha\beta}^\gamma}{\partial u_\lambda}\mathbf{r}_\gamma + \Gamma_{\alpha\beta}^\gamma \mathbf{r}_{\gamma\lambda} + \frac{\partial b_{\alpha\beta}}{\partial u_\lambda}\mathbf{n} + b_{\alpha\beta}\mathbf{n}_\lambda.$$

Observe that

$$\mathbf{n} \cdot \mathbf{n} = 1,$$

and hence

$$\mathbf{n}_\alpha \cdot \mathbf{n} = 0,$$

so that we denote,

$$\mathbf{n}_\alpha = c_\alpha^\gamma \mathbf{r}_\gamma.$$

Observe that,

$$\mathbf{n}_\alpha \cdot \mathbf{r}_\rho = c_\alpha^\gamma \mathbf{r}_\gamma \cdot \mathbf{r}_\rho = c_\alpha^\gamma g_{\gamma\rho},$$

and

$$\mathbf{n}_\alpha \cdot \mathbf{r}_\rho = -b_{\alpha\rho},$$

and from this and

$$g_{\gamma\rho} g^{\rho\tau} = \delta_\gamma^\tau,$$

we get

$$
\begin{aligned}
\mathbf{n}_\alpha \cdot \mathbf{r}_\rho g^{\rho\tau} &= -b_{\alpha\rho} g^{\rho\tau} \\
&= -b_\alpha^\tau \\
&= c_\alpha^\lambda g_{\lambda\rho} g^{\rho\tau} \\
&= c_\alpha^\tau,
\end{aligned}
\tag{8.202}
$$

so that

$$c_\alpha^\tau = -b_\alpha^\tau.$$

We obtain thus the formula of Weingarten:

$$\mathbf{n}_\alpha = -b_\alpha^\beta \mathbf{r}_\beta,$$

where

$$b_\alpha^\beta = g^{\rho\beta} b_{\alpha\rho}.$$

Retaking the earlier expression we had obtained,

$$\mathbf{r}_{\alpha\beta\lambda} = \frac{\partial \Gamma_{\alpha\beta}^\gamma}{\partial u_\lambda} \mathbf{r}_\gamma + \Gamma_{\alpha\beta}^\gamma \mathbf{r}_{\gamma\lambda} + \frac{\partial b_{\alpha\beta}}{\partial u_\lambda} \mathbf{n} + b_{\alpha\beta} \mathbf{n}_\lambda.$$

so that,

$$\mathbf{r}_{\alpha\beta\lambda} = \frac{\partial \Gamma_{\alpha\beta}^{\gamma}}{\partial u_{\lambda}}\mathbf{r}_{\gamma}$$

$$+\Gamma_{\alpha\beta}^{\gamma}\left[\Gamma_{\gamma\lambda}^{\rho}\mathbf{r}_{\rho} + b_{\gamma\lambda}\mathbf{n}\right]$$

$$+\frac{\partial b_{\alpha\beta}}{\partial u_{\lambda}}\mathbf{n}$$

$$-b_{\alpha\beta}b_{\lambda}^{\tau}\mathbf{r}_{\tau}, \qquad\qquad (8.203)$$

that is,

$$\mathbf{r}_{\alpha\beta\lambda} = \left[\frac{\partial \Gamma_{\alpha\beta}^{\rho}}{\partial u_{\lambda}} + \Gamma_{\alpha\beta}^{\gamma}\Gamma_{\gamma\lambda}^{\rho} - b_{\alpha\beta}b_{\lambda}^{\rho}\right]\mathbf{r}_{\rho}$$

$$+\left[\Gamma_{\alpha\beta}^{\rho}b_{\rho\lambda} + \frac{\partial b_{\alpha\beta}}{\partial u_{\lambda}}\right]\mathbf{n}. \qquad\qquad (8.204)$$

Interchanging β and λ, we obtain

$$\mathbf{r}_{\alpha\lambda\beta} = \left[\frac{\partial \Gamma_{\alpha\lambda}^{\rho}}{\partial u_{\rho}} + \Gamma_{\alpha\lambda}^{\gamma}\Gamma_{\gamma\beta}^{\rho} - b_{\alpha\lambda}b\beta^{\rho}\right]\mathbf{r}_{\rho}$$

$$+\left[\Gamma_{\alpha\lambda}^{\rho}b_{\rho\beta} + \frac{\partial b_{\alpha\lambda}}{\partial u_{\beta}}\right]\mathbf{n}. \qquad\qquad (8.205)$$

The vectors $\mathbf{r}_{\alpha}$, $\mathbf{r}_{\beta}$, and $\mathbf{n}$ are linearly independent, so that since

$$\mathbf{r}_{\alpha\beta\lambda} = \mathbf{r}_{\alpha\lambda\beta},$$

we must have

$$\Gamma_{\alpha\beta}^{\rho}b_{\rho\lambda} - \Gamma_{\alpha\lambda}^{\rho}b_{\rho\beta} + \frac{\partial b_{\alpha\beta}}{\partial u_{\lambda}} - \frac{\partial b_{\alpha\lambda}}{\partial u_{\beta}} = 0.$$

These are the Mainard–Codazzi formulas.
From

$$\mathbf{r}_{\alpha\beta\lambda} - \mathbf{r}_{\alpha\lambda\beta} = 0,$$

equating to zero the equations in $\mathbf{r}_{\alpha}$ and $\mathbf{r}_{\beta}$, we obtain,

$$R_{\alpha\lambda\beta}^{\rho} = b_{\alpha\beta}b_{\lambda}^{\rho} + b_{\alpha\lambda}b_{\beta}^{\rho},$$

where

$$R^\rho_{\alpha\lambda\beta} = \frac{\partial\Gamma^\rho_{\alpha\beta}}{\partial u_\lambda} - \frac{\partial\Gamma^\rho_{\alpha\lambda}}{\partial u_\beta}$$

$$+\Gamma^\gamma_{\alpha\beta}\Gamma^\rho_{\gamma\lambda} - \Gamma^\gamma_{\alpha\lambda}\Gamma^\rho_{\gamma\beta}. \tag{8.206}$$

This tensor is called the mixed Riemann curvature tensor.
The tensor

$$R_{\tau\alpha\lambda\beta} = g_{\rho\tau}R^\rho_{\alpha\lambda\beta},$$

is called the Riemann covariant curvature tensor.
Observe that

$$R^\rho_{\alpha\lambda\beta} = b_{\alpha\beta}b^\rho_\lambda - b_{\alpha\lambda}b^\rho_\beta,$$

so that

$$R_{\tau\alpha\lambda\beta} = g_{\rho\tau}R^\rho_{\alpha\lambda\beta}$$

$$= g_{\rho\tau}(b_{\alpha\beta}b^\rho_\lambda - b_{\alpha\lambda}b^\rho_\beta)$$

$$= b_{\alpha\beta}b_{\lambda\tau} - b_{\alpha\lambda}b_{\beta\tau}. \tag{8.207}$$

In particular,

$$R_{1212} = b_{22}b_{11} - b_{21}b_{21}$$

$$= b_{11}b_{22} - (b_{12})^2$$

$$= b. \tag{8.208}$$

Under the context of this last discussion, we have,

Theorem 8.23.1 (Gauss, Egregium Theorem) *Considering the discussion in the above last lines, the Gaussian curvature of the surface S in question is given by,*

$$K = \frac{R_{1212}}{g}.$$

Proof Recall that earlier we had obtained,

$$K = \frac{b}{g},$$

so that from this and

$$R_{1212} = b,$$

we have,

$$K = \frac{R_{1212}}{g}.$$

This completes the proof.

Chapter 9
Integration in $\mathbb{R}^n$

9.1 Integration on Blocks

This chapter develops results concerning the Riemann integration of functions defined in $\mathbb{R}^n$. We address in detail the standard necessary and sufficient condition of zero Lebesgue measure for the set of discontinuities for the Riemann integrability of a scalar function defined in $\mathbb{R}^n$. Other topics such as change of variables are also presented in detail. In this chapter we develop the integration theory in $\mathbb{R}^n$. In the final sections we establish necessary and sufficient conditions for Riemann integrability. The main references for this chapter are [2, 9, 10].

9.2 First Definitions and Results

We start with some fundamental definitions.

Definition 9.2.1 An n-dimensional block $B \subset \mathbb{R}^n$ is a set expressed by

$$B = \prod_{j=1}^{n} [a_j, b_j] = [a_1, b_1] \times \cdots \times [a_n, b_n],$$

where

$$[a_j, b_j] \subset \mathbb{R}$$

is a nonempty, closed, bounded interval, $\forall j \in \{1, \ldots, n\}$.

The volume of B, denoted by $vol(B)$, is defined by

$$vol(B) = \prod_{j=1}^{n} (b_j - a_j) = (b_1 - a_1) \cdots (b_n - a_n).$$

Definition 9.2.2 (Partition) Let $[a, b] \subset \mathbb{R}$ be an interval. We define a partition P of $[a, b]$, as any set of the form:

$$P = \{x_0 = a, x_1, \ldots, x_m = b\},$$

where $x_{i-1} < x_i \; \forall i \in \{1, \ldots, m\}$. Given a block

$$B = \prod_{j=1}^{n} [a_j, b_j] \subset \mathbb{R}^n$$

we define a partition of B, denoted by P, by

$$P = P_1 \times P_2 \times \cdots \times P_n$$

where P_j is a partition of

$$I_j = [a_j, b_j], \; \forall \{1, \ldots, n\}.$$

For $P_j = \{a_j = x_0, x_1, \ldots, x_m = b_j\}$ we define its norm, denoted by $|P_j|$, by

$$|P_j| = \max\{x_k - x_{k-1}, \; k \in \{1, \ldots, m\}\}.$$

Finally, the norm of $P = P_1 \times \cdots \times P_n$ is defined by

$$|P| = \max\{|P_j|, \; j \in \{1, \ldots, n\}\}.$$

Here we highlight that a partition P of B divides B into sub-blocks.

Definition 9.2.3 (Lower and Upper Sums) Let $f : B_0 \to \mathbb{R}$ be a bounded function, where $B_0 \subset \mathbb{R}^n$ is a compact block.

Let P be a partition of B_0. For each sub-block B of B_0 (related to P), we define,

$$m_B = \inf\{f(\mathbf{x}) : \mathbf{x} \in B\},$$

and

$$M_B = \sup\{f(\mathbf{x}) : \mathbf{x} \in B\}.$$

We also define the lower sum of f relating P, denoted by s_f^P, as

$$s_f^P = \sum_{B \in P} m_B vol(B),$$

and the upper sum of f relating P, denoted by S_f^P, as

$$S_f^P = \sum_{B \in P} M_B vol(B).$$

Here $B \in P$ means that B is a sub-block generated by the partition P of B_0. Observe that

$$m \leq m_B \leq M_B \leq M, \ \forall B \in P,$$

where

$$m = \inf\{f(\mathbf{x}) \ : \ \mathbf{x} \in B_0\},$$

and

$$M = \sup\{f(\mathbf{x}) \ : \ \mathbf{x} \in B_0\},$$

so that

$$m Vol(B_0) \leq s_f^P \leq S_f^P \leq M Vol(B_0)$$

$\forall P$ partition of B_0.

Theorem 9.2.4 *Let $B_0 \subset \mathbb{R}^n$ be a nonempty compact block. Let P, Q be partitions of B_0 such that $P \subset Q$*
Let $f : B_0 \to \mathbb{R}$ be a bounded function.
Under such hypotheses,

$$s_f^P \leq s_f^Q \leq S_f^Q \leq S_f^P.$$

Proof Let us denote generically the sub-blocks relating P by B and those relating Q by B'.

Observe that since $P \subset Q$, for each $B' \in Q$ there exists $B \in P$ such that $B' \subset B$,

Also, if $B' \subset B$, then $m_B \leq m_{B'}$ and $M_{B'} \leq M_B$.

Thus,

$$s_f^P = \sum_{B \in P} m_B Vol(B)$$

$$= \sum_{B \in P} m_B \left(\sum_{B' \subset B} Vol(B') \right)$$

$$= \sum_{B \in P} \left(\sum_{B' \subset B} m_B Vol(B') \right)$$

$$\leq \sum_{B \in P} \left(\sum_{B' \subset B} m_{B'} Vol(B') \right)$$

$$= \sum_{B' \in Q} m_{B'} Vol(B')$$

$$= s_f^Q \tag{9.1}$$

Similarly, we may obtain $S_f^P \geq S_f^Q$.

Since the inequality $s_f^Q \leq S_f^Q$ is obvious, the proof is complete.

Corollary 9.2.5 *Let $f : B_0 \to \mathbb{R}$ be a bounded function where B_0 is a compact block. Let P, Q be partitions of B_0.*

Under such hypotheses,

$$s_f^P \leq S_f^Q.$$

Proof Observe that $P \subset P \cup Q$, and $Q \subset P \cup Q$, so that from the last theorem

$$s_f^P \leq s_f^{P \cup Q} \leq S_f^{P \cup Q} \leq S_f^Q,$$

so that

$$s_f^P \leq S_f^Q, \; \forall P, Q, \text{ partitions of } B_0.$$

Definition 9.2.6 Let $B_0 \subset \mathbb{R}^n$ be a compact block. Let $f : B_0 \to \mathbb{R}$ be such that $|f(\mathbf{x})| < K$, $\forall x \in B_0$ for some $K > 0$.

The we define the lower integral of f on B_0, denoted by $\underline{I}$, by

$$\underline{I} = \sup\{s_f^P \; : \; P \text{ is a partition of } B_0\},$$

and the upper integral of f on B_0, denoted by $\overline{I}$, by

$$\overline{I} = \inf\{S_f^P : P \text{ is a partition of } B_0\}.$$

Finally, we say that f is Riemann integrable on B_0 if

$$\underline{I} = \overline{I},$$

and we denote

$$\underline{I} = \overline{I} = I = \int_{B_0} f(\mathbf{x})\, d\mathbf{x}.$$

In such a case $I = \int_{B_0} f(\mathbf{x})\, d\mathbf{x}$ is said to be the Riemann integral of f on B_0.

Theorem 9.2.7 *Let $B_0 \subset \mathbb{R}^n$ be a compact block. Let $f : B_0 \to \mathbb{R}$ be such that $|f(\mathbf{x})| < K$, $\forall \mathbf{x} \in B_0$ for some $K > 0$. Under such hypotheses, f is Riemann integrable if and only if for each $\varepsilon > 0$ there exists a partition P of B_0 such that*

$$S_f^P - s_f^P < \varepsilon.$$

Proof First, we shall prove the condition sufficiency. Suppose the condition is valid. Let $\varepsilon > 0$. Thus, from the condition there exists a partition P of B_0 such that

$$S_f^P - s_f^P < \varepsilon.$$

Observe that

$$s_f^P \leq \underline{I} \leq \overline{I} \leq S_f^P.$$

Hence,

$$\overline{I} - \underline{I} \leq S_f^P - s_f^P < \varepsilon.$$

Since $\varepsilon > 0$ is arbitrary, we obtain

$$\overline{I} = \underline{I}.$$

Reciprocally, suppose that

$$\overline{I} = \underline{I} \equiv I.$$

Suppose given a new and not relabeled $\varepsilon > 0$. Thus there exists partitions P_1 and P_2 of B_0 such that

$$s_f^{P_1} > I - \frac{\varepsilon}{2},$$

and

$$S_f^{P_2} < I + \frac{\varepsilon}{2},$$

Thus,

$$-s_f^{P_1} < -I + \frac{\varepsilon}{2},$$

so that

$$S_f^{P_2} - s_f^{P_1} < I - I + \frac{\varepsilon}{2} + \frac{\varepsilon}{2} = \varepsilon.$$

Observe that

$$s_f^{P_1} \leq s_f^{P_1 \cup P_2} \leq S_f^{P_1 \cup P_2} \leq S_f^{P_2},$$

so that

$$S_f^{P_1 \cup P_2} - s_f^{P_1 \cup P_2} \leq S_f^{P_2} - s_f^{P_1} < \varepsilon.$$

Hence, denoting $P = P_1 \cup P_2$, we obtain

$$S_f^{P} - s_f^{P} < \varepsilon.$$

Therefore, the condition is necessary.
    The proof is complete.

9.3 Riemann Integral Properties

Theorem 9.3.1 *Let $B_0 \subset \mathbb{R}^n$ be a compact block. Let $f_1 : B_0 \to \mathbb{R}$ and $f_2 : B_0 \to \mathbb{R}$ be Riemann integrable functions. Under such hypotheses,*

$$f \equiv f_1 + f_2$$

is also Riemann integrable and,

$$\int_{B_0} (f_1(\mathbf{x}) + f_2(\mathbf{x})) \, d\mathbf{x} = \int_{B_0} f_1(\mathbf{x}) \, d\mathbf{x} + \int_{B_0} f_2(\mathbf{x}) \, d\mathbf{x}.$$

Proof Let $\varepsilon > 0$ be given.

Denoting,

$$I_1 = \int_{B_0} f_1(\mathbf{x}) \, d\mathbf{x},$$

$$I_2 = \int_{B_0} f_2(\mathbf{x}) \, d\mathbf{x},$$

$$\underline{I} = \sup\{s_f^P \; : \; P \text{ is a partition of } B_0\},$$

and

$$\overline{I} = \inf\{S_f^P \; : \; P \text{ is a partition of } B_0\},$$

there exist partitions P_1 and P_2 of B_0 such that

$$I_1 - \frac{\varepsilon}{4} < s_{f_1}^{P_1} \leq S_{f_1}^{P_1} < I_1 + \frac{\varepsilon}{4},$$

and

$$I_2 - \frac{\varepsilon}{4} < s_{f_2}^{P_2} \leq S_{f_2}^{P_2} < I_2 + \frac{\varepsilon}{4}.$$

Define $Q = P_1 \cup P_2$. Hence,

$$I_1 + I_2 - \frac{\varepsilon}{2}$$

$$< s_{f_1}^{P_1} + s_{f_2}^{P_2}$$

$$\leq s_{f_1}^{Q} + s_{f_2}^{Q}$$

$$\leq s_{f_1+f_2}^{Q}$$

$$\leq \underline{I} \leq \overline{I}$$

$$\leq S_{f_1+f_2}^{Q}$$

$$\leq S_{f_1}^{Q} + S_{f_2}^{Q}$$

$$\leq S_{f_1}^{P_1} + S_{F_2}^{P_2}$$

$$< I_1 + I_2 + \frac{\varepsilon}{2}, \tag{9.2}$$

so that

$$\overline{I} - \underline{I} < \varepsilon.$$

Since $\varepsilon > 0$ is arbitrary, we may conclude that

$$\overline{I} = \underline{I} \equiv I = \int_{B_0} f(\mathbf{x}) \, d\mathbf{x}.$$

Moreover, from this and (9.2), we obtain

$$I_1 + I_2 - \frac{\varepsilon}{2} < I < I_1 + I_2 + \frac{\varepsilon}{2},$$

that is,

$$|I - (I_1 + I_2)| < \frac{\varepsilon}{2},$$

Since $\varepsilon > 0$ is arbitrary we have,

$$I = I_1 + I_2,$$

that is

$$\int_{B_0} f(\mathbf{x}) \, d\mathbf{x} = \int_{B_0} f_1(\mathbf{x}) \, d\mathbf{x} + \int_{B_0} f_2(\mathbf{x}) \, d\mathbf{x}.$$

The proof is complete.

Theorem 9.3.2 *Let $B_0 \subset \mathbb{R}^n$ be a compact block. Let $f : B_0 \to \mathbb{R}$ be a Riemann integrable function.*
Then $-f$ is Riemann integrable and

$$\int_{B_0} (-f(\mathbf{x})) \, d\mathbf{x} = -\int_{B_0} f(\mathbf{x}) \, d\mathbf{x}.$$

Proof Suppose given $\varepsilon > 0$. Thus there exists a partition P of B_0 such that

$$I - \frac{\varepsilon}{2} < s_f^P \le S_f^P < I + \frac{\varepsilon}{2}.$$

Hence,

$$-I - \frac{\varepsilon}{2} < -S_f^P \le -s_f^P < -I + \frac{\varepsilon}{2},$$

so that denoting

$$\underline{I_1} = \sup\{s_{(-f)}^P \ : \ P \text{ is a partition of } B_0\},$$

and

$$\overline{I_1} = \inf\{S^P_{(-f)} \; : \; P \text{ is a partition of } B_0\},$$

we have

$$- I - \frac{\varepsilon}{2} < s^P_{(-f)} \leq \underline{I_1} \leq \overline{I_1} \leq S^P_{(-f)} < -I + \frac{\varepsilon}{2}, \qquad (9.3)$$

that is,

$$\overline{I_1} - \underline{I_1} < \varepsilon.$$

Since $\varepsilon > 0$ is arbitrary, we have

$$\overline{I_1} = \underline{I_1} \equiv I_1 = \int_{B_0} (-f(\mathbf{x})) \, d\mathbf{x}.$$

From this and (9.3) we obtain,

$$-I - \frac{\varepsilon}{2} < I_1 < -I + \frac{\varepsilon}{2},$$

so that

$$|I_1 - (-I)| < \frac{\varepsilon}{2}.$$

Since $\varepsilon > 0$ is arbitrary, we finally obtain,

$$I_1 = -I,$$

that is,

$$\int_{B_0} (-f(\mathbf{x})) \, d\mathbf{x} = - \int_{B_0} f(\mathbf{x}) \, d\mathbf{x}.$$

The proof is complete.

Theorem 9.3.3 *Let $B_0 \subset \mathbb{R}^n$ be a compact block. Let $f : B_0 \to \mathbb{R}$ be Riemann integrable and let $c \in \mathbb{R}$. Under such hypotheses cf is Riemann integrable and*

$$\int_{B_0} cf(\mathbf{x}) \, d\mathbf{x} = c \int_{B_0} f(\mathbf{x}) \, d\mathbf{x}.$$

Proof Assume first $c > 0$.

The case $c = 0$ is immediate and the case $c < 0$ will be dealt at the end of this proof. Let $\varepsilon > 0$ be given. Thus, there exists a partition P of B_0 such that

$$I - \frac{\varepsilon}{2c} < s_f^P \leq S_f^P < I\frac{\varepsilon}{2c}.$$

Hence,

$$cI - \frac{\varepsilon}{2} < cs_f^P \leq cS_f^P < cI + \frac{\varepsilon}{2}.$$

Therefore, denoting,

$$\underline{I_1} = \sup\{s_{(cf)}^P \ : \ P \text{ is a partition of } B_0\},$$

and

$$\overline{I_1} = \inf\{S_{(cf)}^P \ : \ P \text{ is a partition of } B_0\},$$

we have

$$cI - \frac{\varepsilon}{2} < s_{(cf)}^P \leq \underline{I_1} \leq \overline{I_1} \leq S_{(cf)}^P < cI + \frac{\varepsilon}{2}. \tag{9.4}$$

so that

$$\overline{I_1} - \underline{I_1} < \varepsilon.$$

Since $\varepsilon > 0$ is arbitrary, we obtain,

$$\overline{I_1} = \underline{I_1} \equiv I_1 = \int_{B_0} cf(\mathbf{x}) \, d\mathbf{x}.$$

From this and (9.4), we have,

$$cI - \frac{\varepsilon}{2} < I_1 < cI + \frac{\varepsilon}{2},$$

that is,

$$|I_1 - cI| < \frac{\varepsilon}{2}.$$

Finally, since $\varepsilon > 0$ is arbitrary, we obtain

$$I_1 = cI,$$

that is,

$$\int_{B_0} cf(\mathbf{x})\, d\mathbf{x} = c \int_{B_0} f(\mathbf{x})\, d\mathbf{x}.$$

Now suppose $c < 0$. From this last result and the last theorem, we have

$$-\int_{B_0} cf(\mathbf{x})\, d\mathbf{x} = \int_{B_0} (-c)f(\mathbf{x})\, d\mathbf{x} = -c \int_{B_0} f(\mathbf{x})\, d\mathbf{x},$$

so that

$$\int_{B_0} cf(\mathbf{x})\, d\mathbf{x} = c \int_{B_0} f(\mathbf{x})\, d\mathbf{x}.$$

The proof is complete.

Theorem 9.3.4 *Let $B_0 \subset \mathbb{R}^n$ be a compact block. Let $f : B_0 \to \mathbb{R}$ be a Riemann integrable function such that*

$$m \le f(x) \le M, \quad \forall x \in B_0,$$

for some $m, M \in \mathbb{R}$.

Let $g : [m, M] \to \mathbb{R}$ be a continuous function on $[m, M]$.
Under such hypotheses, $(g \circ f) : B_0 \to \mathbb{R}$ is Riemann integrable on B_0, where

$$(g \circ f)(\mathbf{x}) = g(f(\mathbf{x})), \quad \forall x \in B_0.$$

Proof Let $\varepsilon > 0$ be given. Observe that since $[m, M]$ is compact, g is uniformly continuous on $[m, M]$. Choose $K > 0$ such that

$$|g(t)| < K, \quad \forall t \in [m, M].$$

Thus, there exists $0 < \delta < \frac{\varepsilon}{4K}$ such that if $s, t \in [m, M]$ and $|s - t| < \delta$, then

$$|g(s) - g(t)| < \frac{\varepsilon}{2Vol(B_0)}.$$

Also, since f is integrable, there exists a partition P of B_0 such that

$$I - \frac{\delta^2}{2} < s_f^P \le S_f^P < I - \frac{\delta^2}{2},$$

where

$$I = \int_{B_0} f(\mathbf{x})\, d\mathbf{x}.$$

Therefore,

$$S_f^P - s_f^P = \sum_{B \in P}^{n} (M_B - m_B) Vol(B) < \delta^2,$$

where we denote,

$$m_B = \inf\{f(\mathbf{x}) \; : \; \mathbf{x} \in B\},$$

$$M_B = \sup\{f(\mathbf{x}) \; : \; \mathbf{x} \in B\},$$

$$m_B^* = \inf\{g(f(\mathbf{x})) \; : \; \mathbf{x} \in B\},$$

$$M_B^* = \sup\{g(f(\mathbf{x})) \; : \; \mathbf{x} \in B\}.$$

Denote by α the set of blocks $B \in P$ such that $M_B - m_B < \delta$. Denote by β the set of blocks $B \in P$ such that $M_B - m_B \geq \delta$.

Observe that

$$\delta \sum_{B \in \beta} Vol(B) \leq \sum_{B \in \beta} (M_B - m_B) Vol(B) < \delta^2,$$

and hence,

$$\sum_{B \in \beta} Vol(B) < \delta < \frac{\varepsilon}{4K}.$$

Observe that if $B \in \alpha$, then $M_B - m_B < \delta$ and thus

$$M_B^* - m_B^* < \frac{\varepsilon}{2Vol(B_0)}.$$

Therefore,

$$S_{(g \circ f)}^P - s_{(g \circ f)}^P = \sum_{B \in P} (M_B^* - m_B^*) Vol(B)$$

$$= \sum_{B \in \alpha} (M_B^* - m_B^*) Vol(B) + \sum_{B \in \beta} (M_B^* - m_B^*) Vol(B)$$

$$< \frac{\varepsilon}{2Vol(B_0)} \sum_{B \in \alpha} Vol(B) + 2K \sum_{B \in \beta} Vol(B)$$

$$< \frac{\varepsilon Vol(B_0)}{2Vol(B_0)} + \frac{2K\varepsilon}{4K}$$

$$= \frac{\varepsilon}{2} + \frac{\varepsilon}{2}$$

$$= \varepsilon. \tag{9.5}$$

Summarizing,

$$S^P_{(g \circ f)} - S^P_{(g \circ f)} < \varepsilon.$$

Since $\varepsilon > 0$ is arbitrary, from Theorem 9.2.7 we may conclude that $(g \circ f)$ is Riemann integrable.

The proof is complete.

Proposition 9.3.5 *Let $B_0 \subset \mathbb{R}^n$ be a compact block. Let $f_1, f_2 : B_0 \to \mathbb{R}$ be Riemann integrable functions. Under such hypotheses:*

1. *$f_1 \cdot f_2$ is Riemann integrable.*
2. *$|f_1|$ is Riemann integrable and*

$$\left| \int_{B_0} f_1(\mathbf{x}) \, d\mathbf{x} \right| \leq \int_{B_0} |f_1(\mathbf{x})| \, d\mathbf{x}.$$

Proof

1. With $g(t) = t^2$, from the last theorem and previous results, we have that

$$f_1 \cdot f_2 = \frac{(f_1 + f_2)^2 - (f_1 - f_2)^2}{4}$$

   is Riemann integrable.
2. With $g(t) = |t|$, from the last Theorem $|f_1| = (g \circ f_1)$ is Riemann integrable. Moreover since

$$\pm f_1(\mathbf{x}) \leq |f_1(\mathbf{x})|, \forall \mathbf{x} \in B_0,$$

   we obtain

$$\pm \int_{B_0} f_1(\mathbf{x}) \, d\mathbf{x} \leq \int_{B_0} |f_1(\mathbf{x})| \, d\mathbf{x}$$

   so that

$$\left| \int_{B_0} f_1(\mathbf{x}) \, d\mathbf{x} \right| \leq \int_{B_0} |f_1(\mathbf{x})| \, d\mathbf{x}.$$

The proof is complete.

Theorem 9.3.6 *Let $B_0 \subset \mathbb{R}^n$ be a compact block. Let $f : B_0 \to \mathbb{R}$ be a continuous function on B_0.*

Under such hypotheses f is Riemann integrable on B_0.

Proof Let $\varepsilon > 0$. Since f is continuous on B_0 and B_0 is compact, we have that f is uniformly continuous on B_0. Thus, there exists $\delta > 0$ such that if $\mathbf{x}, \mathbf{y} \in B_0$ and $|\mathbf{x} - \mathbf{y}| < \delta$, then

$$|f(\mathbf{y}) - f(\mathbf{x})| < \frac{\varepsilon}{2Vol(B_0)}.$$

Let P be a partition of B_0 such that $0 < |P| < \delta/\sqrt{n}$. Hence,

$$|\mathbf{x} - \mathbf{y}| < \delta, \quad \forall \mathbf{x}, \mathbf{y} \in B, \ \forall B \in P.$$

Thus,

$$M_B - m_B \le \frac{\varepsilon}{2Vol(B_0)}, \quad \forall B \in P,$$

so that

$$
\begin{aligned}
S_f^P - s_f^P &= \sum_{B \in P} (M_B - m_B)Vol(B) \\
&\le \sum_{B \in P} \frac{\varepsilon}{2Vol(B_0)} Vol(B) \\
&= \frac{\varepsilon Vol(B_0)}{2Vol(B_0)} \\
&= \frac{\varepsilon}{2} \\
&< \varepsilon.
\end{aligned}
\tag{9.6}
$$

Since $\varepsilon > 0$ is arbitrary, from this and Theorem 9.2.7, f is integrable on B_0.

9.4   A Criterion of Riemann Integrability

In this section we present a condition necessary and sufficient for Riemann integrability. We start with the definition of exterior and Lebesgue measures.

Definition 9.4.1 (Exterior Measure in $\mathbb{R}^n$) Given a set $A \subset \mathbb{R}^n$, we define its exterior measure denoted by $m^*(A)$ by

$$m^*(A) = \inf \left\{ \sum_{n=1}^{\infty} Vol(B_n) \; : \; A \subset \cup_{i=1}^{n} B_n \right\},$$

where B_n is an open block $\forall n \in \mathbb{N}$.

Definition 9.4.2 (Measurable Set) A set $E \subset \mathbb{R}^n$ is said to be Lebesgue measurable if for each $A \subset \mathbb{R}^n$ we have

$$m^*(A) = m^*(A \cap E) + m^*(A \cap E^c),$$

where

$$E^c = \mathbb{R}^n \setminus E = \{\mathbf{x} \in \mathbb{R}^n \; : \; \mathbf{x} \notin E\}.$$

Thus, if E is Lebesgue measurable, we shall define $m(E) = m^*(E)$ where $m(E)$ is said to be the Lebesgue measure of E.

At this point we present some results relating the exterior measure.

Proposition 9.4.3 *Let* $A \subset B \subset \mathbb{R}^n$. *Under such assumptions, we have*

$$m^*(A) \leq m^*(B).$$

Proof If $m^*(B) = +\infty$ the result follows immediately, since in such a case

$$m^*(A) \leq +\infty = m^*(B).$$

Thus, assume $m^*(B) = \alpha < +\infty$.

Let $\varepsilon > 0$ be given.

Since $\alpha + \varepsilon > \alpha = m^*(B)$, there exists a sequence of open blocks $\{B_m\}$ such that

$$B \subset \cup_{m=1}^{\infty} B_m$$

and

$$\alpha \leq \sum_{m=1}^{\infty} Vol(B_m) < \alpha + \varepsilon.$$

Therefore,

$$A \subset B \subset \cup_{m=1}^{\infty} B_m,$$

so that

$$m^*(A) \le \sum_{m=1}^{\infty} Vol(B_m) < \alpha + \varepsilon.$$

Hence,

$$m^*(A) < \alpha + \varepsilon.$$

Since $\varepsilon > 0$ is arbitrary, we may infer that

$$m^*(A) \le \alpha = m^*(B).$$

The proof is complete.

Also very important is the next result.

Proposition 9.4.4 *Let $A, B \subset \mathbb{R}^n$. Under such hypotheses, we have*

$$m^*(A \cup B) \le m^*(A) + m^*(B).$$

Proof If $m^*(A) = +\infty$ or $m^*(B) = +\infty$, the result follows immediately, since in such a case,

$$m^*(A \cup B) \le +\infty = m^*(A) + m^*(B).$$

Thus, suppose $m^*(A) = \alpha_1 < +\infty$ and $m^*(B) = \alpha_2 < +\infty$.
Let $\varepsilon > 0$.
Since $\alpha_1 + \varepsilon/2 > \alpha_1$ and $\alpha_2 + \varepsilon/2 > \alpha_2$ there exist sequences $\{B_m\}$ and $\{C_m\}$ of open blocks such that

$$A \subset \cup_{m=1}^{\infty} B_m \text{ and } B \subset \cup_{m=1}^{\infty} C_m,$$

and also,

$$\alpha_1 \le \sum_{m=1}^{\infty} Vol(B_m) < \alpha_1 + \varepsilon/2 \text{ and } \alpha_2 \le \sum_{m=1}^{\infty} Vol(C_m) < \alpha_2 + \varepsilon/2.$$

Observe that

$$A \cup B \subset [\cup_{m=1}^{\infty} B_m] \cup [\cup_{m=1}^{\infty} C_m],$$

so that

$$m^*(A \cup B) \leq \sum_{m=1}^{\infty} Vol(B_m) + \sum_{m=1}^{\infty} Vol(C_m)$$
$$< \alpha_1 + \varepsilon/2 + \alpha_2 + \varepsilon/2$$
$$= m^*(A) + m^*(B) + \varepsilon. \tag{9.7}$$

Since $\varepsilon > 0$ is arbitrary, we may infer that

$$m^*(A \cup B) \leq m^*(A) + m^*(B).$$

This completes the proof.

Proposition 9.4.5 *Let* $\{E_j\}_{j \in \mathbb{N}} \subset \mathbb{R}^n$ *be a sequence of sets. Under such assumptions,*

$$m^*(\cup_{j=1}^{\infty} E_j) \leq \sum_{j=1}^{\infty} m^*(E_j).$$

Proof If $m^*(E_j) = +\infty$ for some $j \in \mathbb{N}$, then the result follows immediately, since in such a case,

$$m^*(\cup_{j=1}^{\infty} E_j) \leq +\infty = \sum_{j=1}^{\infty} m^*(E_j).$$

Thus, assume

$$m^*(E_j) = \alpha_j < +\infty, \ \forall j \in \mathbb{N}.$$

Since $\alpha_j + \frac{\varepsilon}{2^j} > \alpha_j$, there exists a sequence $\{B_m^j\}_{m \in \mathbb{N}}$ of open blocks such that

$$E_j \subset \cup_{m=1}^{\infty} B_m^j,$$

and

$$\alpha_j \leq \sum_{m=1}^{\infty} Vol(B_m^j) < \alpha_j + \frac{\varepsilon}{2^j},$$

$\forall j \in \mathbb{N}$.
     Therefore,

$$\cup_{j=1}^{\infty} E_j \subset \cup_{j=1}^{\infty} [\cup_{m=1}^{\infty} B_m^j],$$

and

$$m^*(\cup_{j=1}^\infty E_j) \le \sum_{j=1}^\infty \left(\sum_{m=1}^\infty Vol(B_m^j) \right)$$

$$< \sum_{j=1}^\infty \left(\alpha_j + \frac{\varepsilon}{2^j} \right)$$

$$= \sum_{j=1}^\infty \alpha_j + \sum_{j=1}^\infty \frac{\varepsilon}{2^j}$$

$$= \sum_{j=1}^\infty \alpha_j + \varepsilon. \tag{9.8}$$

Hence,

$$m^*(\cup_{j=1}^\infty E_j) < \sum_{j=1}^\infty m^*(E_j) + \varepsilon.$$

Since $\varepsilon > 0$ is arbitrary, we may infer that

$$m^*(\cup_{j=1}^\infty E_j) \le \sum_{j=1}^\infty m^*(E_j).$$

The proof is complete.

Remark 9.4.6 We shall show that if $m^*(E) = 0$, then E is Lebesgue measurable. Indeed, let $E \subset \mathbb{R}^n$ be such that $m^*(E) = 0$.
Let $A \subset \mathbb{R}$. Since $A \cap E \subset E$ we obtain $m^*(A \cap E) = 0$.
Thus

$$m^*(A \cap E) + m^*(A \cap E^c) = m^*(A \cap E^c) \le m^*(A). \tag{9.9}$$

Since $m^*(B \cup C) \le m^*(B) + m^*(C), \forall B, C \subset \mathbb{R}^n$ we obtain

$$m^*(A) = m^*((A \cap E) \cup (A \cap E^c)) \le m^*(A \cap E) + m^*(A \cap E^c),$$

so that from this and (9.9) we get

$$m^*(A) = m^*(A \cap E) + m^*(A \cap E^c), \quad \forall A \subset \mathbb{R}^n,$$

and thus E is Lebesgue measurable whenever $m^*(E) = 0$.

Also we recall that in this case

$$0 = m^*(E) = \inf\left\{\sum_{n=1}^{\infty} Vol(B_n) \;:\; A \subset \cup_{i=1}^{\infty} B_n\right\},$$

so that for each $\varepsilon > 0$ we may find a sequence $\{B_n\}$ of open blocks such that

$$E \subset \cup_{n=1}^{\infty} B_n$$

and

$$\sum_{n=1}^{\infty} Vol(B_n) < \varepsilon.$$

Theorem 9.4.7 *Let $B_0 \subset \mathbb{R}^n$ be a compact block. Let $f : B_0 \to \mathbb{R}$ be a bounded function on B_0. Denote by A the set where f is not continuous.*

Suppose $m(A) = 0$.

Under such hypotheses, f is Riemann integrable on B_0

Proof From the hypotheses $m(A) = 0$ and there exists $K > 0$ such that

$$|f(\mathbf{x})| < K, \;\; \forall \mathbf{x} \in B_0.$$

Let $\varepsilon > 0$. Thus there exists a sequence of open blocks $\{B_m\}$ such that $A \subset \cup_{m=1}^{\infty} B_m$ and

$$\sum_{m=1}^{\infty} Vol(B_m) < \frac{\varepsilon}{4K2^n},$$

so that since B_m is open, $Vol(B_m) > 0, \forall m \in \mathbb{N}$.

Define $\tilde{B} = B_0 \setminus A$.

So f is continuous on $\tilde{B}$.

Thus, for each $\mathbf{x} \in \tilde{B}$, there exists $\delta_x > 0$ such that if $\mathbf{y} \in B_0$ and $|\mathbf{y} - \mathbf{x}| < \delta_x$, then

$$|f(\mathbf{y}) - f(\mathbf{x})| < \frac{\varepsilon}{4Vol(B_0)}. \tag{9.10}$$

Observe that

$$\tilde{B} \subset \cup_{\mathbf{x} \in \tilde{B}} B_{\frac{\delta_x}{2}}(\mathbf{x}).$$

Hence,

$$B_0 = A \cup \tilde{B} \subset [\cup_{m=1}^{\infty} B_m] \cup [\cup_{\mathbf{x} \in \tilde{B}} B_{\frac{\delta_x}{2}}].$$

Since B_0 is compact, there exist $m_1, \ldots, m_k \in \mathbb{N}$ and $\mathbf{x}_1, \ldots, \mathbf{x}_p \in \tilde{B}$ such that

$$B_0 \subset [\cup_{l=1}^{k} B_{ml}] \cup [\cup_{j=1}^{p} B_{\frac{\delta_{x_j}}{2}} (\mathbf{x}_j)].$$

Define

$$C = \min\{\delta_{x_j}/2 \; : \; j \in \{1, \ldots, p\}\},$$

and denoting,

$$B_{ml} = (a_1, b_1) \times \cdots \times (a_n, b_n),$$

define

$$b_{ml} = \min\{(b_i - a_i) : i \in \{1, \ldots, n\}\}.$$

Define also,

$$D = \min\{b_{ml} \; : \; l \in \{1, \ldots, k\}\},$$

and

$$\delta = \min\{C, D/2\}.$$

Let P be a partition of B_0 such that

$$|P| < \frac{\delta}{\sqrt{n}}.$$

Denote, generically, by B the blocks of P which intersects $\cup_{l=1}^{k} B_{ml}$ and by B' the remaining blocks of P.

Thus for $B' \in P$ we have

$$B' \cap [\cup_{l=1}^{k} B_{ml}] = \emptyset.$$

Let $B \in P$.

Hence,

$$B \cap [\cup_{l=1}^{k} B_{ml}] \neq \emptyset.$$

Therefore, there exists $\tilde{l} \in \{1, \ldots, k\}$ such that

$$B \cap B_{m\tilde{l}} \neq \emptyset.$$

Let E_l be the block with the same center as B_{ml}, with all faces parallel to the corresponding faces of B_{ml}, however, with dimensions multiplied by 2, relating the corresponding dimensions of B_{ml}.

Hence,

$$Vol(E_l) = 2^n Vol(B_{ml}),$$

and moreover, since the greatest distances between two pints of B are $\delta < D/2$, we obtain

$$E_{\bar{l}} \supset B_{m\bar{l}} \cup B,$$

so that

$$\cup_{B \in P} B \subset \cup_{l=1}^{k} E_l,$$

and therefore,

$$\sum_{B \in P} Vol(B) = Vol(\cup_{B \in P} B)$$

$$\leq Vol(\cup_{l=1}^{k} E_l)$$

$$\leq \sum_{l=1}^{k} Vol(E_l)$$

$$\leq 2^n \sum_{l=1}^{k} B_{ml}$$

$$\leq 2^n \sum_{m=1}^{\infty} Vol(B_m)$$

$$< \frac{2^n \varepsilon}{4K 2^n}$$

$$= \frac{\varepsilon}{4K}. \tag{9.11}$$

Therefore,

$$\sum_{B \in P} Vol(B) < \frac{\varepsilon}{4K}. \tag{9.12}$$

Let $B' \in P$.

Denote by $\mathbf{x}'$ the center of B'.

Observe that since $B' \cap \cup_{l=1}^{k} B_{ml} = \emptyset$, we have

$$B' \subset \cup_{j=1}^{p} B_{\frac{\delta x_j}{2}}(\mathbf{x}_j).$$

So, there exists $j_0 \in \{1, \ldots, p\}$ such that

$$x' \in B_{\frac{\delta x_{j_0}}{2}}(\mathbf{x}_{j_0}).$$

Observe that since $|P| < \frac{\delta}{\sqrt{n}}$ we recall that the greatest distance between two points of B' is δ.

Let $\mathbf{x} \in B'$.

Thus,

$$\begin{aligned}
|\mathbf{x} - \mathbf{x}_{j_0}| &= |\mathbf{x} - \mathbf{x}' + \mathbf{x}' - \mathbf{x}_{j_0}| \\
&\leq |\mathbf{x} - \mathbf{x}'| + |\mathbf{x}' - x_{j_0}| \\
&< \delta + \frac{\delta_{x_{j_0}}}{2} \\
&\leq \frac{\delta_{x_{j_0}}}{2} + \frac{\delta_{x_{j_0}}}{2} \\
&= \delta_{x_{j_0}}.
\end{aligned} \tag{9.13}$$

Hence from this and (9.10) we obtain

$$|f(\mathbf{x}) - f(\mathbf{x}_{j_0})| < \frac{\varepsilon}{4Vol(B_0)}.$$

Similarly, if $\mathbf{y} \in B'$, then

$$|f(\mathbf{y}) - f(\mathbf{x}_{j_0})| < \frac{\varepsilon}{4Vol(B_0)},$$

so that

$$\begin{aligned}
|f(\mathbf{x}) - f(\mathbf{y})| &= |f(\mathbf{x}) - f(\mathbf{x}_{j_0}) + f(\mathbf{x}_{j_0}) - f(\mathbf{y})| \\
&\leq |f(\mathbf{x}) - f(\mathbf{x}_{j_0})| + |f(\mathbf{x}_{j_0}) - f(\mathbf{y})| \\
&< \frac{\varepsilon}{4Vol(B_0)} + \frac{\varepsilon}{4Vol(B_0)} \\
&= \frac{\varepsilon}{2Vol(B_0)}
\end{aligned} \tag{9.14}$$

So, denoting,

$$M_{B'} = \sup\{f(\mathbf{x}) \; : \; \mathbf{x} \in B'\},$$

$$m_{B'} = \inf\{f(\mathbf{x}) \; : \; \mathbf{x} \in B'\},$$

from (9.14), we obtain

$$M_{B'} - m_{B'} \le \frac{\varepsilon}{2Vol(B_0)}. \qquad (9.15)$$

On the other hand, if $B \in P$, denoting

$$M_B = \sup\{f(\mathbf{x}) \; : \; \mathbf{x} \in B\}$$

and

$$m_B = \inf\{f(\mathbf{x}) \; : \; \mathbf{x} \in B\},$$

since

$$|f(\mathbf{x})| < K, \forall \mathbf{x} \in B_0,$$

we have

$$M_B - m_B \le 2K, \; \forall B \in P. \qquad (9.16)$$

Thus, from (9.12), (9.15), and (9.16), we obtain,

$$
\begin{aligned}
S_f^P - s_f^P &= \sum_{B \in P}(M_B - m_B)Vol(B) + \sum_{B' \in P}(M_{B'} - m_{B'})Vol(B') \\
&\le 2K \sum_{B \in P} Vol(B) + \frac{\varepsilon}{2Vol(B_0)} \sum_{B' \in P} Vol(B') \\
&< 2K \frac{\varepsilon}{4K} + \frac{\varepsilon}{2Vol(B_0)} Vol(B_0) \\
&= \frac{\varepsilon}{2} + \frac{\varepsilon}{2} \\
&= \varepsilon.
\end{aligned}
\qquad (9.17)
$$

Therefore,

$$S_f^P - s_f^P < \varepsilon.$$

Since $\varepsilon > 0$ is arbitrary, from Theorem 9.2.7, f is Riemann integrable.

9.4.1 Oscillation

Definition 9.4.8 Let $B_0 \subset \mathbb{R}^n$ be a compact block. Let $f : B_0 \subset \mathbb{R}^n$ be a bounded function and let $A \subset B_0$. We define the oscillation of f on A denoted by $\omega_f(A)$ by

$$\omega_f(A) = \sup\{|f(\mathbf{x}) - f(\mathbf{y})| \ : \ \mathbf{x}, \mathbf{y} \in A\}.$$

Let $\mathbf{x} \in B_0$. We define the oscillation of f at $\mathbf{x}$ denoted by $\omega_f(\mathbf{x})$ by,

$$\omega_f(\mathbf{x}) = \lim_{\delta \to 0^+} \omega_f(B_\delta(\mathbf{x}) \cap B_0).$$

Remark 9.4.9 Defining

$$g_x(\delta) = \omega_f(B_{\delta(\mathbf{x})} \cap B_0)$$

we have that g_x is nondecreasing in $\delta > 0$, so that

$$
\begin{aligned}
\omega_f(\mathbf{x}) &= \lim_{\delta \to 0^+} \omega_f(B_\delta(\mathbf{x}) \cap B_0) \\
&= \lim_{\delta \to 0^+} g_x(\delta),
\end{aligned}
\tag{9.18}
$$

is well defined as a lateral limit of a monotone function as

$$\delta \to 0^+.$$

Theorem 9.4.10 *Let $B_0 \subset \mathbb{R}^n$ be a compact block. Let $f : B_0 \to \mathbb{R}$ be such that $|f(\mathbf{x})| < K$, , $\forall \mathbf{x} \in [a, b]$ for some $K > 0$. Let $\mathbf{x} \in [a, b]$. Under such hypotheses, f is continuous at $\mathbf{x}$ if, and only if,*

$$\omega_f(\mathbf{x}) = 0.$$

Proof Suppose f is continuous at $\mathbf{x} \in B_0$. Let $\varepsilon > 0$ be given. Thus there exists $\delta_0 > 0$ such that if $\mathbf{y} \in B_0$ and $|\mathbf{y} - \mathbf{x}| < \delta_0$, then

$$|f(\mathbf{y}) - f(\mathbf{x})| < \varepsilon.$$

Thus,

$$g_x(\delta) = \sup\{|f(\mathbf{y}) - f(\mathbf{z})| \ : \ \mathbf{y}, \mathbf{z} \in B_{\delta(\mathbf{x})} \cap B_0\} < \varepsilon, \text{ if } 0 < \delta < \delta_0,$$

so that

$$\omega_f(\mathbf{x}) = \lim_{\delta \to 0^+} g_x(\delta) \le \varepsilon.$$

Since $\varepsilon > 0$ is arbitrary, we may conclude that,

$$\omega_f(\mathbf{x}) = 0.$$

Reciprocally, suppose $\omega_f(\mathbf{x}) = 0$. Let a new and not relabeled $\varepsilon > 0$ be given. Thus there exists $\delta_0 > 0$ such that if $0 < \delta < \delta_0$, then $g_x(\delta) < \varepsilon$. Thus, for $\delta = \delta_0/2$ we have that $|f(\mathbf{y}) - f(\mathbf{x})| < \varepsilon, \forall \mathbf{y} \in B_\delta(\mathbf{x}) \cap B_0$.

We may conclude that

$$\lim_{\mathbf{y} \to \mathbf{x}} f(\mathbf{y}) = f(\mathbf{x}),$$

so that f is continuous at $\mathbf{x}$.

The proof is complete.

Theorem 9.4.11 *Let $B_0 \subset \mathbb{R}^n$ be a compact block. Let $f :\to \mathbb{R}$ be such that $|f(\mathbf{x})| < K, \forall \mathbf{x} \in B_0$ for some $K > 0$. Suppose f is Riemann integrable. Let $A \subset B_0$ be the subset in which f is not continuous. Under such hypotheses, $m(A) = 0$.*

Proof Observe that $\mathbf{x} \in A$ if, and only if, f is not continuous at $\mathbf{x}$. In such a case, from the last theorem, $\omega_f(\mathbf{x}) > \frac{1}{m}$ for some $m \in \mathbb{N}$. For each $k \in \mathbb{N}$, define

$$B_k = \left\{ \mathbf{x} \in B_0 \ : \ \omega_f(\mathbf{x}) > \frac{1}{k} \right\}.$$

Then

$$A = \cup_{k=1}^{\infty} B_k.$$

Let $k \in \mathbb{N}$. We shall prove that $m^*(B_k) = 0$.

Indeed, suppose given $\varepsilon > 0$. Since f is integrable, there exists a partition P of B_0 of $[a, b]$ such that

$$S_f^P - s_f^P < \varepsilon.$$

Denote generically by B the blocks of P which intersects B_k and by B' the remaining blocks.

Thus, if for $B \in P$ we have $B \cap B_k \neq \emptyset$, we have also

$$M_B - m_B > \frac{1}{k},$$

where

$$M_B = \sup\{f(\mathbf{x}) \ : \ \mathbf{x} \in B\},$$

and

$$m_B = \inf\{f(\mathbf{x}) \ : \ \mathbf{x} \in B\}.$$

On the other hand, since for each B' we have $B' \cap B_k = \emptyset$ and

$$[\cup_{B \in P}] \cup [\cup_{B' \in P}] = B_0 \supset B_k,$$

we also obtain

$$\cup_{B \in P} B \supset B_k,$$

so that

$$\sum_{B \in P} Vol(B) = Vol(\cup_{B \in P} B) \geq m^*(B_k).$$

Therefore,

$$\varepsilon > S_f^P - s_f^P$$
$$\geq \sum_{B \in P} (M_B - m_B) Vol(B)$$
$$> \frac{1}{k} \sum_{B \in P} Vol(B)$$
$$\geq \frac{1}{k} m^*(B_k), \tag{9.19}$$

and hence,

$$m^*(B_k) \leq k\varepsilon.$$

Since $\varepsilon > 0$ is arbitrary, we obtain,

$$m^*(B_k) = 0, \ \forall k \in \mathbb{N}.$$

Finally,

$$m^*(A) = m^*(\cup_{k=1}^{\infty} B_k) \leq \sum_{k=1}^{\infty} m^*(B_k) = 0,$$

so that

$$m(A) = m^*(A) = 0.$$

The proof is complete.

9.5 Riemann Sums

Definition 9.5.1 Let $B_0 \subset \mathbb{R}^n$ be a compact block and let $f : B_0 \to \mathbb{R}$ be a bounded function.

Let P be a partition of B_0. Denote by B the sub-blocks of P.

Thus we define a Riemann sum of f relating P, denoted by R_f^P, by

$$R_f^P = \sum_{B \in P} f(\mathbf{x}_B) Vol(B),$$

for some $\mathbf{x}_B \in B$, $\forall B \in P$.

Definition 9.5.2 Let $B_0 \in \mathbb{R}^n$ be a compact block and let $f : B_0 \to \mathbb{R}$ be a bounded function. We say that I is the limit of R_f^P as $|P| \to 0$, if for each $\varepsilon > 0$, there exists $\delta > 0$ such that $0 < |P| < \delta$ then

$$|R_f^P - I| < \varepsilon,$$

for any Riemann sum of f relating P.

In such a case we denote:

$$\lim_{|P| \to 0} R_f^P = I.$$

Theorem 9.5.3 *Let $B_0 \in \mathbb{R}^n$ be a compact block and let $f : B_0 \to \mathbb{R}$ be a bounded function, that is, assume there exists $K > 0$ such that*

$$|f(\mathbf{x})| < K, \ \forall \mathbf{x} \in B_0.$$

Suppose f is Riemann integrable.
Under such hypotheses,

$$\lim_{|P| \to 0} R_f^P = I = \int_{B_0} f(\mathbf{x}) \, d\mathbf{x}.$$

Reciprocally, suppose there exists $I \in \mathbb{R}$ such that

$$\lim_{|P| \to 0} R_f^P = I.$$

Under such hypotheses, f is Riemann integrable, and

$$I = \int_{B_0} f(\mathbf{x}) \, d\mathbf{x}.$$

Proof Assume f is Riemann integrable. Let $\varepsilon > 0$. Thus there exists a partition P of B_0 such that

$$I - \frac{\varepsilon}{4} < s_f^P \le S_f^P < I + \frac{\varepsilon}{4}.$$

where

$$I = \int_{B_0} f(\mathbf{x}) \, d\mathbf{x}.$$

Choose $\delta > 0$ such that

$$0 < \delta < \min\left\{\frac{\varepsilon}{4KA_1}, \frac{|P|}{2}, 1\right\},$$

where A_1 denotes the sum of areas of all faces of all blocks of P. Let P_1 be a partition of B_0 such that $|P_1| < \delta$.

Let us denote generically, by B the blocks of P and by B' the blocks of P_1 Also, denote by B'_α the blocks of P_1 which are totally contained in some block B of P, and by B'_β the remaining blocks of P_1.

Thus,

$$S_f^P - s_f^P = \sum_{B \in P} (M_B - m_B) Vol(B)$$

$$\ge \sum_{B \in P} \left(\sum_{B'_\alpha \subset B} (M_{B'_\alpha} - m_{B'_\alpha}) Vol(B'_\alpha) \right)$$

$$= \sum_{B'_\alpha \in P_1} (M_{B'_\alpha} - m_{B'_\alpha}) Vol(B'_\alpha), \tag{9.20}$$

where we have denoted

$$M_B = \sup\{f(\mathbf{x}) \ : \ \mathbf{x} \in B\},$$

$$m_B = \inf\{f(\mathbf{x}) \ : \ \mathbf{x} \in B\},$$

$$M_{B_\alpha} = \sup\{f(\mathbf{x}) \ : \ \mathbf{x} \in B_\alpha\},$$

$$m_{B_\alpha} = \inf\{f(\mathbf{x}) \ : \ \mathbf{x} \in B_\alpha\}.$$

Therefore, we may write:

$$S_f^{P_1} - s_f^{P_1} = \sum_{B_\alpha' \in P_1} (M_{B_\alpha'} - m_{B_\alpha'}) Vol(B_\alpha')$$

$$+ \sum_{B_\beta' \in P_1} (M_{B_\beta'} - m_{B_\beta'}) Vol(B_\beta')$$

$$\leq S_f^P - s_f^P + \sum_{B_\beta' \in P_1} 2K Vol(B_\beta')$$

$$\leq S_f^P - s_f^P + 2K A_1 \delta$$

$$\leq \frac{\varepsilon}{2} + \frac{\varepsilon}{2}$$

$$= \varepsilon. \tag{9.21}$$

Hence

$$S_f^{P_1} - s_f^{P_1} < \varepsilon.$$

Finally, observe that,

$$s_f^{P_1} \leq R_f^{P_1} \leq S_f^{P_1},$$

and

$$-S_f^{P_1} \leq -I \leq -s_f^{P_1},$$

so that

$$-S_f^{P_1} + s_f^{P_1} \leq R_f^{P_1} - I \leq S_f^{P_1} - s_f^{P_1},$$

and thus,

$$|R_f^{P_1} - I| < S_f^{P_1} - s_f^{P_1} < \varepsilon, \forall \text{ partition } P_1 \text{ such that} |P_1| < \delta.$$

Therefore, we may write,

$$\lim_{|P_1| \to 0} R_f^{P_1} = I = \int_{B_0} f(\mathbf{x}) \, d\mathbf{x}.$$

Conversely, suppose

$$\lim_{|P| \to 0} R_f^P = I.$$

Suppose given a not relabeled $\varepsilon > 0$. Thus there exists $\delta > 0$ such that $0 < |P| < \delta$ then

$$I - \frac{\varepsilon}{2} < R_f^P < I + \frac{\varepsilon}{2},$$

for each Riemann sum relating P.

Choosing a particular P such that $0 < |P| < \delta$, since s_f^P and S_f^P are also arbitrarily close to Riemann sums, we have,

$$I - \frac{\varepsilon}{2} < s_f^P \leq \underline{I_1} \leq \overline{I_1} \leq S_f^P < I + \frac{\varepsilon}{2},$$

where

$$\underline{I_1} = \sup\{s_f^P \ : \ P \text{ is a partition of } B_0\},$$

and

$$\overline{I_1} = \inf\{S_f^P \ : \ P \text{ is partition of } B_0\},$$

Thus we have got,

$$\overline{I_1} - \underline{I_1} \leq \varepsilon,$$

and since $\varepsilon > 0$ is arbitrary, we obtain

$$\underline{I_1} = \overline{I_1} \equiv I_1 = \int_{B_0} f(\mathbf{x}) \, d\mathbf{x},$$

that is, f is Riemann integrable.

From this and above,

$$I - \frac{\varepsilon}{2} < I_1 < I + \frac{\varepsilon}{2},$$

that is,

$$|I - I_1| < \frac{\varepsilon}{2}.$$

Since $\varepsilon > 0$ is arbitrary, we have got,

$$I = I_1 = \int_{B_0} f(\mathbf{x}) \, d\mathbf{x}.$$

This completes the proof.

9.6 Applications, Integration in $\mathbb{R}^2$, and $\mathbb{R}^3$

In this section we develop in some detail the integration in $\mathbb{R}^2$ and $\mathbb{R}^3$.

9.6.1 Iterative Double Integration

Proposition 9.6.1 *Let $B_0 = [a, b] \times [c, d] \subset \mathbb{R}^2$ be a compact block, where $a < b$ and $c < d$.*
Let $f : B_0 \to \mathbb{R}$ a bounded function on B_0.
Let $P = P_1 \times P_2$ be a partition of B_0, where

$$P_1 = \{x_0 = a, x_1, \ldots, x_{m_1} = b\},$$

and

$$P_2 = \{y_0 = c, y_1, \ldots, y_{m_2} = d\}.$$

Select

$$y_j^* \in [y_{j-1}, y_j], \quad \forall j \in \{1, \ldots, m_2\},$$

and denote

$$B_{ij} = [x_{i-1}, x_i] \times [y_{j-1}, y_j],$$

$$m_{ij} = \inf\{f(x, y) \ : \ (x, y) \in B_{ij}\},$$

$$\tilde{m}_{ij} = \inf\{f(x, y_j^*) \ : \ (x, y_j^*) \in B_{ij}\},$$

$$M_{ij} = \sup\{f(x, y) \ : \ (x, y) \in B_{ij}\},$$

$$\tilde{M}_{ij} = \sup\{f(x, y_j^*) \ : \ (x, y_j^*) \in B_{ij}\},$$

and,

$$s_f^{P_1 \times P_2} = \sum_{i=1}^{m_1} \sum_{j=1}^{m_2} m_{ij} A(B_{i,j}),$$

$$\tilde{s}_f^{P_1 \times P_2} = \sum_{i=1}^{m_1} \sum_{j=1}^{m_2} \tilde{m}_{ij} A(B_{ij}),$$

$$S_f^{P_1 \times P_2} = \sum_{i=1}^{m_1} \sum_{j=1}^{m_2} M_{ij} A(B_{ij}),$$

$$\tilde{S}_f^{P_1 \times P_2} = \sum_{i=1}^{m_1} \sum_{j=1}^{m_2} \tilde{M}_{ij} A(B_{ij}),$$

and finally,

$$A(B_{ij}) = (x_i - x_{i-1})(y_i - y_{i-1}), \ \forall i \in \{1, \ldots, m_1\}, \ \forall j \in \{1, \ldots, m_2\}.$$

Under such hypotheses and notation,

$$s_f^{P_1 \times P} \leq \tilde{s}_f^{P_1 \times P_2} \leq \tilde{S}_f^{P_1 \times P_2} \leq S_f^{P_1 \times P_2}, \tag{9.22}$$

Proof Clearly

$$m_{ij} \leq \tilde{m}_{ij} \leq \tilde{M}_{ij} \leq M_{ij},$$

from which the result immediately follows.

Remark 9.6.2 We have already proven that

$$s_f^P \leq S_f^Q, \ \forall \ P, Q \ \text{partitions of } B_0.$$

In particular, concerning the last definitions and proposition, if P_a and P_b are partitions of $[a, b]$, and P_2 is a partition of $[c, d]$, then $P = P_a \times P_2$ and $Q = P_a \times P_2$ are partitions of $B_0 = [a, b] \times [c, d]$ so that

$$s_f^{P_a \times P_2} \leq S_f^{P_b \times P_2}.$$

Moreover, from this and from (9.22), we have:

$$s_f^{P_1 \times P_2} \leq \tilde{s}_f^{P_1 \times P_2}$$

$$\leq \sup_{P_1} \tilde{s}_f^{P_1 \times P_2}$$

$$\leq \sum_{j=1}^{m_2} \underline{\int_a^b} f(x, y_j^*) \, dx \, \Delta y_j$$

$$\leq \sum_{j=1}^{m_2} \overline{\int_a^b} f(x, y_j^*) \, dx \, \Delta y_j$$

$$= \inf_{P_1} \tilde{S}_f^{P_1 \times P_2}$$

$$\leq \tilde{S}_f^{P_1 \times P_2}$$

$$\leq S_f^{P_1 \times P_2}. \tag{9.23}$$

Theorem 9.6.3 *Let $f : B_0 = [a, b] \times [c, d] \to \mathbb{R}$ be a continuous function, where $a < b$ and $c < d$.*

Under such hypotheses

$$\int \int_{B_0} f(x, y) \, dx dy = \int_c^d \left[\int_a^b f(x, y) \, dx \right] dy$$

$$= \int_a^b \left[\int_c^d f(x, y) \, dy \right] dx. \tag{9.24}$$

Proof Let $\varepsilon > 0$.

Thus there exists a partition $P = P_1 \times P_2$ of B_0 such that

$$I - \frac{\varepsilon}{2} < s_f^{P_1 \times P_2} \leq S_f^{P_1 \times P_2} \leq I + \frac{\varepsilon}{2}.$$

Denote

$$P_2 = \{y_0 = c, y_1, \ldots, y_{m_2} = d\},$$

and choose

$$y_j^* \in [y_{j-1}, y_j], \ \forall j \in \{1, \ldots, m_2\}.$$

Also denote,

$$g(y) = \int_a^b f(x, y) \, dx.$$

Observe that $g = [c, d] \to \mathbb{R}$ is continuous (this follows from the uniform continuity of f on B_0) so that from above

$$I - \frac{\varepsilon}{2} \leq s_f^{P_1 \times P_2}$$

$$\leq \sum_{j=1}^{m_2} \int_a^b f(x, y_j^*) \, dx \Delta y_j$$

$$\leq \sum_{j=1}^{m_2} \overline{\int}_a^b f(x, y_j^*)\, dx\, \Delta y_j$$

$$\leq S_f^{P_1 \times P_2}$$

$$\leq I + \frac{\varepsilon}{2}. \qquad\qquad (9.25)$$

From this we obtain

$$\left| \sum_{j=1}^{m_2} \overline{\int}_a^b f(x, y_j^*)\, dx\, \Delta y_j - \sum_{j=1}^{m_2} \underline{\int}_a^b f(x, y_j^*)\, dx\, \Delta y_j \right| < \varepsilon.$$

Since $\varepsilon > 0$ is arbitrary, we have,

$$\sum_{j=1}^{m_2} \overline{\int}_a^b f(x, y_j^*)\, dx\, \Delta y_j = \sum_{j=1}^{m_2} \underline{\int}_a^b f(x, y_j^*)\, dx\, \Delta y_j \equiv \sum_{j=1}^{m_2} \int_a^b f(x, y_j^*)\, dx\, \Delta y_j \equiv R_g^{P_2}.$$

Thus we have obtained

$$|I - R_g^{P_2}| < \varepsilon,$$

and since $\{y_j^*\}$ is arbitrary, we obtain

$$|I - s_g^{P_2}| < \varepsilon$$

and

$$|I - S_g^{P_2}| < \varepsilon,$$

so that

$$I - \varepsilon \leq s_g^{P_2} \leq \int_c^d g(y)\, dy \leq S_g^{P_2} \leq I + \varepsilon,$$

which means,

$$\left| I - \int_c^d g(y)\, dy \right| < \varepsilon.$$

Since $\varepsilon > 0$ is arbitrary, we have got,

$$I = \iint_{B_0} f(x, y)\, dx dy = \int_c^d g(y)\, dy = \int_c^d \left[\int_a^b f(x, y)\, dx \right] dy.$$

Similarly, we may prove that

$$I = \int\int_{B_0} f(x, y)\, dxdy = \int_a^b \left[\int_c^d f(x, y)\, dy \right] dx.$$

The proof is complete.

9.6.2 Integration on More General Regions

Let $f : D \subset \mathbb{R}^2 \to \mathbb{R}$ be continuous functions, where,

$$D = \{(x, y) \in \mathbb{R}^2 \ : \ a \le x \le b \text{ and } y_1(x) \le y \le y_2(x)\},$$

where y_1 and $y_2 : [a, b] \to \mathbb{R}$ are piecewise smooth functions, so that

$$m(\partial D) = 0,$$

that is, the Lebesgue measure in $\mathbb{R}^2$ of the boundary ∂D of D is zero.

Let $[c, d] \subset \mathbb{R}$ be such that

$$D \subset [a, b] \times [c, d] \equiv B_0.$$

Define $\tilde{f} : B_0 \to \mathbb{R}$, by

$$\tilde{f}(x, y) = \begin{cases} f(x, y) & \text{if } (x, y) \in D \\ 0 & \text{if } (x, y) = B_0 \setminus D. \end{cases} \tag{9.26}$$

Thus, the set of discontinuities of $\tilde{f}$ is contained in ∂D, for which we have

$$m(\partial D) = 0.$$

We may conclude that $\tilde{f}$ is Riemann integrable, so that from the exposed in the last section. We may also define:

$$I = \int\int_D f(x, y)\, dxdy$$

$$\equiv \int\int_{B_0} \tilde{f}(x, y)\, dxdy$$

$$= \int_a^b \left[\int_c^d \tilde{f}(x, y)\, dy \right] dx$$

$$= \int_a^b \left[\int_{y=y_1(x)}^{y=y_2(x)} f(x, y) \, dy \right] dx$$

$$= \int_a^b \left[\int_{y=y_1(x)}^{y=y_2(x)} f(x, y) \, dy \right] dx. \tag{9.27}$$

The reason is because we may only infer that

$$\int \int_{B_0} \tilde{f}(x, y) \, dxdy = \int_a^b \left[\int_c^d \tilde{f}(x, y) \, dy \right] dy,$$

and not

$$\int \int_{B_0} \tilde{f}(x, y) \, dxdy = \int_a^b \left[\int_c^d \tilde{f}(x, y) \, dy \right] dx,$$

is because $\tilde{f}$ may not be continuous on ∂D.

9.7   Change of Variables in the Double Integral

Consider the integral

$$I = \int \int_D f(x, y) \, dxdy,$$

where D is a simple region of $\mathbb{R}^2$.

Consider also $C^1(D_0)$ class mappings $X : D_0 \subset \mathbb{R}^2 \to \mathbb{R}$ and $Y : D_0 \subset \mathbb{R}^2 \to$ \ and the change of variables given by:

$$x = X(u, v),$$

and

$$y = Y(u, v), \forall (u, v) \in D_0.$$

Let us denote,

$$\mathbf{r}(u, v) = (X(u, v), Y(u, v)),$$

where we assume $\mathbf{r} : D_0 \to D$ to be a bijection.

Let $u, v \in D_0$. Consider the elementary area $\Delta\tilde{A} = (u, u + \Delta u) \times (v, v + \Delta v)$, whose Lebesgue measure is

$$\Delta\tilde{A}(u, v) = \Delta u \Delta v.$$

Denote $\Delta A = \mathbf{r}(\Delta\tilde{A})$.
Observe that

$$\Delta r_u = \mathbf{r}(u + \Delta u, v) - \mathbf{r}(u, v) = \frac{\partial \mathbf{r}(u, v)}{\partial u}\Delta u + o(\Delta u) \approx \frac{\partial \mathbf{r}(u, v)}{\partial u}\Delta u,$$

and

$$\Delta r_v = \mathbf{r}(u, v + \Delta v) - \mathbf{r}(u, v) = \frac{\partial \mathbf{r}(u, v)}{\partial v}\Delta v + o(\Delta v) \approx \frac{\partial \mathbf{r}(u, v)}{\partial v}\Delta v,$$

so that denoting

$$\Delta A(u, v) = m(\Delta A),$$

we have

$$\Delta A(u, v) \approx |\Delta\tilde{r}_u \times \Delta\tilde{r}_v|,$$

where

$$\Delta\tilde{r}_u = (\Delta r_u, 0) \in \mathbb{R}^3,$$

and

$$\Delta\tilde{r}_v = (\Delta r_v, 0) \in \mathbb{R}^3,$$

so that

$$\Delta\tilde{r}_u = (X_u, Y_u, 0)\Delta u,$$

and

$$\Delta\tilde{r}_v = (X_v, Y_v, 0)\Delta v.$$

Thus,

$$\Delta\tilde{r}_u \times \Delta\tilde{r}_v = \begin{vmatrix} \mathbf{i} & \mathbf{j} & \mathbf{k} \\ X_u & Y_u & 0 \\ X_v & Y_v & 0 \end{vmatrix} \Delta u \Delta v \qquad (9.28)$$

and hence

$$\Delta\tilde{r}_u \times \Delta\tilde{r}_v = 0\mathbf{i} + 0\mathbf{j} + (X_u Y_v - X_v Y_u)\Delta u \Delta v \mathbf{k},$$

so that

$$\Delta A(u, v) \approx |\Delta\tilde{r}_u \times \Delta\tilde{r}_v| = |X_u Y_v - X_v Y_u|\Delta u \Delta v,$$

or in its differential form

$$dA(u, v) = |X_u Y_v - X_v Y_u|du dv = |J(u, v)|\, du dv,$$

where the Jacobian $J(u, v) = \frac{\partial(X,Y)}{\partial(u,v)} = \begin{vmatrix} X_u & Y_u \\ X_v & Y_v \end{vmatrix} = X_u Y_v - X_v Y_u.$

Finally, we may infer that

$$
\begin{aligned}
I &= \int\int_D f(x, y)\, dx dy \\
&= \int\int_{D_0} f(X(u, v), Y(u, v))\, dA(u, v) \\
&= \int\int_{D_0} f(X(u, v), Y(u, v))|J(u, v)|\, du dv.
\end{aligned}
\tag{9.29}
$$

9.7.1   A More Formal Proof of the Change of Variables Formula in $\mathbb{R}^2$

The content of this subsection may be found in a similar form in [2], where more details may be found.

Consider the integral

$$I = \int\int_D dx dy,$$

where D is a simple region of $\mathbb{R}^2$.

Consider also $C^1(D_0)$ class mappings $X : D_0 \subset \mathbb{R}^2 \to \mathbb{R}$ and $Y : D_0 \subset \mathbb{R}^2 \to \mathbb{R}$ and the change of variables given by

$$x = X(u, v),$$

and

$$y = Y(u, v), \forall (u, v) \in D_0.$$

We want to formally prove that

$$I = \int_D dx dy = \int_{D_0} |J(u, v)| \, du dv.$$

Observe that from the Green's formula with $Q(x, y) = x$ and $P(x, y) = 0$, we obtain

$$I = \int_D dx \, dy$$

$$= \int \int_D \left(\frac{\partial Q}{\partial x} - \frac{\partial P}{\partial y} \right) dx \, dy$$

$$= = \int_{\partial D} P \, dx + Q \, dy$$

$$= \int_{\partial D} x \, dy. \tag{9.30}$$

On the other hand

$$J(u, v) = \frac{\partial X}{\partial u} \frac{\partial Y}{\partial v} - \frac{\partial X}{\partial v} \frac{\partial Y}{\partial u}$$

$$= \frac{\partial X}{\partial u} \frac{\partial Y}{\partial v} + X \frac{\partial^2 Y}{\partial u \partial v}$$

$$- X \frac{\partial^2 Y}{\partial v \partial u} - \frac{\partial X}{\partial v} \frac{\partial Y}{\partial u}$$

$$= \frac{\partial}{\partial u} \left(X \frac{\partial Y}{\partial v} \right) - \frac{\partial}{\partial v} \left(X \frac{\partial Y}{\partial u} \right). \tag{9.31}$$

From the Green's Theorem with $\tilde{Q}(u, v) = X \frac{\partial Y}{\partial v}$ and $\tilde{P}(u, v) = X \frac{\partial Y}{\partial u}$, we obtain,

$$\int \int_{D_0} J(u, v) \, du \, dv = = \int \int_{D_0} \left(\frac{\partial \tilde{Q}}{\partial u} - \frac{\partial \tilde{P}}{\partial v} \right) du \, dv$$

$$= \int_{\partial D_0} \tilde{P} \, du + \tilde{Q} \, dv$$

$$= \int_{\partial D_0} X \frac{\partial Y}{\partial u} \, du + X \frac{\partial Y}{\partial v} \, dv. \tag{9.32}$$

Thus, it suffices to prove that,

$$\int_D x \, dy = \int_{\partial D_0} X \frac{\partial Y}{\partial u} \, du + X \frac{\partial Y}{\partial v} \, dv.$$

So, consider a parametrization $\mathbf{r}_0 : [a, b] \to \mathbb{R}^2$ of ∂D_0 that is,

$$\mathbf{r}_0(t) = U(t)\mathbf{i} + V(t)\mathbf{j},$$

and the corresponding parametrization $\mathbf{r} : [a, b] \to \mathbb{R}^2$ of ∂D, given by

$$\mathbf{r}(t) = X(U(t), V(t))\mathbf{i} + Y(U(t), V(t))\mathbf{j}.$$

Observe that

$$\mathbf{r}'(t) = \left(\frac{\partial X}{\partial u}U'(t) + \frac{\partial X}{\partial v}V'(t)\right)\mathbf{i} + \left(\frac{\partial Y}{\partial u}U'(t) + \frac{\partial Y}{\partial v}V'(t)\right)\mathbf{j},$$

so that

$$\int_{\partial D} x \, dy = \int_a^b X[U(t), V(t)] \left(\frac{\partial Y}{\partial u}U'(t) + \frac{\partial Y}{\partial v}V'(t)\right) dt$$

$$= \int_{\partial D_0} X\frac{\partial Y}{\partial u} \, du + X\frac{\partial Y}{\partial v} \, dv. \tag{9.33}$$

This completes the proof.

9.7.2   Integration in $\mathbb{R}^3$ (Triple Integration)

Let $B_0 = [a, b] \times [c, d] \times [e, f]$ be a compact block in $\mathbb{R}^3$.
Let $f : B_0 \to \mathbb{R}$ be a continuous function on B_0.
Under such hypothesis, similar to the case in $\mathbb{R}^2$, we may show that

$$\int\int\int_{B_0} f(x, y, z) \, dxdydz = \int_e^f \left[\int_c^d \left[\int_a^b f(x, y, z) \, dx\right] dy\right] dz,$$

and in particular

$$\int\int\int_{B_0} f(x, y, z) \, dxdydz = \int_a^b \left[\int_c^d \left[\int_e^f f(x, y, z) \, dz\right] dy\right] dx.$$

9.7.3   Triple Integral on More General Domains

Let $V \subset \mathbb{R}^3$ be a bounded, simply connected set, such that there exist C^1 class
functions $z_1, z_2 : D \subset \mathbb{R}^2 \to \mathbb{R}$ and such that D is a simple region of $\mathbb{R}^2$ (Fig. 9.1),

Fig. 9.1 A simple region V in $\mathbb{R}^3$

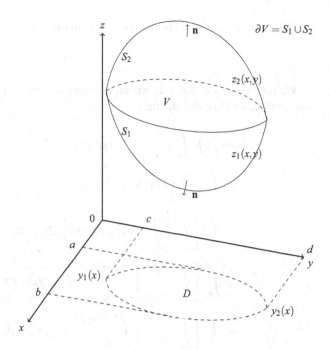

that is, there exist piecewise smooth functions $y_1, y_2 : [a, b] \to \mathbb{R}$ such that

$$D = \{(x, y) \in \mathbb{R}^2 : y_1(x) \le y \le y_2(x) \text{ and } a \le x \le b\},$$

so that

$$V = \{(x, y, z) \in \mathbb{R}^3 : z_1(x, y) \le z \le z_2(x, y) \text{ and } (x, y) \in D\},$$

that is,

$$V = \{(x, y, z) \in \mathbb{R}^3 : z_1(x, y) \le z \le z_2(x, y), \ y_1(x) \le y \le y_2(x), \ a \le x \le b\}.$$

Observe that we have assume $m(\partial V) = 0$, that is, the Lebesgue measure in $\mathbb{R}^3$ of the boundary of V is zero. Also, since V is bounded, there exists a compact block $[a, b] \times [c, d] \times [e, f]$ such that

$$V \subset B_0 \to \mathbb{R}.$$

Define $\tilde{f} : B_0 \to \mathbb{R}$, by

$$\tilde{f}(x, y, y) = \begin{cases} f(x, y, z) & \text{if } (x, y, z) \in V \\ 0 & \text{if } (x, y, z) = B_0 \setminus V. \end{cases} \tag{9.34}$$

Thus, the set of discontinuities of $\tilde{f}$ is contained in ∂V, for which we have

$$m(\partial V) = 0.$$

We may conclude that $\tilde{f}$ is Riemann integrable, so that from the exposed in the last section, we may also define

$$
\begin{aligned}
I &= \int \int \int_V f(x, y, z) \, dx dy dz \\
&\equiv \int \int \int_{B_0} \tilde{f}(x, y, z) \, dx dy dz \\
&= \int_a^b \left[\int_c^d \left[\int_e^f \tilde{f}(x, y, z) \, dz \right] dy \right] dx \\
&= \int_a^b \left[\int_{y=y_1(x)}^{y=y_2(x)} \left[\int_{z=z_1(x,y)}^{z=z_2(x,y)} f(x, y, z) \, dz \right] dy \right] dx \\
&= \int_a^b \left[\int_{y=y_1(x)}^{y=y_2(x)} \left[\int_{z=z_1(x,y)}^{z=z_2(x,y)} f(x, y, z) \, dz \right] dy \right] dx. \qquad (9.35)
\end{aligned}
$$

The reason because we may only infer that

$$\int \int \int_{B_0} \tilde{f}(x, y, z) \, dx dy dz = \int_a^b \left[\int_c^d \left[\int_e^f \tilde{f}(x, y, z) \, dz \right] dy \right] dy,$$

and not

$$\int \int \int_{B_0} \tilde{f}(x, y, z) \, dx dy dz = \int_a^b \left[\int_c^d \left[\int_e^f \tilde{f}(x, y, z) \, dz \right] dy \right] dx,$$

is because $\tilde{f}$ may not be continuous on ∂V.

9.8   Change of Variables in the Triple Integral

Consider the integral

$$I = \int \int \int_V f(x, y, z) \, dx dy dz,$$

where V is a simple region of $\mathbb{R}^3$.

Consider also $C^1(V_0)$ class mappings $X : V_0 \subset \mathbb{R}^3 \to \mathbb{R}$, $Y : V_0 \subset \mathbb{R}^3 \to \mathbb{R}$ and $Z : V_0 \subset \mathbb{R}^3 \to \mathbb{R}$, and the change of variables given by:

$$x = X(u, v, w), \quad y = Y(u, v, w), \quad z = Z(u, v, w), \quad \forall (u, v, w) \in V_0,$$

where we assume

$$\mathbf{r} : V_0 \to V$$

to be a bijection, where,

$$\mathbf{r}(u, v, w) = (X(u, v, w), Y(u, v, w), Z(u, v, w)).$$

Let $(u, v, w) \in V_0$. Consider the elementary volume $\Delta \tilde{V} = (u, u + \Delta u) \times (v, v + \Delta v) \times (w, w + \Delta w)$, whose Lebesgue measure is

$$\Delta \tilde{V}(u, v, w) = \Delta u \Delta v \Delta w.$$

Denote $\Delta V = \mathbf{r}(\Delta \tilde{V})$.
Observe that

$$\begin{aligned}
\Delta r_u &= \mathbf{r}(u + \Delta u, v, w) - \mathbf{r}(u, v, w) \\
&= \frac{\partial \mathbf{r}(u, v, w)}{\partial u} \Delta u + o(\Delta u) \\
&\approx \frac{\partial \mathbf{r}(u, v, w)}{\partial u} \Delta u,
\end{aligned} \tag{9.36}$$

$$\begin{aligned}
\Delta r_v &= \mathbf{r}(u, v + \Delta v, w) - \mathbf{r}(u, v, w) \\
&= \frac{\partial \mathbf{r}(u, v, w)}{\partial v} \Delta v + o(\Delta v) \\
&\approx \frac{\partial \mathbf{r}(u, v, w)}{\partial v} \Delta v,
\end{aligned} \tag{9.37}$$

and,

$$\begin{aligned}
\Delta r_w &= \mathbf{r}(u, v, w + \Delta w) - \mathbf{r}(u, v, w) \\
&= \frac{\partial \mathbf{r}(u, v, w)}{\partial w} \Delta w + o(\Delta w) \\
&\approx \frac{\partial \mathbf{r}(u, v, w)}{\partial w} \Delta w,
\end{aligned} \tag{9.38}$$

so that denoting

$$\Delta V(u, v, w) = m(\Delta V),$$

we have

$$\Delta V(u, v, w) \approx |\Delta r_u \cdot (\Delta r_v \times \Delta r_w)|,$$

where,

$$\Delta r_u = (X_u, Y_u, Z_u)\Delta u,$$

$$\Delta r_v = (X_v, Y_v, Z_v)\Delta v.$$

and

$$\Delta r_v = (X_w, Y_w, Z_w)\Delta w.$$

Thus,

$$(\Delta r_u \cdot (\Delta r_v \times \Delta r_w)) = \begin{vmatrix} X_u & Y_u & Z_u \\ X_v & Y_v & Z_v \\ X_w & Y_w & Z_w \end{vmatrix} \Delta u \Delta v \Delta w \qquad (9.39)$$

and hence

$$\Delta V(u, v, w) \approx |J(u, v, w)|\Delta u \Delta v \Delta w,$$

$$J(u, v, w) = \begin{vmatrix} X_u & Y_u & Z_u \\ X_v & Y_v & Z_v \\ X_w & Y_w & Z_w \end{vmatrix} \qquad (9.40)$$

or in its differential form

$$dV(u, v, w) = |J(u, v, w)| \, dudvdw.$$

Finally, we may infer that

$$I = \iiint_V f(x, y, z) \, dxdydz$$

$$= \iiint_{V_0} f(X(u, v, w), Y(u, v, w), Z(u, v, w)) \, dV(u, v, w)$$

$$= \iiint_{V_0} f(X(u, v, w), Y(u, v, w), Z(u, v, w))|J(u, v, w)| \, dudvdw. \quad (9.41)$$

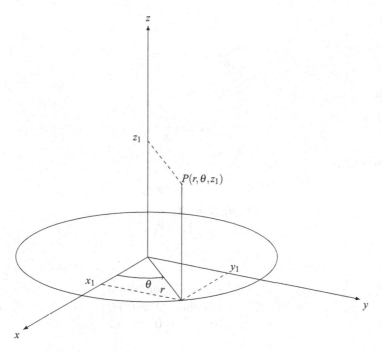

Fig. 9.2 Cylindrical coordinates (r, θ, z_1) for the point $P(x_1 = r\cos(\theta), y_1 = r\sin(\theta), z_1) \in \mathbb{R}^3$

9.8.1   Integration on $\mathbb{R}^3$ Through Cylindrical Coordinates

For the Cartesian coordinates $(x, y, z) \in \mathbb{R}^3$, consider the change of variables defined by,

$$x = X(r, \theta, z) = r\cos(\theta),$$

$$y = Y(r, \theta, z) = r\sin(\theta),$$

$$z = Z(r, \theta, z) = z.$$

Thus, (r, θ, z) are the cylindrical coordinates associated with (x, y, z) (Fig. 9.2). Observe that

$$r^2[\cos(\theta)]^2 + r^2[\sin(\theta)]^2 = x^2 + y^2,$$

so that

$$r = \sqrt{x^2 + y^2}.$$

Also,

$$\frac{\sin(\theta)}{\cos(\theta)} = \tan(\theta) = \frac{y}{x},$$

so that,

$$\theta = \arctan\left(\frac{y}{x}\right).$$

Observe that

$$J(r, \theta, z) = \begin{vmatrix} X_r & Y_r & Z_r \\ X_\theta & Y_\theta & Z_\theta \\ X_z & Y_z & Z_z \end{vmatrix} \tag{9.42}$$

so that

$$= J(r, \theta, z) = \begin{vmatrix} \cos(\theta) & \sin(\theta) & 0 \\ -r\sin(\theta) & r\cos(\theta) & 0 \\ 0 & 0 & 1 \end{vmatrix} = r \tag{9.43}$$

that is,

$$J(r, \theta, z) = r$$

Thus, in cylindrical coordinates, we have,

$$I = \int \int \int_V f(x, y, z)\, dx\, dy\, dz$$

$$= \int \int \int_{V_0} f(X(u, v, w), Y(u, v, w), Z(u, v, w))\, dV(u, v, w)$$

$$= \int \int \int_{V_0} f(X(u, v, w), Y(u, v, w), Z(u, v, w))|J(u, v, w)|\, du\, dv\, dw$$

$$= \int \int \int_{V_0} f(r\cos(\theta), r\sin(\theta), z)\, r\, dr\, d\theta\, dz. \tag{9.44}$$

Fig. 9.3 Spherical
coordinates (r, θ, ϕ)

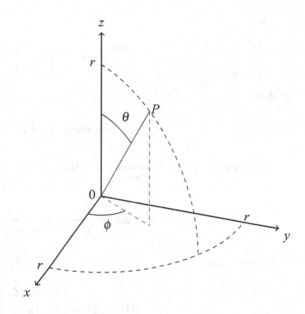

9.8.2 Integration on $\mathbb{R}^3$ in Spherical Coordinates

For the Cartesian coordinates of a point $P = (x, y, z) \in \mathbb{R}^3$, consider the change of variables defined by,

$$x = X(r, \theta, \phi) = r \sin(\theta) \cos(\phi),$$

$$y = Y(r, \theta, \phi) = r \sin(\theta) \sin(\phi),$$

$$z = Z(r, \theta, \phi) = r \cos(\theta).$$

Thus, (r, θ, ϕ) are the spherical coordinates associated with $P = (x, y, z)$, where $0 \leq \theta \leq \pi$ and $0 \leq \phi \leq 2\pi$ (Fig. 9.3).
Observe that

$$r^2 = r^2[\sin(\theta)]^2[\cos(\theta)]^2 + r^2[\sin(\theta)]^2[\sin(\phi)]^2 + r^2[\cos(\theta)]^2 = x^2 + y^2 + y^2,$$

so that

$$r = \sqrt{x^2 + y^2 + z^2}.$$

Also,

$$\frac{\sin(\phi)}{\cos(\phi)} = \tan(\phi) = \frac{y}{x},$$

so that,

$$\phi = \arctan\left(\frac{y}{x}\right).$$

Also, $\cos(\theta) = \frac{z}{r} = \frac{z}{\sqrt{x^2+y^2+z^2}}$, so that,

$$\theta = \arccos\left(\frac{z}{\sqrt{x^2 + y^2 + z^2}}\right).$$

Observe that

$$J(r, \theta, \phi) = \begin{vmatrix} X_r & Y_r & Z_r \\ X_\theta & Y_\theta & Z_\theta \\ X_\phi & Y_\phi & Z_\phi \end{vmatrix} \tag{9.45}$$

so that

$$= J(r, \theta, \phi) = \begin{vmatrix} \sin(\theta)\cos(\phi) & \sin(\theta)\sin(\phi) & \cos(\phi) \\ r\cos(\theta)\cos(\phi) & r\cos(\theta)\sin(\phi) & -r\sin(\theta) \\ -r\sin(\theta)\sin(\phi) & r\sin(\theta)\cos(\phi) & 0 \end{vmatrix} = r^2\sin\theta \tag{9.46}$$

that is,

$$J(r, \theta, z) = r^2\sin(\theta)$$

Thus, in spherical coordinates, we have,

$$I = \int\int\int_V f(x, y, z)\, dx\, dy\, dz$$

$$= \int\int\int_{V_0} f(X(u, v, w), Y(u, v, w), Z(u, v, w))\, dV(u, v, w)$$

$$= \int\int\int_{V_0} f(X(u, v, w), Y(u, v, w), Z(u, v, w))|J(u, v, w)|\, du\, dv\, dw$$

$$= \int\int\int_{V_0} f(r\sin(\theta)\cos(\phi), r\sin(\theta)\sin(\phi), r\cos(\theta))\, r^2\,\sin(\theta)\, dr\, d\theta\, d\phi.$$

Exercises 9.8.1

1. Let $B_0 \subset \mathbb{R}^n$ be a nonempty compact block and let $f : B_0 \to \mathbb{R}$ be such that

$$f(\mathbf{x}) = f(x_1, \ldots, x_n) = \begin{cases} 0, & \text{if } x_j \in \mathbb{Q}, \ \forall j \in \{1, \ldots, n\}, \\ 1, & \text{if } x_j \in \mathbb{R} \setminus \mathbb{Q}, \ \text{for some } j \in \{1, \ldots, n\}. \end{cases}$$

   Show that f is not Riemann integrable on B_0.

2. Let $B_0 \subset \mathbb{R}^n$ be a nonempty compact block. Define $A = \{x_1, \ldots, x_k\} \subset B_0$ and let $f : B_0 \to \mathbb{R}$ be such that

$$f(\mathbf{x}) = \begin{cases} 0, & \text{if } \mathbf{x} \in A, \\ 1, & \text{if } \mathbf{x} \in B_0 \setminus A \end{cases}$$

   Show that f is Riemann integrable on B_0 and calculate

$$\int_{B_0} f(\mathbf{x}) \, d\mathbf{x}.$$

3. Let $B_0 \subset \mathbb{R}^n$ be a nonempty compact block. Let $A = \{x_m\}_{m \in \mathbb{N}} \subset B_0$ be a sequence such that $\lim_{m \to \infty} x_m = x_0 \in \mathbb{R}^n$.
   Define $B = A \cup \{x_0\}$ and let $f : B_0 \to \mathbb{R}$ be such that

$$f(\mathbf{x}) = \begin{cases} 0, & \text{if } \mathbf{x} \in B, \\ 1, & \text{if } \mathbf{x} \in B_0 \setminus B \end{cases}$$

   Show that f is Riemann integrable on B_0 and calculate

$$\int_{B_0} f(\mathbf{x}) \, d\mathbf{x}.$$

4. Let $D \subset \mathbb{R}^3$ be an open nonempty set. Let $f : D \to \mathbb{R}$ be a continuous function.
   Assume

$$\int_V f(\mathbf{x}) \, d\mathbf{x} = 0$$

   for each compact set $V \subset D$ such that ∂V is of C^1 class.
   Under such hypotheses, show that

$$f(\mathbf{x}) = 0, \ \forall \mathbf{x} \in D.$$

5. Let $D \subset \mathbb{R}^n$ be an open, bounded, nonempty set which the boundary ∂D is such that $m(\partial D) = 0$. Let $f : \overline{D} \to \mathbb{R}$ be a continuous function on D, such

that

$$f(\mathbf{x}) \geq 0, \ \forall \mathbf{x} \in D.$$

Suppose

$$\int_{\underline{D}} f(\mathbf{x}) \, d\mathbf{x} = 0.$$

Under such hypotheses, show that

$$f(\mathbf{x}) = 0, \ \forall \mathbf{x} \in D.$$

6. Let $B_0 \subset \mathbb{R}^n$ be a nonempty compact block and let $f : B_0 \to \mathbb{R}$ be a bounded function. Let $I_1 \in \mathbb{R}$ be such that for each $\varepsilon > 0$ there exists a partition P_ε of B_0 such that

$$|s_f^P - I_1| < \varepsilon,$$

for all partition P such that $P_\varepsilon \subset P$.
   Show that

$$I_1 = \underline{I},$$

where $\underline{I}$ denotes the lower integral of f on B_0.

7. Let $A, B_0 \subset \mathbb{R}^n$ be blocks. Suppose that $A \subset B_0$ and f is Riemann integrable on B_0.
   Prove that a restriction f to A, denoted by $f|_A$, is Riemann integrable on A.

8. Let B_0 be a nonempty compact block and let $f : B_0 \to \mathbb{R}$ be a Riemann integrable function on B_0.
   Show that the set

$$G_f = \{(\mathbf{x}, f(\mathbf{x})) \ : \ \mathbf{x} \in B_0\},$$

has zero measure in $\mathbb{R}^{n+1}$.

9. Let B_0 be a nonempty compact block. Show that each Riemann integrable function $f : B_0 \to \mathbb{R}$ may be written as the difference of two nonnegative Riemann integrable functions.

10. Let $D \subset \mathbb{R}^n$ be a compact path connected set. More specifically, assume that for each $\mathbf{A}, \mathbf{B} \in D$ there exists a piecewise smooth continuous function $r :$ $[a, b] \to D$ such that $\mathbf{r}(a) = \mathbf{A}$ and $\mathbf{r}(b) = \mathbf{B}$. Let $f : D \to \mathbb{R}$ be a continuous function on D. Let $\gamma \in [\alpha, \beta]$, where

$$\alpha = \min\{f(\mathbf{x}) \ : \ \mathbf{x} \in D\}$$

and

$$\beta = \max\{f(\mathbf{x}) \ : \ \mathbf{x} \in D\}.$$

Prove that there exists $\mathbf{x}_0 \in D$ such that

$$f(\mathbf{x}_0) = \gamma.$$

11. Let $f : D \subset \mathbb{R}^n \to \mathbb{R}$ be a continuous function on D, where such a set is compact and path connected, as specified in the last exercise. Suppose ∂D is of C^1 class. Show that there exists $\mathbf{x}_0 \in D$ such that

$$\int_D f(\mathbf{x}) \, d\mathbf{x} = f(\mathbf{x}_0) \, Vol(D).$$

12. Let $f : D \subset \mathbb{R}^n \to \mathbb{R}$ be a continuous function where D is an open set. Let $\mathbf{x}_0 \in D$. Show that

$$\lim_{r \to 0} \frac{\int_{B_r(\mathbf{x}_0)} f(\mathbf{x}) \, d\mathbf{x}}{Vol(B_r(\mathbf{x}_0))} = f(\mathbf{x}_0).$$

13. Calculate

$$I = \int \int_D xy \sqrt[4]{\left(\frac{x^4}{9} + \frac{y^4}{25}\right)} \, dxdy,$$

where

$$D = \{(x, y) \in \mathbb{R}^2 \ : \ x^4/9 + y^4/25 \le 1, \ x \ge 0, \ y \ge 0\}.$$

14. Calculate

$$I = \int \int_D e^{\sqrt{2x^2 + 3y^2}} \, dxdy,$$

where

$$D = \{(x, y) \in \mathbb{R}^2 \ : \ 2x^2 + 3y^2 \le 1\}.$$

15. Find the volume of the solid in the first octant bounded by the plans $x = 0$, $y = 0$, $z = 0$, $z = x$ and by the surface $z = 1 - y^2$.

16. Calculate

$$I = \int \int \int_D x^2 \, dxdydz,$$

where

$$D = \{(x, y, z) \in \mathbb{R}^3 \; : \; x^2/a^2 + y^2/b^2 + z^2/c^2 \le 1\}.$$

17. Find the volume of the solid corresponding to the intersection of spheres $x^2 + y^2 + z^2 \le R^2$ and $x^2 + y^2 + z^2 \le 2Rz$, where $R > 0$.

18. Let $A, B \subset \mathbb{R}^n$ be nonempty compact blocks. Let $f : A \to \mathbb{R}$ and $g : B \to \mathbb{R}$ nonnegative bounded functions. Define $\varphi : A \times B \to \mathbb{R}$ by

$$\varphi(\mathbf{x}, \mathbf{y}) = f(\mathbf{x}) \cdot g(\mathbf{y}).$$

Prove that

$$\overline{\int}_{A \times B} \varphi(\mathbf{z}) \, d\mathbf{z} = \overline{\int}_{A} f(\mathbf{x}) \, d\mathbf{x} \cdot \overline{\int}_{B} g(\mathbf{y}) \, d\mathbf{y}.$$

19. Let $D \subset \mathbb{R}^n$ be a nonempty compact set such that $m(\partial D) = 0$.
    Let $f, g : D \to \mathbb{R}$ be Riemann integrable functions. Prove that

$$\left(\int_D f(\mathbf{x}) g(\mathbf{x}) \, d\mathbf{x} \right)^2 \le \int_D f(\mathbf{x})^2 \, d\mathbf{x} \cdot \int_D g(\mathbf{x})^2 \, d\mathbf{x}.$$

20. Let $A \subset \mathbb{R}^n$ be a nonempty compact block. Let $f : A \to \mathbb{R}$ be a function such that

$$0 \le f(\mathbf{x}) \le M, \forall \mathbf{x} \in A.$$

Consider the set

$$C(f) = \{(\mathbf{x}, y) \in A \times [0, M] \; : \; 0 \le y \le f(\mathbf{x})\}.$$

Show that

$$\overline{\int}_{A} f(\mathbf{x}) \, d\mathbf{x} = \overline{\int}_{A \times M} \chi_{C(f)}(\mathbf{x}, y) \, d\mathbf{x} \, dy,$$

where

$$\chi_{C(f)}(\mathbf{x}, y) = \begin{cases} 1, & \text{if } (\mathbf{x}, y) \in C(f) \\[2mm] 0, & \text{if } (\mathbf{x}, y) \notin C(f). \end{cases} \tag{9.47}$$

Establish an analogous result for the lower integral and conclude, justifying your answer, that a necessary and sufficient condition for the integrability of f is that $m(\partial C(f)) = 0$ in $\mathbb{R}^{n+1}$.

21. For $a > 0$ let $B_n(a)$ be the volume of region

$$\{(x_1, x_2, \ldots, x_n\} \in \mathbb{R}^n \mid |x_1| + |x_2| + \cdots + |x_n| \le a\}.$$

(a) Prove that

$$B_n(a) = a^n B_n(1).$$

(b) Prove that

$$B_n(1) = \frac{2}{n} B_{n-1}(1).$$

Hint: Observe that

$$|x_1| + \cdots + |x_n| \le 1, \text{ if and only if, } |x_1| + \cdots + |x_{n-1}| \le 1 - |x_n| \text{ and } |x_n| \le 1\}.$$

Thus

$$B_n(1) = \int_{-1}^{1} \left(\int \cdots \int_{|x_1| + \cdots + |x_{n-1}| \le 1 - |x_n|} dx_1 \cdots dx_{n-1} \right) dx_n.$$

(c) Prove that

$$B_n(1) = \frac{2^n}{n!}$$

and that

$$\lim_{n \to \infty} B_n(1) = 0.$$

Chapter 10
Topics on Vector Calculus and Vector Analysis in $\mathbb{R}^n$

In this chapter we address the main definitions and results for vector analysis in $\mathbb{R}^n$. This part of the text comprises topics such as differential forms in surfaces in $\mathbb{R}^n$, including the volume form for a surface in $\mathbb{R}^n$ and the Green, Gauss, and Stokes theorems in both standard calculus and abstract versions. Indeed we develop an abstract version of the Stokes theorem and recover the classical Divergence and Stokes theorems from such a general approach. We finish the chapter with an introduction to Riemannian geometry.

The main references for this chapter are [1, 2, 5, 10, 12, 14].

We start by recalling some basic definitions.

10.1 Curves in $\mathbb{R}^3$

Definition 10.1.1 (Curve in $\mathbb{R}^3$) A smooth curve in $\mathbb{R}^3$ is defined by the graph of a vectorial one variable function

$$\mathbf{r} : [a, b] \to \mathbb{R}^3,$$

where

$$\mathbf{r}(t) = x(t)\mathbf{i} + y(t)\mathbf{j} + z(t)\mathbf{k}$$

is such that x', y' and z' are continuous on $[a, b]$.

Definition 10.1.2 (Piecewise Smooth Curve) Let $\mathbf{r} : [a, b] \to \mathbb{R}^3$ be a continuous function on $[a, b]$. Assume there exists a partition $P = \{t_0 = a, t_1, \ldots, t_m = b\}$ of $[a, b]$, such that $\mathbf{r}'(t)$ is continuous on (t_{i-1}, t_i), $\forall i \in \{1, \ldots, m\}$. Moreover,

© Springer International Publishing AG, part of Springer Nature 2018
F. S. Botelho, *Real Analysis and Applications*,
https://doi.org/10.1007/978-3-319-78631-5_10

assume that $\mathbf{r}'(a+)$, $\mathbf{r}'(b-)$ and $\mathbf{r}'(t\pm)$ exists on (a, b). In such a case we say that $\mathbf{r}$ is a piecewise smooth vectorial one variable function whose graph defines a piecewise smooth curve.

Finally, here we have denoted,

$$\mathbf{r}'(a+) = \lim_{h \to 0^+} \frac{\mathbf{r}(a+h) - \mathbf{r}(a)}{h},$$

$$\mathbf{r}'(b-) = \lim_{h \to 0^-} \frac{\mathbf{r}(b+h) - \mathbf{r}(b)}{h},$$

$$\mathbf{r}'(t+) = \lim_{h \to 0^+} \frac{\mathbf{r}(t+h) - \mathbf{r}(t)}{h}, \ \forall t \in (a, b),$$

$$\mathbf{r}'(t-) = \lim_{h \to 0^-} \frac{\mathbf{r}(t+h) - \mathbf{r}(t)}{h}, \ \forall t \in (a, b).$$

Remark 10.1.3 Similarly we may define a piecewise smooth curve in $\mathbb{R}^n$, just replacing $\mathbb{R}^3$ by $\mathbb{R}^n$ in the last definition.

10.2   Line Integrals

Definition 10.2.1 (Line Integral) Let $\mathbf{F} : D \subset \mathbb{R}^3 \to \mathbb{R}^3$ be a continuous vectorial function and let C be a piecewise smooth curve defined by $\mathbf{r} : [a, b] \to \mathbb{R}^3$.

We define the line integral of $\mathbf{F}$ over C, denoted by

$$I = \int_C \mathbf{F} \cdot d\mathbf{r},$$

as

$$I = \int_C \mathbf{F} \cdot d\mathbf{r} = \int_a^b \mathbf{F}(\mathbf{r}(t)) \cdot \mathbf{r}'(t) \, dt.$$

Example 10.2.2 Let $\mathbf{F} : \mathbb{R}^3 \to \mathbb{R}^3$ be given by

$$\mathbf{F}(x, y, z) = x^2 \mathbf{i} + y^3 \mathbf{j} + z \mathbf{k},$$

and $\mathbf{r} : [0, 1] \to \mathbb{R}^3$ be given by

$$\mathbf{r}(t) = t^3 \mathbf{i} + t \mathbf{j} + t^2 \mathbf{k},$$

so that

$$\mathbf{r}(t) = x(t)\mathbf{i} + y(t)\mathbf{j} + z(t)\mathbf{k}$$

where

$$x(t) = t^3, \ y(t) = t \text{ and } z(t) = t^2.$$

Thus

$$d\mathbf{r}(t) = \mathbf{r}'(t) \, dt = 3t^2 \, dt\mathbf{i} + dt\mathbf{j} + 2t \, dt\mathbf{k},$$

so that

$$
\begin{aligned}
I &= \int_C \mathbf{F} \cdot d\mathbf{r} \\
&= \int_0^1 \mathbf{F}(\mathbf{r}(t)) \cdot \mathbf{r}'(t) \, dt \\
&= \int_0^1 \mathbf{F}(x(t), y(t), z(t)) \cdot \mathbf{r}'(t) \, dt \\
&= \int_0^1 [(t^3)^2\mathbf{i} + t^3\mathbf{j} + t^2\mathbf{k}] \cdot [3t^2 \, dt\mathbf{i} + dt\mathbf{j} + 2t \, dt\mathbf{k}] \\
&= \int_0^1 [3t^8 + t^3 + 2t^3] \, dt \\
&= [3t^9/9 + t^4/4 + 2t^4/4]_0^1 \\
&= 3/9 + 1/4 + 2/4 \\
&= 1/3 + 1/4 + 1/2 \\
&= (4 + 3 + 6)/12 \\
&= 13/12.
\end{aligned}
\tag{10.1}
$$

10.3 On the Green Theorem in $\mathbb{R}^2$

Definition 10.3.1 (Path Connected Sets in $\mathbb{R}^n$ and Simple Sets in $\mathbb{R}^2$) An open set $V \subset \mathbb{R}^n$ is said to be (path) connected if for which $\mathbf{A}$ and $\mathbf{B} \in V$ such that $\mathbf{A} \neq \mathbf{B}$, there exists a one variable vectorial piecewise smooth function $\mathbf{r} : [a, b] \to V$ such that $\mathbf{r}(a) = \mathbf{A}$ and $\mathbf{r}(b) = \mathbf{B}$.

An open set $V \subset \mathbb{R}^n$ is said to be simply connected if it is connected and, for each piecewise smooth closed curve C such that $C \subset V$, we have that the region interior to C is all contained in V. We could say that a simply connected set is a connected one which has no holes.

If a set $V \subset \mathbb{R}^n$ is connected but not simply connected, it is said to be multiply connected. We could say that a multiply connected set is a connected one with one or more holes.

Finally, we say that a region $D \subset \mathbb{R}^2$ is simple if $\circ D$ is simply connected and there exists piecewise smooth functions, y_1, y_2, x_1, x_2 and $a, b, c, d \in \mathbb{R}$ such that

$$D = \{(x, y) \in \mathbb{R}^2 : a \le x \le b, \text{ and } y_1(x) \le y \le y_2(x)\},$$

and, at the same time,

$$D = \{(x, y) \in \mathbb{R}^2 : c \le y \le d, \text{ and } x_1(y) \le x \le x_2(y)\}.$$

Theorem 10.3.2 (Green Theorem in the Plane) *Let $D \subset \mathbb{R}^2$ be a simple region in the plane and let $P, Q : D \subset \mathbb{R}^2 \to \mathbb{R}$ be C^1 class functions on D, that is, continuous and with derivatives of order up to one continuous on D.*

Under such hypotheses,

$$\int\int_D \left(\frac{\partial Q}{\partial x} - \frac{\partial P}{\partial y} \right) dx \, dy = \oint_{\partial D} P \, dx + Q \, dy,$$

where the line integrals on closed curves, from now and on, must be always considered in the counter-clockwise sense, unless otherwise indicated.

Proof Observe that

$$D = \{(x, y) \in \mathbb{R}^2 : a \le x \le b \text{ and } y_1(x) \le y \le y_2(x)\}.$$

Let us denote $\partial D = \partial D_1 \cup \partial D_2$, where ∂D_1 corresponds to the boundary portion relating y_1 and ∂D_2 corresponds to the boundary portion relating y_2.

Define,

$$I_1 = \int\int_D \frac{\partial P}{\partial y} \, dx \, dy,$$

so that

$$I_1 = \int_{x=a}^{x=b} \left[\int_{y=y_1(x)}^{y=y_2(x)} \frac{\partial P}{\partial y} \, dy \right] dx$$

$$= \int_{x=a}^{x=b} [P(x, y)]_{y=y_1(x)}^{y=y_2(x)} \, dx$$

$$= \int_a^b P(x, y_2(x)) \, dx - \int_a^b P(x, y_1(x)) \, dx$$

$$= -\int_{\partial D_2} P\, dx - \int_{\partial D_1} P\, dx$$

$$= -\oint_{\partial D_1 \cup \partial D_2} P\, dx$$

$$= -\oint_{\partial D} P\, dx. \tag{10.2}$$

Summarizing,

$$I_1 = \int\int_D \frac{\partial P}{\partial y}\, dx\, dy = -\oint_{\partial D} P\, dx.$$

Similarly, we may show that

$$\int\int_D \frac{\partial Q}{\partial x}\, dx\, dy = \oint_{\partial D} Q\, dy,$$

so that

$$\int\int_D \left(\frac{\partial Q}{\partial x} - \frac{\partial P}{\partial y}\right) dx\, dy = \oint_{\partial D} P\, dx + Q\, dy.$$

This completes the proof.

Remark 10.3.3 At this point, we highlight the Green Theorem may be easily extended for more general regions, such as those which comprise unions of simple regions. The idea is just to apply the previous result to each simple part. The next step is to add the partial results to obtain the final one.

10.4 Exact Differential Forms

Definition 10.4.1 Let $P, Q, R : D \subset \mathbb{R}^3$, where D is an open set.
We say that the differential form

$$P\, dx + Q\, dy + R\, dz$$

is exact, if there exists a function $U : D \to \mathbb{R}$ of C^1 class such that

$$P = \frac{\partial U}{\partial x}, \quad Q = \frac{\partial U}{\partial y} \quad R = \frac{\partial U}{\partial z}.$$

In such a case we denote,

$$dU = P\,dx + Q\,dy + R\,dz$$
$$= \frac{\partial U}{dx}\,dx + \frac{\partial U}{dy}\,dy + \frac{\partial U}{dz}\,dz. \tag{10.3}$$

Also, the function U in question is said to be the potential associated with the field

$$\mathbf{F} = P\mathbf{i} + Q\mathbf{j} + R\mathbf{k}.$$

Theorem 10.4.2 *Let $P, Q, R : D \rightarrow \mathbb{R}$ be continuous functions, where D is an open connected set. Under such hypotheses, the following conditions are equivalent:*

1. *The form $P\,dx + Q\,dy + R\,dz$ is exact.*
2. *The line integral*

$$\int_C P\,dx + Q\,dy + R\,dz,$$

   where C connects the points $\mathbf{A}$ and $\mathbf{B} \in D$ depends only on such initial and final points $\mathbf{A}$ and $\mathbf{B}$, but not on the piecewise smooth curve C connecting them.
3. *We have*

$$\oint_C P\,dx + Q\,dy + R\,dz = 0,$$

   for all every piecewise smooth closed curve C contained in D.

Proof

- 1 implies 2:
  Suppose there exists $U : D \subset R^3 \rightarrow \mathbb{R}$ of C^1 class such that:

$$P = \frac{\partial U}{\partial x}, \quad Q = \frac{\partial U}{\partial y} \quad R = \frac{\partial U}{\partial z}.$$

  Let $\mathbf{A}$ and $\mathbf{B} \in D$. Let C be a piecewise smooth curve connecting $\mathbf{A}$ and $\mathbf{B}$. Suppose that C is defined by $\mathbf{r} : [a, b] \rightarrow \mathbb{R}^3$ where,

$$\mathbf{r}(t) = x(t)\mathbf{i} + y(t)\mathbf{j} + z(t)\mathbf{k},$$

  and such that $\mathbf{r}(a) = \mathbf{A}$ and $\mathbf{r}(b) = \mathbf{B}$.
  Let $t_1, t_2, \ldots, t_{m-1}$ be the points where $\mathbf{r}'$ is not continuous and denote $t_0 = a$, $t_m = b$.

Thus,

$$I = \int_C P \, dx + Q \, dy + R \, dz$$

$$= \sum_{i=0}^{m-1} \int_{t_i}^{t_{i+1}} \left(\frac{\partial U}{\partial x} \, dx + \frac{\partial U}{\partial y} \, dy + \frac{\partial U}{\partial z} \, dz \right)$$

$$= \sum_{i=0}^{m-1} \int_{t_i}^{t_{i+1}} \left(\frac{\partial U(\mathbf{r}(t))}{\partial x} \, x'(t) \, dt + \frac{\partial U(\mathbf{r}(t))}{\partial y} \, y'(t) \, dt + \frac{\partial U(\mathbf{r}(t))}{\partial z} \, z'(t) \, dt \right)$$

$$= \sum_{i=0}^{m-1} \int_{t_i}^{t_{i+1}} \frac{dU(\mathbf{r}(t))}{dt} \, dt$$

$$= \sum_{i=0}^{m-1} [U(\mathbf{r}(t))]_{t_i}^{t_{i+1}}$$

$$= \sum_{i=0}^{m-1} U(\mathbf{r}(t_{i+1})) - U(\mathbf{r}(t_i))$$

$$= U(\mathbf{r}(b)) - U(\mathbf{r}(a))$$

$$= U(\mathbf{B}) - U(\mathbf{A}). \tag{10.4}$$

- 2 implies 1:

  Suppose the line integral on D depends only on the initial point $\mathbf{A}$ and final point $\mathbf{B}$, but not on the piecewise smooth curve connecting them.

  Choose $\mathbf{A} \in D$.

  Let $\mathbf{P} \in D$, and since the integral in question does not depend on the piecewise smooth curve connecting $\mathbf{A}$ and $\mathbf{P}$, we may define

$$U(\mathbf{P}) = \int_{\mathbf{A}}^{\mathbf{P}} P \, dx + Q \, dy + R \, dz.$$

Thus, denoting $\mathbf{P} = (x, y, z)$ and $\mathbf{P}_1 = (x + \Delta x, y, z)$ we get,

$$\frac{U(\mathbf{P}_1) - U(\mathbf{P})}{\Delta x}$$

$$= \frac{\int_{\mathbf{A}}^{\mathbf{P}_1}(P \, dx + Q \, dy + R \, dz) - \int_{\mathbf{A}}^{\mathbf{P}}(P \, dx + Q \, dy + R \, dz)}{\Delta x}$$

$$= \frac{\int_{\mathbf{A}}^{\mathbf{P}}(P \, dx + Q \, dy + R \, dz) + \int_{\mathbf{P}}^{\mathbf{P}_1}(P \, dx + Q \, dy + R \, dz) - \int_{\mathbf{A}}^{\mathbf{P}}(P \, dx + Q \, dy + R \, dz)}{\Delta x}$$

$$= \frac{\int_{\mathbf{P}}^{\mathbf{P}_1}(P \, dx + Q \, dy + R \, dz)}{\Delta x}. \tag{10.5}$$

Since D is open, we may consider, for a sufficiently small Δx, the path between $\mathbf{P}$ and $\mathbf{P}_1$, defined by $\mathbf{r}_1 : [0, 1] \to \mathbb{R}^3$, given by:

$$\mathbf{r}_1(t) = (1 - t)\mathbf{P} + t\mathbf{P}_1$$
$$= (x + t\Delta x, y, z), \tag{10.6}$$

so that from the mean value theorem for integrals, we have

$$\frac{U(\mathbf{P}_1) - U(\mathbf{P})}{\Delta x}$$
$$= \frac{\int_0^1 P(x + t\Delta x)\Delta x \, dt}{\Delta x}$$
$$= P(x + \tilde{t}\Delta x), \tag{10.7}$$

for some $\tilde{t} \in (0, 1)$.

Hence, letting $\Delta x \to 0$, we obtain,

$$\frac{U(x + \Delta x, y, z) - U(x, y, z)}{\Delta x}$$
$$= \frac{U(\mathbf{P}_1) - U(\mathbf{P})}{\Delta x}$$
$$= P(x + \tilde{t}\Delta x) \to P(x), \text{ as } \Delta x \to 0, \tag{10.8}$$

so that

$$\frac{\partial U(x, y, z)}{\partial x} = \lim_{\Delta x \to 0} \frac{U(x + \Delta x, y, z) - U(x, y, z)}{\Delta x}$$
$$= P(x), \tag{10.9}$$

Similarly, we may show that

$$\frac{\partial U}{\partial y} = Q,$$

and

$$\frac{\partial U}{\partial z} = R,$$

and conclude that the form in question is exact.

- 2 implies 3:

  Suppose again the integral

$$I = \int_C P \, dx + Q \, dy + R \, dz,$$

depends only on the initial point **A** and final point **B** but not on the piecewise smooth curve C connecting **A** and **B**.

Let C be a piecewise smooth closed curve contained in D. Choose **A** and **B** $\in C$, so that **A** $\neq$ **B**.

Observe that we may denote

$$C = C_1 \cup (-C_2)$$

where C_1 is a part of C connecting **A** and **B** and $-C_2$ is the other part of C, from **B** to **A**.

From the hypotheses

$$\int_{C_1} P\,dx + Q\,dy + R\,dz = \int_{C_2} P\,dx + Q\,dy + R\,dz,$$

that is,

$$\int_{C_1} P\,dx + Q\,dy + R\,dz - \int_{C_2} P\,dx + Q\,dy + R\,dz$$

$$= \int_{C_1} P\,dx + Q\,dy + R\,dz + \int_{-C_2} P\,dx + Q\,dy + R\,dz$$

$$= \int_{C_1 \cup (-C_2)} P\,dx + Q\,dy + R\,dz$$

$$= \int_{C} P\,dx + Q\,dy + R\,dz$$

$$= 0 \tag{10.10}$$

- 3 implies 2:
  Suppose now

$$\oint_{C} P\,dx + Q\,dy + R\,dz = 0,$$

for any piecewise smooth closed curve.

Choose **A**, **B** $\in D$.

Let C_1 and C_2 be two piecewise smooth curves connecting **A** and **B**. Hence,

$$C = C_1 \cup (-C_2),$$

is a piecewise smooth closed curve.

From the hypotheses,

$$\oint_{C_1 \cup (-C_2)} P\, dx + Q\, dy + R\, dz = 0,$$

that is,

$$\int_{C_1} P\, dx + Q\, dy + R\, dz + \int_{-C_2} P\, dx + Q\, dy + R\, dz$$

$$= \int_{C_1} P\, dx + Q\, dy + R\, dz - \int_{C_2} P\, dx + Q\, dy + R\, dz = 0, \quad (10.11)$$

so that

$$\int_{C_1} P\, dx + Q\, dy + R\, dz = \int_{C_2} P\, dx + Q\, dy + R\, dz,$$

that is, the integral does not depend on the piecewise smooth curve connecting **A** and **B**.

This completes the proof.

10.5   Surface Integrals

Consider a surface $S \subset \mathbb{R}^3$ defined by the equation $z = f(x, y)$, that is,

$$S = \{(x, y, z) \in \mathbb{R}^3 \ : \ (x, y) \in D \text{ and } z = f(x, y)\},$$

where $D \subset \mathbb{R}^2$ is a simple region and $f : D \to \mathbb{R}$ is a C^1 class function.
Thus, S is defined through the equation,

$$z - f(x, y) = 0,$$

that is, denoting

$$F(x, y, z) = z - f(x, y).$$

by

$$F(x, y, z) = 0.$$

Let $x_0 = (x_0, y_0, z_0) \in S$, and let C be a smooth curve contained in S and which contains x_0.

Assume that C is defined by a C^1 class function $\mathbf{r} : [a, b] \to \mathbb{R}^3$ where

$$\mathbf{r}(t) = x(t)\mathbf{i} + y(t)\mathbf{j} + z(t)\mathbf{k}.$$

From the hypotheses, since $C \subset S$ we have,

$$F(\mathbf{r}(t)) = 0,$$

that is,

$$F(x(t), y(t), z(t)) = 0, \forall t \in [a, b].$$

Observe that there exists $t_0 \in [a, b]$ such that

$$\mathbf{r}(t_0) = x_0\mathbf{i} + y_0\mathbf{j} + z_0\mathbf{k},$$

so that

$$F(\mathbf{r}(t_0)) = F(x_0, y_0, z_0) = 0.$$

On the other hand, since

$$F(x(t), y(t), z(t)) = 0, \forall t \in [a, b],$$

we obtain

$$\frac{dF(x(t), y(t), z(t))}{dt} = 0,$$

that is,

$$\frac{\partial F(\mathbf{r}(t))}{\partial x} x'(t) + \frac{\partial F(\mathbf{r}(t))}{\partial y} y'(t) + \frac{\partial F(\mathbf{r}(t))}{\partial z} z'(t) = 0,$$

so that in particular in $t = t_0$ we get,

$$\frac{\partial F(\mathbf{r}(t_0))}{\partial x} x'(t_0) + \frac{\partial F(\mathbf{r}(t_0))}{\partial y} y_0'(t) + \frac{\partial F(\mathbf{r}(t_0))}{\partial z} z'(t_0) = 0,$$

that is,

$$\nabla F(\mathbf{r}(t_0)) \cdot \mathbf{r}'(t_0) = 0.$$

Thus, $\nabla F(x_0, y_0, z_0)$ is orthogonal to $\mathbf{r}'(t_0)$ which is a direction tangent to C and in particular, tangent to S at (x_0, y_0, z_0).

Since C is an arbitrary curve, we may say that in such a specific sense, $\nabla F(x_0, y_0, z_0)$ is orthogonal to any direction tangent to S at (x_0, y_0, z_0), so that is, indeed, orthogonal to the tangent plane to S at (x_0, y_0, z_0) (in fact we do not provide all details on that at this point).

Therefore, the tangent plane to S at (x_0, y_0, z_0) has, as a possible normal direction, the vector

$$\begin{aligned} \mathbf{N} &= \nabla F(x_0, y_0, z_0) \\ &= (-f_x(x_0, y_0, z_0), -f_y(x_0, y_0, z_0), 1), \end{aligned} \tag{10.12}$$

and the corresponding unit normal vector $\mathbf{n}$ is given by:

$$\mathbf{n} = \frac{\mathbf{N}}{|\mathbf{N}|} = \frac{(-f_x, -f_y, 1)}{\sqrt{f_x^2 + f_y^2 + 1}},$$

where the functions in question must be considered at the point (x_0, y_0, z_0).

Let γ be the angle between $\mathbf{n}$ and the $\mathbf{k} = (0, 0, 1)$ axis.

Thus,

$$\begin{aligned} \cos(\gamma) &= \frac{\mathbf{n} \cdot \mathbf{k}}{|\mathbf{n}||\mathbf{k}|} \\ &= \frac{(-f_x, -f_y, 1)}{\sqrt{f_x^2 + f_y^2 + 1}} \cdot (0, 0, 1) \\ &= \frac{1}{\sqrt{f_x^2 + f_y^2 + 1}}, \end{aligned} \tag{10.13}$$

so that,

$$\cos(\gamma) = \frac{1}{\sqrt{f_x^2 + f_y^2 + 1}}.$$

10.5.1   Surface Areas

Consider now the problem of calculating the area of a surface S defined by

$$S = \{(x, y, z) \in \mathbb{R}^3 \; : \; (x, y) \in D \text{ and } z = f(x, y)\},$$

where $D \subset \mathbb{R}^2$ is a simple region and $f : D \to \mathbb{R}$ is a C^1 class function.

The elementary surface area ΔS corresponding to the rectangle $(x, x + \Delta x) \times (y, y + \Delta y) \subset D$, whose measure is $\Delta A = \Delta x \Delta y$, is approximately given by

$$\Delta S \approx \frac{\Delta x \Delta y}{\cos \gamma}$$

where γ is the angle between the normal $\mathbf{n}$ and the axis $\mathbf{k}$ at $(x, y) \in D$.

So, in its differential form, we may obtain

$$dS = \frac{dx\, dy}{\cos(\gamma)} = \sqrt{f_x(x, y)^2 + f_y(x, y)^2 + 1}\, dx\, dy,$$

so that we define, the area of S, denoted by $A(S)$, by

$$A(S) = \int\int_S dS = \int\int_D \sqrt{f_x^2 + f_y^2 + 1}\, dx\, dy.$$

10.5.2   Parametric Equations for a Surface

A surface S may be defined by:

$$S = \{\mathbf{r}(u, v) \; : \; (u, v) \in D\},$$

where $D \subset \mathbb{R}^2$ and

$$\mathbf{r}(u, v) = X(u, v)\mathbf{i} + Y(u, v)\mathbf{j} + Z(u, v)\mathbf{k}.$$

As an example, consider the sphere of radius $R > 0$ surface S, where

$$x = R \sin(\theta) \cos(\phi),$$

$$y = R \sin(\theta) \sin(\phi),$$

and,

$$z = R \cos(\theta).$$

Thus,

$$S = \{\mathbf{r}(\theta, \phi) \; : \; 0 \le \theta \le \pi, \; 0 \le \phi \le 2\pi\},$$

that is

$$\mathbf{r}(\theta, \phi) = X(\theta, \phi)\mathbf{i} + Y(\theta, \phi)\mathbf{j} + Z(\theta, \phi)\mathbf{k},$$

where,

$$X(\theta, \phi) = R\sin(\theta)\cos(\phi),$$

$$Y(\theta, \phi) = R\sin(\theta)\sin(\phi),$$

and,

$$Z(\theta, \phi) = R\cos(\theta).$$

10.5.3   Area Calculus Through Parametric Equations

Consider a surface

$$S = \{\mathbf{r}(u, v) \; : \; (u, v) \in D\},$$

where $\mathbf{r}$ is of C^1 class and $D \subset \mathbb{R}^2$ is a simple region, and where,

$$\mathbf{r}(u, v) = X(u, v)\mathbf{i} + Y(u, v)\mathbf{j} + Z(u, v)\mathbf{k}.$$

Consider also a rectangle $(u, u + \Delta u) \times (v, v + \Delta v)$, of dimensions Δu and Δv resulting from the discretization in u and v relating D.

Observe that for Δu and Δv sufficiently small, we have,

$$\begin{aligned}
\Delta r_u &= \mathbf{r}(u + \Delta u, v) - \mathbf{r}(u, v) \\
&= \frac{\partial \mathbf{r}(u, v)}{\partial u}\Delta u + o(\Delta u) \\
&\approx \frac{\partial \mathbf{r}(u, v)}{\partial u}\Delta u,
\end{aligned} \tag{10.14}$$

and

$$\begin{aligned}
\Delta r_v &= \mathbf{r}(u, v + \Delta v) - \mathbf{r}(u, v) \\
&= \frac{\partial \mathbf{r}(u, v)}{\partial v}\Delta v + o(\Delta v) \\
&\approx \frac{\partial \mathbf{r}(u, v)}{\partial v}\Delta v.
\end{aligned} \tag{10.15}$$

In the surface, the corresponding area to the planar region $(u, u+\Delta) \times (v, v+\Delta v)$ is

$$\Delta S \approx |\Delta r_u \times \Delta r_v|,$$

where, from above,

$$\Delta r_u \approx (X_u(u, v), Y_u(u, v), Z_u(u, v))\Delta u,$$

and

$$\Delta r_v \approx (X_v(u, v), Y_v(u, v), Z_v(u, v))\Delta v.$$

Therefore,

$$\Delta S \approx \left| \frac{\partial \mathbf{r}(u, v)}{\partial u} \times \frac{\partial \mathbf{r}(u, v)}{\partial v} \right| \Delta u \Delta v,$$

or in its differential form,

$$dS = \left| \frac{\partial \mathbf{r}(u, v)}{\partial u} \times \frac{\partial \mathbf{r}(u, v)}{\partial v} \right| dudv.$$

Thus, we define the area of S, denoted by $A(S)$, as

$$A(S) = \int \int_D dS = \int \int_D \left| \frac{\partial \mathbf{r}(u, v)}{\partial u} \times \frac{\partial \mathbf{r}(u, v)}{\partial v} \right| dudv.$$

In particular for

$$S = \{(x, y, z) \in \mathbb{R}^3 \ : \ z = f(x, y) \text{ and } (x, y) \in D\},$$

we may write:

$$\mathbf{r}(x, y) = (x, y, f(x, y)),$$

so that

$$\mathbf{r}_x(x, y) = (1, 0, f_x(x, y)),$$

$$\mathbf{r}_y(x, y) = (0, 1, f_y(x, y)),$$

and thus,

$$\mathbf{r}_x \times \mathbf{r}_y = \begin{vmatrix} \mathbf{i} & \mathbf{j} & \mathbf{k} \\ 1 & 0 & f_x \\ 0 & 1 & f_y \end{vmatrix}. \tag{10.16}$$

We may obtain,

$$\mathbf{r}_x \times \mathbf{r}_y = -f_x \mathbf{i} - f_y \mathbf{j} + \mathbf{k},$$

so that

$$|\mathbf{r}_x \times \mathbf{r}_y| = \sqrt{f_x^2 + f_y^2 + 1}.$$

From this we finally have

$$A(S) = \int\int_S dS$$

$$= \int\int_D |\mathbf{r}_x \times \mathbf{r}_y| \, dx \, dy$$

$$= \int\int_D \sqrt{f_x^2 + f_y^2 + 1} \, dx \, dy, \tag{10.17}$$

so that we have re-obtained the earlier formula.

10.6   Divergence Theorem

Definition 10.6.1 (Divergent) Let $\mathbf{F} : D \subset \mathbb{R}^3 \rightarrow \mathbb{R}^3$ be a vectorial field of C^1 class, that is,

$$\mathbf{F}(x, y, z) = P(x, y, z)\mathbf{i} + Q(x, y, z)\mathbf{j} + R(x, y, z)\mathbf{k}.$$

We define the divergent of $\mathbf{F}$, denoted by $div(\mathbf{F})$, by

$$div(\mathbf{F}) = \frac{\partial P}{\partial x} + \frac{\partial Q}{\partial y} + \frac{\partial R}{\partial z}.$$

Definition 10.6.2 (Simple Region in $\mathbb{R}^3$) We say that $V \subset \mathbb{R}^3$ is a simple volume or region of $\mathbb{R}^3$ if there exists C^1 class functions

$$z_1, z_2 : D \subset \mathbb{R}^2 \rightarrow \mathbb{R},$$

$$y_1, y_2 : D_1 \subset \mathbb{R}^2 \rightarrow \mathbb{R},$$

and

$$x_1, x_2 : D_2 \subset \mathbb{R}^2 \rightarrow \mathbb{R}$$

where D, D_1, D_2 are simple regions of $\mathbb{R}^2$, such that V admits concomitantly the following representations:

$$V = \{(x, y, z) \in \mathbb{R}^3 : z_1(x, y) \le z \le z_2(x, y) \text{ and } (x, y) \in D\},$$

$$V = \{(x, y, z) \in \mathbb{R}^3 : y_1(x, z) \le y \le y_2(x, z) \text{ and } (x, z) \in D_1\},$$

and

$$V = \{(x, y, z) \in \mathbb{R}^3 \ : \ x_1(y, z) \le x \le x_2(y, z) \text{ and } (y, z) \in D_2\}.$$

Theorem 10.6.3 (Divergence (Gauss)) *Let* $\mathbf{F} : V \subset \mathbb{R}^3 \to \mathbb{R}$ *be a vectorial field of class* C^1, *where* V *is a simple volume of* $\mathbb{R}^3$.
  Under such hypotheses,

$$\int\int\int_V div(\mathbf{F}) \, dx \, dy \, dz = \int\int_{\partial V} (\mathbf{F} \cdot \mathbf{n}) \, dS,$$

where ∂V *denotes the boundary of* V *and* $\mathbf{n}$ *the relating outward unit normal field.*

Proof Observe that

$$I_1 = \int\int\int_V \frac{\partial R}{\partial z} \, dx dy dz$$

$$= \int\int_D \left(\int_{z=z_1(x,y)}^{z=z_2(x,y)} \frac{\partial R}{\partial z} \, dz \right) dx dy$$

$$= \int\int_D R(x, y, z)_{z=z_1(x,y)}^{z=z_2(x,y)} \, dx dy$$

$$= \int\int_D (R(x, y, z_2(x, y)) - R(x, y, z_1(x, y))) \, dx dy. \qquad (10.18)$$

Thus, denoting by $\gamma(x, y)$ the angle between $\mathbf{n}(x, y)$ and $\mathbf{k}$ at (x, y), we have

$$I_1 = \int\int_D \left(R(x, y, z_2(x, y)) \frac{\cos(\gamma)}{\cos(\gamma)} - R(x, y, z_1(x, y)) \frac{\cos(\gamma)}{\cos(\gamma)} \right) dx dy,$$

that is,

$$I_1 = \int\int_D \left(R(x, y, z_2(x, y)) \frac{\cos(\gamma)}{\cos(\gamma)} + R(x, y, z_1(x, y)) \frac{\cos(\gamma)}{(-\cos(\gamma))} \right) dx dy$$

$$= \int\int_{S_2} R \cos(\gamma) \, dS + \int\int_{S_1} R \cos(\gamma) \, dS, \qquad (10.19)$$

where,

$$S_2 = \{(x, y, z) \in \mathbb{R}^3 \ : \ z = z_2(x, y) \text{ and } (x, y) \in D\},$$

and

$$S_1 = \{(x, y, z) \in \mathbb{R}^3 \ : \ z = z_1(x, y) \text{ and } (x, y) \in D\}.$$

Observe that

$$dS = \frac{dxdy}{\cos(\gamma)}$$

in S_2 and

$$dS = \frac{dxdy}{-\cos(\gamma)}$$

in S_1.

Therefore,

$$I_1 = \int\int_{S_1 \cup S_2} R\cos(\gamma)\, dS$$

$$= \int\int_S R\cos(\gamma)\, dS. \qquad (10.20)$$

Similarly, denoting by α and β the angle between $\mathbf{n}$ and $\mathbf{i}$, and between $\mathbf{n}$ and $\mathbf{j}$, respectively, we may show that,

$$I_2 = \int\int\int_V \frac{\partial P}{\partial x}\, dxdydz = \int\int_S P\cos(\alpha)\, dS,$$

and

$$I_3 = \int\int\int_V \frac{\partial Q}{\partial y}\, dxdydz = \int\int_S Q\cos(\beta)\, dS.$$

Summarizing, we have obtained:

$$\int\int\int_V div(\mathbf{F})\, dx\, dy\, dz = \int\int\int_V \left(\frac{\partial P}{\partial x} + \frac{\partial Q}{\partial y} + \frac{\partial R}{\partial z}\right) dxdydz$$

$$= \int\int_S (P\cos(\alpha) + Q\cos(\beta) + R\cos(\gamma))\, dS$$

$$= \int\int_S (P\mathbf{i} + Q\mathbf{j} + R\mathbf{k}) \cdot (\cos(\alpha)\mathbf{i} + \cos(\beta)\mathbf{j}$$

$$+ \cos(\gamma)\mathbf{k})\, dS$$

$$= \int\int_S \mathbf{F} \cdot \mathbf{n}\, dS, \qquad (10.21)$$

where $S = \partial V$ is the boundary of V. The proof is complete.

Remark 10.6.4 At this point, we highlight that the Divergence Theorem may be easily extended for more general regions, such as those which comprise unions of simple regions. The idea is just to apply the previous result to each simple part. The next step is to add the partial results to obtain the final one.

10.7 The Stokes Theorem in $\mathbb{R}^3$

Definition 10.7.1 (Curl) Let $\mathbf{F} : V \subset \mathbb{R}^3 \to \mathbb{R}^3$ be a vectorial field of C^1 class, where V is an open set.

We define the $curl(\mathbf{F})$ on V, by

$$curl(\mathbf{F}) = \begin{vmatrix} \mathbf{i} & \mathbf{j} & \mathbf{k} \\ \frac{\partial}{\partial x} & \frac{\partial}{\partial y} & \frac{\partial}{\partial z} \\ P & Q & R \end{vmatrix}, \tag{10.22}$$

where we have denoted

$$\mathbf{F} = P\mathbf{i} + Q\mathbf{j} + R\mathbf{k}.$$

Thus,

$$curl(\mathbf{F}) = \left(\frac{\partial R}{\partial y} - \frac{\partial Q}{\partial z} \right) \mathbf{i} + \left(\frac{\partial P}{\partial z} - \frac{\partial R}{\partial x} \right) \mathbf{j} + \left(\frac{\partial Q}{\partial x} - \frac{\partial P}{\partial y} \right) \mathbf{k},$$

on V.

Theorem 10.7.2 (Stokes) *Let S be a surface of C^1 class, where,*

$$S = \{(x, y, z) \in \mathbb{R}^3 \ : \ z = z(x, y) \text{ and } (x, y) \in D\},$$

where $D \subset \mathbb{R}^2$ is a simple region.

Denote by $\mathbf{n}$ the unit outward normal to S.

Let $\mathbf{F} : V \subset \mathbb{R}^3 \to \mathbb{R}^3$ be a C^1 class vectorial field, where V is open and $S \subset V$. Under such hypotheses,

$$\int\int_S curl(\mathbf{F}) \cdot \mathbf{n} \, dS = \oint_C \mathbf{F} \cdot d\mathbf{r},$$

where C is the boundary of S, that is,

$$C = \{(x, y, z) \in \mathbb{R}^3 \ : \ z = z(x, y) \text{ and } (x, y) \in \partial D\},$$

where ∂D is the boundary of D.

Proof Observe that since on S we have $z = z(x, y)$, we may infer that, on S,

$$dz = z_x \, dx + z_y \, dy,$$

so that

$$
\begin{aligned}
I &= \oint_C \mathbf{F} \cdot d\mathbf{r} \\
&= \oint_C P \, dx + Q \, dy + R \, dz \\
&= \oint_C P \, dx + Q \, dy + R(z_x \, dx + z_y \, dy),
\end{aligned}
\tag{10.23}
$$

Thus,

$$
\begin{aligned}
I &= \oint_C (P + Rz_x) \, dx + (Q + Rz_y) \, dy \\
&= \oint_{\partial D} (P(x, y, z(x, y)) + R(x, y, z(x, y))z_x(x, y)) \, dx \\
&\quad + (Q(x, y, z(x, y)) + R(x, y, z(x, y))z_y(x, y)) \, dy \\
&= \oint_{\partial D} \tilde{P} \, dx + \tilde{Q} \, dy,
\end{aligned}
\tag{10.24}
$$

where

$$
\tilde{P}(x, y) = P(x, y, z(x, y)) + R(x, y, z(x, y))z_x(x, y),
$$

and

$$
\tilde{Q}(x, y) = Q(x, y, z(x, y)) + R(x, y, z(x, y))z_y(x, y).
$$

From the Green Theorem in $\mathbb{R}^2$, we have,

$$
\begin{aligned}
I &= \oint_{\partial D} \tilde{P} \, dx + \tilde{Q} \, dy \\
&= \int\int_D \left(\frac{\partial \tilde{Q}}{\partial x} - \frac{\partial \tilde{P}}{\partial y} \right) dx dy \\
&= \int\int_D [Q_x + Q_z z_x + (R_x + R_z z_x)z_y + Rz_{yx}] \, dx dy \\
&\quad - \int\int_D [P_y + P_z z_y + (R_y + R_z z_y)z_x + Rz_{xy}] \, dx dy \\
&= \int\int_D [(R_y - Q_z)(-z_x) + (P_z - R_x)(-z_y) + (Q_x - P_y)] \, dx dy
\end{aligned}
$$

$$= \int \int_D [(R_y - Q_z)\mathbf{i} + (P_z - R_x)\mathbf{j} + (Q_x - P_y)\mathbf{k}] \cdot [-z_x\mathbf{i} - z_y\mathbf{j} + \mathbf{k}]\, dxdy$$

$$= \int \int_D [curl(\mathbf{F})] \cdot \frac{[-z_x\mathbf{i} - z_y\mathbf{j} + \mathbf{k}]}{\sqrt{z_x^2 + z_y^2 + 1}} \sqrt{z_x^2 + z_y^2 + 1}\, dxdy$$

$$= \int \int_S curl(\mathbf{F}) \cdot \mathbf{n}\, dS. \tag{10.25}$$

This completes the proof.

10.8 Surfaces in $\mathbb{R}^n$

Definition 10.8.1 (A Simple Smooth Surface) By a simple smooth surface in $\mathbb{R}^n$ we shall understand the graph of a C^1 class function $\mathbf{r} : D \subset \mathbb{R}^m \to \mathbb{R}^n$ such that $1 \le m < n$ and D is a simply connected set with a regular boundary ∂D (at least C^1 class).

Hence, we may denote

$$M = \{\mathbf{r}(u_1, \ldots, u_m) \in \mathbb{R}^n : \mathbf{u} = (u_1, \ldots, u_m) \in D\},$$

where

$$\mathbf{r}(\mathbf{u}) = X_1(\mathbf{u})\mathbf{e}_1 + X_2(\mathbf{u})\mathbf{e}_2 + \cdots + X_n(\mathbf{u})\mathbf{e}_n,$$

where,

$$\mathbf{e}_i = (0, \ldots, 0, 1, 0, \ldots, 0)$$

with value 1 at the ith entry, and entry 0 otherwise.

We assume $\partial M = \mathbf{r}(\partial D)$, where ∂M denotes the boundary of M.

Definition 10.8.2 (Surface in a More General Form) We define a surface $M \subset \mathbb{R}^n$ of dimension $1 \le m < n$ by a set of the form

$$M = \{\mathbf{r}_\alpha(U_\alpha) : U_\alpha \subset D\},$$

where $\cup_{\alpha \in A} U_\alpha = D$ so that

$$M = \cup_{\alpha \in A} \mathbf{r}_\alpha(U_\alpha),$$

where A is a set of indices, U_α is open $\forall \alpha \in A$, and $\mathbf{r}_\alpha$ is a C^1 class function of

$$\mathbf{u}_\alpha = (u_1^\alpha, \ldots, u_m^\alpha) \in U_\alpha \subset D \subset \mathbb{R}^m$$

concerning an appropriate local system of coordinates.

Moreover we assume that $\mathbf{r}_\alpha$ is a bijection on $\mathbf{r}_\alpha(U_\alpha)$.

Also, if $U_\alpha \cap U_\beta \neq \emptyset$, and if $\tilde{u} \in D$ is such that corresponds to $\mathbf{u}_\alpha \in U_\alpha$ and $\mathbf{u}_\beta \in U$, then

$$\mathbf{r}_\alpha(\mathbf{u}_\alpha) = \mathbf{r}_\beta(\mathbf{u}_\beta),$$

and moreover

$$\mathbf{r}_\alpha \circ \mathbf{r}_\beta^{-1} : \mathbf{r}_\beta(U_\alpha \cap U_\beta) \to \mathbf{r}_\alpha(U_\alpha \cap U_\beta),$$

is a C^1 class mapping, $\forall \alpha, \beta \in A$.

10.9   The Tangent Space

We shall denote generically,

$$M = \cup_{\alpha \in A} \mathbf{r}_\alpha(U_\alpha) \equiv \mathbf{r}(D).$$

To simplify the notation, we shall also denote as the meaning is clear $\mathbf{u}_\alpha = \mathbf{u}$ and

$$\mathbf{r}_\alpha(\mathbf{u}_\alpha) \equiv \mathbf{r}(\mathbf{u}),$$

emphasizing we are in fact referring to a particular system of coordinates relating U_α.

Definition 10.9.1 (The Tangent Space to M) Let $p = \mathbf{r}(\mathbf{u}) \in M$. We define the tangent space to M at p, denoted by $T_p(M)$, by

$$T_p(M) = \left\{ \sum_{i=1}^m \alpha_i \frac{\partial \mathbf{r}(\mathbf{u})}{\partial u_i} \ : \ \alpha_1, \ldots, \alpha_m \in \mathbb{R} \right\}.$$

We shall assume

$$\left\{ \frac{\partial \mathbf{r}(\mathbf{u})}{\partial u_i} \right\}_{i=1}^m$$

is linearly independent $\forall p \in M$.

Remark 10.9.2 We recall that the dual space of a real vector space V is formally defined as the set of all linear continuous functionals (in this context, in fact real functions) defined on V. From the Riesz representation theorem, considering the specific case where $V \subset \mathbb{R}^n$, given a linear continuous functional $F : V \to \mathbb{R}$ there exists $\boldsymbol{\alpha} \in V$ such that

$$F(\mathbf{u}) = \boldsymbol{\alpha} \cdot \mathbf{u}, \ \forall \mathbf{u} \in V.$$

In such a case, as for any Hilbert space, we say that $V^* \approx V$, that is the dual space V^* is indeed identified with V.

In the general case, from elementary linear algebra, given a basis $(\mathbf{e}_1, \ldots, \mathbf{e}_m)$ for V, we may obtain a corresponding dual basis for V^*, given by $F_1, \ldots, F_m \in V^*$, such that

$$F_i(\mathbf{e}_j) = \delta_{ij},$$

where

$$\delta_{ij} = \begin{cases} 1 & \text{if } i = j \\ 0 & \text{otherwise.} \end{cases} \tag{10.26}$$

In the next lines we present the formal details on these last statements.

Theorem 10.9.3 *Let V be a real m-dimensional vector space and let $\{\mathbf{v}_1, \ldots, \mathbf{v}_m\}$ be a basis for V*

Under such hypotheses, there exists a unique basis $\{F_1, \ldots, F_m\}$ for V^* such that

$$F_i(v_j) = \delta_{ij},$$

where

$$\delta_{ij} = \begin{cases} 1 & \text{if } i = j \\ 0 & \text{if } i \neq j, \forall i, j \in \{1, \ldots, m\} \end{cases} \tag{10.27}$$

Proof For each $i \in \{1, \ldots, m\}$, define $F_i : V \to \mathbb{R}$ by

$$F_i(\mathbf{v}) = a_i,$$

where,

$$\mathbf{v} = \sum_{j=1}^{m} a_j \mathbf{v}_j.$$

Therefore,

$$F_i(\mathbf{v}_j) = \delta_{ij}, \ \forall i, j \in \{1, \ldots, m\}.$$

Now we are going to show that such $\{F_i\}$ are unique.

Let $\tilde{F}_1, \ldots, \tilde{F}_m \in V^*$ be such that $\tilde{F}_i(\mathbf{v}_j) = \delta_{ij}, \ \forall i, j \in \{1, \ldots, m\}$.

Let $\mathbf{v} = \sum_{j=1}^{m} a_j \mathbf{v}_j \in V$.

Thus, fixing $i \in \{1, \ldots, m\}$, we have,

$$\tilde{F}_i(\mathbf{v}) = \tilde{F}_i\left(\sum_{j=1}^{m} a_j \mathbf{v}_j\right)$$

$$= \sum_{j=1}^{m} a_j \tilde{F}_i(\mathbf{v}_j)$$

$$= \sum_{j=1}^{m} a_j \delta_{ij}$$

$$= a_i$$

$$= F_i(\mathbf{v}). \tag{10.28}$$

Therefore,

$$\tilde{F}_i(\mathbf{v}) = F_i(\mathbf{v}), \ \forall \mathbf{v} \in V,$$

so that,

$$\tilde{F}_i = F_i, \ \forall i \in \{1, \ldots, m\}.$$

We may conclude that the $F_1, \ldots, F_m \in V^*$ in question are unique.

At this point, we are going to show that $F_1, \ldots, F_m \in V^*$ are linearly independent.

Suppose that $b_1, \ldots, b_m \in \mathbb{R}$ are such that

$$F = \sum_{i=1}^{m} b_i F_i = \mathbf{0}.$$

Thus,

$$0 = F(\mathbf{v}_j) = \sum_{i=1}^{m} b_i F_i(\mathbf{v}_j) = \sum_{i=1}^{m} b_i \delta_{ij} = b_i, \ \forall i \in \{1, \ldots, m\}.$$

Hence, $\{F_1, \ldots, F_m\}$ is a linearly independent set.

To finish the proof, we are going to show that $\{F_1, \ldots, F_m\}$ spans V^*.
Let $S \in V^*$.
Denote $b_i = S(\mathbf{v}_i), \forall i \in \{1, \ldots, m\}$.
Define

$$F = \sum_{i=1}^{m} b_i F_i \in V^*.$$

Thus

$$F(\mathbf{v}_j) = \sum_{i=1}^{m} b_i F_i(\mathbf{v}_j) = \sum_{i=1}^{m} b_i \delta_{ij} = b_j.$$

Let $\mathbf{v} = \sum_{j=1}^{m} a_j \mathbf{v}_j$.
Thus,

$$S(\mathbf{v}) = S\left(\sum_{j=1}^{m} a_j \mathbf{v}_j\right) = \sum_{j=1}^{m} a_j S(\mathbf{v}_j) = \sum_{j=1}^{m} a_j b_j.$$

On the other hand,

$$F(\mathbf{v}) = F\left(\sum_{j=1}^{m} a_j \mathbf{v}_j\right) = \sum_{j=1}^{m} a_j F(\mathbf{v}_j) = \sum_{j=1}^{m} a_j b_j.$$

Hence,

$$F(\mathbf{v}) = S(\mathbf{v}), \forall \mathbf{v} \in V,$$

so that

$$S = F = \sum_{i=1}^{m} b_i F_i.$$

Thus, $\{F_1, \ldots, F_m\}$ spans V^*, so that it is a basis for V^*.
This completes the proof.

Theorem 10.9.4 *Let $M \subset \mathbb{R}^n$ be a m dimensional surface and let $f : M \to \mathbb{R}$ be C^1 class function.*

Let $p = \mathbf{r}(\mathbf{u}) \in M$. Under such hypotheses, we may associate with such a function f, a functional

$$F : T_p(M) \to \mathbb{R}$$

where, for each

$$\mathbf{v} = v_i \frac{\partial \mathbf{r}(\mathbf{u})}{\partial u_i},$$

we have

$$F(\mathbf{v}) = F\left(v_i \frac{\partial \mathbf{r}(\mathbf{u})}{\partial u_i}\right)$$

$$= \lim_{\varepsilon \to 0} \frac{(f \circ \mathbf{r})(\{u_i + \varepsilon v_i\}) - (f \circ \mathbf{r})(\{u_i\})}{\varepsilon}$$

$$= df(\mathbf{v}). \tag{10.29}$$

Conversely, if $F : T_p(M) \to \mathbb{R}$ is a continuous linear functional, that is, if there exists $\alpha \in \mathbb{R}^m$ such that

$$F(\mathbf{v}) = \alpha \cdot \left(v_i \frac{\partial \mathbf{r}(\mathbf{u})}{\partial u_i}\right), \ \forall \mathbf{v} \in T_p(M),$$

then there exists $f : M \to \mathbb{R}$ of C^1 class such that,

$$F(\mathbf{v}) \equiv df(\mathbf{v}), \ \forall \mathbf{v} \in T_p(M).$$

Proof For $f \in C^1(M)$ and $p \in M$ in question, define $F : T_p(M) \to \mathbb{R}$ by

$$F(\mathbf{v}) = df(\mathbf{v}),$$

for all $\mathbf{v} = v_i \frac{\partial \mathbf{r}(\mathbf{u})}{\partial u_i} \in T_p(M)$.
    Hence,

$$F(\mathbf{v}) = df(\mathbf{v})$$

$$= \lim_{\varepsilon \to 0} \frac{(f \circ \mathbf{r})(\{u_i + \varepsilon v_i\}) - (f \circ \mathbf{r})(\{u_i\})}{\varepsilon}$$

$$= \sum_{j=1}^{n} \frac{\partial (f \circ \mathbf{r})(u)}{\partial X_j} \frac{\partial X_j(\mathbf{u})}{\partial u_i} v_i$$

$$= \alpha \cdot \left(v_i \frac{\partial \mathbf{r}(\mathbf{u})}{\partial u_i}\right), \tag{10.30}$$

where

$$\alpha_j = \frac{\partial (f \circ \mathbf{r})(\mathbf{u})}{\partial X_j}, \ \forall j \in \{1, \ldots, n\}.$$

We may conclude that F is continuous and linear on $T_p(M)$, that is,

$$F \in T_p(M)^*.$$

Conversely, assume $F : T_p(M) \to \mathbb{R}$ is linear and continuous, so that by the Riesz representation theorem, there exists $\alpha \in \mathbb{R}^n$ such that

$$F(v) = \alpha \cdot \left(v_i \frac{\partial \mathbf{r}(\mathbf{u})}{\partial u_i} \right), \ \forall \mathbf{v} \in T_p(M).$$

Define $f : M \to \mathbb{R}$ by

$$f(\mathbf{w}) = \alpha \cdot \mathbf{w}, \ \forall \mathbf{w} \in M.$$

Thus $f(\mathbf{r}(u)) = \alpha \cdot \mathbf{r}(\mathbf{u}) = \alpha_j X_j(\mathbf{u})$.
From this,

$$
\begin{aligned}
df(\mathbf{v}) &= \lim_{\varepsilon \to 0} \frac{\alpha_j X_j(\{u_i + \varepsilon v_i\}) - \alpha_j X_j(\{u_i\})}{\varepsilon} \\
&= \alpha_j \frac{\partial X_j(\mathbf{u})}{\partial u_i} v_i \\
&= \alpha \cdot \left(v_i \frac{\partial \mathbf{r}(\mathbf{u})}{\partial u_i} \right) \\
&= F(\mathbf{v}).
\end{aligned}
\tag{10.31}
$$

This completes the proof.

Corollary 10.9.5 *In an appropriate sense, the corresponding dual basis to $T_p(M)^*$, to the primal basis for $T_p(M)$*

$$\left\{ \frac{\partial \mathbf{r}(\mathbf{u})}{\partial u_i} \right\}_{i=1}^m,$$

is given by $\{dw_i(\mathbf{r}(\mathbf{u}))\}$ where $\{w_i(\mathbf{r}(u))\} = \mathbf{r}^{-1}(\mathbf{r}(\mathbf{u})) = \{u_i\}$, so that we could denote such basis for $T_p(M)^*$ by

$$\{du_1, \ldots, du_m\}.$$

Proof Let us denote

$$\left\{ \tilde{\mathbf{e}}_i = \frac{\partial \mathbf{r}(\mathbf{u})}{\partial u_i} \right\}_{i=1}^m,$$

and

$$\tilde{\mathbf{v}}_j = (0, 0, \ldots, 0, 1, 0, \ldots, 0),$$

that is, value 1 at the jth entry and value 0 at the remaining ones.
   Hence

$$
\begin{aligned}
du_i(\tilde{\mathbf{e}}_j) &= dw_i(\mathbf{r}(\mathbf{u}))(\tilde{\mathbf{e}}_j) \\
&= \lim_{\varepsilon \to 0} \frac{w_i(\mathbf{r}(\mathbf{u} + \varepsilon\tilde{\mathbf{v}}_j)) - w_i(\mathbf{r}(\mathbf{u}))}{\varepsilon} \\
&= \frac{\partial w_i(r(\mathbf{u}))}{\partial u_j} \\
&= \frac{\partial u_i}{\partial u_j} = \delta_{ij}.
\end{aligned}
\tag{10.32}
$$

Summarizing, we have obtained

$$du_i(\tilde{\mathbf{e}}_j) = \delta_{ij},$$

so that in the context of last theorem, $\{du_i, \ldots, du_m\}$ is a basis for $T_p(M)^*$.
   The proof is complete.

10.10   Vector Fields

Definition 10.10.1 Let $M \subset \mathbb{R}^n$ be an m-dimensional surface that is, $1 \leq m < n$.
   Generically denoting

$$M = \{\mathbf{r}(\mathbf{u}) \; : \; \mathbf{u} \in D\},$$

let $p = \mathbf{r}(\mathbf{u}) \in M$ and consider the set

$$\left\{ \frac{\partial \mathbf{r}(\mathbf{u})}{\partial u_i} \right\}_{i=1}^m,$$

be a basis for $T_p(M)$.
   We define a vector field

$$X : M \to \{T_p(M), \; p \in M\},$$

by

$$X = \{X_p, \; p \in M\},$$

where point-wise

$$X_p(\mathbf{u}) = \sum_{i=1}^{m} X_{p_i}(\mathbf{u}) \frac{\partial \mathbf{r}(\mathbf{u})}{\partial u_i}.$$

Moreover, for each $p \in M$ we define the operator

$$\tilde{X}_p : C^1(M) \to C^1(D)$$

where

$$X_p = \sum_{i=1}^{m} X_{p_i} \frac{\partial \mathbf{r}(\mathbf{u})}{\partial u_i},$$

by

$$\begin{aligned}
\tilde{X}_p(f) &= df(X_p) \\
&= \lim_{\varepsilon 0} \frac{(f \circ \mathbf{r})(\{u_i + \varepsilon X_{p_i}(\mathbf{u})\}) - (f \circ \mathbf{r})(\{u_i\})}{\varepsilon} \\
&= \frac{\partial (f \circ \mathbf{r})(\mathbf{u})}{\partial X_j} \frac{\partial X_j(\mathbf{u})}{\partial u_i} X_{p_i}(\mathbf{u}) \\
&= \nabla f(\mathbf{r}(u)) \cdot \left(X_{p_i}(\mathbf{u}) \frac{\partial \mathbf{r}(\mathbf{u})}{\partial u_i} \right).
\end{aligned} \tag{10.33}$$

Moreover, we denote by $\mathscr{X}(M)$ the set of C^1 class vector fields on M.

Definition 10.10.2 (The Exterior Product) Let V be a real vector space and V^* its dual one. Let $(F_1, \ldots, F_k) \in V^* \times V^* \times \cdots \times V^*$ (k times) and $(\mathbf{v}_1, \mathbf{v}_2, \ldots, \mathbf{v}_k) \in V \times V \times \cdots \times V$ (k times). We define point-wise the exterior product $(F_1 \wedge F_2 \wedge \cdots \wedge F_k)(\mathbf{v}_1, \mathbf{v}_2, \ldots, \mathbf{v}_k)$, by

$$(F_1 \wedge F_2 \wedge \cdots \wedge F_k)(\mathbf{v}_1, \mathbf{v}_2, \ldots, \mathbf{v}_k) = \det\{F_i(\mathbf{v}_j)\}.$$

Remark 10.10.3 Concerning a C^1 class m-dimensional surface $M \subset \mathbb{R}^n$, where $1 \le m < n$ and

$$M = \{\mathbf{r}(\mathbf{u}) \; : \; \mathbf{u} \in D\},$$

for a vector field $X \in \mathscr{X}(M)$ where

$$X = X_{p_i} \frac{\partial \mathbf{r}(u)}{\partial u_i}$$

and $f : M \to \mathbb{R}$ of C^1 class, we shall denote,

$$X \cdot f = df(X),$$

where point-wisely,

$$(X \cdot f)(\mathbf{r}(\mathbf{u})) = df(X)[\mathbf{r}(\mathbf{u})]$$
$$= \lim_{\varepsilon \to 0} \frac{(f \circ \mathbf{r})(\{u_i + \varepsilon X_{p_i}\}) - (f \circ \mathbf{r})(\mathbf{u})}{\varepsilon}, \qquad (10.34)$$

where $p = \mathbf{r}(\mathbf{u})$.

Definition 10.10.4 (Lie Bracket) Let $M \subset \mathbb{R}^n$ be a C^1 class m-dimensional surface where $1 \le m < n$.

Let $X, Y \in \tilde{\mathscr{X}}(M)$, where $\tilde{\mathscr{X}}(M)$ denotes the set of the $C^\infty(M) = \cap_{k \in \mathbb{N}} C^k(M)$ class vector fields. We define the Lie bracket of X and Y denoted by $[X, Y]$ by,

$$[X, Y] = (X \cdot Y_i - X \cdot Y_i)\frac{\partial \mathbf{r}(\mathbf{u})}{\partial u_i},$$

where

$$X = X_i \frac{\partial \mathbf{r}(\mathbf{u})}{\partial u_i}$$

and

$$Y = Y_i \frac{\partial \mathbf{r}(\mathbf{u})}{\partial u_i}.$$

Theorem 10.10.5 *Let $M \subset \mathbb{R}^n$ be a C^1 class m-dimensional surface where $1 \le m < n$. Let $X, Y \in \tilde{\mathscr{X}}(M)$ and let $f \in C^2(M)$.*

Under such hypotheses,

$$[X, Y] \cdot f = X \cdot (Y \cdot f) - Y \cdot (X \cdot f).$$

Proof Observe that

$$X \cdot (Y \cdot f) = X \cdot (df(Y))$$
$$= X \cdot \left(\frac{\partial (f \circ r)(\mathbf{u})}{\partial u_i} Y_i\right)$$
$$= \left[d\left(\frac{\partial (f \circ r)(\mathbf{u})}{\partial u_i} Y_i\right)\right](X)$$
$$= \frac{\partial^2 (f \circ \mathbf{r})(\mathbf{u})}{\partial u_j \partial u_i} Y_i X_j + \frac{\partial (f \circ \mathbf{r})(\mathbf{u})}{\partial u_i} \frac{\partial Y_i}{\partial u_j} X_j. \qquad (10.35)$$

Similarly,

$$Y \cdot (X \cdot f) = \frac{\partial^2 (f \circ \mathbf{r})(\mathbf{u})}{\partial u_j \partial u_i} X_i Y_j + \frac{\partial (f \circ r)(\mathbf{u})}{\partial u_i} \frac{\partial X_i}{\partial u_j} Y_j,$$

so that

$$X \cdot (Y \cdot f) - Y \cdot (X \cdot f)$$
$$= \frac{\partial (f \circ r)(\mathbf{u})}{\partial u_i} \frac{\partial Y_i}{\partial u_j} X_j - \frac{\partial (f \circ r)(\mathbf{u})}{\partial u_i} \frac{\partial X_i}{\partial u_j} Y_j. \qquad (10.36)$$

On the other hand,

$$[X, Y] \cdot f = \left[(X \cdot Y_i - Y \cdot X_i) \frac{\partial \mathbf{r}(u)}{\partial u_i} \right] \cdot f. \qquad (10.37)$$

Observe that

$$\left[(X \cdot Y_i) \frac{\partial \mathbf{r}(u)}{\partial u_i} \right] \cdot f = df \left[(X \cdot Y_i) \frac{\partial \mathbf{r}(u)}{\partial u_i} \right]$$
$$= \frac{\partial (f \circ \mathbf{r})(\mathbf{u})}{\partial u_i} (X \cdot Y_i)$$
$$= \frac{\partial (f \circ \mathbf{r})(\mathbf{u})}{\partial u_i} dY_i(X)$$
$$= \frac{\partial (f \circ \mathbf{r})(\mathbf{u})}{\partial u_i} \frac{\partial Y_i}{\partial u_j} X_j. \qquad (10.38)$$

Similarly,

$$\left[(X \cdot Y_i) \frac{\partial \mathbf{r}(u)}{\partial u_i} \right] \cdot f$$
$$= \frac{\partial (f \circ \mathbf{r})(\mathbf{u})}{\partial u_i} \frac{\partial X_i}{\partial u_j} Y_j. \qquad (10.39)$$

Therefore,

$$[X, Y] \cdot f = \left[(X \cdot Y_i - Y \cdot X_i) \frac{\partial \mathbf{r}(u)}{\partial u_i} \right] \cdot f$$
$$= \frac{\partial (f \circ \mathbf{r})(\mathbf{u})}{\partial u_i} \frac{\partial Y_i}{\partial u_j} X_j - \frac{\partial (f \circ \mathbf{r})(\mathbf{u})}{\partial u_i} \frac{\partial X_i}{\partial u_j} Y_j$$
$$= X \cdot (Y \cdot f) - Y \cdot (X \cdot f). \qquad (10.40)$$

The proof is complete.

Exercises 10.10.6

1. Let $M \subset \mathbb{R}^n$ be a C^1 class m-dimensional surface where $1 \leq m < n$ and let $X, Y, Z \in \tilde{\mathscr{X}}(M)$. Show that

(a) $[\alpha X + \beta Y, Z] = \alpha[X, Z] + \beta[Y, Z]$, $\forall \alpha, \beta \in \mathbb{R}$,

(b) $[X, \alpha Y + \beta Z] = \alpha[X, Y] + \beta[X, Z]$, $\forall \alpha, \beta \in \mathbb{R}$.

(c) Antisymmetry:

$$[X, Y] = -[Y, X],$$

(d) Jacob Identity:

$$[[X, Y], Z] + [[Y, Z], X] + [[Z, X], Y] = 0.$$

(e) Leibnitz Rule:

$$[fX, gY] = fg[X, Y] + f(X \cdot g)Y - g(Y \cdot f)X, \quad \forall f, g \in C^2(M).$$

At this point we introduce the definition of Lie Algebra.

Definition 10.10.7 (Lie Algebra) A Lie Algebra is a vector space V for which is defined an antisymmetric bilinear form $[\cdot, \cdot] : V \times V \to V$, which satisfies the Jacob identity.

Remark 10.10.8 Observe that from the exposed above, $\tilde{\mathscr{X}}(M)$ is a Lie Algebra with the Lie bracket

$$[X, Y] = (X \cdot Y_i - Y \cdot X_i) \frac{\partial \mathbf{r}(\mathbf{u})}{\partial u_i}.$$

Definition 10.10.9 (Integral Curve) Let $M \subset \mathbb{R}^n$ be a C^1 class m-dimensional surface where $1 \leq m < n$. Let $X \in \mathscr{X}(M)$.

Let A curve $(\mathbf{r} \circ \mathbf{u}) : I = (-\varepsilon, \varepsilon) \to M$ is said to be an integral curve for X on I around $p = \mathbf{r}(u(0))$, if

$$\frac{d\mathbf{r}(\mathbf{u}(t))}{dt} = X(\mathbf{u}(t)), \quad \forall t \in I.$$

Observe that in such a case:

$$\frac{d\mathbf{r}(\mathbf{u}(t))}{dt} = \frac{\partial \mathbf{r}(\mathbf{u}(t))}{\partial u_i} \frac{du_i(t)}{dt} = X(\mathbf{u}(t))$$

$$= X_i(\mathbf{u}(t)) \frac{\partial \mathbf{r}(\mathbf{u}(t))}{\partial u_i}, \tag{10.41}$$

so that

$$\frac{du_i(t)}{dt} = X_i(\mathbf{u}(t)). \tag{10.42}$$

10.11   The Generalized Derivative

At this section we introduce the definition of Lie derivative. We start with a preliminary definition, namely, the generalized derivative.

Definition 10.11.1 (Generalized Derivative) Let $M \subset \mathbb{R}^n$ be a C^1 class m-dimensional surface where $1 \leq m < n$. Let $X \in \mathscr{X}(M)$. Let $\mathbf{r}(\mathbf{u}(t))$ be the point-wise representation of the integral curve of X around $p = \mathbf{r}(\mathbf{u}(0))$.

Let $Y \in \tilde{\mathscr{X}}(M)$ and $f \in C^2(M)$. We define the generalized derivative of f relating Y, along the direction X, at the point $p = \mathbf{r}(\mathbf{u}(0))$, denoted by $D_X Y(f)(p)$, by

$$D_X Y(f) = \frac{d[Y(\mathbf{u}(t)) \cdot f(\mathbf{r}(\mathbf{u}(t)))]}{dt}\Big|_{t=0}.$$

Theorem 10.11.2 *Let $M \subset \mathbb{R}^n$ be a C^1 class m-dimensional surface where $1 \leq m < n$. Let $X \in \mathscr{X}(M)$. Let $\mathbf{r}(\mathbf{u}(t))$ be the point-wise representation of the integral curve of X around $p = \mathbf{r}(\mathbf{u}(0))$.*
Let $Y \in \mathscr{X}(M)$ and $f \in C^2(M)$.
Under such assumptions,

$$D_X Y(f)(p) = X \cdot (Y \cdot f)(p).$$

Proof Observe that,

$$D_X Y(f)(p) = \frac{d[df(Y)]}{dt}\Big|_{t=0}$$

$$= \frac{d}{dt}\left[\frac{\partial(f \circ \mathbf{r})(\mathbf{u}(t))}{\partial u_i} Y_i(\mathbf{u}(t))\right]_{t=0}$$

$$= \left[\frac{\partial^2(f \circ \mathbf{r})(\mathbf{u}(t))}{\partial u_j \partial u_i} \frac{du_j(t)}{dt} Y_i(\mathbf{u}(t))\right]_{t=0}$$

$$+ \left[\frac{\partial(f \circ \mathbf{r})(\mathbf{u}(t))}{\partial u_i} \frac{\partial Y_i(\mathbf{u}(t))}{\partial u_j} \frac{du_j(t)}{dt}\right]_{t=0}$$

$$= \left[\frac{\partial^2(f \circ \mathbf{r})(\mathbf{u}(t))}{\partial u_j \partial u_i} X_j(\mathbf{u}(t)) Y_i(\mathbf{u}(t))\right]_{t=0}$$

$$+ \left[\frac{\partial(f \circ \mathbf{r})(\mathbf{u}(t))}{\partial u_i} \frac{\partial Y_i(\mathbf{u}(t))}{\partial u_j} X_j(\mathbf{u}(t))\right]_{t=0}$$

$$= X \cdot (Y \cdot f)(p). \tag{10.43}$$

Definition 10.11.3 (Derivative of $Y \in \mathscr{X}(M)$ on the Direction of $X \in \mathscr{X}(M)$)
Let $M \subset \mathbb{R}^m$ be an m-dimensional surface where $1 \leq m < n$. Let $X, Y \in \mathscr{X}(M)$.
Let $p = \mathbf{r}(\mathbf{u}(0))$ where $\mathbf{r}(\mathbf{u}(t))$ is the integral curve of X about $p = \mathbf{r}(\mathbf{u}_0)$, where
$\mathbf{u}_0 = \mathbf{u}(0)$.

We define the derivative of Y on the direction X at the point $p = \mathbf{r}(\mathbf{u}_0)$, denoted
by

$$(D_X Y)(p),$$

by

$$(D_X Y)(p) = \lim_{t \to 0} \frac{Y(\mathbf{u}(t)) - Y(\mathbf{u}(0))}{t} = \frac{dY(\mathbf{u}(t))}{dt}\Big|_{t=0}.$$

Remark 10.1 Observe that

$$Y(\mathbf{u}(t)) = Y_i(\mathbf{u}(t)) \frac{\partial \mathbf{r}(\mathbf{u}(t))}{\partial u_i},$$

so that

$$(D_X Y)(p) = \frac{\partial Y_i(\mathbf{u}(0))}{\partial u_j} \frac{du_j(0)}{dt} \frac{\partial \mathbf{r}(\mathbf{u}(0))}{\partial u_i} + Y_i(\mathbf{u}(0)) \frac{\partial^2 \mathbf{r}(\mathbf{u}(0))}{\partial u_i \partial u_j} \frac{du_j(0)}{dt}$$

$$= \frac{\partial Y_i(\mathbf{u}(0))}{\partial u_j} X_j(\mathbf{u}(0)) + Y_i(\mathbf{u}(0)) \frac{\partial^2 \mathbf{r}(\mathbf{u}(0))}{\partial u_i \partial u_j} X_j(\mathbf{u}(0))$$

$$= dY_i(X)(\mathbf{u}_0) \frac{\partial \mathbf{r}(\mathbf{u}_0)}{\partial u_i} + Y_i(\mathbf{u}_0) X_j(\mathbf{u}_0) \frac{\partial^2 \mathbf{r}(\mathbf{u}_0)}{\partial u_i \partial u_j}. \qquad (10.44)$$

Summarizing,

$$(D_X Y)(p) = dY_i(X)(\mathbf{u}_0) \frac{\partial \mathbf{r}(\mathbf{u}_0)}{\partial u_i} + Y_i(\mathbf{u}_0) X_j(\mathbf{u}_0) \frac{\partial^2 \mathbf{r}(\mathbf{u}_0)}{\partial u_i \partial u_j}.$$

Definition 10.11.4 (Lie Derivative of a Vector Field) Let $M \subset \mathbb{R}^n$ be a C^1 class
m-dimensional surface where $1 \leq m < n$. Let $X \in \tilde{\mathscr{X}}(M)$. Let $\mathbf{r}(\mathbf{u}(t))$ be the
point-wise representation of the integral curve of X around $p = \mathbf{r}(\mathbf{u}(0))$.

Let $Y \in \tilde{\mathscr{X}}(M)$ and $f \in C^2(M)$. We define the Lie derivative of Y, along the
direction X, at $f \in C^2(M)$ at the point $p = \mathbf{r}(\mathbf{u}(0))$, denoted by $L_X Y(f)(p)$, by

$$L_X Y(f)(p) = X \cdot (Y \cdot f)(p) - Y \cdot (X \cdot f)(p).$$

Observe that in such a case

$$L_X Y(f)(p) = ([X, Y] \cdot f)(p), \ \forall f \in C^2(M),$$

so that we denote simply,

$$L_X Y = [X, Y].$$

Relating the statements of this last definition, consider the following exercises.

Exercises 10.11.5
1. Prove that

$$L_X[Y, Z] = [L_X, Y, Z] + [Y, L_X Z], \ \forall X, Y, Z \in \tilde{\mathscr{X}}(M).$$

2. Prove that

$$L_X(L_Y)Z - L_Y(L_X)Z = L_{[X,Y]}Z, \ \forall X, Y, Z \in \tilde{\mathscr{X}}((M)).$$

10.11.1   On the Integral Curve Existence

In this section we present some well-known results about the existence of solutions for a nonlinear system of ordinary differential equations. At this point we recall, in a more general fashion, the definition of integral curve.

Let $M \subset \mathbb{R}^n$ be a C^1 class m-dimensional surface, where $1 \le m < n$, where

$$M = \{\mathbf{r}(\mathbf{u}) \ : \ \mathbf{u} \in D\},$$

where $D \subset \mathbb{R}^m$ is a connected set such that ∂D is of C^1 class.
Let $X \in \mathscr{X}(M)$, that is

$$X = X_i(\mathbf{u}) \frac{\partial \mathbf{r}(\mathbf{u})}{\partial u_i}.$$

Let $\mathbf{u}_0 \in D^\circ$ and $p = \mathbf{r}(\mathbf{u}_0)$. An integral curve for X around p is a curve

$$(\mathbf{r} \circ u) : I \to M,$$

where I is a closed interval, such that $t_0 \in I^\circ$,

$$\mathbf{r}(\mathbf{u}(0)) = r(\mathbf{u}_0),$$

and

$$\frac{d\mathbf{r}(\mathbf{u}(t))}{dt} = X(\mathbf{u}(t)),$$

that is

$$\frac{\partial \mathbf{r}(\mathbf{u}(t))}{\partial u_i} \frac{du_i(t)}{\partial t} = X_i(\mathbf{u}(t)) \frac{\partial \mathbf{r}(\mathbf{u}(t))}{\partial u_i},$$

so that in a component wise fashion,

$$\frac{du_i(t)}{\partial t} = X_i(\mathbf{u}(t)), \ \forall i \in \{1, \ldots m\}.$$

We present now the results concerning the existence of solutions for the ordinary differential equation systems in question.

Theorem 10.11.6 *Let $V \subset \mathbb{R}^m$ be an open set. Let $\hat{X} : V \to \mathbb{R}^m$ be a continuous function such that*

$$|\hat{X}(\mathbf{u}) - \hat{X}(\mathbf{v})| \le K|\mathbf{u} - \mathbf{v}|, \ \forall \mathbf{u}, \mathbf{v} \in V,$$

for some $K > 0$.
Let $\mathbf{u}_0 \in V$ and let $r > 0$ be such that $B_r(\mathbf{u}_0) \in V$, where

$$B_r(\mathbf{u}_0) = \{\mathbf{u} \in \mathbb{R}^m \ : \ |\mathbf{u} - \mathbf{u}_0| < r\}.$$

Let $C > 0$ be such that

$$|\hat{X}(\mathbf{u})| < C, \ \forall \mathbf{u} \in B_r(\mathbf{u}_0).$$

Let $t_0 \in \mathbb{R}$ and let $\alpha = \min\{1/K, r/C\}$. Under such hypotheses there exists a curve $\mathbf{u} : I_\alpha \to B_r(\mathbf{u}_0)$ such that

$$\begin{cases} \frac{d\mathbf{u}(t)}{dt} = \hat{X}(\mathbf{u}(t)) \\ \mathbf{u}(t_0) = \mathbf{u}_0, \end{cases} \tag{10.45}$$

where,

$$I_\alpha = [t_0 - \alpha, t_0 + \alpha].$$

Proof Observe that

$$\begin{cases} \frac{d\mathbf{u}(t)}{dt} = \hat{X}(\mathbf{u}(t)) \\ \mathbf{u}(t_0) = \mathbf{u}_0, \end{cases} \tag{10.46}$$

is equivalent to

$$\mathbf{u}(t) = \mathbf{u}_0 + \int_{t_0}^t \hat{X}(\mathbf{u}(s)) \, ds.$$

Define the sequence of functions $\{\mathbf{u}_n : I_\alpha \to \mathbb{R}^m\}$ by

$$\mathbf{u}_1(t) = \mathbf{u}_0,$$

$$\mathbf{u}_{n+1}(t) = \mathbf{u}_0 + \int_{t_0}^t \hat{X}(\mathbf{u}_n(s)) \, ds, \ \forall t \in I_\alpha, \ \forall n \in \mathbb{N}.$$

First, we shall prove by induction that

$$\mathbf{u}_n(t) \in B_r(\mathbf{u}_0), \ \forall n \in \mathbb{N}, \ t \in I_\alpha.$$

Clearly

$$\mathbf{u}_1(t) \in B_r(\mathbf{u}_0), \forall t \in I_\alpha.$$

Suppose

$$\mathbf{u}_n(t) \in B_r(\mathbf{u}_0), \ \forall t \in I_\alpha.$$

Thus,

$$|\hat{X}(\mathbf{u}_n(t))| < C, \ \forall t \in I_\alpha,$$

so that

$$\begin{aligned}
|\mathbf{u}_{n+1}(t) - \mathbf{u}_0| &\leq \left| \int_{t_0}^t |\hat{X}(\mathbf{u}_n(s))| \, ds \right| \\
&< C|t - t_0| \\
&\leq C\alpha \\
&\leq C\frac{r}{C} \\
&= r,
\end{aligned} \tag{10.47}$$

so that

$$\mathbf{u}_{n+1}(t) \in B_r(\mathbf{u}_0), \ \forall t \in I_\alpha.$$

The induction is complete, that is,

$$\mathbf{u}_n(t) \in B_r(\mathbf{u}_0), \ \forall t \in I_\alpha, \ \forall n \in \mathbb{N}.$$

Moreover, for $t \geq t_0$ we have

$$|\mathbf{u}_{n+1}(t) - \mathbf{u}_n(t)| \leq \int_{t_0}^{t} |\hat{X}(\mathbf{u}_n(s)) - \hat{X}(\mathbf{u}_{n-1}(s))| \, ds$$

$$\leq \int_{t_0}^{t} K |\mathbf{u}_n(s) - \mathbf{u}_{n-1}(s)| \, ds$$

$$\leq K \int_{t_0}^{t} \int_{t_0}^{s} |\hat{X}(\mathbf{u}_{n-1}(s_1)) - \hat{X}(\mathbf{u}_{n-2}(s_1))| \, ds_1 \, ds$$

$$\leq K^2 \int_{t_0}^{t} \int_{t_0}^{s} |\mathbf{u}_{n-1}(s_1) - \mathbf{u}_{n-2}(s_1)| \, ds_1 \, ds$$

$$\leq K^2 \int_{t_0}^{t} \int_{t_0}^{s} \int_{t_0}^{s_1} |\hat{X}(\mathbf{u}_{n-2}(s_2)) - \hat{X}(\mathbf{u}_{n-3}(s_2))| \, ds_2 \, ds_1 \, ds$$

$$\leq K^3 \int_{t_0}^{t} \int_{t_0}^{s} \int_{t_0}^{s_1} |\mathbf{u}_{n-2}(s_2) - \mathbf{u}_{n-3}(s_2)| \, ds_2 \, ds_1 \, ds. \quad (10.48)$$

Proceeding inductively in this fashion, and recalling that point-wise $\mathbf{u}_1(t) = \mathbf{u}_0$, we finally would obtain

$$|\mathbf{u}_{n+1}(t) - \mathbf{u}_n(t)| \leq K^n \int_{t_0}^{t} \int_{t_0}^{s} \int_{t_0}^{s_1} \cdots \int_{t_0}^{s_{n-1}} |\mathbf{u}_2(s_n) - \mathbf{u}_0| \, ds_n \, ds_{n_1} \cdots ds_1 \, ds$$

$$\leq K^n r \int_{t_0}^{t} \int_{t_0}^{s} \int_{t_0}^{s_1} \cdots \int_{t_0}^{s_{n-1}} \cdot ds_n \, ds_{n-1} \cdots ds_1 \, ds$$

$$\leq K^n r \frac{|t - t_0|^{n+1}}{(n+1)!}. \quad (10.49)$$

The same estimate is valid for $t \leq t_0$.

Therefore,

$$|\mathbf{u}_{n+1}(t) - \mathbf{u}_n(t)| \leq r K^n |t - t_0|^n \frac{|t - t_0|}{(n+1)!}$$

$$\leq r K^n \alpha^n \frac{|t - t_0|}{(n+1)!}$$

$$\leq r \frac{|t - t_0|}{(n+1)!}$$

$$\to 0, \quad \text{uniformly as } n \to \infty. \quad (10.50)$$

Fix $n, p \in \mathbb{N}$.

From the last inequality

$$|\mathbf{u}_{n+p}(t) - \mathbf{u}_n(t)|$$

$$= |\mathbf{u}_{n+p}(t) - \mathbf{u}_{n+p-1}(t) + \mathbf{u}_{n+p-1}(t)$$

$$-\mathbf{u}_{n+p-2}(t) + \mathbf{u}_{n+p-2}(t) + \cdots + \mathbf{u}_{n+1}(t) - \mathbf{u}_n(t)|$$

$$\leq |\mathbf{u}_{n+p}(t) - \mathbf{u}_{n+p-1}(t)| + |\mathbf{u}_{n+p-1}(t) - \mathbf{u}_{n+p-2}(t)| + \cdots + |\mathbf{u}_{n+1}(t) - \mathbf{u}_n(t)|$$

$$\leq r|t - t_0| \sum_{k=1}^{p} \frac{1}{(n+k)!}$$

$$\leq r|t - t_0| \sum_{k=(n+1)}^{\infty} \frac{1}{k!}. \tag{10.51}$$

Let us denote $\{a_k\} = \frac{1}{k!}$. Observe that

$$\lim_{k\to\infty} \frac{a_{k+1}}{a_k} = \lim_{k\to\infty} \frac{k!}{(k+1)!} = \lim_{k\to\infty} \frac{1}{k+1} = 0.$$

Hence,

$$\sum_{k=1}^{\infty} \frac{1}{k!}$$

is converging so that

$$\sum_{k=(n+1)}^{\infty} \frac{1}{k!} \to 0, \text{ as } n \to \infty.$$

From this and (10.51), $\{\mathbf{u}_n\}$ is a uniformly Cauchy sequence of continuous functions, which uniformly converges to some continuous $\mathbf{u} : I_\alpha \to \overline{B}_r(\mathbf{u}_0)$, such that

$$\mathbf{u}(t) = \mathbf{u}_0 + \int_{t_0}^{t} \hat{X}(\mathbf{u}(s))\, ds,$$

so that

$$\begin{cases} \frac{d\mathbf{u}(t)}{dt} = \hat{X}(\mathbf{u}(t)) \\ \mathbf{u}(t_0) = \mathbf{u}_0. \end{cases} \tag{10.52}$$

The proof of uniqueness of $\mathbf{u}$ is left as an exercise.

Theorem 10.11.7 (Gronwall's Inequality) *Let $f, g : [a, b] \to \mathbb{R}$ be continuous and nonnegative functions. Suppose there exists $A > 0$ such that*

$$f(t) \leq A + \int_a^t f(s)g(s) \, ds, \quad \forall t \in [a, b].$$

Under such hypotheses,

$$f(t) \leq A e^{\int_a^t g(s) \, ds}, \quad \forall t \in [a, b].$$

Proof Define $h : [a, b] \to \mathbb{R}$ by

$$h(t) = A + \int_a^t f(s)g(s) \, ds.$$

From the hypotheses, $h(t) > 0$ and

$$f(t) \leq h(t), \quad \forall t \in [a, b].$$

Moreover,

$$h'(t) = f(t)g(t) \leq h(t)g(t), \quad \forall t \in [a, b].$$

Therefore,

$$\frac{h'(t)}{h(t)} \leq g(t), \quad \forall t \in [a, b],$$

so that

$$\frac{d \ln(h(t))}{dt} \leq g(t), \quad \forall t \in [a, b],$$

and hence

$$\ln(h(t)) - \ln(A) \leq \int_a^t g(s) \, ds, \quad \forall t \in [a, b],$$

that is,

$$\ln(h(t)/A) \leq \int_a^t g(s) \, ds,$$

so that

$$\frac{h(t)}{A} \leq e^{\int_a^t g(s) \, ds},$$

that is,

$$f(t) \le h(t) \le Ae^{\int_a^t g(s)\, ds}, \ \forall t \in [a, b].$$

The proof is complete.

Theorem 10.11.8 *Let $V \subset \mathbb{R}^m$ be an open set. Let $\hat{X} : V \to \mathbb{R}^m$ be a continuous function such that*

$$|\hat{X}(\mathbf{u}) - \hat{X}(\mathbf{v})| \le K|\mathbf{u} - \mathbf{v}|, \ \forall \mathbf{u}, \mathbf{v} \in V,$$

for some $K > 0$.

Let $\mathbf{u}_0 \in V$ and let $r > 0$ be such that $B_r(\mathbf{u}_0) \in V$, where

$$B_r(\mathbf{u}_0) = \{\mathbf{u} \in \mathbb{R}^m \ : \ |\mathbf{u} - \mathbf{u}_0| < r\}.$$

Let $C > 0$ be such that

$$|\hat{X}(\mathbf{u})| < C, \ \forall \mathbf{u} \in B_r(\mathbf{u}_0).$$

Let $F_t(\mathbf{u}_0)$ denote the unique integral curve $\mathbf{u} : I_\alpha \to \mathbb{R}$ such that

$$\begin{cases} \frac{d\mathbf{u}(t)}{dt} = \hat{X}(\mathbf{u}(t)) \\ \mathbf{u}(0) = \mathbf{u}_0, \end{cases} \tag{10.53}$$

where $\alpha = \min\{1/K, r/C\}$, and $I_\alpha = [-\alpha, \alpha]$.

Under such hypotheses there exist $r_1 > 0$ and $\varepsilon > 0$ such that for each $\mathbf{v}_0 \in B_r(\mathbf{u}_0)$ there exists an integral $\mathbf{v} : [-\varepsilon, \varepsilon] \to \mathbb{R}^m$ such that

$$\begin{cases} \frac{d\mathbf{v}(t)}{dt} = \hat{X}(\mathbf{v}(t)) \\ \mathbf{v}(0) = \mathbf{v}_0. \end{cases} \tag{10.54}$$

Moreover,

$$|\mathbf{u}(t) - \mathbf{v}(t)| \le e^{K|t|}|\mathbf{u}_0 - \mathbf{v}_0|, \forall t \in [-\varepsilon, \varepsilon].$$

Proof Define $r_1 = r/2$ and $\varepsilon = \min\{1/K, r/(2C)\}$.

Let $\mathbf{v}_0 \in B_r(\mathbf{u}_0)$. Thus $B_{r_1}(\mathbf{v}_0) \subset B_r(\mathbf{u}_0)$, so that

$$|\hat{X}(\mathbf{v})| \le C, \ \forall \mathbf{v} \in B_{r_1}(\mathbf{u}_0).$$

From Theorem 10.11.6, with $\mathbf{v}_0$ in place of $\mathbf{u}_0$, r_1 in place of r, 0 in place of t_0, and ε in place of α, there exists a curve $\mathbf{v} : I_\varepsilon \to \mathbb{R}^m$, where $I_\varepsilon = [-\varepsilon, \varepsilon]$, such that

$$\begin{cases} \frac{d\mathbf{v}(t)}{dt} = \hat{X}(\mathbf{v}(t)) \\ \mathbf{v}(0) = \mathbf{v}_0. \end{cases} \tag{10.55}$$

Now define

$$f(t) = |\mathbf{u}(t) - \mathbf{v}(t)|.$$

Thus,

$$\begin{aligned}
f(t) &= |\mathbf{u}(t) - \mathbf{v}(t)| \\
&= \left| \int_0^t (\hat{X}(\mathbf{u}(s)) - \hat{X}(\mathbf{v}(s))) \, ds + \mathbf{u}_0 - \mathbf{v}_0 \right| \\
&\leq \left| \int_0^t K |\mathbf{u}(s) - \mathbf{v}(s)| \, ds \right| + |\mathbf{u}_0 - \mathbf{v}_0| \\
&\leq |\mathbf{u}_0 - \mathbf{v}_0| + K \int_0^{|t|} f(s) \, ds.
\end{aligned} \tag{10.56}$$

From the Gronwall's inequality with $A = |\mathbf{u}_0 - \mathbf{v}_0|$ and $g(t) = K$ we obtain,

$$\begin{aligned}
|\mathbf{u}(t) - \mathbf{v}(t)| &= f(t) \\
&\leq h(t) \\
&= |\mathbf{u}_0 - \mathbf{v}_0| + K \int_0^{|t|} f(s) \, ds \\
&\leq A e^{\int_0^{|t|} K \, ds} \\
&= |\mathbf{u}_0 - \mathbf{v}_0| e^{K|t|}.
\end{aligned} \tag{10.57}$$

The proof is complete.

10.12   Differential Forms

We start with the following preliminary result.

Theorem 10.12.1 (Partition of Unity) *Let $K \subset \mathbb{R}^n$ be a compact set such that*

$$K \subset \cup_{i=1}^m V_i,$$

where $V_i \subset \mathbb{R}^n$ is bounded and open, $\forall i \in \{1, \dots, m\}$.
Under such hypotheses, there exist functions $h_1, \dots, h_m : \mathbb{R}^n \to \mathbb{R}$ such that

$$\sum_{i=1}^m h_i(u) = 1, \ \forall u \in K,$$

$$0 \leq h_i(u) \leq 1, \ \forall u \in \mathbb{R}^n$$

and

$$h_i \in C_c(V_i),$$

that is, h_i is continuous and with compact support contained in V_i, $\forall i \in \{1, \ldots, m\}$.

We recall that the support of h_i denoted by $supp \, h_i$ is defined by

$$supp \, h_i = \overline{\{u \in \mathbb{R}^n : h_i(u) \neq 0\}},$$

$\forall i \in \{1, \ldots, m\}$.

Proof Let $u \in K \subset \cup_{i=1}^m V_i$.

Thus, there exists $j \in \{1, \ldots, m\}$ such that $u \in V_j$.

Since V_j is open, there exists $r_u > 0$ such that

$$\overline{B}_{r_u}(u) \subset V_j.$$

Observe that

$$K \subset \cup_{u \in K} B_{r_u}(u).$$

Since K is compact, there exists $u_1, \ldots, u_N \in K$ such that

$$K \subset \cup_{j=1}^N B_{r_j}(u_j),$$

where we have denoted $r_{u_j} = r_j$, $\forall j \in \{1, \ldots, N\}$.

For each $i \in \{1, \ldots, m\}$ define $\tilde{W}_i$ as the union of all $\overline{B}_{r_j}(u_j)$ which are contained in V_i.

For each $i \in \{1, \ldots, m\}$, select also an open set W_i such that

$$\overline{\tilde{W}}_i \subset W_i \subset \overline{W}_i \subset V_i.$$

At this point, define $g_i : \mathbb{R}^n \to \mathbb{R}$, by

$$g_i(u) = \frac{d(u, W_i^c)}{d(u, W_i^c) + d(u, \tilde{W}_i)},$$

where generically, for $B \subset \mathbb{R}^n$ we define

$$d(u, B) = \inf\{|u - v| : v \in B\}.$$

Observe that

$$g_i(u) = 1, \forall u \in \tilde{W}_i,$$

$$0 \leq g_i(u) \leq 1, \forall u \in \mathbb{R}^n$$

and

$$g_i(u) = 0, \forall u \in \overline{W_i^c},$$

so that

$$\text{supp } g_i \subset V_i, \ \forall i \in \{1, \ldots, m\}.$$

Define

$$h_1 = g_1,$$

$$h_2 = (1 - g_1)g_2,$$

$$h_3 = (1 - g_1)(1 - g_2)g_3,$$

$$\vdots$$

$$h_m = (1 - g_1)(1 - g_2) \cdots (1 - g_{m-1})g_m.$$

Hence

$$0 \leq h(u) \leq 1, \forall u \in \mathbb{R}^n,$$

and $h_i \in C_c(V_i)$, $\forall i \{1, \ldots, m\}$.

Now we are going to show by induction that

$$h_1 + h_2 + \cdots + h_j = 1 - (1 - g_1)(1 - g_2) \cdots (1 - g_j), \ \forall j \in \{1, \ldots, m\}.$$

For $j = 1$, we have $h_1 = g_1 = 1 - (1 - g_1)$.

Suppose that for $2 \leq j < m$ we have

$$h_1 + h_2 + \cdots + h_j = 1 - (1 - g_1)(1 - g_2) \cdots (1 - g_j).$$

Thus,

$$h_1 + h_2 + \cdots + h_j + h_{j+1}$$
$$= 1 - (1 - g_1)(1 - g_2) \cdots (1 - g_j) + (1 - g_1)(1 - g_2) \cdots (1 - g_j)g_{j+1}$$
$$= 1 - (1 - g_1)(1 - g_2) \cdots (1 - g_{j+1}). \tag{10.58}$$

The induction is complete so that, in particular, we have obtained

$$h_1 + h_2 + \cdots + h_m = 1 - (1 - g_1)(1 - g_2) \cdots (1 - g_m). \tag{10.59}$$

Hence, if $u \in K$, then $u \in K \subset \cup_{j=1}^N B_{r_j}(u_j) = \cup_{i=1}^m \tilde{W}_i$ so that $u \in \tilde{W}_j$, for some $j \in \{1, \ldots, m\}$.

Thus, $g_j(u) = 0$, so that from this and (10.59), we obtain

$$h_1 + \cdots + h_m(u) = 1, \forall u \in K.$$

The proof is complete.

Remark 10.12.2 Concerning this last theorem, the set $\{h_1, \ldots, h_m\}$ is said to be a partition of unity subordinate to $V_1, \ldots, V_m$ and related to K.

At this point, we recall to have already established that $\{du_i\}_{i=1}^m$ is the dual basis for $T_p(M)^*$ corresponding to the basis

$$\left\{ \frac{\partial \mathbf{r}(\mathbf{u})}{\partial u_i} \right\}_{i=1}^m,$$

of $T_p(M)$.

Observe that for $f \in C^1(M)$ and

$$\mathbf{v} = v_i \frac{\partial \mathbf{r}(\mathbf{u})}{\partial u_i} \in T_p(M),$$

we have,

$$df(\mathbf{v}) = \frac{\partial f(\mathbf{r}(\mathbf{u}))}{\partial X_j} \frac{\partial X_j(\mathbf{u})}{\partial u_i} v_i. \tag{10.60}$$

On the other hand, denoting

$$\tilde{\mathbf{e}}_j = \frac{\partial \mathbf{r}(\mathbf{u})}{\partial u_j},$$

$$du_i(\mathbf{v}) = du_i \left(\sum_{j=1}^m v_j \tilde{\mathbf{e}}_j \right)$$

$$= v_j du_i(\tilde{\mathbf{e}}_j) = v_j \delta_{ij} = v_i. \tag{10.61}$$

From these last results we obtain,

$$df(\mathbf{v}) = \frac{\partial f(\mathbf{r}(\mathbf{u}))}{\partial X_j} \frac{\partial X_j(\mathbf{u})}{\partial u_i} v_i$$

$$= \frac{\partial f(\mathbf{r}(\mathbf{u}))}{\partial X_j} \frac{\partial X_j(\mathbf{u})}{\partial u_i} du_i(\mathbf{v})$$

$$= \frac{\partial f(\mathbf{r}(\mathbf{u}))}{\partial u_i} du_i(\mathbf{v}). \tag{10.62}$$

We have just got the differential expression

$$df = \frac{\partial f(\mathbf{r}(\mathbf{u}))}{\partial u_i} du_i,$$

that is,

$$df = \frac{\partial f(\mathbf{r}(\mathbf{u}))}{\partial X_j} \frac{\partial X_j(\mathbf{u})}{\partial u_i} du_i,$$

so that

$$df = \frac{\partial f(\mathbf{r}(\mathbf{u}))}{\partial X_j} dX_j(\mathbf{u}).$$

Hence, in this context, through $\{dX_j(\mathbf{u})\}$, we may express a set which spans $T_p(M)^*$.

Definition 10.12.3 Let M be a C^1 class manifold and let $p = \mathbf{r}(\mathbf{u})$. We define a 1-form as an element ω of $T_p(M)^*$, expressed by

$$\omega = \sum_{k=1}^{n} \omega_k(\mathbf{u}) dX_k(\mathbf{u}).$$

We define a k-form by

$$\omega = \sum_{I} \omega_I(\mathbf{u}) dX_I,$$

where ω_I is C^1 class function and $dX_I = dX_{i_1}(\mathbf{u}) \wedge \cdots \wedge dX_{i_k}(\mathbf{u})$, and

$$I = (i_1, \ldots, i_k)$$

is any collection of k indices $i_j \in \{1, \ldots, n\}$.

Definition 10.12.4 (Differential of a Form) For a 1-form

$$\omega = \sum_{k=1}^{n} \omega_k(\mathbf{u}) dX_k(\mathbf{u}),$$

we define its exterior differential denoted by $d\omega$ by

$$d\omega = \sum_{k=1}^{n} d\omega_k(\mathbf{u}) \wedge dX_k(\mathbf{u}),$$

that is,

$$d\omega = \sum_{k=1}^{n} \sum_{j=1}^{m} \frac{\partial \omega_k(\mathbf{u})}{\partial u_j} \, du_j \wedge dX_k(\mathbf{u}).$$

10.13 Integration of Differential Forms

In this section first we present a discussion about the subject, before presenting the main result, namely, the Stokes Theorem on its general form.

Let $M \subset \mathbb{R}^n$ be an m-dimensional C^1 class surface with a boundary ∂M, where generically,

$$M = \{\mathbf{r}(\mathbf{u}) \ : \ \mathbf{u} \in D\},$$

and

$$\partial M = \mathbf{r}(\partial D).$$

Here we assume D, ∂D and therefore M to be compact. Let $\omega = \omega_I \, dX_I$ be a k-form and consider the problem of calculating the integral:

$$I = \int_M \omega_I \, dX_I.$$

At this point, we suppose the manifold M is oriented, in the sense that, considering a canonical system

$$\{\mathbf{e}_1, \ldots, \mathbf{e}_n\} \subset \mathbb{R}^n,$$

and

$$\left\{ \frac{\partial \mathbf{r}(\mathbf{u})}{\partial u_i} \right\}_{i=1}^{m},$$

we may select linearly independent normal C^1 class fields, point-wise denoted by $\mathbf{n}_j(\mathbf{u})$, $\forall j \in \{1, \cdots, n - m\}$, orthogonal to

$$\left\{ \frac{\partial \mathbf{r}(\mathbf{u})}{\partial u_i} \right\}_{i=1}^{m},$$

and such that, for each $\mathbf{u} \in D$, $\mathbb{R}^n$ is spanned by

$$\left\{ \{\mathbf{n}_j(\mathbf{u})\}_{j=1}^{n-m}, \left\{ \frac{\partial \mathbf{r}(\mathbf{u})}{\partial u_i} \right\}_{i=1}^{m} \right\}.$$

Moreover, we assume that the matrix to change from this local basis

$$\left\{ \{ \mathbf{n}_j(\mathbf{u}) \}_{j=1}^{n-m} , \left\{ \frac{\partial \mathbf{r}(u)}{\partial u_i} \right\}_{i=1}^{m} \right\}$$

to the canonical one of $\mathbb{R}^n$, has always positive (or always negative) determinant, $\forall \mathbf{u} \in D$.

Let $p \in \partial \Omega$. Thus $\mathbf{u}_p = \mathbf{r}^{-1}(p) \in \partial D$.

We also assume that for an appropriate local system of coordinates compatible with the orientation, there exists a C^1 class function point-wise indicated by $g^p(u_1, \ldots, u_m)$ such that for a rectangle $B_p = \prod_{k=1}^{m} [\alpha_k^p, \beta_k^p]$ we have,

$$D \cap B_p = \{(u_1, \ldots, u_m) \in \mathbb{R}^m \ : \ \alpha_k^p \le u_k \le \beta_k^p, \ \forall k \in \{1, \ldots, m-1\},$$

$$\text{and } \alpha_m^p \le g^p(u_1, \ldots, u_m) \le u_m \le \beta_m^p \}. \tag{10.63}$$

Thus,

$$\cup_{p \in M} B_p^\circ \supset \partial D,$$

and since ∂D is compact, There exists $p_1, \ldots, p_s \in M$ such that

$$\partial D \subset \cup_{l=1}^{s} B_{p_l}^\circ.$$

Define

$$B_{p_0} = D^\circ \setminus \cup_{l=1}^{s} B_{p_l}.$$

Select a partition of unit

$$\{\rho_{p_l}\}_{l=0}^{s}$$

subordinate to

$$\{B_{p_0}^\circ, B_{p_1}^\circ, \ldots, B_{p_s}^\circ \}.$$

Observe that

$$\text{supp}\{\rho_{p_l}\} \subset B_{p_l}^\circ, \ \forall l \in \{0, \ldots, s\},$$

$$0 \le \rho_{p_l}(\mathbf{u}) \le 1, \ \forall l \in \{0, \ldots, s\}, \ \mathbf{u} \in D,$$

$$\sum_{l=0}^{s} \rho_{p_l} = 1, \forall \mathbf{u} \in D.$$

Let us first consider the specific general case in which:

$$\omega = \sum_{j=1}^{m} (-1)^{j+1} f_j(\mathbf{u}) \, du_1 \wedge du_2 \cdots \wedge \hat{du}_j \wedge \cdots \wedge du_m,$$

where $f_j : D \to \mathbb{R}$ are C^1 class functions $\forall j \in \{1, \ldots, m\}$, and where $\hat{du}_j$ means that the term du_j is absent in the products in question.

Observe that

$$d\omega = \sum_{j=1}^{m} \frac{\partial f_j(\mathbf{u})}{\partial u_j} du_1 \wedge \cdots \wedge du_m.$$

Denote also,

$$\mathbf{s}_k = \frac{\partial \mathbf{r}(\mathbf{u})}{\partial u_k} \, du_k,$$

$$\mathbf{S}_k = (\mathbf{s}_1, \mathbf{s}_2, \ldots, \hat{\mathbf{s}}_k, \ldots, \mathbf{s}_m),$$

$$\mathbf{S} = (\mathbf{s}_1, \mathbf{s}_2, \ldots, \mathbf{s}_m),$$

Thus,

$$\int_M d\omega = \int_D \sum_{j=1}^{m} \frac{\partial f_j(\mathbf{u})}{\partial u_j} (du_1 \wedge \cdots \wedge du_m)(\mathbf{S}), \tag{10.64}$$

and hence,

$$\int_M d\omega$$

$$= \int_D \sum_{j=1}^{m} (-1)^{j+1} \frac{\partial f_j(\mathbf{u})}{\partial u_j} (du_j \wedge (du_1 \wedge du_2 \cdots \wedge \hat{du}_j \wedge \cdots \wedge du_m))(\mathbf{S})$$

$$= \int_D \sum_{j=1}^{m} (-1)^{j+1} \frac{\partial [(\sum_{l=1}^{s} \rho_{pl}) f_j(\mathbf{u})]}{\partial u_j} (du_j \wedge (du_1 \wedge du_2 \cdots \wedge \hat{du}_j \wedge \cdots \wedge du_m))(\mathbf{S})$$

$$= \sum_{l=1}^{s} \int_D \sum_{j=1}^{m} \left[(-1)^{j+1} \frac{\partial (\rho_{pl}(\mathbf{u}) f_j(\mathbf{u}))}{\partial u_j} du_j \right] du_1 du_2 \cdots \hat{du}_j \cdots du_m$$

$$= \sum_{l=1}^{s} \int_{\partial B_l} \sum_{j=1}^{m-1} (-1)^{j+1} [(\rho_{pl} f_j)(u_1, \ldots, \beta_j^l, \ldots, u_m)]$$

$$-(\rho_{p_l} f_j)(u_1, \ldots, \alpha^l_j, \ldots, u_m))]du_1 du_2 \cdots \hat{du}_j \cdots du_m$$

$$+ \int_{\partial B_l} (-1)^{m+1} (\rho_{p_l} f_m)(u_1, \ldots, u_{m-1}, \beta^l_m) \, du_1 \cdots du_{m-1}$$

$$- \int_{\partial D \cap B_l} (-1)^{m+1} (\rho_{p_l} f_m)(u_1, \ldots, u_{m-1}, g^l(u_1, \ldots, u_{m-1}) \, du_1 \cdots du_{m-1}$$

$$= 0 - \sum_{l=1}^{s} \int_{\partial D \cap B_l} (-1)^{m+1} (\rho_{p_l} f_m)(u_1, \ldots, u_{m-1}, g^l(u_1, \ldots, u_{m-1})) \, du_1 \cdots du_{m-1}$$

$$= - \int_{\partial D} (-1)^{m+1} \left(\sum_{l=1}^{s} \rho_{p_l} f_m \right) (u_1, \ldots, u_{m-1}, g^l(u_1, \ldots, u_{m-1})) \, du_1 \cdots du_{m-1}$$

$$= \int_{\partial D} \left(\sum_{l=1}^{s} \rho_{p_l}(\mathbf{u}) \right) \sum_{j=1}^{m} (-1)^{j+1} f_j(\mathbf{u}) du_1 du_2 \cdots \hat{du}_j \cdots du_m$$

$$= \int_{\partial D} \sum_{j=1}^{m} (-1)^{j+1} f_j(\mathbf{u}) du_1 du_2 \cdots \hat{du}_j \cdots du_m, \tag{10.65}$$

so that

$$\int_M d\omega = \int_{\partial D} \sum_{j=1}^{m} (-1)^{j+1} f_j(\mathbf{u}) du_1 du_2 \cdots \hat{du}_j \cdots du_m$$

$$= \int_{\partial D} \sum_{j=1}^{m} (-1)^{j+1} f_j(\mathbf{u})(du_1 \wedge du_2 \cdots \wedge \hat{du}_j \wedge \cdots \wedge du_m)(\mathbf{S}_j)$$

$$= \int_{\partial M} \omega. \tag{10.66}$$

10.14   A Simple Example to Illustrate the Integration Process

We would emphasize that, at the end of last section, when the wedge product is absent, the differential forms and integrations are relating the usual calculus sense.

In the next example, we see how to connect the general differential form to a usual one, also in the ordinary calculus sense.

So, let $M \subset \mathbb{R}^n$ be C^1 class surface given by:

$$M = \{\mathbf{r}(\mathbf{u}) \; : \; \mathbf{u} \in D\},$$

where, for simplicity $D = [a, b] \times [c, d]$.

Consider the integral

$$I = \int_M f \, dX_1 \wedge dX_2$$

$$= \int_D f(\mathbf{u})(dX_1(\mathbf{u}) \wedge dX_2(\mathbf{u})) \left(\frac{\partial \mathbf{r}(\mathbf{u})}{\partial u_1} \, du_1, \frac{\partial \mathbf{r}(\mathbf{u})}{\partial u_2} \, du_2 \right) \qquad (10.67)$$

where, $X_1 : D \rightarrow \mathbb{R}$ and $X_2 : D \rightarrow \mathbb{R}$ are C^1 class functions.

Observe that

$$dX_1(\mathbf{u}) = \frac{\partial X_1(\mathbf{u})}{\partial u_1} \, du_1 + \frac{\partial X_1(\mathbf{u})}{\partial u_2} \, du_2$$

and

$$dX_2(\mathbf{u}) = \frac{\partial X_2(\mathbf{u})}{\partial u_1} \, du_1 + \frac{\partial X_2(\mathbf{u})}{\partial u_2} \, du_2$$

Consider the elementary rectangle

$$(u_1, u_1 + \Delta u_1) \times (u_2, u_2 + \Delta u_2),$$

which has as its sides the vectors

$$(\Delta u_1, 0) \in \mathbb{R}^2,$$

and

$$(0, \Delta u_2) \in \mathbb{R}^2.$$

Define

$$\tilde{\mathbf{s}}_1 = \frac{\partial \mathbf{r}(u)}{\partial u_1} \Delta u_1,$$

and

$$\tilde{\mathbf{s}}_2 = \frac{\partial \mathbf{r}(u)}{\partial u_2} \Delta u_2,$$

Considering the basis

$$\left\{ \frac{\partial \mathbf{r}(u)}{\partial u_1}, \frac{\partial \mathbf{r}(u)}{\partial u_2} \right\},$$

all to simplify the notation we shall denote

$$\tilde{\mathbf{s}}_1 = (\Delta u_1, 0),$$

and

$$\tilde{\mathbf{s}}_2 = (0, \Delta u_2).$$

Recall that, for the general case, for

$$\mathbf{v} = v_i \frac{\partial \mathbf{r}(u)}{\partial u_i}.$$

we have

$$du_j(\mathbf{v}) = v_j, \forall j \in \{1, \ldots, m\}.$$

Recall also that from Definition 10.10.2,

$$(F_1 \wedge \cdots \wedge F_k)(\mathbf{v}_1, \ldots, \mathbf{v}_k) = \det\{F_i(\mathbf{v}_j)\}_{i,j=1}^k.$$

Let us evaluate

$$(dX_1(\mathbf{u}) \wedge dX_2(\mathbf{u})) \left(\frac{\partial \mathbf{r}(\mathbf{u})}{\partial u_1} \Delta u_1, \frac{\partial \mathbf{r}(\mathbf{u})}{\partial u_2} \Delta u_2 \right)$$

$$= (dX_1(\mathbf{u}) \wedge dX_2(\mathbf{u}))[(\tilde{\mathbf{s}}_1, \tilde{\mathbf{s}}_2)]$$

$$= (dX_1(\mathbf{u}) \wedge dX_2(\mathbf{u}))[(\Delta u_1, 0), (0, \Delta u_2)]$$

$$= \left[\left(\frac{\partial X_1(\mathbf{u})}{\partial u_1} du_1 + \frac{\partial X_1(\mathbf{u})}{\partial u_2} du_2 \right) \wedge \left(\frac{\partial X_2(\mathbf{u})}{\partial u_1} du_1 + \frac{\partial X_2(\mathbf{u})}{\partial u_2} du_2 \right) \right]$$

$$((\Delta u_1, 0), (0, \Delta u_2))$$

$$= \left[\left(\frac{\partial X_1(\mathbf{u})}{\partial u_1} \frac{\partial X_2(\mathbf{u})}{\partial u_2} du_1 \wedge du_2 \right) + \left(\frac{\partial X_1(\mathbf{u})}{\partial u_2} \frac{\partial X_2(\mathbf{u})}{\partial u_1} du_2 \wedge du_1 \right) \right]$$

$$((\Delta u_1, 0), (0, \Delta u_2))$$

$$= \left(\frac{\partial X_1(\mathbf{u})}{\partial u_1} \frac{\partial X_2(\mathbf{u})}{\partial u_2} du_1 \wedge du_2 \right) ((\Delta u_1, 0), (0, \Delta u_2))$$

$$+ \left(\frac{\partial X_1(\mathbf{u})}{\partial u_2} \frac{\partial X_2(\mathbf{u})}{\partial u_1} du_2 \wedge du_1 \right) ((\Delta u_1, 0), (0, \Delta u_2))$$

$$= \frac{\partial X_1(\mathbf{u})}{\partial u_1} \frac{\partial X_2(\mathbf{u})}{\partial u_2} \begin{vmatrix} du_1(\Delta u_1, 0) & du_1(0, \Delta u_2) \\ du_2(\Delta u_1, 0) & du_2(0, \Delta u_2) \end{vmatrix}$$

$$+ \frac{\partial X_1(\mathbf{u})}{\partial u_2} \frac{\partial X_2(\mathbf{u})}{\partial u_1} \begin{vmatrix} du_2(\Delta u_1, 0) & du_2(0, \Delta u_2) \\ du_1(\Delta u_1, 0) & du_1(0, \Delta u_2) \end{vmatrix}$$

$$= \frac{\partial X_1(\mathbf{u})}{\partial u_1} \frac{\partial X_2(\mathbf{u})}{\partial u_2} \begin{vmatrix} \Delta u_1 & 0 \\ 0 & \Delta u_2 \end{vmatrix}$$

$$+ \frac{\partial X_1(\mathbf{u})}{\partial u_2} \frac{\partial X_2(\mathbf{u})}{\partial u_1} \begin{vmatrix} 0 & \Delta u_2 \\ \Delta u_1 & 0 \end{vmatrix}$$

$$= \left(\frac{\partial X_1(\mathbf{u})}{\partial u_1} \frac{\partial X_2(\mathbf{u})}{\partial u_2} - \frac{\partial X_1(\mathbf{u})}{\partial u_2} \frac{\partial X_2(\mathbf{u})}{\partial u_1} \right) \Delta u_1 \Delta u_2. \tag{10.68}$$

So, to summarize,

$$(dX_1(\mathbf{u}) \wedge dX_2(\mathbf{u}))[(\Delta u_1, 0), (0, \Delta u_2)]$$

$$= \left(\frac{\partial X_1(\mathbf{u})}{\partial u_1} \frac{\partial X_2(\mathbf{u})}{\partial u_2} - \frac{\partial X_1(\mathbf{u})}{\partial u_2} \frac{\partial X_2(\mathbf{u})}{\partial u_1} \right) \Delta u_1 \Delta u_2, \tag{10.69}$$

or in its differential form:

$$(dX_1(\mathbf{u}) \wedge dX_2(\mathbf{u}))[(du_1, 0), (0, du_2)]$$

$$= \left(\frac{\partial X_1(\mathbf{u})}{\partial u_1} \frac{\partial X_2(\mathbf{u})}{\partial u_2} - \frac{\partial X_1(\mathbf{u})}{\partial u_2} \frac{\partial X_2(\mathbf{u})}{\partial u_1} \right) du_1 du_2. \tag{10.70}$$

Observe that this last differential form is one in the usual sense calculus, which is also used in the final usual integration process.

10.15   Volume (Area) of a Surface

Consider the problem of calculating the volume defined by the vectors

$$\mathbf{v}_1, \mathbf{v}_2, \ldots, \mathbf{v}_m \in \mathbb{R}^n$$

where $1 \leq m \leq n$.

Definition 10.15.1 Given $\mathbf{v}_1, \mathbf{v}_2 \in \mathbb{R}^n$ we shall define the volume (in fact area) defined by $\mathbf{v}_1$ and $\mathbf{v}_2$, denoted by V_2, by

$$V_2 = |\mathbf{v}_1||\mathbf{v}_2|| \sin \theta|,$$

where θ is the angle between $\mathbf{v}_1$ and $\mathbf{v}_2$, which is given by:

$$\cos \theta = \frac{\mathbf{v}_1 \cdot \mathbf{v}_2}{|\mathbf{v}_1||\mathbf{v}_2|},$$

so that

$$V_2 = |\mathbf{v}_1||\mathbf{v}_2|\sqrt{1 - \cos^2\theta}$$

$$= |\mathbf{v}_1||\mathbf{v}_2|\sqrt{1 - \frac{(\mathbf{v}_1 \cdot \mathbf{v}_2)^2}{|\mathbf{v}_1|^2|\mathbf{v}_2|^2}}$$

$$= \sqrt{|\mathbf{v}_1|^2|\mathbf{v}_2|^2 - (\mathbf{v}_1 \cdot \mathbf{v}_2)^2}, \tag{10.71}$$

that is,

$$V_2^2 = |\mathbf{v}_1|^2|\mathbf{v}_2|^2 - (\mathbf{v}_1 \cdot \mathbf{v}_2)^2$$

$$= \begin{vmatrix} \mathbf{v}_1 \cdot \mathbf{v}_1 & \mathbf{v}_1 \cdot \mathbf{v}_2 \\ \mathbf{v}_2 \cdot \mathbf{v}_1 & \mathbf{v}_2 \cdot \mathbf{v}_2 \end{vmatrix}$$

$$\equiv g_2, \tag{10.72}$$

so that,

$$V_2 = \sqrt{g_2},$$

where

$$g_2 = \det\{\mathbf{v}_i \cdot \mathbf{v}_j\}_{i,j=1}^2.$$

Definition 10.15.2 Let $\mathbf{v}_1, \ldots, \mathbf{v}_m \subset \mathbb{R}^n$ be nonzero vectors, where $2 \leq m \leq n$.

For $2 \leq s < m$, we shall define inductively the volume defined by $\mathbf{v}_1, \mathbf{v}_2, \ldots, \mathbf{v}_{s+1}$, by

$$V_2^2 = g_2,$$

and

$$V_{k+1}^2 = V_k^2|\tilde{\mathbf{v}}_{k+1}|^2, \forall k \in \{2, \ldots, s\},$$

where

$$\tilde{\mathbf{v}}_{k+1} = \mathbf{v}_{k+1} - \sum_{j=1}^k \tilde{a}_j \mathbf{v}_j,$$

and where

$$\{\tilde{a}_j\} = \arg\min\left\{ \left|\mathbf{v}_{k+1} - \sum_{j=1}^k a_j \mathbf{v}_j\right|^2 : a_1, \ldots, a_k \in \mathbb{R}\right\}$$

Theorem 10.15.3 *Under the statements of last definition, we have:*

$$V_k^2 = g_k, \forall k \in \{2, \ldots, s+1\},$$

where

$$g_k = \det\{\mathbf{v}_i \cdot \mathbf{v}_j\}_{i,j=1}^k.$$

Proof We prove the result by induction.

For $k = 2$ we have already:

$$V_2^2 = g_2.$$

For $2 < k < s + 1$ assume

$$V_k^2 = g_k.$$

It suffices, to complete the induction, to show that

$$V_{k+1}^2 = g_{k+1}.$$

By definition,

$$V_{k+1}^2 = V_k^2 |\tilde{\mathbf{v}}_{k+1}|^2,$$

where

$$\tilde{\mathbf{v}}_{k+1} = \mathbf{v}_{k+1} - \sum_{j=1}^k \tilde{a}_j \mathbf{v}_j,$$

and where

$$\{\tilde{a}_j\} = \arg\min\left\{ \left| \mathbf{v}_{k+1} - \sum_{j=1}^k a_j \mathbf{v}_j \right|^2 \; : \; a_1, \ldots, a_k \in \mathbb{R} \right\}$$

For this last optimization problem, the extremal necessary conditions are given by:

$$\left(\mathbf{v}_{k+1} - \sum_{j=1}^k \tilde{a}_j \mathbf{v}_j \right) \cdot \mathbf{v}_s = 0, \forall s \in \{1, \ldots, k\},$$

so that

$$\{\mathbf{v}_s \cdot \mathbf{v}_j\}\{\tilde{a}_j\} = \{\mathbf{v}_{k+1} \cdot \mathbf{v}_s\},$$

that is,

$$\{\tilde{a}_j\} = \{\mathbf{v}_s \cdot \mathbf{v}_j\}^{-1}\{\mathbf{v}_{k+1} \cdot \mathbf{v}_s\}.$$

Observe that

$$g_{k+1} = \{\mathbf{v}_i \cdot \mathbf{v}_j\}_{i,j=1}^{k+1}$$

$$= \begin{vmatrix} \mathbf{v}_1 \cdot \mathbf{v}_1 & \mathbf{v}_1 \cdot \mathbf{v}_2 & \cdots & \mathbf{v}_1 \cdot \mathbf{v}_{k+1} \\ \mathbf{v}_2 \cdot \mathbf{v}_1 & \mathbf{v}_2 \cdot \mathbf{v}_2 & \cdots & \mathbf{v}_2 \cdot \mathbf{v}_{k+1} \\ \vdots & \vdots & \ddots & \vdots \\ \mathbf{v}_{k+1} \cdot \mathbf{v}_1 & \mathbf{v}_{k+1} \cdot \mathbf{v}_2 & \cdots & \mathbf{v}_{k+1} \cdot \mathbf{v}_{k+1} \end{vmatrix} \qquad (10.73)$$

so that,

$$g_{k+1} = \begin{vmatrix} \mathbf{v}_1 \cdot \mathbf{v}_1 & \mathbf{v}_1 \cdot \mathbf{v}_2 & \cdots & \mathbf{v}_1 \cdot (\mathbf{v}_{k+1} - \sum_{j=1}^{k} \tilde{a}_j \mathbf{v}_j) \\ \mathbf{v}_2 \cdot \mathbf{v}_1 & \mathbf{v}_2 \cdot \mathbf{v}_2 & \cdots & \mathbf{v}_2 \cdot (\mathbf{v}_{k+1} - \sum_{j=1}^{k} \tilde{a}_j \mathbf{v}_j) \\ \vdots & \vdots & \ddots & \vdots \\ \mathbf{v}_{k+1} \cdot \mathbf{v}_1 & \mathbf{v}_{k+1} \cdot \mathbf{v}_2 & \cdots & \mathbf{v}_{k+1} \cdot (\mathbf{v}_{k+1} - \sum_{j=1}^{k} \tilde{a}_j \mathbf{v}_j) \end{vmatrix} \qquad (10.74)$$

that is,

$$g_{k+1} = \begin{vmatrix} \mathbf{v}_1 \cdot \mathbf{v}_1 & \mathbf{v}_1 \cdot \mathbf{v}_2 & \cdots & 0 \\ \mathbf{v}_2 \cdot \mathbf{v}_1 & \mathbf{v}_2 \cdot \mathbf{v}_2 & \cdots & 0 \\ \vdots & \vdots & \ddots & \vdots \\ \mathbf{v}_{k+1} \cdot \mathbf{v}_1 & \mathbf{v}_{k+1} \cdot \mathbf{v}_2 & \cdots & \mathbf{v}_{k+1} \cdot (\mathbf{v}_{k+1} - \sum_{j=1}^{k} \tilde{a}_j \mathbf{v}_j) \end{vmatrix}. \qquad (10.75)$$

From this, since

$$\mathbf{v}_s \cdot \left(\mathbf{v}_{k+1} - \sum_{j=1}^{k} \tilde{a}_j \mathbf{v}_j \right) = 0, \forall s \in \{1, \dots, k\},$$

we obtain,

$$g_{k+1} = \begin{vmatrix} \mathbf{v}_1 \cdot \mathbf{v}_1 & \mathbf{v}_1 \cdot \mathbf{v}_2 & \cdots & 0 \\ \mathbf{v}_2 \cdot \mathbf{v}_1 & \mathbf{v}_2 \cdot \mathbf{v}_2 & \cdots & 0 \\ \vdots & \vdots & \ddots & \vdots \\ \mathbf{v}_{k+1} \cdot \mathbf{v}_1 & \mathbf{v}_{k+1} \cdot \mathbf{v}_2 & \cdots & (\mathbf{v}_{k+1} - \sum_{j=1}^{k} \tilde{a}_j \mathbf{v}_j) \cdot (\mathbf{v}_{k+1} - \sum_{j=1}^{k} \tilde{a}_j \mathbf{v}_j) \end{vmatrix}$$

so that

$$
g_{k+1} = \begin{vmatrix}
\mathbf{v}_1 \cdot \mathbf{v}_1 & \mathbf{v}_1 \cdot \mathbf{v}_2 & \cdots & 0 \\
\mathbf{v}_2 \cdot \mathbf{v}_1 & \mathbf{v}_2 \cdot \mathbf{v}_2 & \cdots & 0 \\
\vdots & \vdots & \ddots & \vdots \\
\mathbf{v}_{k+1} \cdot \mathbf{v}_1 & \mathbf{v}_{k+1} \cdot \mathbf{v}_2 & \cdots & \tilde{\mathbf{v}}_{k+1} \cdot \tilde{\mathbf{v}}_{k+1}
\end{vmatrix}
\tag{10.76}
$$

from which, calculating the determinant through the last column, we have,

$$
g_{k+1} = g_k |\tilde{\mathbf{v}}_{k+1}|^2.
$$

This completes the induction.

The proof is complete.

Let us turn our attention to the problem of calculating the volume of a manifold.

Consider a m-dimensional C^1 class surface $M \subset \mathbb{R}^n$, where, $1 \le m \le n$ and

$$
M = \{ \mathbf{r}(\mathbf{u}) \ : \ \mathbf{u} \in D \}.
$$

We define, the volume of M, denoted by V, by:

$$
V = \int_M dM,
$$

where dM will be specified in the next lines.

We may also denote,

$$
V = \int_D dM(\mathbf{u}).
$$

At this point we address the problem of finding $dM(\mathbf{u})$.

Let $\mathbf{u} = (u_1, \dots, u_m) \in D$, and let $p = \mathbf{r}(\mathbf{u})$.

Consider the m-dimensional elementary rectangle

$$
(u_1, u_1 + \Delta u_1) \times (u_2, u_2 + \Delta u_2) \times \cdots \times (u_m, u_m + \Delta u_m).
$$

We are going to obtain the corresponding approximate volume in the surface, which we denote by

$$
\Delta M(u).
$$

Observe that,

$$
\Delta \mathbf{r}_{u_j} = \mathbf{r}(u_1, \ldots, u_j + \Delta u_j, \ldots, u_m) - \mathbf{r}(u_1, \ldots, u_j, \ldots, u_m)
$$

$$
= \frac{\partial \mathbf{r}(\mathbf{u})}{\partial u_j} \Delta u_j + o(\Delta u_j)
$$

$$
\approx \frac{\partial \mathbf{r}(\mathbf{u})}{\partial u_j} \Delta u_j, \tag{10.77}
$$

for Δu_j sufficiently small.

Hence, $\Delta M(\mathbf{u})$ is approximately defined by the set of vectors

$$
\{\Delta \mathbf{r}_{u_1}, \ldots, \Delta \mathbf{r}_{u_m}\} \subset \mathbb{R}^n,
$$

where, $\Delta \mathbf{r}_{u_j} \approx \frac{\partial \mathbf{r}(\mathbf{u})}{\partial u_j} \Delta u_j$, from the last theorem, we obtain,

$$
\Delta M(\mathbf{u}) \approx \sqrt{g} \Delta u_1 \Delta u_2 \cdots \Delta u_m,
$$

where

$$
g = \det\{g_{ij}\}_{i,j=1}^m,
$$

and where

$$
\mathbf{g}_i \equiv \frac{\partial \mathbf{r}(\mathbf{u})}{\partial u_j}, \forall i \in \{1, \ldots, m\},
$$

and finally

$$
g_{ij} = \mathbf{g}_i \cdot \mathbf{g}_j, \ \forall i, j \in \{1, \ldots, m\}.
$$

So, in its differential form, we could write,

$$
dM(\mathbf{u}) = \sqrt{g} \, du_1 \cdots du_m,
$$

so that up to local coordinates and concerning partition of unity, we have

$$
\int_M dM = \int_D dM(\mathbf{u})
$$

$$
= \int_D \sqrt{g} \, du_1 \cdots du_m. \tag{10.78}
$$

10.16 Change of Variables: The General Case

Let $D \subset \mathbb{R}^n$ be a compact block (or even a more general compact region analogous to a simple one in $\mathbb{R}^3$). Let $f : D \to \mathbb{R}$ be a continuous function.

Consider the integral I, where

$$I = \int_D f(\mathbf{x})\, d\mathbf{x} = \int_D f(x_1, \ldots, x_n)\, dx_1 \cdots dx_n.$$

Consider also the change in variables, given by the C^1 class functions $X_1, \ldots, X_n : D_0 \to \mathbb{R}$, where we denote

$$\mathbf{r}(\mathbf{u}) = X_1(\mathbf{u})\mathbf{e}_1 + \cdots + X_n(\mathbf{u})\mathbf{e}_n.$$

We assume $\mathbf{r} : D_0 \to D$ to be a bijection, where $\{\mathbf{e}_1, \ldots, \mathbf{e}_n\}$ denotes the canonical basis for $\mathbb{R}^n$ and

$$\mathbf{u} = (u_1, \ldots, u_n) \in D_0 \subset \mathbb{R}^n.$$

More specifically, the change of variables is given by,

$$x_1 = X_1(\mathbf{u}), \ldots, x_n = X_n(\mathbf{u}).$$

At this point we shall show that

$$\left| \det\left\{ \frac{\partial X_i(\mathbf{u})}{\partial u_j} \right\} \right|^2 = g,$$

where

$$g = \det\{g_{ij}\}$$

and

$$g_{ij} = \frac{\partial \mathbf{r}(\mathbf{u})}{\partial u_i} \cdot \frac{\partial \mathbf{r}(\mathbf{u})}{\partial u_j},$$

so that

$$
\begin{aligned}
g_{ij} &= \frac{\partial \mathbf{r}(\mathbf{u})}{\partial u_i} \cdot \frac{\partial \mathbf{r}(\mathbf{u})}{\partial u_j} \\
&= \sum_{k=1}^{n} \frac{\partial X_k}{\partial u_i} \frac{\partial X_k}{\partial u_j} \\
&= \sum_{k=1}^{n} (X_k)_{u_i} (X_k)_{u_j}.
\end{aligned}
\tag{10.79}
$$

Observe that

$$\left\{\frac{\partial X_i(\mathbf{u})}{\partial u_j}\right\}^T \left\{\frac{\partial X_i(\mathbf{u})}{\partial u_j}\right\}$$

$$= \left\{\begin{array}{cccc} (X_1)_{u_1} & (X_2)_{u_1} & \cdots & (X_n)_{u_1} \\ (X_1)_{u_2} & (X_2)_{u_2} & \cdots & (X_n)_{u_2} \\ \vdots & \vdots & \ddots & \vdots \\ (X_1)_{u_n} & (X_2)_{u_n} & \cdots & (X_n)_{u_n} \end{array}\right\} \cdot \left\{\begin{array}{cccc} (X_1)_{u_1} & (X_1)_{u_2} & \cdots & (X_1)_{u_n} \\ (X_2)_{u_1} & (X_2)_{u_2} & \cdots & (X_2)_{u_n} \\ \vdots & \vdots & \ddots & \vdots \\ (X_n)_{u_1} & (X_n)_{u_2} & \cdots & (X_n)_{u_n} \end{array}\right\} \quad (10.80)$$

so that

$$\left\{\frac{\partial X_i(\mathbf{u})}{\partial u_j}\right\}^T \left\{\frac{\partial X_i(\mathbf{u})}{\partial u_j}\right\}$$

$$= \left\{\sum_{k=1}^{n}(X_k)_{u_i}(X_k)_{u_j}\right\}$$

$$= \left\{\frac{\partial \mathbf{r}(\mathbf{u})}{\partial u_i} \cdot \frac{\partial \mathbf{r}(\mathbf{u})}{\partial u_j}\right\}$$

$$= \{g_{ij}\}. \quad (10.81)$$

Hence,

$$\left|\det\left\{\frac{\partial X_i(\mathbf{u})}{\partial u_j}\right\}\right|^2 = \det\{g_{ij}\} = g.$$

Thus,

$$\left|\det\left\{\frac{\partial X_i(\mathbf{u})}{\partial u_j}\right\}\right| = \sqrt{g},$$

so that from the exposed in the last two sections, we have,

$$I = \int_D f(x_1, \ldots, x_n)\, dx_1 \cdots dx_n$$

$$= \int_{D_0} f(X_1(\mathbf{u}), \ldots, X_n(\mathbf{u}))\, \sqrt{g}\, du_1 \cdots du_n$$

$$= \int_{D_0} f(X_1(\mathbf{u}), \ldots, X_n(\mathbf{u}))\, \left|\det\left\{\frac{\partial X_i(\mathbf{u})}{\partial u_j}\right\}\right|\, du_1 \cdots du_n. \quad (10.82)$$

10.17 The Stokes Theorem

In this subsection we present the Stokes theorem:

Theorem 10.17.1 *Let $M \subset \mathbb{R}^n$ be a C^1 class oriented compact m-dimensional surface with a boundary ∂M, where $1 \leq m < n$.*
   Let $\omega = \omega_I \, dX_I$ be a $(m-1)$-form. Under such hypotheses,

$$\int_M d\omega = \int_{\partial M} \omega.$$

Proof The proof follows from the discussion at Sect. 10.13.
   Indeed, denoting

$$\mathbf{s}_k = \frac{\partial \mathbf{r}(\mathbf{u})}{\partial u_k} \, du_k, \ \forall k \in \{1, \ldots, m\},$$

and

$$\mathbf{S}_j = (\mathbf{s}_1, \ldots, \hat{\mathbf{s}}_j, \ldots, \mathbf{s}_m),$$

where $\hat{\mathbf{s}}_j$ means the absence of such a term in the concerning list of vectors, we may infer that the general form

$$\sum_{j=1}^m (\omega_I(\mathbf{u}) d X_I(\mathbf{u}))(\mathbf{S}_j)$$

would stand for:

$$\sum_{j=1}^m (-1)^{j+1} h_j(\mathbf{u}) \, (du_1 \wedge \cdots \wedge \hat{du}_j \wedge \cdots \wedge du_m)(\mathbf{S}_j),$$

$$= \sum_{j=1}^m (-1)^{j+1} h_j(\mathbf{u}) \, du_1 \cdots \hat{du}_j \cdots du_m, \tag{10.83}$$

for appropriate C^1 class functions $h_j, \forall j \in \{1, \ldots, m\}$.

10.17.1 Recovering the Classical Results on Vector Calculus in $\mathbb{R}^3$ from the General Stokes Theorem

* Recovering the standard stokes Theorem in $\mathbb{R}^3$. Let $S \subset \mathbb{R}^3$ be a C^1 class surface with a boundary $C = \partial S$, where

$$S = \{(x, y, z) \in \mathbb{R}^3 \ : \ z = z(x, y) \text{ and } (x, y) \in D\},$$

where $D \subset \mathbb{R}^2$ is a simple region.

Consider the form $\omega = P\,dx + Q\,dy + R\,dz$, where we denote $\mathbf{F} = P\mathbf{i} + Q\mathbf{j} + R\mathbf{z}$, and where $P, Q, R : V \supset S \to \mathbb{R}$ are C^1 class scalar functions.

From the Stokes Theorem 10.17.1 we have

$$\int_C \omega = \int_S d\omega,$$

that is,

$$\int_C P\,dx + Q\,dy + R\,dz = \int_S d\omega$$

$$= \int_S (dP \wedge dx + dQ \wedge dy + dR \wedge dz)$$

$$= \int_S \left(\frac{\partial P}{\partial dx}\,dx + \frac{\partial P}{\partial y}\,dy + \frac{\partial P}{\partial z}\,dz \right) \wedge dx$$

$$+ \left(\frac{\partial Q}{\partial x}\,dx + \frac{\partial Q}{\partial y}\,dy + \frac{\partial Q}{\partial z}\,dz \right) \wedge dy$$

$$+ \left(\frac{\partial R}{\partial x}\,dx + \frac{\partial R}{\partial y}\,dy + \frac{\partial R}{\partial z}\,dz \right) \wedge dz$$

$$= \int_S (Q_x - P_y)\,dx \wedge dy + \int_S (R_y - Q_z)\,dy \wedge dz$$

$$+ (P_z - R_x)\,dz \wedge dx. \tag{10.84}$$

We recall that in S, $z = z(x, y)$, so that in the context of informal calculus language, we have

$$dz = z_x\,dx + z_y\,dy.$$

Also,

$$dx \wedge dy = dxdy,$$

$$dy \wedge dz = dy \wedge (z_x dx + z_y dy) = -z_x dxdy,$$

$$dz \wedge dx = (z_x dx + z_y dy) \wedge dx = -z_y dxdy,$$

so that from (10.85), we obtain,

$$\int_C P\,dx + Q\,dy + R\,dz$$

$$= \int_D (Q_x - P_y)\,dxdy + \int_D (R_y - Q_z)\,(-z_x)\,dxdy \wedge dz$$

$$+(P_z - R_x)\,(-z_y)\,dxdy$$

$$= \int_D (R_y - Q_z)\,(-z_x)\,dxdy + (P_z - R_x)\,(-z_y)\,dxdy + \int_D (Q_x - P_y)\,dxdy$$

$$= \int_D curl(\mathbf{F}) \cdot [-z_x\mathbf{i} - z_y\mathbf{j} + \mathbf{k}]\,dxdy$$

$$= \int_D curl(\mathbf{F}) \cdot \frac{[-z_x\mathbf{i} - z_y\mathbf{j} + \mathbf{k}]}{\sqrt{z_x^2 + z_y^2 + 1}}\sqrt{z_x^2 + z_y^2 + 1}\,dxdy$$

$$= \int_D curl(\mathbf{F}) \cdot \mathbf{n}\sqrt{z_x^2 + z_y^2 + 1}\,dxdy$$

$$= \int_S curl(\mathbf{F}) \cdot \mathbf{n}\,dS, \tag{10.85}$$

where $\mathbf{n}$ is unit outward normal relating S.

So, to summarize we have obtained,

$$\int_C \mathbf{F} \cdot d\mathbf{r} = \int_S curl(\mathbf{F}) \cdot \mathbf{n}\,dS,$$

which is the standard Stokes Theorem result in $\mathbb{R}^3$.

Remark 10.17.2 We have used some informality here.

In fact, in a more rigorous fashion, we should have:

$$\mathbf{r}(x, y) = (x, y, z(x, y)),$$

and

$$(dx \wedge dy)(\mathbf{r}_x(x, y)\,dx, \mathbf{r}_y(x, y)\,dy)$$

$$= (dx \wedge dy)((1, 0, z_x)\,dx, (0, 1, z_y)\,dy)$$

$$= dx\,dy, \tag{10.86}$$

and with more details,

$$(dy \wedge dz)(\mathbf{r}_x(x, y)\,dx, \ \mathbf{r}_y(x, y)\,dy)$$

$$= (dy \wedge dz)((1, 0, z_x)\,dx, (0, 1, z_y)\,dy)$$

$$= (dy \wedge (z_x dx + z_y dy))((1, 0, z_x)\,dx, \ (0, 1, z_y)\,dy)$$

$$= z_x(dy \wedge dx)((1, 0, z_x)\,dx, (0, 1, z_y)\,dy)$$

$$+ z_y(dy \wedge dy)((1, 0, z_x)\,dx, (0, 1, z_y)\,dy)$$

$$= z_x \begin{vmatrix} dy(dx, 0, z_x dx) & dy(0, dy, z_y \, dy) \\ dx(dx, 0, z_x dx) & dx(0, dy, z_y \, dy) \end{vmatrix}$$

$$= z_x \begin{vmatrix} 0 & dy \\ dx & 0 \end{vmatrix}$$

$$= -z_x dx dy. \tag{10.87}$$

A similar remark is valid for

$$dz \wedge dx.$$

- Recovering the Divergence Theorem in $\mathbb{R}^3$:

  Let $V \subset \mathbb{R}^3$ be a simple volume with a boundary $S = \partial V$. Let $\mathbf{F} = P\mathbf{i} + Q\mathbf{j} + R\mathbf{k}$ where $P, Q, R : V \to \mathbb{R}$ are C^1 class scalar functions.

  From the last item, we have

$$(dx \wedge dy)((1, 0, z_x) \, dx, (0, 1, z_y) \, dy) = dx dy,$$

$$(dy \wedge dz)(((1, 0, z_x) \, dx, (0, 1, z_y) \, dy)) = -z_x dx dy,$$

$$(dz \wedge dx)(((1, 0, z_x) \, dx, (0, 1, z_y) \, dy)) = -z_y dx dy,$$

so that,

$$(P \, dy \wedge dz + Q dz \wedge dx + R dx \wedge dy)((dx, 0), (0, dy))$$

$$= P(-z_x) \, dx dy + Q(-z_y) \, dx dy + R \, dx dy$$

$$= (P\mathbf{i} + Q\mathbf{j} + R\mathbf{k}) \cdot (-z_x\mathbf{i} - z_y\mathbf{j} + \mathbf{k}) \, dx dy$$

$$= \mathbf{F} \cdot \frac{(-z_x\mathbf{i} - z_y\mathbf{j} + \mathbf{k})}{\sqrt{z_x^2 + z_y^2 + 1}} \sqrt{z_x^2 + z_y^2 + 1} \, dx dy$$

$$= \mathbf{F} \cdot \mathbf{n} \sqrt{z_x^2 + z_y^2 + 1} \, dx dy$$

$$= \mathbf{F} \cdot \mathbf{n} \, dS, \tag{10.88}$$

where

$$\mathbf{n} = \frac{(-z_x\mathbf{i} - z_y\mathbf{j} + \mathbf{k})}{\sqrt{z_x^2 + z_y^2 + 1}},$$

is the unit outward normal relating S and

$$dS = \sqrt{z_x^2 + z_y^2 + 1} \, dx dy.$$

Consider, with some informality here, the form

$$\omega = P\,dy \wedge dz + Qdz \wedge dx + Rdx \wedge dy = \mathbf{F} \cdot \mathbf{n}\,dS.$$

From the Stokes Theorem 10.17.1, we have

$$\int_S \omega = \int_V d\omega.$$

Observe that

$$\int_V d\omega = \int_V (dP \wedge (dy \wedge dz)) + (dQ \wedge (dz \wedge dx)) + (dR \wedge (dx \wedge dy))$$

$$= \int_V \left(\frac{\partial P}{\partial x}\,dx + \frac{\partial P}{\partial y}\,dy + \frac{\partial P}{\partial z}\,dz \right) \wedge (dy \wedge dz)$$

$$+ \left(\frac{\partial Q}{\partial x}\,dx + \frac{\partial Q}{\partial y}\,dy + \frac{\partial Q}{\partial z}\,dz \right) \wedge (dz \wedge dx)$$

$$+ \left(\frac{\partial R}{\partial x}\,dx + \frac{\partial R}{\partial y}\,dy + \frac{\partial R}{\partial z}\,dz \right) \wedge (dx \wedge dy)$$

$$= \int_V \left(\frac{\partial P}{\partial x} + \frac{\partial Q}{\partial y} + \frac{\partial R}{\partial z} \right) dx \wedge dy \wedge dz$$

$$= \int_V div(\mathbf{F})\,dx\,dy\,dz. \tag{10.89}$$

Joining the pieces we obtain

$$\int_S \omega = \int_S (\mathbf{F} \cdot \mathbf{n})\,dS$$

$$= \int_S P\,dy \wedge dz + Qdz \wedge dx + Rdx \wedge dy$$

$$= \int_V div(\mathbf{F})\,dx\,dy\,dz, \tag{10.90}$$

which is the standard Gauss Divergence Theorem.

Exercises 10.17.3

1. Find the domain of the one variable vectorial functions $\mathbf{r}$ indicated below.

(a)

$$\mathbf{r}(t) = \frac{1}{t^2 + 1}\mathbf{i} + \sqrt{(t-1)(t+3)}\mathbf{j},$$

(b)

$$\mathbf{r}(t) = \ln(t^2 - 16)\mathbf{i} + \sqrt{t^2 + 2t - 15}\mathbf{j} + \tan(t + 1)\mathbf{k},$$

(c)

$$\mathbf{r}(t) = \sqrt{25 - t^2}\mathbf{i} + \sqrt{t^2 + 2t - 8}\mathbf{j}.$$

2. Through the formula

$$\frac{dy}{dx} = \frac{dy/dt}{dx/dt},$$

calculate the derivatives of the functions defined by the parametric equations indicated,

(a)

$$\mathbf{r}(t) = \frac{e^t}{1 + e^t}\mathbf{i} + t^2 \ln(t)\mathbf{j},$$

(b)

$$\mathbf{r}(t) = \frac{\cos(t)}{5 + \sin(t)}\mathbf{i} + \ln(\sqrt{t^4 + t^2})\mathbf{j}.$$

3. Let $\mathbf{r} : \mathbb{R} \setminus \{-1\} \to \mathbb{R}^2$ be defined by

$$\mathbf{r}(t) = \frac{t}{t + 1}\mathbf{i} + \ln(t^2 + 1)\mathbf{j}.$$

Find the equation of the tangent line to the graph of the curve defined by $\mathbf{r}$ at the point corresponding to $t = 1$.

4. Let $\mathbf{r}, \mathbf{s} : \mathbb{R} \to \mathbb{R}^2$ be defined by

$$\mathbf{r}(t) = t\mathbf{i} + t^2\mathbf{j}$$

and

$$\mathbf{s}(t) = (t^2 + t)\mathbf{i} + t^3\mathbf{j}.$$

Calculate the angle between $\mathbf{r}'(t)$ and $\mathbf{s}'(t)$ at the point corresponding to $t = 1$.

5. Let $\mathbf{r} : \mathbb{R} \to \mathbb{R}^3$ be defined by

$$\mathbf{r}(t) = \frac{2t}{1 + t^2}\mathbf{i} + \frac{1 - t^2}{1 + t^2}\mathbf{j} + \mathbf{k}.$$

Show that the angle between $\mathbf{r}(t)$ and $\mathbf{r}'(t)$ is constant.

6. A vectorial function $\mathbf{r}$ satisfies the equation,

$$t\mathbf{r}'(t) = \mathbf{r}(t) + t\mathbf{A}, \ \forall t > 0$$

where

$$\mathbf{A} \in \mathbb{R}^3.$$

Suppose that $\mathbf{r}(1) = 2\mathbf{A}$. Calculate $\mathbf{r}''(1)$ and $\mathbf{r}(3)$ as functions of $\mathbf{A}$.

7. Find a function $\mathbf{r} : (0, +\infty) \to \mathbb{R}^3$ such that

$$\mathbf{r}(x) = xe^x\mathbf{A} + \frac{1}{x}\int_1^x \mathbf{r}(t) \, dt.$$

where $\mathbf{A} \in \mathbb{R}^3$, $\mathbf{A} \neq \mathbf{0}$.

8. Through the Green Theorem, calculate the areas of the regions D, where,

(a)

$$D = \{(x, y) \in \mathbb{R}^2 \ : \ x^2 + y^2 \leq 1 \text{ and } y \geq 1/2\}.$$

(b)

$$D = \{(x, y) \in \mathbb{R}^2 \ : \ x^2 + y^2 \leq 1 \text{ and } -1/2 \leq y \leq \sqrt{3}/2\}.$$

(c)

$$D = \{(x, y) \in \mathbb{R}^2 \ : \ x^2 + y^2 \leq 1 \text{ and } 0 \leq x \leq 1/2\}.$$

9. Calculate the area of surface S, where

$$S = \left\{(x, y, z) \in \mathbb{R}^3 \ : \ x^2 + y^2 + z^2 = 1 \text{ and } \frac{1}{2} \leq z \leq \frac{\sqrt{3}}{2}\right\}.$$

10. Calculate the area of surface S, where

$$S = \left\{(x, y, z) \in \mathbb{R}^3 \ : \ x^2 + y^2 + z^2 = 1 \text{ and } \frac{-\sqrt{3}}{2} \leq z \leq \frac{1}{2}\right\}.$$

11. Calculate the area of surface S, where

$$S = \left\{(x, y, z) \in \mathbb{R}^3 \ : \ z^2 = x^2 + y^2 \text{ and } x^2 + y^2 \leq 2ax\right\},$$

where $a \in \mathbb{R}$.

12. Calculate $I = \int \int_S x \, dS$, where

$$S = \left\{ (x, y, z) \in \mathbb{R}^3 \ : \ x^2 + y^2 = R^2 \text{ and } |z| \le 1 \right\}.$$

13. Through the Divergence Theorem, calculate $I = \int \int_S (y\mathbf{j} + z\mathbf{k}) \cdot \mathbf{n} \, dS$, where

$$S = \left\{ (x, y, z) \in \mathbb{R}^3 \ : \ x = \sqrt{R^2 - y^2 - z^2} \text{ and } x \ge \frac{\sqrt{3}R}{2} \right\},$$

where $R > 0$.

14. Through the Divergence Theorem, calculate $I = \int \int_S \mathbf{F} \cdot \mathbf{n} \, dS$ where

$$S = \{(x, y, z) \in \mathbb{R}^3 \ : \ x^2 + y^2 + z^2 = 2R_0 x \text{ and } z \ge 0\}$$

and where $\mathbf{F} = x^2\mathbf{i} + y^2\mathbf{j} + z^2\mathbf{k}$ and $R_0 > 0$.

15. Let $u : V \to \mathbb{R}$ be a scalar field and let $\mathbf{F} : V \to \mathbb{R}^3$ be a vectorial one, where $V \subset \mathbb{R}^3$ is open u, $\mathbf{F}$ are of C^1 class. Show that

$$div(u\mathbf{F}) = (\nabla u) \cdot \mathbf{F} + u \, (div\mathbf{F}).$$

16. Let $u, v : V \to \mathbb{R}$ be C^2 class scalar fields, where $V \subset \mathbb{R}^3$ is open and its closure is simple. Defining

$$\nabla^2 u = \frac{\partial^2 u}{\partial x^2} + \frac{\partial^2 u}{\partial y^2} + \frac{\partial^2 u}{\partial z^2}$$

show that $div(\nabla u) = \nabla^2 u$ and prove the Green identities,

(a)

$$\int \int \int_V (v\nabla^2 u + \nabla v \cdot \nabla u) \, dV = \int \int_S v(\nabla u \cdot \mathbf{n}) \, dS$$

where $S = \partial V$ (that is, S is the boundary of V).

(b)

$$\int \int \int_V (v\nabla^2 u - u\nabla^2 v) \, dV = \int \int_S \left(v\frac{\partial u}{\partial \mathbf{n}} - u\frac{\partial v}{\partial \mathbf{n}} \right) dS,$$

where $S = \partial V$ and $\frac{\partial u}{\partial \mathbf{n}} = \nabla u \cdot \mathbf{n}$.

17. Let $u : V \to \mathbb{R}$, $\mathbf{F} : V \to \mathbb{R}^3$ be C^2 class fields on the open set $V \subset \mathbb{R}^3$. Prove that $curl(\nabla u) = \mathbf{0}$ and $div(curl(\mathbf{F})) = 0$, on V.

18. Let $D \subset \mathbb{R}^2$ be a simple region. Let $u, v \in C(\overline{D}) \cap C^1(D)$. Prove that

$$\int\int_D u v_x \, dxdy = \int_{\partial D} uv \, dy - \int\int_D u_x v \, dxdy,$$

and

$$\int\int_D u v_y \, dxdy = \int_{\partial D} uv \, dx - \int\int_D u_y v \, dxdy,$$

19. Let $V \subset \mathbb{R}^3$ be an open region bounded by a closed surface of C^1 class. Through the first Green identity, prove the uniqueness of solution of the Dirichlet problem,

$$\begin{cases} \nabla^2 u = f, \quad \text{in} V \\ u = u_0, \quad \text{on } \partial V, \end{cases} \tag{10.91}$$

where $f : V \to \mathbb{R}$ is continuous and $u_0 : \partial V \to \mathbb{R}$ is also continuous.
   Also through the first Green identity, prove that

(a) for the Neumann problem

$$\begin{cases} \nabla^2 u = f, \quad \text{in } V \\ \frac{\partial u}{\partial \mathbf{n}} = u_0, \quad \text{on } \partial V, \end{cases} \tag{10.92}$$

   to have a solution, it is necessary that

$$\int\int\int_V f \, dxdydz = \int\int_{\partial V} u_0 \, dS.$$

   Hint: Consider $v \equiv 1$ in the first Green identity.
(b) Prove that any two solutions of the Neumann problem differ by a constant.

20. Let $V \subset \mathbb{R}^3$ be a simple region. Let $\mathbf{F} : V \to \mathbb{R}^3$ be a vectorial field of C^1 class.
   Let $\mathbf{x}_0 \in V^\circ$. Show that

$$div(\mathbf{F}(\mathbf{x}_0)) = \lim_{r \to 0} \frac{\int\int_{\partial B_r(\mathbf{x}_0)} \mathbf{F} \cdot \mathbf{n} \, dS}{Vol(B_r(\mathbf{x}_0))},$$

where $\mathbf{n}$ denotes unit outward normal field to $B_r(\mathbf{x}_0)$.

21. Let $V \subset \mathbb{R}^3$ be a simple region. Let $f : V \to \mathbb{R}$ be a scalar field of C^2 class. Let $\mathbf{x}_0 \in V^\circ$. Through the first Green identity, show that

$$\nabla^2 f(\mathbf{x}_0) = \lim_{r \to 0} \frac{\int \int_{\partial B_r(\mathbf{x}_0)} \frac{\partial f}{\partial \mathbf{n}} \, dS}{Vol(B_r(\mathbf{x}_0))},$$

where $\mathbf{n}$ denotes the unit outward normal field to $B_r(\mathbf{x}_0)$.

22. Let $M \subset \mathbb{R}^n$ be a m-dimensional surface of C^2 class, where $1 \leq m < n$.

Let $X, Y, Z \in \tilde{\mathscr{X}}(M)$, where $\tilde{\mathscr{X}}(M)$ denotes the set of tangential vector fields of C^∞ class defined on M.

Show that

(a) $[\alpha X + \beta Y, Z] = \alpha[X, Z] + \beta[Y, Z]$, $\forall \alpha, \beta \in \mathbb{R}$,

(b) $[X, \alpha Y + \beta Z] = \alpha[X, Y] + \beta[X, Z]$, $\forall \alpha, \beta \in \mathbb{R}$.

(c) Antisymmetry:

$$[X, Y] = -[Y, X],$$

(d) Jacob Identity:

$$[[X, Y], Z] + [[Y, Z], X] + [[Z, X], Y] = 0.$$

(e) Leibnitz rule:

$$[fX, gY] = fg[X, Y] + f(X \cdot g)Y - g(Y \cdot f)X, \ \forall f, g \in C^2(M).$$

(f) Recalling that $L_X Y = [X, Y]$, show that

(i)

$$L_X[Y, Z] = [L_X Y, Z] + [Y, L_X Z],$$

(ii)

$$L_X(L_Y)Z - L_Y(L_X)Z = L_{[X,Y]}Z.$$

23. Consider a three-dimensional surface $M \subset \mathbb{R}^4$ defined by

$$M = \{(x, y, z, w) \in \mathbb{R}^4 \ : \ x^2 + y^2 + z^2 + \ln(w) = 1\}.$$

(a) Defining $\mathbf{u} = (u_1, u_2, u_3) = (x, y, z)$, write M in the form,

$$M = \{\mathbf{r}(\mathbf{u}) \in \mathbb{R}^4 \ : \ \mathbf{u} \in \mathbb{R}^3\}.$$

(b) Let $p = \mathbf{r}(\mathbf{u})$. Obtain the tangent space and equation of the hyper-plan tangent to M at p.

(c) For $f : M \to \mathbb{R}$, $X, Y \in \mathscr{X}(M)$ such that

$$f(\mathbf{r}(\mathbf{u})) = x^2 + e^{x^2 y} + (\sin(x^2 + z^2))^3 + w(x, y, z),$$

$$X(x, y, z) = e^x \frac{\partial \mathbf{r}(x, y, z)}{\partial x} + y \frac{\partial \mathbf{r}(x, y, z)}{\partial y} + (x + z^2)^2 \frac{\partial \mathbf{r}(x, y, z)}{\partial z},$$

and

$$Y(x, y, z) = (\sin(xy))^2 \frac{\partial \mathbf{r}(x, y, z)}{\partial x} + (\cos(x^2 + y^2))^3 \frac{\partial \mathbf{r}(x, y, z)}{\partial y}$$

$$+ e^{x+z^2} \frac{\partial \mathbf{r}(x, y, z)}{\partial z},$$

for $p = \mathbf{r}(x_0, y_0, z_0)$, calculate

(i) $(X \cdot f)(p)$,
(ii) $(D_X Y)(p)$,
(iii) $[X, Y](p)$,
(iv) $([X, Y] \cdot f)(p)$
(v) Compute numerically the results obtained in the 4 last items at the point $p_0 = \mathbf{r}(x_0, y_0, z_0) = \mathbf{r}(\pi, 0, 1)$.

(d) Let $Z \in \mathscr{X}(M)$, where

$$Z(x, y, z) = (x + y + z) \frac{\partial \mathbf{r}(x, y, z)}{\partial x} + (2x + y + z) \frac{\partial \mathbf{r}(x, y, z)}{\partial y}$$

$$+ (-y + z) \frac{\partial \mathbf{r}(x, y, z)}{\partial z}.$$

Obtain the integral curve $r(\mathbf{u}(t))$ of Z, such that $\mathbf{u}(0) = (1, -1, 0)$.

24. Obtain the differential $dM(x, y, z)$ to calculate the area of the surface $M \subset \mathbb{R}^4$, where

$$M = \{(x, y, z, w) \in \mathbb{R}^4 : e^w = [\sin(x^2 + y)]^3 + z^2 + 5 \text{ and } x^2 + y^2 + z^2 \leq 1\}.$$

25. Consider the surface $M \subset \mathbb{R}^4$ defined by

$$M = \{(x, y, z) \in \mathbb{R}^4 : e^w - x^2 - y^2 - z^2 = 1\}.$$

Write its equation in the form,

$$M = \{\mathbf{r}(x, y, z) : (x, y, z) \in \mathbb{R}^3\},$$

where

$$\mathbf{r}(x, y, z) = X_1(x, y, z)\mathbf{e}_1 + \cdots + X_4(x, y, z)\mathbf{e}_4.$$

Let

$$dX_1 = \frac{\partial X_1}{\partial x}\, dx + \frac{\partial X_1}{\partial y}\, dy + \frac{\partial X_1}{\partial z}\, dz,$$

and

$$dX_4 = \frac{\partial X_4}{\partial x}\, dx + \frac{\partial X_4}{\partial y}\, dy + \frac{\partial X_4}{\partial z}\, dz.$$

(a) Calculate

$$(dX_1 \wedge dX_4)(\mathbf{s}_1, \mathbf{s}_2),$$

where

$$\mathbf{s}_1 = \frac{\partial \mathbf{r}(x, y, z)}{\partial x}\, \Delta x,$$

and

$$\mathbf{s}_2 = \frac{\partial \mathbf{r}(x, y, z)}{\partial y}\, \Delta y,$$

and where $\Delta x,\ \Delta y \in \mathbb{R}$.

(b) Consider the differential form

$$\omega = (w(x, y, z) + x^2 y + z)dX_1 \wedge dX_4 + (w(x, y, z)^2$$
$$+ \sin(x^2 + y) - z^2)dX_1 \wedge dX_2,$$

where

$$dX_2 = \frac{\partial X_2}{\partial x}\, dx + \frac{\partial X_2}{\partial y}\, dy + \frac{\partial X_2}{\partial z}\, dz.$$

Obtain the exterior differential $d\omega$ of ω at $(\mathbf{s}_1, \mathbf{s}_2, \mathbf{s}_3)$, where

$$\mathbf{s}_1 = \frac{\partial \mathbf{r}(x, y, z)}{\partial x}\, \Delta x,$$

$$\mathbf{s}_2 = \frac{\partial \mathbf{r}(x, y, z)}{\partial y}\, \Delta y,$$

and

$$\mathbf{s}_3 = \frac{\partial \mathbf{r}(x, y, z)}{\partial z}\, \Delta z,$$

and where $\Delta x,\ \Delta y,\ \Delta z \in \mathbb{R}$.

26. Let $M \subset \mathbb{R}^n$ be a three-dimensional C^1 class surface, where $n \geq 4$,

$$M = \{\mathbf{r}(\mathbf{u}) = X_i(\mathbf{u})\mathbf{e}_i \ : \ \mathbf{u} \in D\},$$

$D \subset \mathbb{R}^3$ and $\{\mathbf{e}_1, \ldots, \mathbf{e}_n\}$ is the canonical basis for $\mathbb{R}^n$,
Let $\omega = dX_1 \wedge dX_4 \wedge dX_3$ be a 3-form on M, where,

$$dX_1(\mathbf{u}) = \frac{\partial X_1(\mathbf{u})}{\partial u_1} \, du_1 + \frac{\partial X_1(\mathbf{u})}{\partial u_2} \, du_2 + \frac{\partial X_1(\mathbf{u})}{\partial u_3} \, du_3,$$

$$dX_4(\mathbf{u}) = \frac{\partial X_4(\mathbf{u})}{\partial u_1} \, du_1 + \frac{\partial X_4(\mathbf{u})}{\partial u_2} \, du_2 + \frac{\partial X_4(\mathbf{u})}{\partial u_3} \, du_3,$$

and

$$dX_3(\mathbf{u}) = \frac{\partial X_3(\mathbf{u})}{\partial u_1} \, du_1 + \frac{\partial X_3(\mathbf{u})}{\partial u_2} \, du_2 + \frac{\partial X_3(\mathbf{u})}{\partial u_3} \, du_3.$$

Compute

$$(dX_1(\mathbf{u}) \wedge dX_4(\mathbf{u}) \wedge dX_3(\mathbf{u}))(\mathbf{s}_1, \mathbf{s}_2, \mathbf{s}_3),$$

where

$$\mathbf{s}_1 = \frac{\partial \mathbf{r}(\mathbf{u})}{\partial u_1} \, \Delta u_1,$$

$$\mathbf{s}_2 = \frac{\partial \mathbf{r}(\mathbf{u})}{\partial u_2} \, \Delta u_2$$

and

$$\mathbf{s}_3 = \frac{\partial \mathbf{r}(\mathbf{u})}{\partial u_3} \, \Delta u_3.$$

27. Consider the vectorial field $F : \mathbb{R}^3 \to \mathbb{R}^3$ where $\mathbf{F} = x^2\mathbf{i} + y^2\mathbf{j} + (z - x^2)\mathbf{k}$.
Through the Stokes Theorem, calculate

$$I = \int \int_S curl(\mathbf{F}) \cdot \mathbf{n} \, dS$$

where

$$S = \{(x, y, z) \in \mathbb{R}^3 \ : \ z = 8 - x^2 - 2y^2 \text{ and } 2 \leq z \leq 4\}.$$

28. Consider the vectorial field $F : \mathbb{R}^3 \to \mathbb{R}^3$ where $\mathbf{F} = y\mathbf{i} + y\mathbf{j} + 5\mathbf{k}$.

Through the Stokes theorem, calculate

$$I = \int\int_S curl(\mathbf{F}) \cdot \mathbf{n}\, dS$$

where

$$S = \{(x, y, z) \in \mathbb{R}^3 \; : \; z = 16 - x^2 - 3y^2 \text{ and } z \geq y^2 + 2x + y\}.$$

29. Let $D \subset \mathbb{R}^n$ be an open set and let $f : D \to \mathbb{R}$ be a function of C^1 class (therefore differentiable on D).

    Let $\mathbf{x}_0 \in D$ and $\varepsilon > 0$. Prove that there exists $\delta > 0$ such that if $\mathbf{x} \in D$ and $|\mathbf{x} - \mathbf{x}_0| < \delta$, then

    (a)

    $$f(\mathbf{x}) - f(\mathbf{x}_0) = f'(\mathbf{x}_0) \cdot (\mathbf{x} - \mathbf{x}_0) + r(\mathbf{x}),$$

    where

    $$|r(\mathbf{x})| \leq \varepsilon|\mathbf{x} - \mathbf{x}_0|.$$

    (b) Use the previous item to show that if $\mathbf{x}, \; \mathbf{y} \in D$, $|\mathbf{x}-\mathbf{x}_0| < \delta$ and $|\mathbf{y}-\mathbf{x}_0| < \delta$ then

    $$f(\mathbf{y}) - f(\mathbf{x}) = f'(\mathbf{x}_0) \cdot (\mathbf{y} - \mathbf{x}) + r_1(\mathbf{x}, \mathbf{y}),$$

    where

    $$|r_1(\mathbf{x}, \mathbf{y})| \leq 2\varepsilon\delta.$$

30. Let $M \subset \mathbb{R}^n$ be a m-dimensional surface of C^1 class, where $1 \leq m < n$, where

    $$M = \{\mathbf{r}(\mathbf{u}) \; : \; \mathbf{u} \in D \subset \mathbb{R}^m\}.$$

    Let $f \in C^2(M)$ and $X, Y \in \mathscr{X}(M)$.
    In this chapter, we have denoted

    $$X \cdot f = df(X) = \frac{\partial(f \circ \mathbf{r})(\mathbf{u})}{\partial u_i} X_i(\mathbf{u}).$$

    (a) Calculate

    $$X \cdot (Y \cdot f).$$

(b) Show that

$$X \cdot (Y \cdot f) - Y \cdot (X \cdot f) = [X, Y] \cdot f,$$

where

$$[X, Y] = (dY_i(X) - dX_i(Y)) \frac{\partial \mathbf{r}(\mathbf{u})}{\partial u_i}.$$

(c) Consider the three-dimensional manifold $M \subset \mathbb{R}^4$ defined by

$$M = \{(x, y, z, w) \in \mathbb{R}^4 \ : \ e^w - x^6 - y^2 - z^4 = 5\}.$$

(i) Defining $\mathbf{u} = (u_1, u_2, u_3) = (x, y, z)$, write M in the form,

$$M = \{\mathbf{r}(\mathbf{u}) \in \mathbb{R}^4 \ : \ \mathbf{u} \in \mathbb{R}^3\}.$$

(ii) For $f : M \to \mathbb{R}$, $X, Y \in \mathscr{X}(M)$ such that

$$f(\mathbf{r}(\mathbf{u})) = 5y^2 + w(x, y, z),$$

$$X(x, y, z) = \cos(x - y) \frac{\partial \mathbf{r}(x, y, z)}{\partial x} + y^3 \frac{\partial \mathbf{r}(x, y, z)}{\partial y}$$

$$+ (x + z^2)^3 \frac{\partial \mathbf{r}(x, y, z)}{\partial z},$$

and

$$Y(x, y, z) = e^x \frac{\partial \mathbf{r}(x, y, z)}{\partial x} + (\sin(x^2 + y^3))^4 \frac{\partial \mathbf{r}(x, y, z)}{\partial y} + e^{x^3 z} \frac{\partial \mathbf{r}(x, y, z)}{\partial z},$$

for $p = \mathbf{r}(x, y, z)$, calculate

$$([X, Y] \cdot f)(p)$$

31. Consider a three-dimensional surface $M \subset \mathbb{R}^4$ defined by

$$M = \{(x, y, z) \in \mathbb{R}^4 \ : \ \ln(w) - x + 2y^2 - z^3 = 1\}.$$

Write the equation of M in the form,

$$M = \{\mathbf{r}(x, y, z) \ : \ (x, y, z) \in \mathbb{R}^3\},$$

where

$$\mathbf{r}(x, y, z) = X_1(x, y, z)\mathbf{e}_1 + \cdots + X_4(x, y, z)\mathbf{e}_4,$$

(a) Obtain $(dX_1 \wedge dX_2 \wedge dX_4)(\mathbf{s}_1, \mathbf{s}_2, \mathbf{s}_3)$.

(b) Consider the differential form

$$\omega = e^{x^2 + y^5}\, dX_2 + \sin(xy^2)\, dX_1.$$

Obtain the exterior differential $d\omega$ of ω at $(\mathbf{s}_1, \mathbf{s}_2)$, where for the last two sub-items,

$$\mathbf{s}_1 = \frac{\partial \mathbf{r}(x, y, z)}{\partial x}\, \Delta x,$$

$$\mathbf{s}_2 = \frac{\partial \mathbf{r}(x, y, z)}{\partial y}\, \Delta y,$$

and

$$\mathbf{s}_3 = \frac{\partial \mathbf{r}(x, y, z)}{\partial z}\, \Delta z,$$

and where $\Delta x,\ \Delta y,\ \Delta z \in \mathbb{R}$.

10.18   The First and Second Green Identities in a Surface

In this section we develop the first and second Green identities in a surface.

We start with the following definition.

Definition 10.18.1 (Gradient and Divergent in a Surface) Let $M \subset \mathbb{R}^n$ be an m-dimensional C^1 class surface where $1 \leq m < 1$. We denote,

$$M = \{\mathbf{r}(\mathbf{u})\ :\ \mathbf{u} \in \overline{D}\},$$

where D is an open set with a C^1 class boundary.

Let $f : D \to \mathbb{R}$ be a function. We define the gradient of f relating M, denoted by $\nabla f \in T(M)$, point-wise, as

$$\nabla f(\mathbf{u}) = g^{ij} \frac{\partial f(\mathbf{u})}{\partial u_j} \frac{\partial \mathbf{r}(\mathbf{u})}{\partial u_i}.$$

At this point, we recall that

$$\{g^{ij}\} = \{g_{ij}\}^{-1},$$

so that

$$g^{ik}g_{kj} = g_{ik}g^{kj} = \delta_{ij},$$

$\forall i, j \in \{1, \ldots, m\}$.

Also given the vector field $X \in \mathscr{X}(M)$, where point-wisely

$$X(\mathbf{u}) = X_i(\mathbf{u})\frac{\partial \mathbf{r}(\mathbf{u})}{\partial u_i},$$

we define the divergent of X at $\mathbf{u} \in D$, denoted by

$$\mathrm{div}X(\mathbf{u}),$$

by

$$\mathrm{div}X(\mathbf{u}) = \frac{1}{\sqrt{g}}\frac{\partial [X_i(\mathbf{u})\sqrt{g}]}{\partial u_i}.$$

At this point we present the first Green identity.

Theorem 10.18.2 (The First Green Identity) *Let* $M \subset \mathbb{R}^n$ *be compact m-dimensional surface of* C^1 *class with a boundary* ∂M, *where* $1 \leq m < n$.
Here

$$M = \{\mathbf{r}(\mathbf{u}) \; : \; \mathbf{u} \in \overline{D} \subset \mathbb{R}^m\},$$

where D *is an open set with a* C^1 *class boundary* ∂D.
Let $f : D \to \mathbb{R}$ *be a* C^1 *class function and let* $X \in \tilde{\mathscr{X}}(M)$ *be a vector field.*
Under such hypotheses, we have,

$$\int_M X \cdot \nabla f \, dM + \int_M (div(X))f \, dM$$

$$= \int_{\partial M} (-1)^{j+1}X_j f\sqrt{g}du_1 \wedge \cdots \wedge \widehat{du_j} \wedge \cdots \wedge du_m, \qquad (10.93)$$

where the notation $\widehat{du_j}$ *means that such a term is absent in the product in question.*

Proof Observe that

$$X \cdot \nabla f = \left\langle X_i \frac{\partial \mathbf{r}}{\partial u_i}, g^{jk}\frac{\partial f}{\partial u_j}\frac{\partial \mathbf{r}}{\partial u_k} \right\rangle_{\mathbb{R}^m}$$

$$= g^{jk}\left\langle \frac{\partial \mathbf{r}}{\partial u_i}, \frac{\partial \mathbf{r}}{\partial u_k} \right\rangle_{\mathbb{R}^m} X_i \frac{\partial f}{\partial u_j}$$

$$= g^{jk} g_{ik} X_i \frac{\partial f}{\partial u_j}$$

$$= g^{jk} g_{ki} X_i \frac{\partial f}{\partial u_j}$$

$$= \delta_{ji} X_i \frac{\partial f}{\partial u_j}$$

$$= X_i \frac{\partial f}{\partial u_i}. \tag{10.94}$$

From the Stokes theorem, defining

$$\omega = (-1)^{j+1} X_j \, f \, \sqrt{g} \, du_1 \wedge \cdots \widehat{du_j} \cdots \wedge du_m,$$

we have

$$\int_M \partial\omega = \int_{\partial M} \omega,$$

so that

$$\int_M \partial\omega = \int_M \frac{\partial(X_j \, f \, \sqrt{g})}{\partial u_k} (-1)^{j+1} \, du_k \wedge \cdots \widehat{du_j} \cdots \wedge du_m$$

$$= \int_M \frac{\partial(X_j \, f \, \sqrt{g})}{\partial u_j} du_1 \wedge \cdots \wedge du_m$$

$$= \int_{\partial M} \omega$$

$$= \int_{\partial M} (-1)^{j+1} X_j \, f \, \sqrt{g} \, du_1 \wedge \cdots \wedge \widehat{du_j} \wedge \cdots \wedge du_m. \tag{10.95}$$

On the other hand,

$$\int_M \frac{\partial(X_j \, f \, \sqrt{g})}{\partial u_j} du_1 \wedge \cdots \wedge du_m$$

$$= \int_M \left[\frac{\partial(X_j \, \sqrt{g})}{\partial u_j} f + X_j \frac{\partial f}{\partial u_j} \sqrt{g} \right] du_1 \wedge \cdots \wedge du_m$$

$$= \int_M \frac{1}{\sqrt{g}} \frac{\partial(X_j \, \sqrt{g})}{\partial u_j} f \, \sqrt{g} \, du_1 \wedge \cdots \wedge du_m$$

$$+ \int_M X_j \frac{\partial f}{\partial u_j} \sqrt{g} \, du_1 \wedge \cdots \wedge du_m$$

$$= \int_M \mathrm{div} X \; f \sqrt{g} du_1 \wedge \cdots \wedge du_m$$

$$+ \int_M X \cdot \nabla f \sqrt{g} du_1 \wedge \cdots \wedge du_m$$

$$= \int_M (\mathrm{div}(X)) f \; dM + \int_M X \cdot \nabla f \; dM. \tag{10.96}$$

From (10.95) and (10.96), we obtain,

$$\int_M (\mathrm{div}(X)) f \; dM + \int_M X \cdot \nabla f \; dM$$

$$= \int_{\partial M} (-1)^{j+1} X_j \; f \; \sqrt{g} \; du_1 \wedge \cdots \widehat{du_j} \cdots \wedge du_m. \tag{10.97}$$

The proof is complete.

Theorem 10.18.3 (The Second Green Identity) *Let $M \subset \mathbb{R}^n$ be compact m-dimensional surface of C^1 class with a boundary ∂M, where $1 \le m < n$.*
Here

$$M = \{\mathbf{r}(\mathbf{u}) \; : \; \mathbf{u} \in \overline{D} \subset \mathbb{R}^m\},$$

where D is an open set with a C^1 class boundary ∂D.
Let $f, h : D \to \mathbb{R}$ be C^2 class functions.
Denoting generically,

$$\nabla^2 \hat{h} = \mathrm{div} \nabla \hat{h}$$

$$= \frac{1}{\sqrt{g}} \left[\frac{\partial}{\partial u_i} \left(g^{ij} \frac{\partial \hat{h}}{\partial u_j} \sqrt{g} \right) \right], \; \forall \hat{h} \in C^2(D), \tag{10.98}$$

we have,

$$\int_M (\nabla^2 f) h \; dM - \int_M (\nabla^2 h) f \; dM$$

$$= \int_{\partial M} (-1)^{j+1} g^{jk} \left[\left(\frac{\partial f}{\partial u_k} \right) h - \left(\frac{\partial h}{\partial u_k} \right) f \right] \sqrt{g} du_1 \wedge \cdots \widehat{du_j} \cdots \wedge du_m,$$

$$\tag{10.99}$$

where the notation $\widehat{du_j}$ means that such a term is absent in the product in question.

Proof From the first Green identity with $X = \nabla h$ we obtain,

$$\int_M (\nabla h) \cdot (\nabla f)\, dM + \int_M \operatorname{div}(\nabla h) f\, dM$$
$$=$$
$$= \int_{\partial M} (-1)^{j+1} g^{jk} \frac{\partial h}{\partial u_k} f\, \sqrt{g}\, du_1 \wedge \cdots \wedge \widehat{du_j} \wedge \cdots \wedge du_m, \quad (10.100)$$

From this, interchanging the roles of f and h, we obtain,

$$\int_M (\nabla h) \cdot (\nabla f)\, dM + \int_M \operatorname{div}(\nabla f) h\, dM$$
$$= \int_{\partial M} (-1)^{j+1} g^{jk} \frac{\partial f}{\partial u_k} h\, \sqrt{g}\, du_1 \wedge \cdots \wedge \widehat{du_j} \wedge \cdots \wedge du_m, \quad (10.101)$$

From (10.100) and (10.101), we have,

$$\int_M (\nabla^2 h) f\, dM - \int_M (\nabla^2 f) h\, dM$$
$$= \int_{\partial M} (-1)^{j+1} g^{jk} \left[\frac{\partial h}{\partial u_k} f - \frac{\partial f}{\partial u_k} h \right] \sqrt{g}\, du_1 \wedge \cdots \wedge \widehat{du_j} \wedge \cdots \wedge du_m.$$

This completes the proof.

Remark 10.2 From the first Green identity with $f \equiv 1$ we obtain the Gauss (divergence) theorem, which is summarized by the equation,

$$\int_M \operatorname{div} X\, dM = \int_{\partial M} (-1)^{j+1} X_j \sqrt{g}\, du_1 \wedge \cdots \widehat{du_j} \cdots \wedge du_m.$$

10.19   More Topics on Differential Geometry

Let $M \subset \mathbb{R}^n$ be an m-dimensional surface where $1 \le m < n$, and where

$$M = \{\mathbf{r}(\mathbf{u}) \ : \ \mathbf{u} \in D\}.$$

Here $D \subset \mathbb{R}^m$ is a open bounded connected set with ∂D of C^1 class.

Let $\mathbf{u}_0, \mathbf{v}_0 \in D$ and let $\mathbf{A} = \mathbf{r}(\mathbf{u}_0)$ and $\mathbf{B} = \mathbf{r}(\mathbf{v}_0)$, where $\mathbf{A} \ne \mathbf{B}$.

Consider a curve C defined by $(\mathbf{r} \circ \mathbf{u}) : [a, b] \to M$, where $\mathbf{u} : [a, b] \to D$ is of C^1 class, such that

$$(\mathbf{r} \circ \mathbf{u})(a) = \mathbf{A},$$

and

$$(\mathbf{r} \circ \mathbf{u})(b) = \mathbf{B}.$$

Let $P = \{t_0 = a, t_1, \ldots, t_n = b\}$ be a partition of $[a, b]$.
Thus, a first approximation of the length of C is given by,

$$L \approx \sum_{j=1}^{k} |\Delta \mathbf{r}_j|,$$

where $\Delta \mathbf{r}_j = \mathbf{r}(\mathbf{u}(t_j)) - \mathbf{r}(\mathbf{u}(t_{j-1}))$.
Observe that

$$\begin{aligned}
\Delta \mathbf{r}_j &= \mathbf{r}(\mathbf{u}(t_j)) - \mathbf{r}(\mathbf{u}(t_{j-1})) \\
&= \frac{d\mathbf{r}(\mathbf{u}(t_j))}{dt} \Delta t_j + \mathscr{O}(\Delta t_j^2) \\
&\approx \frac{d\mathbf{r}(\mathbf{u}(t_j))}{dt} \Delta t_j.
\end{aligned} \tag{10.102}$$

Hence

$$L \approx \sum_{j=1}^{k} \left| \frac{d\mathbf{r}(\mathbf{u}(t_j))}{dt} \right| \Delta t_j \equiv S_{\mathbf{r}}^{P}.$$

With such an approximation in mind, recalling $\mathbf{u}$ is of C^1 class, we define the length of C, also denoted by L, by

$$\begin{aligned}
L &= \lim_{|P| \to 0} S_{\mathbf{r}}^{P} \\
&= \int_{a}^{b} \left| \frac{d\mathbf{r}(t)}{dt} \right| dt.
\end{aligned} \tag{10.103}$$

Observe also that

$$\begin{aligned}
&\frac{d\mathbf{r}(\mathbf{u}(t))}{dt} \cdot \frac{d\mathbf{r}(\mathbf{u}(t))}{dt} \\
&= \left(\frac{\partial \mathbf{r}(\mathbf{u}(t))}{\partial u_i} \frac{du_i(t)}{dt} \cdot \frac{\partial \mathbf{r}(\mathbf{u}(t))}{\partial u_j} \frac{du_j(t)}{dt} \right) \\
&= \left(\frac{\partial \mathbf{r}(\mathbf{u}(t))}{\partial u_i} \cdot \frac{\partial \mathbf{r}(\mathbf{u}(t))}{\partial u_j} \right) \frac{du_i(t)}{dt} \frac{du_j(t)}{dt} \\
&= g_{ij} \frac{du_i(t)}{dt} \frac{du_j(t)}{dt},
\end{aligned} \tag{10.104}$$

where $\mathbf{g}_i = \frac{\partial \mathbf{r}(\mathbf{u})}{\partial u_i}$, and

$$g_{ij} = \mathbf{g}_i \cdot \mathbf{g}_j, \ \forall i, j \in \{1, \ldots, m\}.$$

Summarizing,

$$L(\mathbf{u}) = \int_a^b \sqrt{g_{ij}(\mathbf{u}(t))\frac{du_i(t)}{dt}\frac{du_j(t)}{dt}} \, dt.$$

At this point, consider the problem of finding $\mathbf{u} : [a, b] \to \mathbb{R}^m$ which minimizes L.

Observe first that to minimize L is equivalent to minimize $J(\mathbf{u})$ where

$$J(\mathbf{u}) = \int_a^b g_{ij}(\mathbf{u}(t))\frac{du_i(t)}{dt}\frac{du_j(t)}{dt},$$

on U_{ad} where

$$U_{ad} = \{\mathbf{u} \in W^{1,2}([a, b]; D) \ : \ \mathbf{u}(a) = \mathbf{u}_0 \text{ and } \mathbf{u}(b) = \mathbf{v}_0\}.$$

Observe that

$$g_{ij}\frac{du_i}{dt}\frac{du_j}{dt} = \frac{d\mathbf{r}(\mathbf{u}(t))}{dt} \cdot \frac{d\mathbf{r}(\mathbf{u}(t))}{dt} = \left|\frac{d\mathbf{r}(t)}{dt}\right|^2 \geq 0,$$

$\forall \mathbf{u} \in U_{ad}$, so that L is a positive functional.

The existence of a minimizer for J is a subtle question, since J may not be convex.

However, a point of local minimum for J satisfies the Euler Lagrange equations,

$$\frac{d^2 u_i}{dt^2} + \Gamma_{jk}^i(\mathbf{u})\frac{du_j}{dt}\frac{du_k}{dt} = 0, \text{ in } [a, b], \ \forall i \in \{1, \ldots, m\},$$

where,

$$\Gamma_{jk}^i(\mathbf{u}) = \frac{1}{2}g^{il}\left(\frac{\partial g_{kl}}{\partial u_j} + \frac{\partial g_{il}}{\partial u_k} - \frac{\partial g_{jk}}{\partial u_l}\right), \forall i, j, k \in \{1, \ldots, m\},$$

are the so-called Christoffel symbols for M.

10.20 Some Topics on Riemannian Geometry

Definition 10.20.1 Let $M \subset \mathbb{R}^n$ be an m-dimensional C^1 class volume where $m = n$. Let us denote by $\mathscr{X}(M)$ the set of C^1 class vector fields on $T(M)$. Under such statements and assumptions, we define an affine connection on M as a map

$$\nabla : \mathscr{X}(M) \times \mathscr{X}(M) \to \mathscr{X}(M)$$

such that

1.

$$\nabla_{fX+gY} Z = f \nabla_X Z + g \nabla_Y Z,$$

2.

$$\nabla_X (Y + Z) = \nabla_X Y + \nabla_X Z,$$

3.

$$\nabla_X (fY) = (X \cdot f) Y + f \nabla_X Y,$$

$\forall X, Y, Z \in \mathscr{X}(M), \ f, g \in C^\infty(M; \mathbb{R})$.

Remark 10.20.2 $\nabla_X Y$ is referred as the covariant derivative of Y along X.

Theorem 10.20.3 *Let $M \subset \mathbb{R}^n$ be an m-dimensional C^1 class volume where $m = n$. Let ∇ be an affine connection on M. Let $p = \mathbf{r}(\mathbf{u}) \in M$ and $X, Y \in \mathscr{X}(M)$. Denoting*

$$X = X_i \frac{\partial \mathbf{r}(\mathbf{u})}{\partial u_i},$$

and

$$Y = Y_i \frac{\partial \mathbf{r}(\mathbf{u})}{\partial u_i},$$

we have,

$$\nabla_X Y = \sum_{i=1}^{m} \left(X \cdot Y_i + \sum_{j,k=1}^{m} \Gamma_{jk}^i X^j Y^k \right) \frac{\partial \mathbf{r}(\mathbf{u})}{\partial u_i} \in T_p(M), \qquad (10.105)$$

where Γ^i_{jk} are defined by the relations,

$$\nabla_{\frac{\partial \mathbf{r}(\mathbf{u})}{\partial u_j}} \frac{\partial \mathbf{r}(\mathbf{u})}{\partial u_k} = \Gamma^i_{jk} \frac{\partial \mathbf{r}(\mathbf{u})}{\partial u_i}.$$

Proof Observe that

$$
\begin{aligned}
\nabla_X Y &= \nabla_{X_i \frac{\partial \mathbf{r}(\mathbf{u})}{\partial u_i}} \left(Y_j \frac{\partial \mathbf{r}(\mathbf{u})}{\partial u_j} \right) \\
&= X_i \nabla_{\frac{\partial \mathbf{r}(\mathbf{u})}{\partial u_i}} \left(Y_j \frac{\partial \mathbf{r}(\mathbf{u})}{\partial u_j} \right) \\
&= X_i \left[\frac{\partial \mathbf{r}(\mathbf{u})}{\partial u_i} \cdot Y_j \right] \frac{\partial \mathbf{r}(\mathbf{u})}{\partial u_j} + X_i Y_j \nabla_{\frac{\partial \mathbf{r}(\mathbf{u})}{\partial u_i}} \frac{\partial \mathbf{r}(\mathbf{u})}{\partial u_j} \\
&= \left[X_i \frac{\partial \mathbf{r}(\mathbf{u})}{\partial u_i} \cdot Y_j \right] \frac{\partial \mathbf{r}(\mathbf{u})}{\partial u_j} + X_i Y_j \Gamma^k_{ij} \frac{\partial \mathbf{r}(\mathbf{u})}{\partial u_k} \\
&= \sum_{i=1}^m \left(X \cdot Y_i + \sum_{j,k=1}^m \Gamma^i_{jk} X_j Y_k \right) \frac{\partial \mathbf{r}(\mathbf{u})}{\partial u_i}.
\end{aligned}
\tag{10.106}
$$

The proof is complete.

Consider now an m-dimensional C^1 class volume $M \subset \mathbb{R}^n$ where $m = n$ and locally

$$M = \{ \mathbf{r}(\mathbf{u}) \ : \ \mathbf{u} \in D \}$$

where $D \subset \mathbb{R}^m$ is a bounded simply connected set with a C^1 class boundary ∂D. Consider also a C^1 class curve $(\mathbf{r} \circ \mathbf{u}) : [a, b] \to \mathbb{R}^n$ where we point-wise denote

$$\mathbf{c}(t) = \mathbf{r}(\mathbf{u}(t)), \ \forall t \in [a, b].$$

Let V be a vector field on $T(M)$ where we also denote:

$$V(t) = V_i(\mathbf{r}(\mathbf{u}(t))) \frac{\partial \mathbf{r}(\mathbf{u}(t))}{\partial u_i}.$$

We define,

$$\frac{DV(t)}{dt} = \nabla_{\dot{\mathbf{c}}(t)} V,$$

Observe that

$$\dot{\mathbf{c}}(t) = \frac{\partial \mathbf{r}(\mathbf{u}(t))}{\partial u_i} \dot{u}_i,$$

so that from the earlier definition,

$$\frac{DV(t)}{dt} = \nabla_{\dot{\mathbf{c}}(t)} V$$

$$= \left[\dot{\mathbf{c}}(t) \cdot V_i(\mathbf{r}(\mathbf{u}(t))) + \sum_{j,k=1}^{m} \Gamma_{jk}^i(\mathbf{u}(t)) \dot{u}_j V_k(\mathbf{r}(\mathbf{u}(t))) \right] \frac{\partial \mathbf{r}(\mathbf{u}(t))}{\partial u_i}.$$

$$(10.107)$$

Also,

$$\dot{\mathbf{c}}(t) \cdot V_i(\mathbf{r}(\mathbf{u})) = \lim_{\varepsilon \to 0} \frac{V_i(\mathbf{r}(\mathbf{u}(t) + \varepsilon\{\dot{u}_j\})) - V_i(\mathbf{r}(\mathbf{u}(t)))}{\varepsilon}$$

$$= \frac{\partial V_i(\mathbf{r}(\mathbf{u}(t)))}{\partial u_j} \dot{\mathbf{u}}_j$$

$$= \frac{d[V_i(\mathbf{r}(\mathbf{u}(t)))]}{dt} \equiv \dot{V}_i(t). \qquad (10.108)$$

Hence,

$$\frac{DV(t)}{dt} = \sum_{i=1}^{m} \left(\dot{V}_i(t) + \Gamma_{jk}^i(\mathbf{r}(\mathbf{u}(t))) \dot{u}_j V_k \right) \frac{\partial \mathbf{r}(\mathbf{u}(t))}{\partial u_i}.$$

Moreover, a vector field V on $T(M)$ as defined along a curve $\mathbf{c} = (\mathbf{r} \circ \mathbf{u})$: $[a, b] \to \mathbb{R}^n$ is said to be parallel along such a curve, if

$$\nabla_{\dot{\mathbf{c}}(t)} V = 0, \quad \text{on } [a, b].$$

Also, the curve defined point-wisely by

$$\mathbf{c}(t) = \mathbf{r}(\mathbf{u}(t)),$$

is said to be geodesic if,

$$\frac{D\dot{\mathbf{c}}(t)}{dt} = 0,$$

where,

$$\dot{\mathbf{c}}(t) = \frac{\partial \mathbf{r}(\mathbf{u}(t))}{\partial u_i} \frac{\partial u_i}{\partial t}$$

$$= \frac{\partial u_i}{\partial t} \frac{\partial \mathbf{r}(\mathbf{u}(t))}{\partial u_i}. \qquad (10.109)$$

So, from

$$\nabla_{\dot{c}(t)}\dot{c}(t) = 0,$$

that is, from (10.107) with

$$V_i = \frac{\partial u_i(t)}{\partial t} = \dot{u}_i,$$

we obtain,

$$\ddot{u}_i + \Gamma^i_{jk}\dot{u}_j\dot{u}_k = 0, \ \forall i \in \{1, \ldots, m\}.$$

These last equations are called the geodesics ones.

Finally, if the connection in question is such that

$$\Gamma^i_{jk} = g^{il}\left(\frac{\partial g_{kl}}{\partial u_j} + \frac{\partial g_{jl}}{\partial u_k} - \frac{\partial g_{jk}}{\partial u_l}\right)$$

such a connection is said to be the Levi-Civita one.

In such a case the geodesics equations here obtained coincide with the necessary conditions obtained through the calculus of variations to minimize the arc length over the surface in question between

$$\mathbf{A} = \mathbf{r}(\mathbf{u}(a))$$

and

$$\mathbf{B} = \mathbf{r}(\mathbf{u}(b)).$$

Definition 10.20.4 (Curvature of a Connection) Let $M \subset \mathbb{R}^n$ be an m-dimensional C^1 class volume, where $m = n$, and let $\nabla \tilde{\mathscr{X}} \times \tilde{\mathscr{X}} \to \tilde{\mathscr{X}}$ be a connection. We define the curvature of ∇ as the map defined by:

$$R(X, Y)Z = \nabla_X\nabla_Y Z - \nabla_Y\nabla_X Z - \nabla_{[X,Y]}Z.$$

Hence R is a way of measuring the commutativity of a connection.

Theorem 10.20.5 *Let $M \subset \mathbb{R}^n$ be an m-dimensional C^1 class volume where $m = n$, let ∇ be a connection and R be the corresponding curvature. Under such hypotheses,*

$$R(fX_1 + gX_2, Y)Z = fR(X_1, Y)Z + gR(X_2, Y)Z, \ \forall X_1, X_2, Y, Z \in \tilde{\mathscr{X}}.$$

Proof Observe that

$$\nabla_{fX_1+gX_2}(\nabla_Y Z) = f\nabla_{X_1}\nabla_Y Z + g\nabla_{X_2}\nabla_Y Z, \qquad (10.110)$$

and,

$$\nabla_Y(\nabla_{fX_1+gX_2}Z)$$
$$= \nabla_Y(f\nabla_{X_1}Z + g\nabla_{X_2}Z)$$
$$= \nabla_Y(f\nabla_{X_1}Z) + \nabla_Y(g\nabla_{X_2}Z)$$
$$= (Y\cdot f)\nabla_{X_1}Z + f\nabla_Y\nabla_{X_1}Z + (Y\cdot g)\nabla_{X_2}Z + g\nabla_Y\nabla_{X_2}Z, \quad (10.111)$$

Also,

$$\nabla_{[fX_1+gX_2,Y]}Z$$
$$= \nabla_{\left((fX_1)\cdot Y_i - Y\cdot(f(X_1)_i)\frac{\partial \mathbf{r}}{\partial u_i} + (Y\cdot(g(X_2)_i - (gX_2)\cdot Y_i)\frac{\partial \mathbf{r}}{\partial u_i}\right)}Z$$
$$= (fX_1)\cdot Y_i - Y\cdot(f(X_1)_i)\nabla_{\frac{\partial \mathbf{r}}{\partial u_i}}Z + (g(X_2)\cdot Y_i - Y\cdot g(X_2)_i)\nabla_{\frac{\partial \mathbf{r}}{\partial u_i}}Z,$$

$$(10.112)$$

so that

$$R(fX_1 + gX_2, Y)Z$$
$$= f\nabla_{X_1}\nabla_Y Z + g\nabla_{X_2}\nabla_Y Z$$
$$\quad -(Y\cdot f)\nabla_{X_1}f - f\nabla_Y(\nabla_{X_1}Z)$$
$$\quad -(Y\cdot g)\nabla_{X_2}Z - g\nabla_Y\nabla_{X_2}Z$$
$$\quad -[(fX_1)\cdot Y_i + (gX_2)\cdot Y_i]\nabla_{\frac{\partial \mathbf{r}}{\partial u_i}}Z$$
$$\quad +[(Y)\cdot(fX_1)_i + (Y)\cdot(gX_2)_i]\nabla_{\frac{\partial \mathbf{r}}{\partial u_i}}Z, \qquad (10.113)$$

Observe that

$$(Y\cdot(f(X_1)_i))\nabla_{\frac{\partial \mathbf{r}}{\partial u_i}}Z$$
$$= d(f(X_1)_i)(Y)\nabla_{\frac{\partial \mathbf{r}}{\partial u_i}}Z$$
$$= df(Y)(X_1)_i\nabla_{\frac{\partial \mathbf{r}}{u_i}}Z + fd(X_1)_i(Y)\nabla_{\frac{\partial \mathbf{r}}{\partial u_i}}Z$$
$$= (Y\cdot f)(X_1)_i\nabla_{\frac{\partial \mathbf{r}}{u_i}}Z + f(Y\cdot(X_1)_i)\nabla_{\frac{\partial \mathbf{r}}{\partial u_i}}Z$$
$$= (Y\cdot f)\nabla_{X_1}Z + f(Y\cdot(X_1)_i)\nabla_{\frac{\partial \mathbf{r}}{\partial u_i}}Z, \qquad (10.114)$$

and similarly,

$$(Y \cdot (g(X_2)_i)) \nabla_{\frac{\partial \mathbf{r}}{\partial u_i}} Z$$

$$= (Y \cdot g) \nabla_{X_2} Z + f(Y \cdot (X_2)_i) \nabla_{\frac{\partial \mathbf{r}}{\partial u_i}} Z. \qquad (10.115)$$

Thus,

$$R(fX_1 + gX_2, Y)Z$$

$$= f \nabla_{X_1} \nabla_Y Z + g \nabla_{X_2} \nabla_Y Z$$

$$\quad - (Y \cdot f) \nabla_{X_1} Z - f \nabla_Y (\nabla X_1 Z)$$

$$\quad - (Y \cdot g) \nabla_{X_2} Z - g \nabla_Y \nabla_{X_2} Z$$

$$\quad - [(fX_1) \cdot Y_i + (gX_2)Y_i] \nabla_{\frac{\partial \mathbf{r}}{\partial u_i}} Z$$

$$\quad + (Y \cdot f) \nabla_{X_1} Z + (Y \cdot g) \nabla_{X_2} Z$$

$$\quad + (f(Y \cdot (X_1)_i) + g(Y \cdot (X_2)_i)) \nabla_{\frac{\partial \mathbf{r}}{\partial u_i}} Z, \qquad (10.116)$$

so that,

$$R(fX_1 + gX_2, Y)Z$$

$$= f(\nabla_{X_1} \nabla_Y Z - \nabla_Y \nabla_{X_1} Z)$$

$$\quad + g(\nabla_{X_2} \nabla_Y Z - \nabla_Y \nabla_{X_2} Z)$$

$$\quad - [(fX_1) \cdot Y_i - f(Y \cdot (X_1)_i)] \nabla_{\frac{\partial \mathbf{r}}{\partial u_i}} Z$$

$$\quad - [(gX_2) \cdot Y_i - g(Y \cdot (X_2)_i)] \nabla_{\frac{\partial \mathbf{r}}{\partial u_i}} Z$$

$$= f(\nabla_{X_1} \nabla_Y Z - \nabla_Y \nabla_{X_1} Z)$$

$$\quad + g(\nabla_{X_2} \nabla_Y Z - \nabla_Y \nabla_{X_2} Z)$$

$$\quad - f[(X_1) \cdot Y_i - Y \cdot ((X_1)_i)] \nabla_{\frac{\partial \mathbf{r}}{\partial u_i}} Z$$

$$\quad - g[(X_2) \cdot Y_i - Y \cdot ((X_2)_i)] \nabla_{\frac{\partial \mathbf{r}}{\partial u_i}} Z$$

$$= f(\nabla_{X_1} \nabla_Y Z - \nabla_Y \nabla_{X_1} Z)$$

$$\quad + g(\nabla_{X_2} \nabla_Y Z - \nabla_Y \nabla_{X_2} Z)$$

$$\quad - f \nabla_{[X_1, Y]} Z - g \nabla_{[X_2, Y]} Z$$

$$= f R(X_1, Y)Z + g R(X_2, Y)Z. \qquad (10.117)$$

Exercise 10.20.6

1. Considering the last definitions and results, prove that

$$R(X, Y)(fZ_1 + gZ_2) = f R(X, Y)Z_1 + g R(X, Y)Z_2, \quad \forall X, Y, Z_1, Z_2 \in \tilde{\mathscr{X}}(M).$$

References

1. R. Abraham, J.E. Marsden, T. Ratiu, *Manifolds, Tensor Analysis and Applications, Applied Mathematical Sciences*, vol. 75, 2nd edn. (Springer, New York, 1988)
2. T.M. Apostol, *Calculus*, vol. 2, 2nd edn. (Wiley India, Noida, 2007)
3. F. Botelho, *Functional Analysis and Applied Optimization in Banach Spaces* (Springer, Cham, 2014)
4. H. Brezis, *Functional Analysis, Sobolev Spaces and Partial Differential Equations* (Springer, New York, 2010)
5. L. Godinho, J. Natario, *An Introduction to Riemannian Geometry, with Applications to Mechanics and Relativity* (Springer, Cham, 2014)
6. J. Jost, *Postmodern Analysis* (Springer, Berlin 2005)
7. S.G. Krantz, H.R. Parks, *The Implicit Function Theorem* (Birkhäuser, Boston, 2003)
8. E. Kreyszig, *Differential Geometry* (Dover Publications, Inc., New York, 2013)
9. E.L. Lima, *Curso de Análise*, vol. 1, 14th edn. (IMPA, Rio de Janeiro, 2014)
10. E.L. Lima, *Curso de Análise*, vol 2, 11th edn. (IMPA, Rio de Janeiro, 2015)
11. S.M. Robinson, Strong regular generalized equations. Math. Oper. Res. **5**, 43–62 (1980)
12. W. Rudin, *Principles of Mathematical Analysis* (McGraw Hill, New York, 1976)
13. E.M. Stein and R. Shakarchi, Fourier Analysis, an Introduction, Princeton University Press, Princeton, New Jersey, 2003.
14. J. Jost, Riemaniann Geometry and Geometric Analysis, 6th edition, Springer-Verlag, Berlin, 2011.

© Springer International Publishing AG, part of Springer Nature 2018
F. S. Botelho, *Real Analysis and Applications*,
https://doi.org/10.1007/978-3-319-78631-5

Index

© Springer International Publishing AG, part of Springer Nature 2018
F. S. Botelho, *Real Analysis and Applications*,
https://doi.org/10.1007/978-3-319-78631-5

Printed in the United States
By Bookmasters